T0201379

SUCCESSFUL WOMEN CERAMIC AND GLASS SCIENTISTS AND ENGINEERS

SUCCESSFUL WOMEN CERAMIC AND GLASS SCIENTISTS AND ENGINEERS

100 Inspirational Profiles

Lynnette D. Madsen, Ph.D.

The American Ceramic Society

WILEY

This book is dedicated to all of the women featured in it—without their enthusiasm and perseverance in science and engineering, successes, and willingness to share their stories, this book would not have been possible.

CONTENTS

THE WOMEN

FOREWORD

It takes thousands of years for wind and water to erode a boulder before it becomes sand. Yet it takes only one person, in a single moment, to heat that sand to a temperature that changes its state from solid to liquid, then to glass. The transformation occurs at a moment when slow and natural progress meets intention.

For centuries, leaders in engineering and science have chipped away at the gendered stereotypes that exist in our fields. These champions have encouraged girls and young women to pursue their interests, led by example as positive role models, and shared the tremendous career possibilities available to anyone in STEM fields. With incremental exposure to these elements, the stereotypes that have limited change are slowly wearing down.

In the following pages, Dr. Lynnette Madsen chronicles 100 remarkable engineers and scientists who have made lasting changes in the world. Some are designing technological solutions to the energy challenges, while others are pioneering more efficient ways to use natural resources. Each of these women is an exceptional example of why intentional leadership in science and engineering has never been more important than it is today.

The Faculty of Applied Science and Engineering at the University of Toronto is one of the top-ranked and most diverse engineering schools in the world. As Dean, I experience the benefit that involving multiple genders, ethnicities, and backgrounds brings to our campus and to industry. Multiple perspectives and multidisciplinary collaboration enhance creativity, which enables engineers and scientists to invent technologies and processes that have never existed before. These can become innovations that create prosperity today and ensure a bright future for the generations of tomorrow.

There has never been a more important time for Dr. Madsen's careful exploration into the professional and personal lives of these 100 trailblazers. Through a number of efforts that increase awareness of the rewarding careers in engineering and science among the public and students of all ages, as well as targeted recruitment efforts, many universities in Canada, the United States, and around rest of the world are experiencing record-high enrolment of female engineering and science students. Through similar strategic outreach initiatives and fostering engaging environments, the numbers of female graduate students and faculty members at these institutions are also on the rise.

The women you will read about in the forthcoming pages should inspire our momentum. Their stories should also remind us just how far we have to go. My intention is to see a world where gender is no longer among the defining features of a success story.

This is why I find Dr. Madsen's commitment to this effort extraordinarily inspiring—her research not only highlights the outstanding and tremendous contributions of female ceramic and glass scientists and engineers, but also gives greater visibility to the boundless potential of women in engineering, science, and beyond.

Each of us can be a catalyst in a girl's journey to science or engineering. Male or female, parent or educator, leader or peer, we each have the ability to illuminate and celebrate the difference women can make in the world through a career in these fields.

When I was growing up in Uruguay, I was fortunate to have several people who set their intention on seeing me reach my potential. First, it was a primary school teacher who encouraged me to "tinker" and take things apart. Then it was my parents who let me continue exploring, even when it meant finding our radio scattered in dozens of pieces across the living room floor. Later, it was a high school physics teacher who encouraged me to pursue engineering at the university level in Venezuela and, later, the United States. Many of these champions, mentors, and advocates were men. Intention sees possibilities through to change, regardless of gender.

Like the other engineers and scientists described in this book, incremental change, engagement, and influence from many different types of people lit my path to engineering. When evolving perceptions of what women could contribute to the field met the intention set by leaders and mentors in my life, it transformed my future. My career has taken me around the world, enabled me to pursue meaningful research, and eventually provided me the privilege and responsibility of leading change in higher education and in industry.

I am honored to introduce Dr. Madsen's book to you, and extend my deepest gratitude for her conviction and dedication to enhancing diversity. If we all join together in actions great and small, we can transform stones into sand, and sand into glass.

DR. CRISTINA AMON
Dean, Applied Science and Engineering,
University of Toronto

PREFACE

Mid-2012, I contacted Wiley about writing a book about women working with ceramics and glass. I wanted to focus on women because they are not only underrepresented in the physical sciences, but also underappreciated. Both of these factors have affected their success. I chose a narrow area of materials science and engineering to make a point— **there are plenty of successful women out there!** Underrepresentation is not the same as complete absence. However, those who have persisted in the field have been held to higher standards and encountered subtle and overt gender-based barriers. If I selected a broader category, it is likely that all of the top women in the materials science would have achieved many great awards or other recognitions already. For example, in 2012 alone, Prof. Mildred Dresselhaus was a co-recipient of the Enrico Fermi Award and was also awarded the Kavli Prize. Although some of the women featured in this book are members of prestigious organizations (e.g., U.S. National Academy of Sciences, U.S. National Academy of Engineering, the World Academy of Ceramics, etc.), others have yet to be included. Many of the women are still on the road to greater success and recognition. **It is my hope that by highlighting them in this book, it will aid in their journey.**

The difficult task was to *select* 100—not to *find* 100—notable women in ceramics and glass, and, at the same time, to get a *diverse* set of women—something I am familiar with in my day job as a Program Director at the National Science Foundation. Obviously, it was easiest for me to find U.S. women from academe—a group that I interact with daily, but my wish was for a broader cross section **to attract young women from around the world to the field and to encourage them to stay.** I want the reader to find someone in this book who reflects who they are and where they want to go. The women featured in this book have shared their career paths, life lessons or important turning points in their lives, what it has meant to be a woman in a nontraditional field, and some general advice.

It was challenging to simultaneously seek diverse women (in terms of geography, work sector, race/ethnicity, etc.) who worked to varying extents in one or more areas of ceramics or glass. Sometimes I found it difficult to appreciate women's careers when their websites were predominately in a language other than English. This limitation may account for the lack of women from some countries.

In the course of writing and selecting the 100 women to feature, I have left out many excellent women. When torn between several excellent choices, I gave a slight preference to selecting more senior women. So, the women in this book are merely illustrative, not exhaustive. Many others could and perhaps should have been included. I want to hear from you and your colleagues—I welcome your input for volume II, the next edition, and/or other articles or books. To obtain a form for submitting a profile, e-mail me at Lynnette@svedbergscience.com.

Despite the number limitation, I think this book can still be considered a first resource for identifying successful women working in the field of ceramics and glass, dispelling the myth that there are no highly successful women in this exciting area of the physical sciences. Women are featured from all over the world, different sectors of the workforce, and reflect a diverse population. The book includes a rare view of women in industry who are less accessible than those in academe (who generally have work-related websites). The stories of these women come alive for the younger generation who seek role models and they remind the more senior populace that there are plenty of women in the field to consider when making nominations for honors, guest speakers, or awards.

LYNNETTE D. MADSEN, PH.D.

ABOUT THE AUTHOR

Dr. Lynnette D. Madsen was raised, educated, and worked for a decade in industry in Canada. Subsequently, she held a faculty position in physics at Linköping University in Sweden. Since 2000, Dr. Madsen has worked at the National Science Foundation (NSF) in the United States as the Program Director for Ceramics. In addition to recommending the distribution of the annual Ceramics Program budget, she has an active independent research program, led new cooperative activities with European researchers in materials, and been part of the driving force in program development and initiatives in nano-technology, manufacturing, sustainability, education, and diversity. Also, she is Vice President of Svedberg Science, Inc.—a small business that provides consulting about scientific research directions and trends. This career trajectory has provided her with a bird's-eye view of the researchers in the ceramics and glass research field, which has informed the approach of this volume. More information about the author is available in her own profile, which appears later in this book.

QUICK GUIDE TO SELECT GROUPS

Many of the women featured are working in academe—they were the easiest to identify due to their publication records and university websites highlighting their work. Other career paths—at the onset of one's career, or later on—can be equally rewarding and beneficial. I have highlighted two groups below.

Government and Nonprofit Organizations

Alida Bellosi (Institute of Science and Technology for Ceramic Materials, National Research Institute in Italy)

Zhili Chen (Communist Party of China)

Melanie W. Cole (Department of Defense)

Bonnie J. Dunbar (National Aeronautics and Space Administration (NASA))

Catherine P. Foley (Commonwealth Scientific and Industrial Research Organisation (CSIRO) in Australia)

Carol M. Jantzen (Savannah River National Laboratory)

Sylvia M. Johnson (NASA)

Gretchen Kalonji (The United Nations Educational, Scientific and Cultural Organization (UNESCO))

Gabrielle G. Long (Argonne National Laboratory)

Lynnette D. Madsen (National Science Foundation)

Tina M. Nenoff (Sandia National Laboratory (SNL))

Julia Phillips (SNL)

Maxine Savitz (National Academies)

Debra R. Rolison (Naval Research Laboratory)

Inna G. Talmy (Naval Surface Warfare Center)

Ellen Williams (Advanced Research Projects Agency—Energy (ARPA-E))

Industry/Business

Uma Chowdhry (DuPont)

Dana Goski (Allied Mineral Products)

Lina M. Echeverria (Corning)
Katharine G. Frase (IBM)
Susan N. Houde-Walter (LaserMax)
Mareike Klee (Philips Research)
Merrilea Mayo (Mayo Enterprises)
Marina R. Pascucci (CeraNova)
Nelly M. Rodriguez (Catalytic Materials)
Eva M. Vogel (Bell Labs)
Wanda Wolny (Meggitt)

Despite the issues/obstacles in the United States, many of the women who are obviously successful in this field now reside in the United States; they account for about half of the women in this book. Nevertheless, women who are included represent 28 other countries.

Australia, Russia, and Asia

Inna P. Borovinskaya (Russia)
Rachel A. Caruso (Australia)
Helen Lai Wa Chan (Hong Kong, China)
Li-Chyong Chen (Taiwan)
Jen-Sue Chen (Taiwan)
Zhili Chen (China)
Catherine P. Foley (Australia)
Huey Hoon Hng (Singapore)
Kazumi Kato (Japan)
Michiko Kusunoki (Japan)
Jueinai R. Kwo (Taiwan)
Hua Kun Liu (Australia)
Soon Ja Park (Korea)
Tanusri Saha-Dasgupta (India)
Wei-Ying Sun (China)
Jackie Ying (Singapore)
Jing Zhu (China)

Europe

Alida Bellosi (Italy)
Serena M. Best (UK)
Dominique Chatain (France)

Ulrike Diebold (Austria)
Alicia Duran Carrera (Spain)
Natalia Dubrovinskaia (Germany)
Mari-Ann Einarsrud (Norway)
Monica Ferraris (Italy)
Maria Veronica Ganduglia-Pirovano (Spain)
Dagmar Gerthsen (Germany)
Clare P. Grey (UK)
Kersti Hermansson (Sweden)
Gretchen Kalonji (France)
Maarit Karppinen (Finland)
Mareike Klee (The Netherlands)
Marija Kosec (Slovenia)
Anne L. Leriche (France)
Claude Levy-Clement (France)
Janina Molenda (Poland)
Beatriz Noheda (The Netherlands)
Ellen Ivers-Tiffee (Germany)
Judith L. MacManus-Driscoll (UK)
Tatiana A. Prikhna (Ukraine)
Nava Setter (Switzerland)
Nicola A. Spaldin (Switzerland)
Paula Maria L.S. Vilarinho (Portugal)
Wanda Wolny (Denmark)
Maria Magdalena Zaharescu (Romania)

The Americas (excluding the USA)

Viola I. Birss (Canada)
Linda F. Nazar (Canada)
Ruth H.G.A. Kiminami (Brazil)
Eliana N.S. Muccillo (Brazil)
Noemi "Betty" Elisabeth Walsöe de Reca (Argentina)

Women of Color in the USA

Unfortunately, there are very few senior/accomplished women of color in the United States today with a glass or ceramics specialization to include in this book; the "pipeline" looks better.

Sossina M. Haile
Helen M. Chan
Deborah D.L. Chung
Uma Chowdhry
Lina M. Echeverria
Rosario Gerhardt
Nelly M. Rodriguez
Lourdes Salamanca-Riba

Women in Academe in the USA

The largest single group is women in academe in the United States. It was difficult to select individuals from the many excellent candidates. Some women are members of the National Academies, many are members of the World Academy of Ceramics, some are deans, and a few are past presidents of The American Ceramic Society.

Engineering and Applied Sciences (or closely related department)
Liesl Folks (University at Buffalo, SUNY)

Elsa Garmire (Dartmouth College)

Jennifer A. Lewis (Harvard University)

Materials Science and Engineering (or closely related department)
Dawn A. Bonnell (University of Pennsylvania)

Helen M. Chan (Lehigh University)

Elizabeth C. Dickey (North Carolina State University)

Mildred S. Dresselhaus (Massachusetts Institute of Technology)

Doreen D. Edwards (Alfred University)

Katherine T. Faber (California Institute of Technology)

Rosario Gerhardt (Georgia Tech)

Sossina M. Haile (Northwestern University)

Linda E. Jones (Alfred University)

Lisa C. Klein (Rutgers, The State University of New Jersey)

Waltraud M. Kriven (University of Illinois at Urbana-Champaign)

Alexandra Navrotsky (University of California, Davis)

Caroline A. Ross (Massachusetts Institute of Technology)

Della M. Roy (Arizona State University)

Lourdes Salamanca-Riba (University of Maryland)

Susan B. Sinnott (University of Florida)

Susanne Stemmer (University of California, Santa Barbara)

Susan Trolier-McKinstry (Penn State University)

Mechanical and Aerospace Engineering (or closely related department)
Emily A. Carter (Princeton University)
Deborah D.L. Chung (University at Buffalo, SUNY)

Optics & Photonics
Kathleen A. Cerqua-Richardson (University of Central Florida)

Physics (or closely related department)
Laura H. Greene (Florida State University)
Karin M. Rabe (Rutgers, The State University of New Jersey)

INTRODUCTION

Why This Book?

The book features 100 women scientists and engineers—all of whom were alive when the book was started.

In the Swedish newspaper, *Ny Teknik*, they described the most influential people in technology in Sweden; so herein, it was not merely the number of publications or citations that brought women to this book, but their positions of influence as well. Women included have at least one and typically several of the following attributes:

- Top recognitions, e.g., elected World Academy of Ceramic academician, elected U.S. National Academy member, winner of a prestigious international award or senior chair position, President of the American Ceramic Society
- A large number of citations (evident by a double-digit h-index)
- Positions of influence or international impact (e.g., astronaut, politician, etc.)
- Leadership, evident through a senior position held (or previously held) at a major company or organization
- Owner of a small business

Selection was not based on personal acquaintance, and in fact, I was "introduced" to many women through the process of writing this book.

The profiles hit a middle ground between the terse *Who's Who* format and a longer essay style. For most of the women featured, there are several pages of text with one or more images. Typically, each profile includes her affiliation, points of contact, a brief biography, an image of her active at her job, highlights of her successes (most cited publications, prestigious recognitions/awards, etc.), and many also include personal pointers to younger scientists finding their way.

For Whom?

The format I selected is based on how I pictured the book being used by the various readers. Many audiences, I envision five, will find the profiles in this volume to be useful.

1. **Women** (students of all ages, postdoctoral associates, new graduates, and mid-career professionals) **and people from other underrepresented groups** may draw inspiration and guidance from the profiles, and see a variety of career paths and options.

2. **Faculty advisors** (of students and postdoctoral associates) **and other mentors** will find this a useful book to share with their protégés (both female and male) to illustrate the diversity of career pathways available.

3. **Conference organizers and search and selection committees** (for awards, appointments, etc.) may use it to aid them in identifying women suitable for various opportunities. These opportunities in turn should enable women to showcase their research results, utilize their skills in new ways, and/or gain recognition for their successes. These profiles will help to overcome one of the challenges associated with the gender difference in professional networks: Women's networks are noted to differ in composition from men's—women's networks are broader in intellectual scope and may not be as instrumental for obtaining recognition in a specific field. Moreover, many women are conditioned to avoid seeking reward or recognition. On the other hand, I hope that committees will see this book as a tool to recruit and promote proactively. For this reason, points of contact have been provided for each woman.

4. **Professional and scientific societies**—both groups focused exclusively on women and organizations that are broader—will have a compilation of female role models.

5. **Professors and scholars of women's studies and/or gender studies** may find these profiles useful for the rich diversity of women in the field and insights about this group of successful scientists and engineers.

The Writing Journey

The idea comes in a flash—you see a way forward past the naysayers and doubtful... the idea is refined in hours to days as you consider what is out there in book form and available on the web. The dialogue with the publisher then takes days to weeks. The review of your book proposal takes additional weeks. Finally, the writing—including the background investigations and requests for additional information—takes months. As with any major endeavor one carries out as a "side activity" rather than vocation, this book evolved over quite some time from concept to reality. Key periods were particularly productive, e.g., during holidays around the end of the calendar year, and the 16-day Federal Government shutdown of 2013 (where I could not, by law, do any official government work until either a continuing resolution or budget was passed). During the government shutdown, the weather fully cooperated—we had several days of very hot weather (which limited what I wanted to do outside) followed by several days of rain . . . all in all, it was a very productive period for me.

The identification of the women and the collection of background information were done without the aid of Facebook or LinkedIn; I am not a member of either. Perhaps it would have been useful to join one or both of these, but I also thought it might also complicate things. As an added bonus, I didn't need to review their Data Use Policy and Statement of Rights and Responsibilities.

When Marija Kosec passed away in December 2012, in the period between the initial idea for this book and signing the contract, I made the decision to ask the oldest women first. And, for the most part, they are a tough group—while a few, e.g., Mildred Dresselhaus, had an extensive website (with sections geared for different age groups), others had retired and had little or no web presence. Della Roy, born in 1926, has the honor of being the oldest woman included in the book; in addition, there are a handful of women who were born in the 1930s. The youngest women in the book were born in the late 1960s and early 1970s. In total, the book covers about a 45-year span.

The first profile, my own, took the longest as I struggled with the format, content, and wording. All of my older pictures were not digital and many of my "action" shots were dark or out of focus. I did wonder how I was going to fare with the entire group, if I was having so much "fun" doing my own profile.

The profiles vary in length, depending upon the amount of information that was available about each woman, the length of her career, and the extent to which each woman, herself, participated in writing the profile. It is this latter element that provides more depth to the profiles in this book than the standard biographies in similar volumes.

There were four key sections I asked each woman to complete:

1. Proudest Career Moment (there were several variations on the title of this section)
2. Challenges
3. On being a woman in this field . . .
4. Words of Wisdom

The women profiled also supplied many images. In addition, I also asked for their help on completing the rest of each profile, in particular I tried to capture a historical biography. You will note there is also a little variation in the section heading of the Cited Papers section; most often I supplied the three most cited papers. All of the sections (and subsections) were entirely optional—most women completed all sections, several chose to omit a few major or minor sections, and a few completed almost nothing. My goal was simply to the keep the format of each profile similar.

There are a number of ways to capture publication impact, each with its strengths and weaknesses. My use of Thomson Reuters' *Web of Science* database for publication and citation data appeared to be a controversial choice by a few of the women included in this book. Some women preferred Elsevier's *Scopus*, others preferred *Google Scholar*, and one woman was committed to *Harzing.com's Publish or Perish*. The h-index for *Google Scholar* is almost always higher than that in *Web of Science*. At the time I was working on this book, in the physical sciences, there is a high regard for *Web of Science*. In addition to the other issues, when I started writing the book, I did not have direct access to *Scopus*. It should be noted that even the *Web of Science* statistics vary based on whether one uses the Core Collection[1] or All Databases and is affected by the time span of the subscription. In most cases, I could choose All Databases plus I allowed each woman to make corrections (so that they captured any name changes or other variations).

Additionally, individual identifiers are included with some publication databases. They provide an avenue to reduce potential problems of author ambiguity within the scholarly research community. For example, *ORCID* provides a persistent digital

[1] The Core Collection includes journal articles in the sciences, social sciences, and arts and humanities; All Databases is broader, including journal articles, patents, websites, conference proceedings, and open access material.

identifier (http://orcid.org/) and Thomson Reuters offers *ResearcherID*; information can be shared between the two platforms. If a *ResearcherID* existed, I noted it in the profile and used it for updates. At the time of writing, joining *ResearcherID* was free and there was no charge or subscription needed to access information (at http://www.researcherid. com/). The statistics are mostly reliable, but sometimes dated (since the lists must be manually updated by the author with new entries). One researcher had a link to her *ResearcherID* on her website (http://www.mse.umd.edu/faculty/salamanca-riba).

There were many connections between the women in these profiles and in the images women supplied. Some of them are provided below:

- Millie Dresselhaus served as the Ph.D. advisor for two women featured in this book: Deborah D.L. Chung and Lourdes Salamanca-Riba.
- Prof. Helen L.W. Chan (Hong Kong)—to whom I was referred by Helen M. Chan in the United States—supplied a picture of her with Prof. Dresselhaus.
- Prof. Maria Zaharescu in Romania supplied a picture of her with Prof. Alex Navrotsky.
- Nava Setter and Gretchen Kalonji both have strong ties to Tanzania.
- Two women featured received their first degrees in chemistry from Bryn Mawr College; however, Dr. Frase was about 20 years later than Dr. Savitz.
- Half a dozen of the women have backgrounds in geology (or related fields) and moved (more) into ceramics and glass, while the opposite trend is true for Gretchen Kalonji who started with a materials science and engineering background and then moved to UNESCO's science division where her new focus included hydrology, geology, oceanography, biodiversity, and climate. Alex Navrotsky maintains strong connections with both geosciences and ceramics.

Various pairs have published together: Liesl Folks and Ulrike Diebold, Susan Sinnott and Beth Dickey, and Jackie Ying and Alex Navrotsky, to name a few. There may be many other research connections (that I missed). Many of the women have won prestigious awards; however, only Della Roy has one named after her.

There was some institutional overlap as well:

- Two of the women featured—Judith L. MacManus-Driscoll and Serena Best— were colleagues at the University of Cambridge (at the time this book was written).
- Dagmar Gerthsen and Ellen Ivers-Tiffee were both at the Karlsruhe Institute of Technology.
- Millie Dresselhaus and Caroline Ross were both at MIT.
- Trudy Kriven and Laura Greene were both at the University of Illinois at Urbana-Champaign, but in different departments; since then, Laura has moved universities. Similarly, Karin Rabe and Lisa Klein were both at Rutgers, again in different departments.
- While I was writing this book, Kathy Faber moved from Northwestern University to California Institute of Technology making her a departmental colleague of Sossina Haile, but shortly thereafter Sossina moved to Northwestern University.

- Liesl Folks and Deborah Chung were both at the University at Buffalo, SUNY, where Folks was a Dean and Chung was a Professor.
- Tina M. Nenoff worked at Sandia National Laboratory (SNL) and Julia Phillips had just retired from there.
- At Alfred University, there were both Linda Jones as a VP and Doreen Edwards as a Dean. However, while this book was being written, Linda Jones moved universities.
- Both Lourdes Salamanca-Riba (Materials Science and Engineering) and Ellen Williams (Physics) were at the University of Maryland, but Ellen now spends more of her time at ARPA-E.

WORDS OF PRAISE FOR THE BOOK

Praise for This Book

I am very excited about your book—such a book will provide "connectivity" and "virtual role models" to help me through my physics education and provide inspiration for my quest for a career in science.

Alex, a 20-year-old female student

In today's highly competitive world, talent, creativity, and innovation comprise the coin of the realm. All nations recognize that science, engineering, and technology drive the global economy and, to win in the marketplace, investment in the next generation of scientists and engineers is critical. Yet, what goes unrecognized is that half of the generational pool is underutilized or, more often, ignored and that half is comprised of brilliant women engineers and scientists. Lynnette Madsen has carefully detailed the careers and biographies of 100 outstanding women from countries around the world to illustrate their magnificent achievements in materials science and ceramics. Selection of women to highlight was difficult, not because of so few but quite the opposite, because of so many who are extraordinarily accomplished in this one area of science and engineering. If other fields were to be similarly highlighted, it is certain that there would be the same challenge: how to select just 100 from so many distinguished scientists and engineers who are women and leaders in their fields. Lynnette has done a great service in writing this book, not just for women, but for society at large, because in the twenty-first century, we can no longer underutilize or ignore that half of the best. When countries like Japan and those in Scandinavia have mandated proportional representation of women in leadership positions in government, academia, and industry, and other countries will follow, we can thank Lynnette Madsen for launching a source book of female talent, creativity, and innovation!

Rita Colwell, Director, United States National Science Foundation (1998–2004)
Distinguished University Professor, University of Maryland, College Park,
and Johns Hopkins Bloomberg School of Public Health, USA

Praise for This Book from the Women Featured

Welcome your brilliant initiative. I think there is a real need of such documentation for the young women who are going forward with their dreams.

Alida Bellosi, Director, The Institute of Science and
Technology for Ceramics (ISTEC), Italy

The questions you ask are rather intriguing; I'll be very interested to see how others have responded when the book comes out…

Helen M. Chan, Professor, Lehigh University, USA

It is a special experience for me to write up these paragraphs. I have been asked similar questions from time to time, but usually answered them in a vague way. Thanks for sharing this opportunity with me so that I have a chance to review my own career.

Jen-Sue Chen, Professor, National Cheng Kung University, Taiwan

I hope that your book will give some inspiration to and have an impact on the young generations.

Li-Chyong Chen, Distinguished Research Fellow and Director,
National Taiwan University, Taiwan

Sounds like a fantastic project.

Ulrike Diebold, Professor, Vienna University of Technology (TU Wien), Austria

I hope it goes really well. It is a really good idea.

Judith Driscoll, Professor, University of Cambridge, UK

You have a noble goal "to inspire others in their career choices and activities."

Natalia Dubrovinskaia, Professor, University of Bayreuth, Germany

Thanks for including me in this work. I really appreciate it. I hope that our stories will inspire more young women to pursue careers in science and engineering.

Doreen Edwards, Dean, Alfred University, USA

I admire you for getting Wiley onboard and having the patience and fortitude to put this together. Any opportunity to convince young women about opportunities in engineering is important to seize.

Katherine Faber, Professor, California Institute of Technology, USA

What an excellent idea to provide inspiration to those of us in the midst of our careers, and to those that will follow.

Liesl Folks, Dean, University of Buffalo, USA

Thanks again for making me reflect on all that has transpired over the years!

Rosario Gerhardt, Professor, Georgia Institute of Technology, USA

The idea for your book is splendid, and it will be very interesting to read the different biographies and experiences of female scientists. Good luck for finishing it soon. (I know that it takes a lot of stamina to get the contributions from all these busy women!)
Dagmar Gerthsen, Professor, Karlsruhe Institute of Technology, Germany

I am so pleased and honored to be invited to be part of this important work. You remind us that we are important to the younger generation. Thank you for your efforts.
Sossina Haile, Professor, Northwestern University, USA

Your book sounds exciting. It is a gift to women scientists, and your efforts are appreciated.
Susan Houde-Walter, CEO and Co-founder, LaserMax, Inc., USA

I think that your initiative is really a great idea. You certainly should be heartily commended for undertaking and completing this magnum opus that will surely serve to as a significant encouragement to the next generation of young women. For them, I thank you from the bottom of my heart. Good job, Lynnette! Well done!
Trudy Kriven, Professor, University of Illinois Urbana-Champaign, USA

Congratulations! This sounds like a very interesting project.
Jennifer Lewis, Professor, Harvard University, USA

Thank you very much for writing this book. It will be very encouraging to other women scientists...and good luck with the final touches of the book.
Lourdes Salamaca-Riba, Professor, University of Maryland, USA

It is great that you are doing this book. It shows that progress has been made, over the past 50 years, in the number of women and positions they are able to hold in engineering and science. I hope this book will encourage more women to enter and remain in scientific and engineering-related careers.
Maxine Savitz, Former Vice-President, National Academy of Engineering,
Washington, DC, USA

I see women scientists' experiences through the microcosm of Bell Labs and the affirmative action effects in the corporate environment of the 1960s onwards. Your questions made me think back on the good times, but as you know there were challenges.
Eva Vogel, Former Distinguished Member of Technical Staff,
Lucent Technologies—Bell Labs, USA

Profile 1

Alida Bellosi

Director, The Institute of Science and Technology for
Ceramics (ISTEC)
National Research Council of Italy (CNR)
Via Granarolo 64
48018 Faenza
Italy

e-mail: alida.bellosi@istec.cnr.it
telephone: +39 0546 699712

Alida Bellosi in the laboratory
in 1980.

Birthplace
Castel Bolognese (in the Province of Ravenna), Italy
Born
January 13, 1952

Publication/Invention Record

>150 publications: h-index 30
4 patents, editor of 7 books

Tags

❖ Administration and
   Leadership
❖ Government
❖ Domicile: Italy
❖ Nationality: Italian
❖ Caucasian
❖ Children: 2

Proudest Career Moment (to date)

Appointment as Director of ISTEC-CNR, Institute of Science and Technology for
Ceramics, belonging to National Research Council of Italy (CNR).

Academic Credentials

Ph.D. (1988) Physical Chemistry, National Competition for Title/Degree, Italy.
M. Degree (1974) Physics, Bologna University, Bologna, Italy.

Research Expertise: oxide and non-oxide-based ceramics (nitrides, borides, carbides,
and related composites), forming and sintering processes, oxidation and corrosion resistance of structural ceramics, joining dissimilar materials, materials design and engineering

Other Interests: evaluating and discussing collaborative programs, editing books and
proceedings, training of young scientists, scientific organization of congresses and schools

Successful Women Ceramic and Glass Scientists and Engineers: 100 Inspirational Profiles. Lynnette D. Madsen.
© 2016 The American Ceramic Society and John Wiley & Sons, Inc. Published 2016 by John Wiley & Sons, Inc.

Key Accomplishments, Honors, Recognitions, and Awards

- Stuijts Award from the European Ceramic Society, 2015
- Elected Fellow of the European Ceramic Society, 2013
- Elected Fellow of the American Ceramic Society, 2013
- Promoter and Organizer of MiMe—Materials in Medicine International Conference, Faenza, 2013
- Promoter and Co-Chair of CERMODEL 2013—Modelling and Simulation Meet Innovation in Ceramics Technology, International Congress, Trento, Italy, 2013
- Chairperson of the Working Group "Research and Development" of the European Ceramic Society, 2010–2013
- Member for the International Advisory Committee of CICECO—Centre for Research in Ceramics and Composite Materials, Aveiro, Portugal, 2008–2011
- Selected by the Italian Ceramic Society for the role of Chairperson of the 2nd International Congress on Ceramics (ICC2), Verona, 2008
- Co-Chair of the Meeting "Ultra-High Temperature Ceramics: Materials for Extreme Environment Applications," Engineering Conferences International, Lake Tahoe, USA, 2008
- Scientific Coordinator for the bilateral agreement between ISTEC and Shanghai Institute of Ceramics, Shanghai, China. Project on processing and characterization of ultra-high-temperature ceramics, 2007–2010
- Scientific Coordinator for the bilateral agreement between ISTEC and Institute of Inorganic Chemistry, Bratislava, Slovakia. Project on the development of structural ceramics, 1997–2007
- Member of the Editorial Board of Materials Letters, Elsevier since 2008
- Member of the Scientific Committee of the Department "Manufacturing Technologies" of CNR, 2006–2011

Alida Bellosi at the Conference "Engineering Ceramics", NATO ASI Series in 1996 in Slovakia.

- Member of the "Technical and Scientific Committee" of ASTER, Emilia Romagna Region, Italy, 2006–2009
- Award for 2nd place in the poster presentation, "Microstructure and properties of porous SiC templates from soft woods" at the "8th International Conference on Ceramic Processing Science" in Hamburg, 2002

- Award for 3rd place in the poster presentation, "Microstructure and properties of ultrafine SiC produced through liquid phase sintering of nanopowders" at the "8th International Conference on Ceramic Processing Science" in Hamburg, 2002
- Award for 2nd place in the presentation, "Fabrication of Al_2O_3–SiC nanocomposites" at the "2002 Conference on Advanced Ceramics and Composites" in Cocoa Beach, 2002
- Coordinator of the Section "Advanced Ceramics Processing Technologies" in the frame of the National Targeted Project on Advanced Materials, 1989–2000
- Evaluator of national and regional projects in Italy, 1998–2013
- Expert of the European Commission for the Evaluation of Proposals, 2001
- Director and Scientific Organizer of the NATO Advanced Research Workshop "Interfacial Science in Ceramic Joining," Bled, Slovenia, 1997
- Congress Chair of the IV European Ceramic Society Conference: Meeting and Exhibition, Riccione, Italy, 1995
- Acknowledgment of the European Ceramic Society for the role of "Congress Chair" of the IV European Ceramic Society Conference: Meeting and Exhibition, Riccione, Italy, 1995

The award ceremony during the 4th International Conference of ECerS. The award is given to Alida Bellosi as Chairperson of the Conference, by the President of the Italian Ceramic Society Prof. Leopoldo Cini (left-hand side) and by the President of the European Ceramic Society Dr. Gian Nicola Babini (right-hand side).

- Co-Chair of Special Session "Modern Applications of Electron and Scanning Probe Microscopy to Ceramics," IV European Ceramic Society Conference, Riccione, Italy, 1995
- Co-Chair of the Short Course "Advanced Techniques for Surface Analyses of Ceramic Materials: Theory and Applications," in the frame of the IV European Ceramic Society Conference, 1995
- Scientific Organizer of the Workshops "Hexagonal Initiative, Italy/Europe" and "Workshop Italy/USA," Rimini, Italy, 1992
- Scientific Organizer of the International Congress CERMAT '92, Rimini, Italy, 1992
- Member of the Italian Delegation at International Bilateral Meeting on New Ceramics, between UK and Italy, promoted by CNR and Materials Commission Science and Engineering Research Council, 1990

Biography

Early Life and Education

Alida Bellosi was raised in Italy by Italian parents; she has two sisters and one brother. They lived in the fertile countryside of Romagna until she was about 25 years old. She received her early education in the public schools and continued with the higher school education at the Liceo Scientifico in Faenza, where she graduated in 1970.

She carried out her undergraduate studies at the University of Bologna, Faculty of Physics, where she received the Degree (equivalent to the current Master's Degree) in Physics, summa cum laude, on July 22, 1974. At that time, doctoral courses/programs had not yet been established in Italy. The Ph.D. program began in 1985—at the start, scientific researchers participated in a National Competition. Based on her qualifications and examination results, Alida Bellosi received her Ph.D. in Physical Chemistry in 1988.

Alida Bellosi is congratulated at the degree ceremony after completion of her thesis dissertation, University of Bologna, 1974.

From 1974 to 1975, she served as a volunteer researcher at the Research Unit National Group on Materials Structure, Institute of Physics, University of Bologna. In 1976, she joined the National Research Council of Italy, in the role of scientific researcher, at the Institute of Science and Technology for Ceramics, in Faenza.

In 1977, she got married. She is the now the mother of two children: Daniela (1981) and Alberto (1985).

Career History

Since 1976, as a researcher at ISTEC, Alida Bellosi has been the coordinator of research projects, responsible for the management of scientific laboratories, and the reference scientist for research contracts. Her technical expertise is in new ceramic materials (oxides, non-oxides, and composites) for structural, electrical, and biomedical applications. She has held some key positions within ISTEC. For example, she has served as Head of research projects on innovative ceramic engineering applications, Head of the Laboratory of Electron Microscopy and Microanalysis, Coordinator of ~30 research contracts and activities with third parties and companies, and Principal Investigator of research projects within national programs and of research groups within European projects. In addition, she has been responsible for collaboration activities with several international institutions and the management–organi-zation–assessment of programs and collaborations at national and international level. Scientific activities on the development and characterization of new materials for innovative applications have been carried out in coordination with other international activities, and accompanied by various initiatives and commitments aimed at ensuring the contribution and participation of ISTEC in the international scenario of new materials.

Alida Bellosi has devoted considerable effort to the implementation of initiatives related to training of students and young scientists at all levels. She has personally contributed to teaching in various venues and has collaborated in the organization of professional courses and schools of specializations on advanced materials. Some significant examples include teaching (seminars/lectures in schools, master classes, and university courses), mentoring (fellows, young researchers, and guests), co-tutoring undergraduates, graduate students, and interns, serving as a member of doctoral dissertation committees, and organizing schools, seminars, workshops, and conferences.

She has carried out numerous initiatives related to the promotion of research and the integration with other local, national, and international contributions, aimed primarily at supporting innovation. She has presented keynote and invited talks at ~40 conferences and international symposia and has acted as co-chair of several congresses.

Alida Bellosi established a strong collaboration network of relationships with many universities, nationally and abroad in the United States, France, Belgium, The Netherlands, Slovak Republic, Venezuela, China, Vietnam, Indonesia, and Japan.

3 Most Cited Publications

Title: Processing and properties of zirconium diboride-based composites
Author(s): Monteverde, F; Bellosi, A; Guicciardi, S
Source: Journal of the European Ceramic Society; volume: 22; issue: 3; pages: 279–288; published: March 2002
Times Cited: 218 (from Web of Science)

Title: Advances in microstructure and mechanical properties of zirconium diboride based ceramics
Author(s): Monteverde, F; Guicciardi, S; Bellosi, A
Source: Materials Science and Engineering A—Structural Materials: Properties, Micro-structure and Processing; volume: 346; issue: 1–2; pages: 310–319; article number: PII S0921-5093(02)00520-8; published: April 15, 2003
Times Cited: 215 (from Web of Science)

Title: Oxidation of ZrB_2-based ceramics in dry air
Author(s): Monteverde, F; Bellosi, A
Source: Journal of the Electrochemical Society; volume: 150; issue: 11; pages: B552–B559; published: November 2003
Times Cited: 135 (from Web of Science)

ResearcherID (B-6167-2014)

Challenges

These are the major challenges in my career:

1. It was challenging to attend university and obtain a degree with honors in the shortest academic time. At that time, it was not easy for young students living in the countryside to attend the university because of the personal sacrifices (associated with the family supporting the necessary costs) and poor transportation; i.e., a daily journey by train was made from home to Bologna early in the morning, which then returned late at night.

2. Finding the right conditions to build a career in science while working at ISTEC-CNR has been challenging. Over the years, I have held positions of head of laboratories, research projects of numerous contracts with companies, and groups of units in several European, national, and regional programs. I have collaborated with research groups in many countries, and I played an active role within the European Ceramic Society.

3. I was appointed Director of ISTEC-CNR: it was a challenge for me to collect the qualifications and merits to be admitted to the competition for this role as Director. The role of Institute Director constitutes the top level in the command structure of the National Research Council of Italy; above, there is only the centralized governance in the CNR headquarters in Rome. CNR is the major public organization for research in Italy. The Institute Director (one hundred positions at national level for one hundred institutes) has full responsibility of the research structure (s)he is leading: for the scientific activities, all of the legal and administrative aspects including personnel, building, goods, and assets, and for the financial management and strategic decisions and policy for the development of the Institute.

On being a woman in this field . . .

In both the roles of a scientific researcher and Institute Director, I did not experience problems based on being a woman in this field. I did not suffer from overt discrimination, nor did I resent it, if sometimes I did not see my efforts recognized thoroughly. The limiting factor for moving forward in my career (more quickly) was having children. Having a family brings responsibilities and commitments that are associated with hard work and many sacrifices: the freedom to manage my own time and the scheduling of my own days has been affected significantly by being a mother. And for some of the critical years as a mother, my job performance and accordingly my career prospects suffered some setbacks. But I have no regrets, it was worth it—I would do it again. I would choose to have a family, even though this extra "burden on my shoulders" has left its mark on me, personally, and on the output from my job. I'm sure I could have done more in my career without the constraints and responsibilities of having a family, but I'm very happy with my choice. I recognize that I have also been lucky since I have had the good fortune to carry out "the unique job I dreamed of" and, at the same time, I also had kids. I have no complaints or discontent for what I may have missed and pathway has been filled with its share of rewards.

Words of Wisdom

I will share my thoughts about a research-focused career path versus an administrative one. Although I had learned a lot from Dr. Babini, the former Director of ISTEC, by closely following his direction for 23 years, when I took over the role of ISTEC (in 2010) I only then fully realized how heavy and burdensome the load was. As Director, I am responsible to ensure the ongoing professional development of staff, to provide the institute with the technological skills it needs. Moreover, it is very demanding to develop strategies to keep pace with the times in the various sectors and lead the frontier of research at the international level. Among the ethical responsibilities, the most engaging and compelling is to ensure maximum standard of safety and protection of personnel and capital goods. Having a career as

A recent image of Alida Bellosi at her desk at ISTEC-CNR.

a researcher is not sufficient preparation to be a good Director of an institute such as ISTEC (where there are ~100 people). After about 4 years in this role as a Director, I feel comfortable and positive. I'd make this choice again, even though my research has taken a back seat. Conducting research satisfies an innate curiosity to study, investigate, compare, discuss, plan, correlate data and theories, propose solutions, formulate models, explain what no one before has attempted to do, compare at the international level, propose innovative ways, etc. However, in the role of Director, it's impossible to maintain an active role in the research: there is no time to collect extensive data, to read or write articles, and to design and follow experimental activities. The daily duties focus on circulars, regulations, reports, financial statements, and contracts. It is rather impossible to have an active presence in the international scientific community; that requires the maximum level of knowledge, being up-to-date, and having an ongoing consistent presence at conferences and meetings. However, the satisfaction comes in a different way. I am glad to have created the conditions for future development of the Institute, to have laid the foundation for the growth of staff and young students, and to have encouraged so many young people in research. It is excellent to see their enthusiasm and interest. In closing, let me say that a researcher has "the best job in the world," while the director has "a continuous and interesting uphill challenge." Some of the perks/ benefits provided to directors make it feel "not so bad."

I want to advise young people looking at a career in research: study a lot, commit, never give up, try to rise above the others, make a name for yourself, travel and open your mind to the experience, listen a lot to senior scientists (at least at the beginning), treasure the advice of someone who has experience, sacrifice something of your private life for professional growth, and believe in what you do. In the lucky case, your road will be TO DO RESEARCH: convince yourself that you had a great fortune and you are a great soul, you have undertaken one of the most exciting existing jobs, though . . . you'll never (probably) be rich in money, but (certainly) very loaded with personality and intellect.

Serena M. Best

Professor
University of Cambridge
Department of Materials Science and
Metallurgy
27, Charles Babbage Road
Cambridge
United Kingdom

e-mail: smb51@cam.ac.uk
telephone: +44 1223 334307

Serena Best cooking in Luxembourg at about age 21.

Birthplace
United Kingdom

Born
January 24, 1964

Publication/Invention Record
>175 publications: h-index40 (Web of Science)

Proudest Career Moment (to date)
My proudest career moment was being elected to a Chair in Cambridge and as Fellow of the FREng.

Academic Credentials
Ph.D. (1990) Bioceramics, University of London, London, United Kingdom (UK).
M.A. (2003) University of Cambridge, Cambridge, UK.
B.Sc. (1986) Materials Science and Technology(2i), University of Surrey, Guildford, Surrey, UK.

Research Expertise: bioceramics, skeletal reconstruction, bioactive glass-ceramic implants and articulating surfaces

Tags
❖ Academe
❖ Domicile: UK
❖ Nationality: UK
❖ Caucasian
❖ Children: 2

Successful Women Ceramic and Glass Scientists and Engineers: 100 Inspirational Profiles. Lynnette D. Madsen.
© 2016 The American Ceramic Society and John Wiley & Sons, Inc. Published 2016 by John Wiley & Sons, Inc.

Key Accomplishments, Honors, Recognitions, and Awards

- Chair, Medical Technologies Community of Practice, Royal Academy of Engineering
- Editorial Board member Royal Society Journal, *Interface*, 2009 to present
- Member, Royal Academy of Engineering, Biomedical Engineering Panel, 2009 to present
- Chair, Biomedical Applications Division and Council member, IoM3, 2005 to present
- Editor, *Journal of Materials Science: Materials in Medicine*, 2004 to present
- Editorial Board member, *Advances in Applied Ceramics*, 2002 to present
- President's Prize from the UK Society for Biomaterials (UKSB), 2015
- Fellowship of the Royal Academy of Engineering (FREng.), 2012
- Fellowship of the Federation of Institutions for Biomaterials Science and Engineering (FBSE), 2012
- External Evaluator, Materials Science, Chalmers University, Sweden, 2012
- Chapman Medal from the Institute of Materials, Minerals and Mining, 2011
- Kroll medal from the Institute of Materials, Minerals and Mining, 2011
- Grant Application Review Panel Member, Finnish Science Academy, 2011
- Elected Academician of World Academy of Ceramics, 2009
- Editorial Board member, *European Cells and Tissues*, 2004–2009
- Keynote Lecture at Materials Chemistry Forum 8 of the Royal Society of Chemistry: Advancing Materials by Chemical Design, 2008
- Global Road Map Editor for Ceramics for Biotechnology and Healthcare, International Ceramic Federation, (Verona, ICC2), 2008
- Council member, Royal Microscopical Society, 2004–2008
- Fellowship of the Institute of Materials Minerals and Mining (FIMMM), 2007
- External Evaluator Nanyang Technological University, Singapore, 2007
- UK—Netherlands Bilateral Consultation on "Regenerative Medicine"—British Embassy, The Hague, 2006
- Chair, Electron Microscopy Committee, Royal Microscopical Society, 2001–2004
- Jean Leray Award from the European Society for Biomaterials, 1999
- Chartered Engineer through the Institute of Materials, Minerals and Mining, 1995
- Founding Member of the UK Society for Biomaterials, 1996

Biography

Early Life and Education

Serena Best completed a first degree in 1986 at Surrey University in Materials Technology before joining Professor Bonfield in the Materials Department at Queen Mary and Westfield College to study for a doctorate in Bioceramics. After she was awarded her Ph.D. in 1990, Professor Best joined the Cookson Group to work at the Cookson Technology Centre in Yarnton, Oxfordshire. The Interdisciplinary Research Centre (IRC) in Biomedical Materials was established toward the end of 1991 and

Professor Best returned to Queen Mary and Westfield College to run the Bioceramics Research Group in the Department. Areas of particular interest in the Group include the synthesis and characterization of novel apatite-based ceramics with the objective of systematically studying the biological and mechanical effects of the addition of carbonate, fluoride, magnesium, silicon, and sodium ions into the hydroxyapatite lattice.

Credit: http://www.msm.cam.ac.uk/ccmm/about/academic/bestbiog.html

Career History

Professor Serena Best was appointed reader in the Department of Materials Science and Metallurgy, University of Cambridge in October 2003 after serving as lecturer since 2000. Now, Professor of Materials Science and a Fellow of St. John's College, together with Prof. Ruth Cameron she directs the Cambridge Centre for Medical Materials. She originally started her medical materials research in the bioceramics field and this work culminated with the spinout of ApaTech. Her interests encompass bioactive ceramics, coatings, composites, and scaffolds for skeletal and soft tissue repair and regeneration. Work in the collagen scaffolds area led to the spinout of a second company, Orthomimetics. She is a Fellow of the Royal Academy of Engineering and also the Institute of Materials, Minerals and Mining, she was recently awarded both the Chapman Medal and the Kroll Medal for her work in her field. She is Editor of the Journal of Materials Science: Materials in Medicine and has been invited to act as a specialist on both national and international panels.

A recent image of Serena Best skiing. (See insert for color version of figure.)

In her spare time, Serena enjoys cookery, house renovation, skiing, and yoga.

Credit: http://www.msm.cam.ac.uk/ccmm/about/academic/bestbiog.html

3 Most Cited Publications

Title: Chemical characterization of silicon-substituted hydroxyapatite
Author(s): Gibson, IR; Best, SM; Bonfield, W
Source: Journal of Biomedical Materials Research; volume: 44; issue: 4; pages: 422–428; doi: 10.1002/(SICI)1097-4636(19990315)44:4<422::AID-JBM8>3.0. CO;2-#; published: March 15, 1999
Times Cited: 292 (from Web of Science)

Title: A comparative study on the in vivo behavior of hydroxyapatite and silicon substituted hydroxyapatite granules
Author(s): Patel, N; Best, SM; Bonfield, W; et al.
Source: Journal of Materials Science-Materials in Medicine; volume: 13; issue: 12; pages: 1199–1206; doi: 10.1023/A:1021114710076; published: December 2002
Times Cited: 247 (from Web of Science)

Title: Characterization of porous hydroxyapatite
Author(s): Hing, KA; Best, SM; Bonfield, W
Source: Journal of Materials Science-Materials in Medicine; volume: 10; issue: 3; pages: 135–145; doi: 10.1023/A:1008929305897; published: March 1999
Times Cited: 199 (from Web of Science)

Challenges

Combining family life with an academic career!

On being a woman in this field . . .

I have enjoyed every stage of my career and feel that being a female academic in the field has brought many more advantages than disadvantages.

Words of Wisdom

I have always been lucky to have a mentor who has pushed me to achieve as much as I can. The only point to remember is that it is sometimes necessary to prioritize the many opportunities that are presented.

Profile 3

Viola I. Birss

Department of Chemistry
University of Calgary
2500 University Drive NW
Calgary, Alberta T2N 1N4
Canada

e-mail: birss@ucalgary.ca
telephone: (403) 220-6432

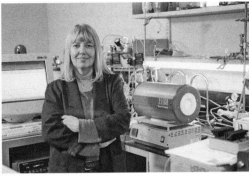

Dr. Viola Birss in the high-temperature solid oxide fuel cell/electrolysis cell lab at the University of Calgary.

Birthplace
Coleman, Alberta, Canada
Born
October 6

Publication/Invention Record
>150 publications: h-index 29
(Web of Science)

Tags
❖ Academe
❖ Domicile: Canada
❖ Nationality: Canadian
❖ Caucasian
❖ Children: 0

Proudest Career Moment (to date)

When I received notice that we had successfully landed the funding (ca. $6 million) to support the pan-Canadian Solid Oxide Fuel Cell (SOFC) Canada Strategic Research Network, after several years of work on my part. The Network was focused on bringing the Canadian SOFC community together under one large research umbrella while also carrying out breakthrough research especially on Ni-based SOFC anodes and also on several new SOFC cell architectures. The competition was very severe, starting with over 50 teams competing for these funds. This was then whittled down to roughly 10, and finally, we learned that we were one of only a few teams that received 5 years of Network funding in Canada. My role in the Network was as the Scientific Director, a position that I held for the lifetime of the Network.

Academic Credentials

Ph.D. (1978) Chemistry, University of Auckland, Auckland, New Zealand.
B.Sc. (1972) Chemistry, University of Calgary, Calgary, Alberta, Canada.

Successful Women Ceramic and Glass Scientists and Engineers: 100 Inspirational Profiles. Lynnette D. Madsen.
© 2016 The American Ceramic Society and John Wiley & Sons, Inc. Published 2016 by John Wiley & Sons, Inc.

Research Expertise: fuel cells, electrolysis cells, electrochemical energy conversion, electrode processes, corrosion, supercapacitors, corrosion and wear resistance of surfaces

Other Interests: supervision and mentoring of students

Key Accomplishments, Honors, Recognitions, and Awards

- Director, Calgary Advanced Energy Storage and Conversion Research Group Technologies (CAESR-Tech), 2013 to present
- One of the founders and now Co-Chair of the Western Canada Fuel Cell Initiative, a group of researchers from the Universities of Calgary, Alberta, BC, Victoria, Saskatchewan, and Manitoba, as well as the Alberta Research Council, 2002 to present
- Mentor and group leader of the Chemistry Women in Science and Engineering Connections Group, 1989 to present
- Canada Research Chair (Tier I)—Fuel Cells and Related Energy Applications, 2004–2011 and 2011–2018
- Founder and Scientific Co-Director of a pan-Canadian network of researchers in the solid oxide fuel cell area (SOFC Canada Strategic Research Network), consisting of over 20 Canadian research groups (at 8 universities, the National Research Council, and Alberta Innovates Technology Futures), roughly 100 graduate students and postdoctoral fellows, and ca. 15 industry partners, 2008–2013
- Fellow of The Royal Society of Canada, 2011
- Alberta Science and Technology Leadership (ASTech) Foundation's Outstanding Leadership in Alberta Technology, Honouree, 2010
- Fellow of the Electrochemical Society Inc., 2007
- Top 40 Alumni, University of Calgary, 2006
- National Science Foundation ADVANCE Distinguished Lectureship, 2005
- Killam Resident Fellowship, July 1–December 31, 2002
- Member of the prestigious Killam Fellowship and Prize Selection Committee, 2000–2003
- Editor of the Canadian Journal of Chemistry, 1993–2001
- Sigma Xi, President of Calgary Chapter, 2000–2001
- Nominated for Graduate Student's Association Teaching Excellence Award, University of Calgary (one of three nominees from Faculty of Science), 2000
- CIC Lecture Award, presented at the University of Sherbrooke, Sherbrooke, Québec, 1998
- Member of the Natural Sciences and Engineering Research Council of Canada (NSERC) Strategic Grant (Energy and Mineral Resources) Selection Panel, 1994–1997
- Faculty of Science, University of Calgary, Excellence in Research Award, 1995
- YWCA Woman of Distinction Award in Science & Technology, Calgary, 1994

- First recipient of the Clara Benson Award of the Canadian Society of Chemistry (sponsored by the Council of Canadian University Chemistry Chairs), 1993
- Member, and then Chair, of the NSERC Grant Selection Committee (Physical and Analytical Chemistry), 1989–1992
- W. Lash Miller Award in Electrochemistry, Canadian Section of the Electrochemical Society Inc., 1985–1987
- Commonwealth Scholar, Canada to New Zealand, 1975–1978

Biography

Early Life and Education

I grew up in a small town (the Crowsnest Pass) in the mountains of southern Alberta, attending the first 5 years of school there, while also enjoying the fact that wilderness was just outside our back door. I then moved with my family to Calgary, which gave us access to better schools, jobs for my parents, etc. I always loved science as I found it stimulating and challenging, especially compared to the social sciences, humanities, etc., and this had great appeal to me. As I approached my university years, I felt that physics was a little too abstract and biology somewhat too descriptive, and felt that chemistry provided just the right mix of everything. Also, I was attracted to the fact that there is usually a "right answer" in the sciences, versus primarily opinions and conjecture, and that is why I chose chemistry as my field as an undergraduate student. Further, I always had a strong concern for the environment, and particularly in identifying clean and sustainable ways of converting, storing, and using energy. This drew me into the field of electrochemistry and materials science, with a particular emphasis on clean energy options, as a graduate student and afterward.

Beyond science, my main interests focus on the outdoors (wilderness hiking, cross-country skiing, etc), certain sports (tennis, other racket sports, bicycling, etc.), and yoga. These activities have often included members of my research group, which helps to build a team and adds collegiality. I also very much enjoy reading, and typically read one or two books per week.

Career History

Dr. Birss was a Commonwealth Scholar who obtained her doctorate in the research area of anodic films on silver electrodes, from the University of Auckland under the supervision of Dr. G.A. Wright. She completed a postdoctoral position with Dr. B.E. Conway at the University of Ottawa where she carried out pioneering work on the supercapacitive properties of hydrous metal oxides, particularly Ru oxide, a project funded by Continental Group Inc. (CGI). Dr. Birss was then hired by CGI as a Research Scientist to continue this work, most of which has been patented, and which then spawned the current interest in supercapacitive devices.

Subsequently, Dr. Birss began her independent career as a Research Scientist with Alcan International Ltd. in Kingston, Ontario, where she developed several new approaches for establishing the susceptibility of Al alloys to pitting and stress corrosion.

She also contributed significantly to the development of a new, high-strength, corrosion-resistant Al–Mg–Si alloy. After 1 year at Alcan, Dr. Birss joined the Chemistry Department at the University of Calgary as an Assistant Professor in 1983, where she was promoted to Associate Professor in 1987 and then to Full Professor in 1991, followed by her appointment, in 2004, as a Canada Research Chair (Tier I, Electrochemistry of Fuel Cells and Related Energy Applications).

Since joining the University of Calgary, Dr. Birss has published close to 160 very high caliber refereed papers, ca. 50 conference proceedings papers, and over 100 research reports for industrial sponsors. Her principal area of research has been focused on understanding and controllably modifying the electrochemical, morphological, chemical, and physical properties of a wide range of films and layers on electrode surfaces. The most significant contributions have included pioneering work on redox-active metal oxides, the development and evaluation of catalytic

Dr. Viola Birss in the materials testing lab at the University of Calgary.

materials and microstructures for fuel cell applications, the protection of Al and Mg alloys from corrosion using a novel anodization method, and, more recently, the development of a highly promising redox-active glucose sensor. Leading research in the low-temperature fuel cell area has included the development of novel approaches to the formation of electrocatalysts for methanol oxidation and the preparation of low-cost non-noble metal electrocatalysts for the reduction of oxygen. In the SOFC area, Dr. Birss' key research achievements include the establishment of the kinetics and mechanisms of fuel oxidation and oxygen reduction reactions using three-electrode electrochemical methods, the development of a number of highly promising new fuel cell materials, and understanding and overcoming SOFC electrode degradation processes.

Dr. Birss has trained over 50 graduate students, over 70 undergraduates, and approximately 30 postdoctoral fellows, most of who are now employed in prestigious positions in the government, industry, and university sectors in Canada and around the world. Dr. Birss has also been extremely well funded from both government and industry sources. She has held a wide range of NSERC grants, funding from the National Research Council (NRC)—NSERC Fuel Cell Program, the Alberta Energy Research Institute, Western Economic Diversification, CANMET, and more. Her collaborations with Canadian companies and knowledge transfer to them continue to be extremely strong. In past years, these have included Alcan International, Allied

Signal, Advanced Measurements, The Electrometallurgy Consortium (which included Alcan, Canadian Electrolytic Zinc, Falconbridge Ltd., Hydro-Québec, INCO Ltd., and QIT-Fer et Titane), ESSO Chemicals, Shell Canada, Travis Chemicals, and Versa Power (previously Global Thermoelectric). Dr. Birss' group is currently in close collaboration on a wide range of electrochemical projects with Ballard Power Systems, Nova Chemicals, and Seal Well, as well as Honeywell Aerospace in the United States, who are now scaling up one of Dr. Birss' electrochemical coating processes for potential commercialization. At present, her research group consists of over 20 individuals, including many highly talented students holding prestigious NSERC and Alberta Ingenuity Fund (AIF) awards, and her research is supported by grants totaling close to $750,000 per year.

Dr. Birss is the founder and Co-Chair (since 2002) of the Western Canada Fuel Cell Initiative (WCFCI), involving over 35 research groups at eight institutions in the four western provinces. The Alberta component of the WCFCI received $2 million of funding over 2 years (2006–2007) under Dr. Birss' leadership. She was then the co-founder and Co-Chair of Solid Oxide Fuel Cells Canada (SOFC Canada), a national umbrella organization involving all relevant groups in the SOFC area within the triple helix of industry, government, and academia in Canada. As of January 2008, Dr. Birss has been the Scientific Co-Director of the SOFC Canada NSERC Strategic Research Network, funded at $5.5 million over 5 years, distributed over 16 research groups at seven universities from across Canada. The Network also involves research groups from both the Alberta Research Council and the NRC, as well as 12 industrial partners. This has led to a very high level of collaboration and multiple student exchange opportunities within Canada.

Dr. Birss' outstanding contributions in the electrochemical area have often been covered by the media. She is also very well known at all levels at NSERC, particularly as a result of the many examples she has brought to their attention of the importance and impact of electrochemistry within Canada. Dr. Birss is recognized internationally for the breadth and depth of her many outstanding accomplishments and contributions to the field of electrochemical science, as well as for significantly advancing the field of electrochemistry within Canada. She has also been the recipient of a number of Calgary-based awards in recognition of her many significant and far-reaching research contributions, her excellence in all aspects of teaching and training, and her dedicated mentorship of young women in science. Dr. Birss is also recognized for the many opportunities she provides for her students to attend and participate at national and international meetings, and her students have often been the recipients of best poster awards at conferences.

3 Most Cited Publications

Title: Conversion of methane by oxidative coupling
Author(s): Amenomiya, Y; Birss, VI; Goledzinowski, M; et al.
Source: Catalysis Reviews: Science and Engineering; volume: 32; issue: 3; pages: 163–227; doi: 10.1080/01614949009351351; published: 1990
Times Cited: 389 (from Web of Science)

Title: High performance PtRuIr catalysts supported on carbon nanotubes for the anodic oxidation of methanol
Author(s): Liao, SJ; Holmes, KA; Tsaprailis, H; et al.
Source: Journal of the American Chemical Society; volume: 128; issue: 11; pages: 3504–3505; doi: 10.1021/ja0578653; published: March 22, 2006
Times Cited: 177 (from Web of Science)

Title: Oxygen reduction at sol-gel derived $La_{0.8}Sr_{0.2}Co_{0.8}Fe_{0.2}O_3$ cathodes
Author(s): Liu, JB; Co, AC; Paulson, S; et al.
Source: Solid State Ionics; volume: 177; issue: 3–4; pages: 377–387; doi: 10.1016/j.ssi.2005.11.005; published: January 31, 2006
Times Cited: 84 (from Web of Science)

Challenges

The biggest challenge in my life and in my career is finding enough time to do everything that I'm interested in. This of course includes work-related tasks and activities, but I would also like to have more time for R&R, as well as time just for reflection. Another challenge has been to successfully mentor the next generation(s) of women so that they stay in the field of science. While I have had many female graduate students and postdoctoral fellows and 5–10 of these have gone on to leadership positions in academics or government, there are many who should have gone further. However, there are still barriers for young women to move into these professions, including the heavy time commitment of jobs in these areas, challenges related to childcare, etc.

On being a woman in this field . . .

While things have improved greatly over the years for women, I can easily recall many of the difficult situations that I have encountered in my career specifically related to being a female. However, now that many of the younger members in the field have spouses who are also scientists and engineers, there is much more recognition of the problems women face and the playing field is being leveled. Even so, it is worrisome that there are so few women in science and engineering, and especially in my field of electrochemistry and materials science.

Words of Wisdom

To be a scientist is to be focused on problem solving. If you like to solve puzzles, this is the right area of pursuit. Science is the right field for a person who likes challenge and also wishes to make a very useful contribution to the world. The future seems to be extremely positive for the employment of scientists, as technology continues to advance. Universities are and will increasingly be facing an extreme shortage of professors, while the opportunities in industry are increasing constantly. There are very few unemployed or underemployed scientists, especially as compared to graduates in the social sciences and humanities. Excellent communication skills are essential for scientists of the future.

Dawn A. Bonnell

Professor of Materials Science & Engineering
and Vice Provost for Research
University of Pennsylvania
Philadelphia, PA 19104-6272
USA

e-mail: bonnell@lrsm.upenn.edu
telephone: (215) 898-7236

Dawn Bonnell as a new Assistant Professor. Credit: University of Pennsylvania.

Birthplace

Detroit, MI, USA

Publication/Invention Record

>200 publications: h-index 34 (Web of Science)
5 patents, Editor of 7 books, 13 book chapters

Tags

❖ Administration and Leadership
❖ Academe
❖ Domicile: USA
❖ Nationality: American
❖ Caucasian
❖ Children: 2

Important Career Moment (to date)

There is no single proudest moment. I take great pride in recognition that comes from my colleagues, especially when it derives from crossing disciplines. Two instances have been memorable. The first instance, occurring early in my career, was the publication of the first atomic-resolution scanning tunneling microscope (STM) image of an oxide surface in Science magazine in 1991. This was both validating and gratifying since I had been warned that this area of research was not viable and the experiments would not be

Dawn Bonnell in her laboratory next to her new scanning tunneling microscope 2004. Credit: University of Pennsylvania. (See insert for color version of figure.)

Successful Women Ceramic and Glass Scientists and Engineers: 100 Inspirational Profiles. Lynnette D. Madsen.
© 2016 The American Ceramic Society and John Wiley & Sons, Inc. Published 2016 by John Wiley & Sons, Inc.

successful. The second instance occurred years later when I heard from the National Science Foundation (NSF) that Penn would receive a nanotechnology center grant. As founding director I knew what our outstanding team of faculty and students could achieve . . . and what they have accomplished is beyond any expectation. To work with these colleagues has been a privilege.

Academic Credentials

Ph.D. (1986) Materials Science and Engineering, University of Michigan, Ann Arbor, Michigan, USA.
M.S. (1984) Engineering Materials, University of Michigan, Ann Arbor, Michigan, USA.
B.S.E. (1983) Materials Science and Engineering, University of Michigan, Ann Arbor, Michigan, USA.

Research Expertise: atomic imaging and local electronic structure of complex oxide surfaces, advancing probes of local properties, interfaces in electronic and plasmonic hybrid nanostructures, ferroelectric nanolithography

Other Interests: increasing scientific literacy in the local community, gender equity in STEM fields

Key Accomplishments, Honors, Recognitions, and Awards

- Elected Fellow of the Materials Research Society (MRS), 2015
- American Vacuum Society (AVS) Nanotechnology Award, 2014
- Henry Robinson Towne Professor of Engineering and Applied Science, 2014
- Elected National Academy of Engineering (NAE) Member for development of atomic-resolution surface probes, and for institutional leadership in nanoscience, 2013
- Sosman Award, The American Ceramic Society (ACerS), 2012
- Distinguished Alumni Award, University of Michigan, 2011
- Racheff Distinguished Lecture, University of Illinois, 2009
- Penn State University Distinguished Lecturer, 2008

Dawn Bonnell, Henry Towne Professor in the Department of Materials Science and Engineering, and Nader Engheta, H. Nedwill Ramsey Professor in the Department of Electrical and Systems Engineering, have been named 2015 MRS Fellows. Credit: University of Pennsylvania, ca. 2008.

- Heilmeier Faculty Research Award, University of Pennsylvania, 2006
- Kliegel Lectureship, California Institute of Technology, 2006
- Elected Fellow of the American Association for the Advancement of Science (AAAS), 2006
- Staudinger/Durrer Medal from Swiss Federal Institute of Technology, ETH (in German: Eidgenössische Technische Hochschule), 2005
- Elected Fellow of AVS, 2005
- Provost Distinguished Lectureship, Case Western Reserve University, 2004
- President AVS: Science and Technology Society, 2003
- Ross Coffin Purdy Award for most significant contribution to the literature, ACerS, 2003
- Trustee AVS: Science and Technology Society, 1999–2002
- Vice President Program, Meetings, and Expositions, ACerS, 2000–2001
- Director AVS Science and Technology Society Board of Directors, 1998–1999
- Visiting Professor, ETH Zurich (Swiss Federal Engineering University), 1997
- Fellow of ACerS, 1996
- Presidential Young Investigators Award, NSF, 1990
- Associated Western Universities Faculty Fellowship, 1990
- Fulbright/German Academic Exchange Service (DAAD) Scholarship to Max-Planck-Institute, Alpha Sigma Mu Honors Society, 1984

Biography

Early Life and Education

Dawn Bonnell grew up in the suburbs of Detroit, the eldest of four children in a close-knit family. It was always the intention (by her and her parents) that she would go to college, but she deferred this goal to start a family. Several years later as a single parent of two toddlers, she found the need for more education in order to support her family. Fortunately, at that time she lived virtually next door to an outstanding university, the University of Michigan.

She began introductory classes and found that she enjoyed chemistry, while most of the class did not. Sensing that this was an indication of a future job market (and after doing a bit of research), she decided to become a chemical engineer. After taking several classes on this track it seemed that the content was more about pumps and pipes than chemistry. In contrast, the required materials science course was replete with chemistry . . . though much of it, solid state. After switching to materials science and doing undergraduate research in ceramics with Prof. Tien, she was hooked on research. The focus of her research was to develop and use knowledge of six-component phase relations to control grain boundary structure and ultimately properties of silicon nitride-based materials. She spent a year at the Max Planck Institute in Stuttgart Germany working for Manfred Ruehle. There, she worked in the best electron microscopy group in the world and mapped composition variations at boundaries. The concept of measuring structures at the atomic level to develop an understanding of materials behavior was compelling. This international experience was transformative, both for her and for her children. On their return to Michigan, she wrote and submitted her doctoral dissertation.

Career History

While finishing her thesis, she was invited to interview at the University of Pennsylvania. Barely ready to interview she considered this as a practice interview. To her surprise, she was offered a position at Penn. Since she was quite impressed with the level of intellectual discourse there, she readily accepted. Shortly afterward she was offered the opportunity to do a postdoc at IBM, working for David Clarke. Penn graciously agreed to wait for her to finish at IBM before starting the faculty position.

The experience at IBM was also transformative, this time in terms of the research focus. The scanning tunneling microscope had been invented at IBM and the scientific world recognized its importance immediately. This instrument provided the ultimate in atomic spatial resolution of structure combined with property analysis. Upon arriving at Penn her goal was clear—she wanted to use the power of this approach (and associated techniques) to advance understanding of ceramic materials. This eventually bifurcated into two primary research directions: the relation of atomic structure at interfaces to properties and the role of surface reconstruc-

Dawn Bonnell in her office at the University of Pennsylvania, 2014. Credit: University of Pennsylvania.

tions on chemical interactions. Based on these two fundamental themes, questions could be addressed that impacted diverse and sometimes, unexpected, applications. The areas of application included chemical catalysis, size-dependent properties of nano contacts in electronics, ferroelectric nanolithography, ferroelectric polarization and molecular adsorption, toughness in fiber-reinforced composites, plasmon-induced hot electron generation, and more. It is an amazing playground within which to explore stimulating ideas.

In 2004, she became the founding director of the Nano/Bio Interface Center (NBIC) at Penn. In this role, she led a multidisciplinary team of colleagues from Penn and partner institutions in the United States, Europe, and Asia with the support of $35M to advance both the science and technology enabled by molecular interactions at bio-interfaces. In addition to research, a new undergraduate and graduate educational program on nanotechnology was developed at Penn. NBIC also established a unique laboratory facility consisting of nine scanning probe-based instrument platforms that enable studies ranging from heart disease interventions to plasmon-based energy solutions to nanosensors.

Most recently she became the Vice Provost for Research at the University of Pennsylvania. In this capacity, she shapes policy and advances administrative initiatives for the University's $800+ million per year research enterprise.

She is an active member of AVS (past president), the ACerS (past vice president), MRS, AAAS, American Physical Society (APS), American Chemical Society (ACS) (past editor of journal ACS Nano), and the NAE.

3 Most Cited Publications

Title: Imaging mechanism of piezoresponse force microscopy of ferroelectric surfaces
Author(s): Kalinin, SV; Bonnell, DA
Source: Physical Review B; volume: 65; issue: 12; article number: 125408; published: March 15, 2002
Times Cited: 222 (from Web of Science)

Title: Local potential and polarization screening on ferroelectric surfaces
Author(s): Kalinin, SV; Bonnell, DA
Source: Physical Review B; volume: 63; issue: 12; article number: 125411; published: March 15, 2001
Times Cited: 170 (from Web of Science)

Title: Controlled fabrication of nanogaps in ambient environment for molecular electronics
Author(s): Strachan, DR; Smith, DE; Johnston, DE; et al.
Source: Applied Physics Letters; volume: 86; issue: 4; article number: 043109; published: January 24, 2005
Times Cited: 152 (from Web of Science)

Challenges

Several challenges arose as a consequence of taking a "non-traditional" path toward a career. For example, due to the 7-year break between high school and the university application process, I didn't have SAT scores with which to compete for scholarships. Hence, no assistance of this sort could be obtained for my undergraduate education. In later years, I encountered challenges regarding research awards/program delineated by age. Since my "professional age" was 7 years younger than my physical age, I was sometimes considered for these when I was no longer eligible. It is satisfying to see that several federal agencies and many institutions have amended policies to address this issue.

Words of Wisdom

To me, curiosity and community are key aspects to a satisfying professional life.

In being continuously curious there is nothing but enjoyment in the inevitable diversions that occur in research.

In caring for and about the community (students, colleagues, and international networks of scientific friends), one receives much satisfaction from mentoring others, while benefiting from the strengths of a healthy community.

Inna Petrovna Borovinskaya

Institute of Structural Macrokinetics and
Materials Science (ISMAN)
Russian Academy of Sciences
Institutskaya Str. 8
Chernogolovka
Moscow Region
Russia

e-mail: inna@ism.ac.ru
telephone: +496-5246205

Inna Borovinskaya when she
was about 30 years old.

Birthplace
Podolsk, Moscow Region, Russia

Born
August 20, 1934

Publication/Invention Record

>200 publications: h-index 14 (Web of Science)
>100 patents

Tags

- ❖ Academe
- ❖ Domicile: Russia
- ❖ Nationality: Russian
- ❖ Caucasian
- ❖ Children: 1

Proudest Career Moment (to date)

The discovery of "The Phenomenon of the Wave Localization of Solid-State Autor-
etarding Reactions," which appeared to be the background of SHS (self-propagating
high-temperature synthesis) and put combustion processes and materials science
together.

Academic Credentials

Dr. Sci. (1989), Institute of Structural Macrokinetics and Materials Science, Russian
Academy of Sciences, Chernogolovka, Russia.
Ph.D. (1972), Institute of Chemical Physics, USSR Academy of Sciences, Chernogo-
lovka, Russia.
Chemical Technologist (1957) Chemistry, Ural Polytechnic Institute, Sverdlovsk
(Yekaterinburg), Russia.

Successful Women Ceramic and Glass Scientists and Engineers: 100 Inspirational Profiles. Lynnette D. Madsen.
© 2016 The American Ceramic Society and John Wiley & Sons, Inc. Published 2016 by John Wiley & Sons, Inc.

Research Expertise: chemistry and experimental diagnostics of mechanisms of self-propagating high-temperature synthesis (SHS), material science and technology of SHS products, and the development of advanced inorganic materials

Other Interests: industrial applications of her work

Key Accomplishments, Honors, Recognitions, and Awards

- A member of a few Scientific Councils: Academic Council (ISMAN), Council on Combustion and Explosion (RAS)
- Editorial boards of journals: "Proceedings of Higher School. Nonferrous Metallurgy," till 2013; "Combustion and Plasmochemistry," since 2002
- Advisory Committee, The International Journal of Self-Propagating High-Temperature Synthesis, 1992 to present
- Vice President of the Scientific Council on Theory and Practice of SHS Processes of the Ministry of Science and Technologies of Russian Federation, during the 1980s and 1990s
- International Ceramics Prize in Basic Research, World Academy of Ceramics, 2008
- Elected Member of the World Academy of Ceramics, 2002
- Awarded an honorary medal "for the discovery of SHS and great contribution in fundamentals, chemistry and technology of SHS" at the IV International Symposium on SHS, Toledo, Spain, 1997
- National Prize Laureate of Armenia, 1980; and Russian Federation, 1996
- Order of Friendship, Russian Federation, 1999
- Honorary Professor of Harbin Institute of Technology, 1995
- Honorary Professor of Uralski Polytechnic Institute, 1993

Biography

Early Life and Education

She was born in 1934 in Podolsk, Moscow Region. In 1957, she graduated from Uralski Polytechnic Institute in Sverdlovsk; she subsequently received Cand. Sci. Degree in 1972 and Dr. Sci. Degree in 1989.

Career History

Adapted from http://www.ism.ac.ru/struct/inna/bor.htm.

She started her professional carrier in a small town in the suburbs of Sverdlovsk. In 1961, she moved to Chernogolovka and began to work in the branch of Institute of Chemical Physics, first on the problems of fluorine organic compounds synthesis, and then, from 1966, she studied the chemistry of condensed systems combustion. Since 1980, she has been the head of the laboratory of self-propagating high-temperature synthesis (SHS) processes. In 1987, together with the group headed by A.G. Merzhanov, she started working in the newly established Institute of Structural Macrokinetics. This laboratory, also headed by I.P. Borovinskaya, was transformed into the SHS Research Center, and a set of new laboratories were established.

Prof. I.P. Borovinskaya is the author of more than 200 scientific works and the holder of more than 100 patents. She is also the author (co-authors are A.G. Merzhanov and V.M. Shkiro) of the scientific discovery "The phenomenon of wave localization of auto-inhibited solid phase reactions" ("The Phenomenon of Solid Flame," State Register of Scientific Discoveries in the USSR, 1984).

Basic results on combustion mechanism of metals in nitrogen, fundamental regularities in filtration combustion of porous solids in reactive gas, synthesis of nitrides (including new compounds and phases), new displays of unstable combustion, for example, front auto-oscillation and spin waves, chemical routes of SHS reactions, development of SHS technologies of simple and composite refractory substances, production of nitride ceramics in SHS gas-static plants, forced SHS densification of hard alloys, etc. belong to I. P. Borovinskaya and her co-authors. Some of her results were realized at an industrial scale in Russia and abroad.

Prof. I.P. Borovinskaya is the head of a section of the Academic Council of ISMAN. Sixteen candidate dissertations were defended under her guidance and eight students were awarded doctorates of science under her supervision.

A recent picture of Inna Borovinskaya.

3 Most Cited Publications

Title: New class of combustion processes
Author(s): Merzhanov, AG; Borovinskaya, IP
Source: Combustion Science and Technology; volume: 10; issue: 5–6; pages: 195–201; published: 1975
Times Cited: 164 (from Web of Science)

Title: Gasless combustion of mixtures of powdered transition metals with boron
Author(s): Borovinskaya, IP; Merzhanov, AG; Novikov, NP; et al.
Source: Combustion, Explosion, and Shock Waves; volume: 10; issue: 1; pages: 2–10; published: 1974
Times Cited: 55 (from Web of Science)

Title: Single-crystal beta-Si_3N_4 fibers obtained by self-propagating high-temperature synthesis
Author(s): Rodriguez, MA; Makhonin, NS; Escrina, JA; et al.

Source: Advanced Materials; volume: 7; issue: 8; pages: 745–747; doi: 10.1002/
adma.19950070815; published: August 1995
Times Cited: 50 (from Web of Science)

Challenges

Sometimes I have not been able to bring to finality some of our interesting ideas because
of the difficulties of their industrial implementation.

On being a woman in this field . . .

One must combine the romance of woman with the acumen of man.

Words of Wisdom

• Be what you are.
• Try to combine your private life with scientific interests.

Profile 6

Emily A. Carter

Founding Director, Andlinger Center
for Energy and the Environment
Gerhard R. Andlinger Professor in Energy
and the Environment
Professor of Mechanical and Aerospace Engineering
& Applied and Computational Mathematics
Department of Mechanical and Aerospace Engineering
Princeton University
Princeton, NJ 08544
USA

Emily Carter at age 32.

e-mail: eac@princeton.edu
telephone: (609) 258-5391

Birthplace

Los Gatos, California, USA

Born

November 28, 1960

Tags

❖ Academe

❖ Domicile: USA

❖ Nationality: American

❖ Caucasian

❖ Children: 3

Publication/Invention Record

>300 publications: h-index 58 (Web of Science)

Proudest Career Moment (to date)

Getting the phone call that I had been elected to the National Academy of Sciences. It simply felt like a tremendous affirmation by the best scientists in the world that all the work I'd done, together with my students and postdocs, to make a difference to the state of knowledge in physical sciences and engineering was being recognized as important, impactful, and meaningful.

Academic Credentials

Ph.D. (1987) Chemistry, California Institute of Technology, Pasadena, California, USA.
B.S. (1982) Chemistry, University of California, Berkeley, California, USA.

Research Expertise: theory, computation: development of efficient and accurate first principles quantum mechanics techniques for electron correlation, embedded

Successful Women Ceramic and Glass Scientists and Engineers: 100 Inspirational Profiles. Lynnette D. Madsen.
© 2016 The American Ceramic Society and John Wiley & Sons, Inc. Published 2016 by John Wiley & Sons, Inc.

correlated wavefunction, and orbital-free density functional theories. Applications are focused entirely on enabling discovery and design of molecules and materials for sustainable energy

Other Interests: increasing the understanding of energy and environmental issues, mentoring, eliminating the gender gap

Key Accomplishments, Honors, Recognitions, and Awards

- Member, Inaugural Editorial Advisory Board, ACS Central Science, 2015 to present
- Member, Editorial Advisory Board of Journal of Physical Chemistry Letters, 2014 to present
- Member, International Advisory Board of the Winton Programme for the Physics of Sustainability, Cambridge University, 2011 to present
- Member, Editorial Advisory Board, Journal of Chemical Theory and Computation, 2010–2018
- Member, Board on Energy and Environmental Systems, National Research Council, National Academy of Sciences, 2014–2017
- Member, SLAC National Accelerator Laboratory Scientific Policy Committee, 2014–2016
- Member, National Science Foundation Mathematical and Physical Sciences Advisory Committee, 2012–2015
- 2015–2016 Joseph O. Hirschfelder Prize in Theoretical Chemistry, Theoretical Chemistry Institute at the University of Wisconsin, Madison, 2015
- Fellow, National Academy of Inventors, 2014
- Malcolm Dole Distinguished Summer Lecturer in Physical Chemistry, Northwestern University, 2014
- Ira Remsen Award, Maryland Section of the American Chemical Society, Johns Hopkins University, 2014
- Linnett Visiting Professor of Chemistry, University of Cambridge, 2014
- Member, Editorial Advisory Board of ChemPhysChem, 2000–2014
- Hoyt C. Hottel Lecturer in Chemical Engineering, Massachusetts Institute of Technology, 2013
- Kenneth S. Pitzer Lecturer, Department of Chemistry, University of California, Berkeley, 2013
- Mathematics of Planet Earth Simons Public Lecturer, Institute for Pure and Applied Mathematics, University of California, Los Angeles, 2013
- Lord Lecturer, Department of Chemistry, Allegheny College, 2013
- Sigillo D'Oro (Golden Sigillum) Medal, Italian Chemical Society, Scuola Normale Superiore, Pisa, Italy, 2013
- Article selected for The Journal of Chemical Physics 80th Anniversary Collection (Chen Huang and Emily A. Carter, "Potential-functional embedding theory for molecules and materials", J. Chem. Phys., 135, 194104 (2011))

- Francis Clifford Phillips Lectureship, Xi Chapter of the Phi Lambda Upsilon National Honorary Chemical Society and the Department of Chemistry, University of Pittsburgh, 2013
- Tedori-Callinan Lectureship, Department of Mechanical Engineering and Applied Mechanics, University of Pennsylvania, 2013
- W. Allan Powell Lectureship, Virginia Section of the American Chemical Society and the University of Richmond, 2013
- Member, Editor-in-Chief Search Committee, Science, 2012–2013
- Chair, DOE-BES Council on Chemical and Biochemical Sciences, 2012–2013
- Docteur Honoris Causa from L'École Polytechnique Fédérale de Lausanne, Switzerland (EPFL), 2012
- Fellow of the American Chemical Society, 2012
- Honorary Mathematical and Physical Sciences Distinguished Lecturer, National Science Foundation, 2012
- Dean's Distinguished Lecture, College of Science and Technology, Temple University, 2012
- Member, Board on Chemical Sciences and Technology, National Research Council, National Academy of Sciences, 2010–2012
- Member, Editorial Board of Modelling and Simulation in Materials Science and Engineering, 2001–2012
- MIT Distinguished Speaker in Computational Science and Engineering, Massachusetts Institute of Technology, 2011
- August Wilhelm von Hofmann Lecture Award, German Chemical Society, 2011
- Jerome B. Cohen Lecturer in Materials Science and Engineering, Northwestern University, 2011
- Ernest Davidson Lecturer in Theoretical Chemistry, University of North Texas, 2011
- Gerhard R. Andlinger Professor in Energy and the Environment, Princeton University, 2011–
- Molecular Foundry Distinguished Lecturer, Lawrence Berkeley National Laboratory, 2010
- Coover Lecturer in Chemistry, Iowa State University, 2010
- Material Simulation Distinguished Lecturer, Penn State University, 2010
- Pelz Memorial Lecturer in Mechanical and Aerospace Engineering, Rutgers University, 2010
- Noyes Lecturer in Physical Chemistry, University of Texas, Austin, 2010
- Member, Editorial Board, Annual Review of Physical Chemistry, 2006–2010
- Member, International Academy of Quantum Molecular Science, 2009
- Member, Advisory Editorial Board of Chemical Physics Letters, 1998–2009
- EaSTChem Visiting Fellow, Universities of Edinburgh and St. Andrews, Scotland, 2008
- Member, National Academy of Sciences, 2008
- Fellow, American Academy of Arts & Sciences, 2008

- Welch Distinguished Lecturer in Chemistry, 2008
- Coulson Lecturer in Theoretical Chemistry, University of Georgia, 2008
- Kivelson Lecturer in Physical Chemistry, University of California, Los Angeles, 2008
- Member, Editor-in-Chief Search Committee, Journal of Chemical Physics, 2007–2008
- Old Dominion Faculty Fellow, Council of the Humanities, Princeton University, 2007–2008
- American Chemical Society Award for Computers in Chemical and Pharmaceutical Research, 2007
- Member, Editorial Advisory Board of Accounts of Chemical Research, 2005–2007
- Member, Editorial Board of SIAM Journal on Multiscale Modeling and Simulation, 2001–2007
- Arthur W. Marks '19 Professor, Princeton University, 2006–2011
- Merck-Frosst Lecturer in Chemistry, Concordia University, 2005
- Chair, American Conference on Theoretical Chemistry, 2005
- Chair, Division of Chemical Physics, American Physical Society, 2004–2005
- Guest Editor, Accounts of Chemical Research special issue on Computational and Theoretical Chemistry, 2004–2005
- Member, Los Alamos National Laboratory Theoretical Division Advisory and Review Committee, 2000–2005
- Fellow of the Institute of Physics, 2004
- Member, Editor-in-Chief Search Committee, Journal of Physical Chemistry, 2003–2004
- Dean's Recognition Award for Research, UCLA, 2002
- McDowell Lecturer in Physical Chemistry, University of British Columbia, 2002
- Member, Editorial Board of Journal of Chemical Physics, 2000–2002
- Member, Editorial Board of the Encyclopedia of Chemical Physics and Physical Chemistry, 1999–2001
- Guest Editor, Journal of Physical Chemistry William A. Goddard issue, 1999–2000
- Fellow of the American Association for the Advancement of Science, 2000
- Member, Editorial Advisory Board of Journal of Physical Chemistry, 1995–2000
- Member, Editorial Advisory Board of Surface Science, 1994–1999
- Fellow of the American Physical Society, 1998
- Hanson-Dow Award for Excellence in Teaching, UCLA, 1998
- Defense Science Study Group Member, 1996–1997
- Dr. Lee Visiting Research Fellowship in the Sciences, Christ Church, Oxford University, England, 1996
- Member, Editorial Advisory Board of Molecular Simulation, 1991–1996
- Peter Mark Memorial Award, American Vacuum Society, 1995
- Fellow of the American Vacuum Society, 1995
- Specialist Editor of Computer Physics Communications, 1993–1994

- Herbert Newby McCoy Research Award, UCLA, 1993
- Medal of the International Academy of Quantum Molecular Science, 1993
- Exxon Faculty Fellowship in Solid State Chemistry, American Chemical Society Inorganic Division Award, 1993
- Glenn T. Seaborg Research Award, UCLA, 1993
- Alfred P. Sloan Research Fellow, 1993–1995
- Camille and Henry Dreyfus Teacher–Scholar Award, 1992–1997
- Union Carbide Innovation Recognition Award, 1990–1991
- Faculty Member of Distinction (Undergraduate Teaching Award), UCLA, 1989–1990
- Union Carbide Innovation Recognition Award, 1989–1990
- Camille and Henry Dreyfus Foundation Distinguished New Faculty Award, 1988–1993
- National Science Foundation Presidential Young Investigator Award, 1988–1993
- SOHIO Fellowship in Catalysis, Caltech, 1986–1987
- International Precious Metals Institute and Gemini Industries Research Grant Award, 1985–1986
- Sigma Xi, Caltech, 1984
- National Science Foundation Predoctoral Fellowship, 1982–1985
- Phi Beta Kappa, UC Berkeley, 1982
- Mabel Kittredge Wilson Prize in Chemistry, UC Berkeley, 1982
- Bruce Howard Memorial Scholar, UC Berkeley, 1981–1982
- Coblentz Society Award for Molecular Spectroscopy, UC Berkeley, 1981
- Mildred Jordan Sharp Torch and Shield Award, UC Berkeley, 1981
- Theodore and Edith Braun Scholar, UC Berkeley, 1979–1980
- Alumni Scholar, UC Berkeley, 1978–1982
- Regents Scholar, University of California, Berkeley, 1978–1982

Biography

Early Life and Education

Professor Carter received her B.S. in Chemistry from UC Berkeley in 1982 (graduating Phi Beta Kappa) and her Ph.D. in Chemistry from Caltech in 1987. She spent a year as a postdoctoral researcher at the University of Colorado, Boulder.

Career History

She spent 16 years on the faculty of University of California, Los Angeles (UCLA) as a Professor of Chemistry and later of Materials Science and Engineering. She moved to Princeton University in 2004. She holds courtesy appointments in Chemistry, Chemical and Biological Engineering, and three interdisciplinary institutes (PICSciE, PRISM, and PEI). The author of over 300 publications, she has delivered more than 480 invited

Emily Carter (at the left) talking with two graduate students, Vincent Ligneres and Kristen Marino, in her office (2006). Credit: Denise Applewhite (Princeton) for the Princeton Weekly Bulletin. (See insert for color version of figure.)

lectures all over the world and serves on numerous international advisory boards spanning a wide range of disciplines. Her scholarly work has been recognized by a number of national and international awards and honors from a variety of entities.

Professor Carter is a theorist/computational scientist first known for her research combining *ab initio* quantum chemistry with molecular dynamics and kinetic Monte Carlo simulations, especially as applied to etching and growth of silicon. Later, she merged quantum mechanics, applied mathematics, and solid-state physics in her linear scaling orbital-free density functional theory (OFDFT) that can treat unprecedented numbers of atoms quantum mechanically, her embedded correlated wavefunction and *ab initio* DFT+U theories that combine quantum chemistry with periodic DFT to treat condensed matter ground and excited electronic states and strongly correlated materials, and her fast algorithms for *ab initio* multi-reference correlated electronic wavefunction methods that permit accurate thermochemical kinetics and excited states to be predicted for large molecules. She was also a pioneer in quantum-based multiscale simulations of materials that eliminate macroscopic empirical constitutive laws and have led to new insights into, e.g., shock Hugoniot behavior of iron and stress-corrosion cracking of steel. Earlier, her doctoral research furnished new understanding into homogeneous and heterogeneous catalysis, while her postdoctoral work presented the condensed matter simulation community with the widely used rare event sampling method known as the Blue Moon Ensemble. Her more recent research into how materials fail due to chemical and mechanical effects furnished concepts for how to optimally protect these materials against failure. Her current research is focused entirely on enabling discovery and design of molecules and materials for sustainable energy, including converting sunlight to electricity and fuels, providing clean electricity from solid oxide fuel cells, clean

Emily Carter (in the middle) talking during a 2008 research group meeting with her graduate student, Donald Johnson; (at the left) graduate student, Peilin Liao, and (to the right) post-doctoral associate, Doron Naveh, and graduate student, Kristen Marino, listen. Credit: Brian Wilson.

and efficient combustion of biofuels, and optimizing lightweight metal alloys for fuel-efficient vehicles and fusion reactor walls.

Credit: http://www.princeton.edu/carter.

3 Most Cited Publications

Title: Constrained reaction coordinate dynamics for the simulation of rare events
Author(s): Carter, EA; Ciccotti, G; Hynes, JT; et al.
Source: Chemical Physics Letters; volume: 156; issue: 5; pages: 472–477; published: April 14, 1989
Times Cited: 497 (from Web of Science)

Title: Solvation dynamics for an ion pair in a polar solvent: time-dependent fluorescence and photochemical charge transfer
Author(s): Carter, EA; Hynes, JT
Source: Journal of Chemical Physics; volume: 94; issue: 9; pages: 5961–5979; doi: 10.1063/1.460431; published: May 1, 1991
Times Cited: 369 (from Web of Science)

Title: Oligoacenes: theoretical prediction of open-shell singlet diradical ground states
Author(s): Bendikov, M; Duong, HM; Starkey, K; et al.
Source: Journal of the American Chemical Society; volume: 126; issue: 24; pages: 7416–7417; doi: 10.1021/ja048919w; published: June 23, 2004
Times Cited: 308 (from Web of Science)

Challenges and on being a woman in this field . . .

When I was just getting started, preconceived notions about women in science were rampant. It seemed, and many women I talked to had similar experiences, that it was assumed by those who did not know you that you could not possibly be competent, whereas men were given the benefit of the doubt. Even worse, it was assumed that you must have been given a break to get to where you are. I took great joy in watching the surprise waft over the faces of men who had assumed I was incompetent, once they heard me speak and heard what I had to say.

The other major challenge is realizing you don't have to choose between career and family (though it isn't easy). It can be done—career and family—and it is absolutely worth it. It is exhausting to do it all but it can be done.

Words of Wisdom

Carve out a unique niche for yourself by getting training in multiple subdisciplines. Work on your self-confidence while still holding yourself to the highest standards of excellence in your work. Be your own most demanding critic, but not to the point of being self-destructive. Your research will be much better for being critical and thorough. Think deeply. Always look for the physical basis behind what you are researching. Don't be

Emily Carter (in the middle) at a 2013 celebration dinner with her family: (starting from left) her husband, Bruce Koel, stepdaughter, Jacqueline Geib, (and to her immediate right) stepson, Brent Koel, and (far right) son, Adam Carter Koel.

afraid to ask questions. That's how you learn and grow. And always consider your audience when trying to convey your ideas to others, either in written or in oral form. That is how you will ensure that your work will have impact and hopefully inspire the work of others. Make a positive, meaningful contribution to the world.

Rachel A. Caruso

Associate Professor & Reader
School of Chemistry
The University of Melbourne
Victoria 3010
Australia

e-mail: rcaruso@unimelb.edu.au
telephone: +61 3 8344-7146

Dr. Rachel Caruso at work in her office.

Tags

* ❖ Administration and Leadership
* ❖ Academe
* ❖ Domicile: Australia
* ❖ Nationality: Australian
* ❖ Caucasian
* ❖ Children: 2

Birthplace

Melbourne, Australia

Publication/Invention Record

>100 publications: h-index 40
2 patents, 4 book chapters

Academic Credentials

Ph.D. (1997), The University of Melbourne, Victoria, Australia.
B.Sc. (1992), The University of Melbourne, Victoria, Australia.

Research Expertise: metal oxides, porous structures, templating techniques, photo-catalysis, photovoltaics

Other Interests: outreach to primary and secondary school students

Key Accomplishments, Honors, Recognitions, and Awards

* Associate Editor of Chemical Communications, 2014 to present
* Associate Professor and Reader, The University of Melbourne, 2011 to present
* Office of the Chief Executive (OCE) Science Leader, Commonwealth Scientific and Industrial Research Organisation (CSIRO) Materials Science and Engineering, 2008 to present

---

Successful Women Ceramic and Glass Scientists and Engineers: 100 Inspirational Profiles. Lynnette D. Madsen.
© 2016 The American Ceramic Society and John Wiley & Sons, Inc. Published 2016 by John Wiley & Sons, Inc.

- Australian Research Council (ARC) Future Fellow, 2010–2014
- Senior Lecturer, The University of Melbourne, 2008–2009
- Hartung Lecturer, The Royal Australian Chemical Institute (RACI), 2008
- ARC Australian Research Fellow, The University of Melbourne, 2003–2008
- Cosmos Bright Sparks Award for Australia's top 10 young scientists, 2007
- Victorian Young Tall Poppy Science Award, 2006
- Centenary Research Fellow, The University of Melbourne, 2003
- Group Leader, Max Planck Institute of Colloids and Interfaces, 2000–2002
- Postdoctoral Fellow, Max Planck Institute of Colloids and Interfaces, Germany, 1999
- Postdoctoral Fellow, Hahn Meitner Institute, Germany, 1998

Biography

Early Life and Education

Rachel Caruso was awarded a Bachelor of Science degree with first-class honors in 1992 from The University of Melbourne. She continued with her research and was awarded a Doctor of Philosophy in 1997. Her doctoral thesis was titled "Colloidal particle formation using sonochemistry."

Career History

Dr. Caruso went to the Hahn-Meitner Institute, Berlin and the Max-Planck Institute of Colloids and Interfaces, Golm-Potsdam for her postdoctoral work. Her research focused on materials chemistry and the fabrication of novel porous materials.

In 2003, she returned to The University of Melbourne as the Inaugural Centenary Research Fellow, securing an ARC Australian Research Fellowship later the same year. Her time has been divided between lecturing at The University of Melbourne and serving as an OCE Science Leader at CSIRO since 2008.

Associate Professor Rachel Caruso, 2010.

3 Most Cited Publications

Title: Nanoengineering of inorganic and hybrid hollow spheres by colloidal templating
Author(s): Caruso, F; Caruso, RA; Mohwald, H
Source: Science; volume: 282; issue: 5391; pages: 1111–1114; published: November 6, 1998
Times Cited: 2960 (from Web of Science)

Title: Magnetic nanocomposite particles and hollow spheres constructed by a sequential layering approach
Author(s): Caruso, F; Spasova, M; Susha, A; et al.

Source: Chemistry of Materials; volume: 13; issue: 1; pages: 109–116; published: January 2001
Times Cited: 484 (from Web of Science)

Title: Mesoporous anatase TiO_2 beads with high surface areas and controllable pore sizes: a superior candidate for high-performance dye-sensitized solar cells
Author(s): Chen, D; Huang, F; Cheng, Y-B; et al.
Source: Advanced Materials; volume: 21; issue: 21; pages: 2206–2210; doi: 10.1002/adma.200802603; published: June 5, 2009
Times Cited: 421 (from Web of Science)

Challenges

Juggling work commitments and family life can be an ongoing challenge and it is really important to get the right balance.

Words of Wisdom

Make sure you are doing what you really enjoy, follow your dreams or passion.

Sources used to create this profile

- Caruso's Homepage, http://www.chemistry.unimelb.edu.au/dr-rachel-a-caruso, accessed December 11, 2014.
- Find an Expert, http://www.findanexpert.unimelb.edu.au/display/person17895, accessed December 11, 2014.
- Caruso's Group Page, http://caruso.chemistry.unimelb.edu.au/, accessed December 11, 2014.
- Web of Science, accessed December 11, 2014.
- CSIRO, http://www.csiro.au/Organisation-Structure/Divisions/CMSE/Surfaces-and-Nanosciences/RachelCaruso.aspx, accessed December 11, 2014.
- Chemical Communications Blog, http://blogs.rsc.org/cc/2014/07/18/chemcomm-introduces-rachel-caruso-as-associate-editor/, accessed December 11, 2014.
- NanoBio Australia, http://www.icbni.com.au/?page=173888&pid=171051, accessed December 11, 2014.

Kathleen A. Cerqua-Richardson

Professor
College of Optics and Photonics
Department of Materials Science and Engineering
University of Central Florida (UCF)
4304 Scorpius Street
Orlando, FL 32816-2700
USA

e-mail: kcr@creol.ucf.edu
telephone: (407) 823-6815

Kathleen Cerqua-Richardson assembling liquid crystalline optics—specifically, circular polarizers, notch filters, and wave plates, fabricated for use in the OMEGA high-power laser system—in a flow hood within the Laboratory for Laser Energetics (LLE) of the University of Rochester, 1986.

Birthplace

Rochester, NY, USA

Born

June 10, 1960

Publication/Invention Record

>250 publications: h-index 30 (Web of Science)
7 patents

Tags

❖ Academe
❖ Domicile: USA
❖ Nationality: American
❖ Caucasian

Proudest Career Moment (to date)

The most memorable aspect of my career to date is being able to see how my mentoring of students has contributed to their success not only as productive engineers and scientists, but also as broadly skilled, cross-discipline-trained, globally aware young professionals with strong interpersonal and cross-cultural skills. This result (which of course takes time to see) has validated my early assumption and hope that crossing over boundaries within related, but distinctly different disciplines (based on "labels") can result in skill sets that lead to unique personal growth, innovation in thinking and working, that often results in new scientific findings.

Successful Women Ceramic and Glass Scientists and Engineers: 100 Inspirational Profiles. Lynnette D. Madsen.
© 2016 The American Ceramic Society and John Wiley & Sons, Inc. Published 2016 by John Wiley & Sons, Inc.

Academic Credentials

Ph.D. (1992) Ceramics, Alfred University, Alfred, NY, USA.
M.Sc. (1988) Glass Science, Alfred University, Alfred, NY, USA.
B.S. (1982) Ceramic Engineering, Alfred University, Alfred, NY, USA.

Research Expertise: optical glass/glass ceramic science and engineering, optical material design, processing, and characterization for use in optical components and systems

Other Interests: international engagement including shared education programs, cooperation with industry

Key Accomplishments, Honors, Recognitions, and Awards

- Member, Board of Fellows, Society of Glass Technology, 2010 to present
- Accreditation Board of Engineering and Technology (ABET) accreditation team, 2000 to present
- Member, Scientific Advisory Board, NSF-ERC on Mid-Infrared Technologies for Health and the Environment (MIRTHE), Princeton University, 2007 to present
- Member, Board of Trustees, Alfred University, 2006 to present
- Academician, World Academy of Ceramics, 2015
- President, American Ceramic Society (ACerS), 2014–2015; President-Elect, 2013–2014; Board of Directors, 2012–2016
- Director, Board of Directors, Society of Photo-Optical Instrumentation Engineers (SPIE), 2012–2015
- U.S. representative, International Commission on Glass [ICG] Coordinating Technical Committee, 2012–2015; Technical Council Member, 2009–2011; Committee member, TC-23 Education, TC-20 Photonics
- I.D. Varshnei Award, Indian Ceramic Society, for outstanding contributions to international glass science and engineering research, 2013
- Associate Editor, International Journal of Applied Glass Science, 2009–2013
- Member, Fellows Committee, Optical Society of America, 2011–2012
- Director, Board of Directors, American Ceramic Society, 2008–2011
- McQueen-Quattlebaum Faculty Achievement Award, Clemson University—College of Engineering and Science (CoES), for faculty accomplishments, achievements, and awards resulting from excellence in scholarly and professional activities, 2011
- Optics Superhero, honored by Edmund Optics for contributions to advances in precision glass molding (PGM) at SPIE Photonics West, San Francisco, CA, 2011
- Member, External Advisory Board, Virginia Institute of Technology, Materials Science and Engineering Department, 2005–2010
- Technical advisor and collaborator, Australian Research Council Centre of Excellence for Ultrahigh-Bandwidth Devices for Optical Systems (CUDOS), Sydney, Australia, 2002–2010

- Award for Faculty Excellence, Clemson University Board of Trustees, 2005, 2007, 2009, 2010
- Fellow, Optical Society of America (OSA), for contributions to the advancement of glass science in optics through teaching and research across institutions and international boundaries, 2009
- Outstanding Educator Award, American Ceramic Society, Ceramic Education Council (CEC), for truly outstanding work and creativity in teaching, in directing student research, or in the general educational process (lectures, publications, etc.) of ceramic educators, 2009
- Member, Tau Beta Pi Honor Society, 2009
- Samuel R. Scholes Lecture and Award, Alfred University, Kazuo Inamori School of Engineering, for international contributions to glass science and engineering research and education, 2008
- Featured lecturer, SPIE Women in Optics Lecture, SPIE Photonics West meeting, San Jose, CA, 2008
- Fellow, Society of Photo-Optical Instrumentation Engineers (SPIE), for distinguished contributions in optical materials and optical science and engineering education, 2008
- Fellow, American Ceramic Society, for longstanding contributions to the advancement of ceramic science and technology to the community, 2006
- Featured Scientist, in Society of Photo-Optical Instrumentation Engineers (SPIE), "Women in Optics," 2006
- Member, President's Engineering Task Force, Alfred University, 2004, 2005
- Member, American Precision Optics Manufacturers Association (APOMA), Executive Board, 1998–2005
- Fellow, Society of Glass Technology, United Kingdom, awarded for outstanding contributions to the international glass science community, 2003
- Research Incentive Award, presented by the UCF Research Council for outstanding productivity in research, 2001
- Young Researcher Symposium Prize, presented at the XIIth International Symposium on Non-Oxide Glasses and Advanced Optical Materials; awarded by the symposium's International Advisory Committee for "outstanding work in the field of non-oxide glass and new optical glass research" to a researcher under the age of 40, 2000
- Member, inaugural Million Dollar Club, University of Central Florida, Office of Sponsored Research, for single year research funding of one million dollars or more, 2000
- Award for Innovative Excellence in Teaching, Learning, and Technology, presented at the 8th National Conference on College Teaching and Learning, 1997
- Profiled, Orlando Science Center, Discovering Women Series, "Seek Out Science Workshop," 1995
- IR-100 Award: "Liquid Crystal Polarizer/Isolator," co-recipient with S.D. Jacobs and K. Marshall, awarded annually by Research and Development Magazine, 1989

Biography

Early Life and Education

I was the kid in the neighborhood who fixed bikes and loved to play baseball with the boys. I was also the first in my family to have a leaning to "all things science" and then to get grades and scholarships that allowed my self-employed parents to send me to college. Living in Rochester, NY, I assumed (as many of my peers did) that we might end up in Rochester working at Kodak and perhaps even doing something that involved optics. My dad didn't work at Kodak, so I had little exposure to the company; however, how far off was our sense of certainty that the once large, historic, and thriving symbol of American industry would be short lived—and that failure to reinvent itself would lead to significant challenges to its employees and the western New York economy.

As a young person I loved science and math but school was not that challenging for me—I welcomed the distraction of playing sports to support my competitive spirit—this made the everyday aspect of school bearable—we had to be in attendance in high school all day to be able to practice and play on the varsity teams. By the time I finished high school, I was playing four sports to keep me "busy" (I lettered in all (4) of them) and was fortunate to have found a Boy Scout-sponsored Explorer Club in Optics (which met weekly at Kodak Park) that allowed me to meet like-minded kids from all over the county that exposed me to the nuts and bolts of what chemists and materials scientists did for a living.

While I didn't appreciate it at the time, the expectation of staying in western NY led to my decision to pursue chemistry at Alfred University, a choice that soon evolved into ceramic engineering (my roommates were having much more fun doing chemistry in Mud Lab, than I was in Organic Lab)! So I changed majors and in addition to processing and characterization of inorganic metal oxides (not flower pots and toilets that most laymen thought were the only applications of ceramics in the late 1970s), I took all the Optics courses I could find in the curriculum. Little did I know that this would be the seed of the future focus of my career. I graduated Alfred with my B.Sc., totally burned out on academics, and after indicating that "I'd had enough and would never go back to school," went on to work at the University of Rochester's Laboratory for Laser Energetics (LLE). That I was hired during the tough economic times of the early 1980s was in itself fortunate (unemployment and interest rates were in the "teens"), but I was most fortunate to be hired by a meticulous, optical engineer—himself a Ph.D. graduate of University of Rochester's prestigious Institute of Optics. Dr. Stephen Jacobs admitted he didn't know (really) what a Ceramic Engineer was or did, but realized someone who could make glass and ceramics might be able to help his other optical engineers do research on novel optical materials. After (3) grueling interviews (where I had to bring in my books to show him what optics I had taken), he hired me and I was off. Overwhelmed by physical and geometric optics, I quickly learned how I best learned—doing first, and then reading the book. This crossover experience, which benefited from an incredible mentor and motivator (who helped me learn that which I didn't know I *needed to know*), was an invaluable part of my early career training and it had a profound impact on my future. While I initially felt ignorant on all things related to the plasma physics experiments going on at LLE at the time associated with the Laser Fusion research there, I immersed and learned all that I could. I also spent time learning how to teach my colleagues that they didn't need to "settle" for the commercially available "flavors" of optical materials on the market—if you knew the chemistry–structure–property relationships that led to those materials, you could design and fabricate them to yield new properties or to enhance existing ones. This was the start of my career as an optical materials scientist.

Career History

Following almost 4 years of work at LLE as an optical materials engineer, I went back to Alfred for a M.Sc. (in Glass Science), undertaking my degree part time while still working at LLE full time. I realized over those 3 years that it was extremely difficult to do a good job at both work and a thesis and thus decided against doing my Ph.D. part time. While I was supervising Ph.D. graduate students and their dissertation research in optics, I still didn't have the piece of paper needed to do that anywhere else. In 1988, I left LLE and went back to Alfred full time for my Ph.D. in Glass Science and Engineering and in 1992 graduated—one of (5) women to graduate with Ph.D.'s in Ceramics that year—equal to the *total number of women Ph.D.'s in Alfred University's* (then) 156-year history.

Having married in 1990 and starting life with my physicist husband and balancing the two Ph.D. career challenge, I took a postdoc with an Optical Materials Group at the Naval Air Defense Laboratory at China Lake, CA. Fortunately for me, the lab did not have the high-temperature melt facilities I needed for my research work, and the Navy agreed to have me on-site at CREOL, a new laser research center established at the University of Central Florida in Orlando in 1986 to support research in electro-optics and lasers. Fortunate because this is where my husband had accepted a Professor position in 1989 moving from NY to FL while I was at Alfred. Thus, I started my work in optical materials—first as a staff scientist postdoc for the Navy, and later as a tenure track professor. I remained at UCF through 2004 until I accepted the position of Director for Clemson University's School of Materials Science and Engineering in Clemson, South Carolina.

Kathleen Cerqua-Richardson speaking to the inaugural meeting of UCF's Women in Lasers and Optics (WiLO) group. Credit: UCF.

I remained at Clemson until 2012 when UCF hired me back to support the nucleus of a critical mass of faculty in Optical Materials. During my time in South Carolina, UCF's effort had expanded from two glass researchers—myself and one other colleague—to now include crystal growth and optical ceramics, a fiber optical processing and characterization facility, and expertise in optical glasses that complement my own in non-oxide, infrared glass and glass ceramic materials. It is during these years that my

group's work with industry and government partners has allowed us to become one of the most recognized efforts in novel infrared optical materials in the world. We maintain anywhere from 10 to 15 students and staff within the glass processing and characterization laboratory (GPCL) at CREOL, persons with expertise across a diverse range of majors including graduate and undergraduate students with funding sources ranging from fundamental to applied research. In late 2013, I bowed to pressure from (3 of) my then current or former students/colleagues as we launched a spin-off company, Irradiance Glass, Inc., to commercialize many of the prototype materials our team has developed over my almost 30-year career. It has been an exciting venture that supports the ongoing need for novel materials with unique function.

A recent image of Kathleen Cerqua-Richardson. Credit: UCF.

3 Most Cited Publications

Title: Chalcogenide photonics
Author(s): Eggleton, BJ; Luther-Davies, B; Richardson, K
Source: Nature Photonics; volume: 5; issue: 3; pages: 141–148; published: March 2011
Times Cited: 335 (from Web of Science)

Title: High-efficiency Bragg gratings in photothermorefractive glass
Author(s): Efimov, OM; Glebov, LB; Glebova, LN; et al.
Source: Applied Optics; volume: 38; issue: 4; pages: 619–627; published: February 1, 1999
Times Cited: 153 (from Web of Science)

Title: Fabrication and characterization of integrated optical waveguides in sulfide chalcogenide glasses
Author(s): Viens, JF; Meneghini, C; Villeneuve, A; et al.
Source: Journal of Lightwave Technology; volume: 17; issue: 7; pages: 1184–1191; published: July 1999
Times Cited: 123 (from Web of Science)

ResearcherID (A-6012-2011)

Challenges

The most challenging part of my career has been in balancing career with family but not in the usual ways. The (two) Ph.D. family (as compared to the often stated challenges with having and raising children) has been the most challenging for me—being able to work nearby my spouse. While in similar fields (we both "do" optics) but at very different professional stages in our careers at various times (due to an almost 20-year age difference), this issue has been the most challenging. My husband's recognition of this and support of me when I needed to change locations/positions to advance allowed me to grow in all kinds of ways. Surprisingly, once I no longer felt intimidated by physicists and others outside of materials science/chemistry, I had no real technical difficulties or respect issues, working across boundaries between Materials and Optics. Nor have I had issues with gender—being a woman in science. While I was often the only woman in the room, I quickly found ways to turn it into "the only glass/materials person" in the room—this made all the difference and it was no longer a gender issue—rather, it became a technical expertise issue where I could "compete" side by side with any other technical discipline. Since there were not many of glass experts around, this quickly led to an opportunity to highlight a skill set that I could use to extend traditional optics in new directions. At that stage, I never looked back.

Words of Wisdom

Find a way to carve out your unique niche—even if you are in a "traditional *whatever*" discipline—find something that you excel at. Through that differentiation, you'll always be able to shine in your own way. Stay honest to yourself—don't necessarily change due a fad or trend—rather, you sometimes need to adapt what you do, to the winds of management or funding agencies. Spin your talent to create a novel solution to a problem that is unique—it's harder to do, but worth it in the end. Know that sometimes you know the problem and have an idea toward the solution, before the funding agency does!

Profile 9

Helen Lai Wa Chan

Professor of Applied Physics
The Hong Kong Polytechnic University
11 Yuk Choi Road
Hung Hom
Hong Kong, People's Republic of China

e-mail: apahlcha@polyu.edu.hk
telephone: (852) 2766-5692

Helen L.W. Chan (at 44) when she first joined the university as a faculty member.

Birthplace
Hong Kong

Publication/Invention Record

>850 publications: h-index 46 (Web of Science)
7 book chapters, 29 patents

Proudest Career Moment (to date)

Receiving the 2002 National Technology Invention Award (2nd class) from the former Premier of China in Beijing.

Academic Credentials

Ph.D. (1987) Materials Science, Macquarie University, New South Wales, Australia.
Cert. Ed. (1976) Education, University of Hong Kong, Pokfulam Road, Hong Kong, China.
M.Phil. (1974) Physics, Chinese University of Hong Kong, Shatin, Hong Kong, China.
B.Sc. (1970) Physics, Chinese University of Hong Kong, Shatin, Hong Kong, China.

Research Expertise: ferroelectrics, dielectrics, composites, photonic and phononic crystals, ultrasonic transducers, acoustic sensors, pyroelectric sensors and arrays

Other Interests: teaching and reading

Tags

❖ Administration and Leadership
❖ Academe
❖ Domicile: Hong Kong
❖ Nationality: Chinese
❖ Asian
❖ Children: 4

Successful Women Ceramic and Glass Scientists and Engineers: 100 Inspirational Profiles. Lynnette D. Madsen.
© 2016 The American Ceramic Society and John Wiley & Sons, Inc. Published 2016 by John Wiley & Sons, Inc.

Key Accomplishments, Honors, Recognitions, and Awards

• Vaisala Award from the Royal Meteorological Society of UK, 2010

Prof. Helen L.W. Chan receiving the Vaisala Award in
June 2011 at Exeter, UK.

• Gold medal in the International Fair of Inventions at Geneva, Switzerland, 2009
• Silver Medal in the 15th National Exhibition of Inventions, Beijing, 2005
• State Technology Invention Award (2nd Class), China, 2002

Helen L.W. Chan receiving the 2002 State Technology
Invention Award handed from the former Premier of
China in Beijing.

- Gold Award in the Seoul International Invention Fair, 2002
- The President's Awards for Outstanding Research and Scholarly Activities, The Hong Kong Polytechnic University (HKPU), 1999 and 2000
- Outstanding Licensing Activities Award, HKPU, 2001
- The Commonwealth Scholarship for Ph.D. study at Macquarie University, 1983–1987
- The Plessey Prize, Sydney, Australia, 1996
- The Hong Kong Government Scholarship for University Studies, 1970–1974

Biography

Early Life and Education

After being awarded a Bachelor of Science degree, Helen took a position as a demonstrator in the Physics Department in the Chinese University of Hong Kong from September 1970 to August 1974. Then, she became a High School teacher at St. Paul's Co-Educational College until August 1980. In February 1981, her family migrated to Sydney, Australia. After her arrival in Sydney, she became a full-time house-wife helping her husband to set up his business and her children to adapt to the new environment. In 1985, she embarked on a Ph.D. program to work on 1-3 piezoceramic/polymer composites for ultrasonic transducer applications.

Prof. Helen L.W. Chan (left) with Prof. M. Dresselhaus from MIT (second from the right) after she received an Honorary Doctorate from The Hong Kong Polytechnic University.

Soon after (in 1987) she completed her Ph.D. studies and was offered a job as a Research Scientist at the Commonwealth Scientific and Industrial Research Organization (CSIRO).

Career History

Her first position was as a Research Scientist in the Division of Applied Physics at CSIRO setting up and managing the "Australian Standards for Medical Ultrasound" Project from August 1987 to August 1991. She then became the Senior Acoustic Engineer for GEC-Marconi Pty in Australia (from September 1991 to September 1992) where she was responsible for the calibration of "Towed Arrays" for underwater acoustics. The towed arrays and other underwater ultrasonic transducers were fabricated using piezoelectric

Helen L.W. Chan at work in the atomic force microscopy (AFM) laboratory.

ceramics as the sensing elements. In October 1992, she moved to The Hong Kong Polytechnic University where she became a Lecturer in the Department of Applied Physics. In August 1997, she was promoted to Associate Professor, and in September 1997 she was promoted again this time to Professor. Subsequently, in September 2003, she was selected as Chair Professor. From July 2005 through November 2013, she served as Head of the Applied Physics Department. Since November 2013, she has been serving as Associate Dean in the Faculty of Applied Science and Textiles.

3 Most Cited Publications

Title: Simple model for piezoelectric ceramic/polymer 1-3 composites used in ultrasonic transducer applications
Author(s): Chan, HLW; Unsworth, J
Source: IEEE Transactions on Ultrasonics, Ferroelectrics, and Frequency Control; volume: 36; issue: 4; pages: 434–441; published: July 1989
Times Cited: 217 (from Web of Science)

Title: Electromechanical and ferroelectric properties of $(Bi_{1/2}Na_{1/2})TiO_3–(Bi_{1/2}K_{1/2})TiO_3–BaTiO_3$ lead-free piezoelectric ceramics
Author(s): Wang, XX; Tang, XG; Chan, HLW
Source: Applied Physics Letters; volume: 85; issue: 1; pages: 91–93; published: July 5, 2004
Times Cited: 216 (from Web of Science)

Title: Diffuse phase transition and dielectric tunability of $Ba(Zr_yTi_{1-y})O_3$ relaxor ferroelectric ceramics
Author(s): Tang, XG; Chew, KH; Chan, HLW
Source: Acta Materialia; volume: 52; issue: 17; pages: 5177–5183; published: October 4, 2004
Times Cited: 209 (from Web of Science)

ResearcherID (B-5633-2014)

On being a woman in this field . . .

I have always maintained a positive attitude and tried to articulate my ideas to my colleagues and coworkers. I do not expect positive discrimination and I am happy to work in a level playing field. I am a good team player and collaborate well with my colleagues. Finding new grand challenge in lead-free ceramics is a major focus I have been concentrated in.

Words of Wisdom

Life-long learning is very important to keep us active and happy.

Profile 10

Helen M. Chan

Materials Science and Engineering Department
Lehigh University
Whitaker Laboratory
5 E. Packer Ave.
Bethlehem, PA 18015
USA

e-mail: hmc0@lehigh.edu
telephone: (610) 758-5554

Helen M. Chan (1990)

Birthplace
London, England

Publication/Invention Record
>175 publications: h-index 37 (Web of Science)
4 patents

Tags
❖ Academe
❖ Administration and Leadership
❖ Domicile: USA
❖ Asian
❖ Children: 2

Proudest Career Moment (to date)

It is difficult to pinpoint a given career moment as my proudest. Firsts, of course, tend to be memorable; the first paper that appears in print, the first proposal funded, the successful graduation of one's first graduate student, and so on. As my career as progressed, I must say I derive great joy from learning about the accomplishments of alums, former graduate students and post-docs, and am very proud to be part of this expanding network of engineers and scientists that makes up the materials community.

Academic Credentials

Ph.D. and D.I.C. (1982) Materials Science & Technology, Imperial College, London, United Kingdom.
B.Sc. (1979) Materials Science (First Class Honors), Imperial College, London, United Kingdom.

Research Expertise: reactive processing to fabricate unique ceramic/metal structures, including cellular and nanopatterned materials; role of dopants and interfacial chemistry on diffusion-limited processes in ceramics

Successful Women Ceramic and Glass Scientists and Engineers: 100 Inspirational Profiles. Lynnette D. Madsen.
© 2016 The American Ceramic Society and John Wiley & Sons, Inc. Published 2016 by John Wiley & Sons, Inc.

Prof. H.M. Chan with graduate students M. Kracum (Left) and K. Anderson (Right).

Other Interests: raising awareness of materials science and engineering in middle and high school students

Key Accomplishments, Honors, Recognitions, and Awards

- Associate Editor for the Journal of the American Ceramic Society, 1999 to present
- Editor of the Journal of Materials Science, 2004 to 2015
- Chair, University of Materials Council, 2011–2012
- Chair, Executive Committee of the Basic Science Division of the American Ceramic Society, 2010–2011
- Chair, Gordon Research Conference for Solid State Studies in Ceramics, 2008
- Fellow of the American Ceramic Society, 2005
- Lehigh University's Eleanor and Joseph F. Libsch Award for excellence in research, 2005
- Named New Jersey Zinc Professor at Lehigh University, 1999
- American Ceramic Society Roland B. Snow award on four separate occasions: 1986, 1990, 1992, and 1999
- Lehigh University "Class of 1961" Professorship for "distinction in teaching, research and service," 1993
- Lehigh University's "Service Teaching Excellence Award" for two consecutive years, 1991 and 1992
- ASM International's Bradley Stoughton Award for outstanding young faculty in the field of Materials Science & Engineering, 1992
- Lehigh Valley Chapter of ASM International "Outstanding Young Member," 1991
- Alfred Noble Robinson Award for "outstanding performance and unusual promise of professional achievement," 1990

Biography

Early Life and Education

Dr. Chan grew up in Northamptonshire, England, her parents having emigrated from Hong Kong. It was a visit to Imperial (with a demonstration involving liquid oxygen) that kindled her interest in the field of Materials. Dr. Chan graduated from Imperial College with a B.Sc. First Class Honors degree in Materials Science, and was the recipient of the Governors' Prize for most outstanding graduate in Materials Science. She was subsequently awarded a Ph.D. and D.I.C. from the Dept. of Materials Science & Technology at Imperial in 1982. To this day she is an avid soccer fan, and London is her favorite city. She would love to have a mineral collection on par with the Smithsonian Natural History Museum, but fears it would be a dust magnet.

Career History

Following appointments as a post-doctoral research associate and research engineer, Dr. Chan joined the Lehigh faculty in 1986. She then took an 18-month leave of absence at the National Institute of Standards and Technology, where she worked in the Mechanical Properties Group of the Ceramics Division. Dr. Chan returned to Lehigh January 1988, and was promoted to the rank of Associate Professor with tenure in 1991 and to the rank of Full Professor in 1995. Dr. Chan served as Chair of the Dept. of Materials Science and Engineering (Lehigh University) 2006–2015. She is included in Thomson ISI's list of highly cited researchers in materials.

Prof. H.M. Chan with graduate students M. Kracum and K. Anderson.

3 Most Cited Publications

Title: Ordering structure and dielectric-properties of undoped and La/Na-doped $Pb(Mg_{1/3}Nb_{2/3})O_3$
Author(s): Chen, J; Chan, HM; Harmer, MP
Source: Journal of the American Ceramic Society; volume: 72; Issue: 4; pages: 593–598; doi: 10.1111/j.1151-2916.1989.tb06180.x; published: April 1989
Times Cited: 430 (from Web of Science)

Title: Mechanical-behavior of alumina silicon-carbide nanocomposites
Author(s): Zhao, JH; Stearns, LC; Harmer, MP; et al.

Source: Journal of the American Ceramic Society; volume: 76; issue: 2; pages: 503–510; doi: 10.1111/j.1151-2916.1993.tb03814.x; published: February 1993
Times Cited: 270 (from Web of Science)

Title: Compensating defects in highly donor-doped $BaTiO_3$
Author(s): Chan, HM; Harmer, MP; Smyth, DM
Source: Journal of the American Ceramic Society; volume: 69; Issue: 6; pages: 507–510; doi: 10.1111/j.1151-2916.1986.tb07453.x Published: June 1986
Times Cited: 179 (from Web of Science)

Challenges

The challenges of being a minority, albeit with regard to gender or some other characteristic, are akin to the difficulties of establishing oneself in a new field. I think we are all familiar with individuals who thrive within the *status quo*, and hence are resistant to change. All communities need leaders in the field who would rather open doors than hold them shut, build bridges rather than moats. Having been fortunate to have been mentored by such special individuals, I strive to follow their example.

Prof. H.M. Chan in her office at Lehigh University.

On Being a Woman in This Field . . .

Without doubt there are individuals (both male and female) who deep down do not believe that a woman can be as good as a man in a technical field. I feel very fortunate in that the number of bigoted individuals that has crossed my path is very small. I do remember though, the time as a graduate student I was in Denmark and was taking a shuttle bus to the conference center. When asked which stop, I mentioned to the driver that I was attending a scientific meeting. He shook his head and muttered in English, "teaching a woman science, one might as well teach a telegraph pole." At the time I was too taken aback to respond, but if I could have that moment again, I would love to hand him a copy of this book!

Words of Wisdom

There isn't any one formula for success (however one chooses to define it), but if I had any words of wisdom, they would be along the lines of "everything I needed to know about work (and life) I learned in a materials lab." For example, *Spills must be cleaned up immediately.* Letting issues drag on in is never conducive to productivity. *There is no such thing as bad data* (provided it is accurate!). As researchers, it is often the unexpected result that leads us in new directions and opens up new fields.

Poor focus – poor micrograph. Try to concentrate on the things that are most important, and allocate your effort accordingly. *Listen to what your TA's have to say*. My advice would be to emulate the habits and characteristics of people who have been successful in your field, but make adaptations with the realization that your own style and skillset may be different. *Having good partners goes a long way toward making the lab more enjoyable*. This of course is self-explanatory, work is not work if you enjoy your interactions with the people around you. Finally, *Does it count toward my final grade?* Our careers and lives are made up of many different facets, in the end the most important grade is the one we assign ourselves.

Dominique Chatain

Directeur de Recherche
National Center of Scientific Research (CNRS)
Marseille 13288
France

e-mail: chatain@cinam.univ-mrs.fr
telephone: +33-660 302 890

Dominique Chatain at 39.

Birthplace
Nancy, France
Born
August 31, 1955

Publication/Invention Record
>100 publications: h-index 27 (Web of Science)
1 book chapter

Tags
❖ Government
❖ Domicile: France
❖ Nationality: French
❖ Caucasian
❖ Children: 0

Pivotal and Proudest Career Moments (to date)

In 1992, I received a small grant that supported a joint program with an American colleague of Carnegie Mellon University. This changed my professional life and opened my eyes to the international world of scientific research.

In 2010, I was very proud to be awarded the Silver Medal of CNRS, in recognition of my scientific achievements.

Academic Credentials

Ph.D. (1983) Physics, Institut National Polytechnique de Grenoble, Grenoble, Isère, France. Engineer's Degree (1979) Metallurgy, Institut National Polytechnique de Grenoble, Grenoble, Isère, France.

Research Expertise: physics and chemistry of wetting; structural, chemical, and energetic properties of surfaces and interfaces; wetting transitions; thermodynamics

---

Successful Women Ceramic and Glass Scientists and Engineers: 100 Inspirational Profiles. Lynnette D. Madsen.
© 2016 The American Ceramic Society and John Wiley & Sons, Inc. Published 2016 by John Wiley & Sons, Inc.

of interfaces of metals and ceramics; multiscale (nano-meso-macro-scale) characterization of interfaces; nano/submicron particles at interfaces and their impact on properties

Other Interests: international collaborations

Key Accomplishments, Honors, Recognitions, and Awards

- Appointment to the CNRS Board for Materials Science, 2012–2016
- Lady Davis Fellowship from the TECHNION, 2014
- Legion of Honneur of the French government for her impacting scientific contributions, 2012
- Head of a department of 20 faculty members/ researchers in my lab, 2007–2011
- Silver Medal of CNRS for impacting contributions to the field of materials science (http://www.cnrs.fr/fr/recherche/prix/docs/argent2010/ChatainD.pdf), 2010
- Leader of an international brainstorming group on interfaces since 2000
- More than 60 plenary or invited talks at international conferences and workshops in Europe and the United States
- Chair and co-chair of more than 20 international conferences

Biography

Early Life and Education

Dominique Chatain was raised in the north of France. She has two younger sisters and one younger brother. She received her early education in the French public school system. After completing her high school diploma (baccalauréat), she was enrolled in a pre-engineering course for 2 years to prepare for the competitive entrance exam to a French

Dominique Chatain in 2010 when she delivered her acceptance speech for the Silver Medal of CNRS.

engineering school. Engineering degrees are among the highest level diplomas in the French system. She was accepted at the Polytechnic Institute of Grenoble where she received an Engineer's Degree in Materials Science.

She completed her doctorate in Physics and Materials Science, and was awarded a Ph.D. from the Polytechnic Institute of Grenoble in 1983.

Career History

She was hired as a CNRS scientific researcher in 1981 in a Materials Science and Engineering laboratory at the University and the Polytechnic Institute of Grenoble. She developed her research in an interdisciplinary manner by merging physical and chemical views of metal-oxide interfaces and wetting. In 1991, she moved to the University of Marseille to a CNRS laboratory with a focus on Crystal Growth research. There she discovered the "nanoworld," and was able to share her expertise in chemistry and thermodynamics with surface physicists. In 2007, the laboratory, which had broadened its scope to include topics in interdisciplinary Materials Science, was reorganized into five departments. Following this reorganization, she led the department of "nano-materials and reactivity" for 4 years. CNRS appointments provide a great deal of freedom, which allowed her to spend a total of about 4 years out of her laboratory for extended visits to other institutions, mainly in the United States.

She is an active member of the American Ceramic Society (ACerS) and the Materials Research Society (MRS).

3 Most Cited Publications

Title: Wettability of monocrystalline alumina by aluminum between its melting point and 1273 K
Author(s): Laurent, V; Chatain, D; Chatillon, C; et al.
Source: Acta Metallurgica; volume: 36; issue: 7; pages: 1797–1803; published: July 1988
Times Cited: 191 (from Web of Science)

Title: Wettability of SiC by aluminum and Al–Si alloys
Author(s): Laurent, V; Chatain, D; Eustathopoulos, N
Source: Journal of Materials Science; volume: 22; issue: 1; pages: 244–250; published: January 1987
Times Cited: 174 (from Web of Science)

Title: Wettability of SiO_2 and oxidized SiC by aluminum
Author(s): Laurent, V; Chatain, D; Eustathopoulos, N
Source: Materials Science and Engineering A—Structural Materials: Properties, Micro-structure and Processing; volume: 135; pages: 89–94; published: March 30, 1991
Times Cited: 132 (from Web of Science)

Challenges

Challenges drive life and often direct your steps toward unexpected tracks. They require you to make choices, which may involve leaving subjects you know for others that you do not. This is the spice of life.

Many moves in my life have been challenging. Among them I should cite the move from Grenoble to Marseille, and the decision to spend part of my professional life in foreign countries. Each move was rewarding through the joy of sharing the passion for my work with friends.

When I discovered that mentorship is rarely benevolent, I decided to follow my own ideas. I have never been interested in following "fashionable trends." Going your own way makes you stand out, and may be uncomfortable in the short term, but is rewarding in the long term, because you are able to generate new insights and create knowledge for others, maybe, but certainly for yourself.

On being a woman in this field . . .

I was supported by my parents to get the best education in a field that I chose. This was unusual for a woman, but my parents had three daughters and the Western World had started to consider that women and men should have equal rights. This period of enlightenment, when I did not need to wonder about my gender, only lasted till the mid-1980s. The 1990s experienced a backlash, where equal rights were sufficiently over-looked that they now have to be enforced by law. However, a woman must still choose between her professional and personal lives. She cannot have both, unlike most of her male colleagues. Thus, only a "superwoman" can succeed in raising children, running her home, and enjoying a professional life.

I had not experienced any gender biases until I started studying for my Ph.D. At that time, I had to face the French research community in Materials Science and Condensed Matter Physics, which was ruled by men who were reluctant to let women achieve positions where they could make their own decisions. I had to fight every time I wanted to follow my own ideas rather than the ones chosen by my closest colleagues.

While France is supposed to be the country of "Egalité," i.e., equality, my gender has always been an issue in my French professional life. Fortunately, I was able to really feel free of this "stigma" when I started to collaborate with U.S. colleagues.

Words of Wisdom

- Diversity is wealth.
- When you make choices, accept them and never look back.
- Do not follow fashion, but make it.
- Trust your instinct in the field for which you feel you have insight. Do not listen to those who will say "it will not work."
- Dare to step into an area you don't know. You can always learn.
- Do not work to get more but to get better.
- Enjoy working with friends; we are so many scientists on this planet that you will surely find a few.
- Fame is not an aim.
- There are so many different people that competing is not a point.
- Success needs a bit of luck, and support from people who recognize your qualities and who believe you have perceptive insights.
- Choose to network with people you respect for their human qualities and avoid opportunists.
- Do not pretend when you don't know and learn from others.

Profile 12

Jen-Sue Chen

Professor of Materials Science and Engineering
National Cheng Kung University
Tainan
Taiwan

e-mail: jenschen@mail.ncku.edu.tw
telephone: 886-6-2757575 ext. 62948

Birthplace

Tainan, Taiwan

Born

November 11, 1966

Publication/Invention Record

>125 publications: h-index 22 (Web of Science)
7 patents, 1 book chapter

Proudest Career Moment (to date)

Jen-Sue Chen at about 30 years old, just after returning to Taiwan for an academic faculty position.

Tags

❖ Academe
❖ Domicile: Taiwan
❖ Nationality: Taiwanese
❖ Asian
❖ Children: 0

To me, the proudest career moment does not link directly to any award or obvious achievement. Similar to many developed countries, in Taiwan, not many students pursue a doctorate degree nowadays. Fortunately, I have quite a few dedicated masters' students and feel content to lead them into the world of research. However, it is a pity that only a few continue their research careers. In March 2014, my masters' degree student, Wen-Hui Cheng, received admission from Caltech for her Ph.D. program. Wen-Hui started her independent study in her junior year as an undergraduate and continued with her master's thesis research in my lab. I was exceedingly happy about her admission from Caltech because it is my alma mater, and Wen-Hui, a young lady now, will be a leading material scientist one day. This is really the best reward for being a teacher and advisor.

Academic Credentials

Ph.D. (1995) Materials Science, California Institute of Technology, Pasadena, California, USA.

Successful Women Ceramic and Glass Scientists and Engineers: 100 Inspirational Profiles. Lynnette D. Madsen.
© 2016 The American Ceramic Society and John Wiley & Sons, Inc. Published 2016 by John Wiley & Sons, Inc.

M.S. (1990) Materials Science, California Institute of Technology, Pasadena, California, USA.
B.Sc. (1988) Materials Science and Engineering, National Tsing Hua University, Hsinchu, Taiwan.

Research Expertise: thin film fabrication of oxides and nitrides; microstructural and chemical characterization; measurement of physical properties (electrical and optical); determination of microelectronic device characteristics (metal-oxide-semiconductor capacitors, thin film transistors, solar cells, and memory devices)

Other Interests: sharing my experiences of studying/traveling abroad with students

Key Accomplishments, Honors, Recognitions, and Awards

- Associate Editor, Materials Chemistry and Physics, Elsevier, 2010 to present
- National Science Council (Taiwan) College Research Award, Advisor, 2001, 2011, 2012
- Outstanding Teaching Award, College of Engineering, National Cheng Kung University, 1999, 2000, 2003, 2011
- Outstanding Teaching Award, National Cheng Kung University, 2007
- Outstanding Research Award, College of Engineering, National Cheng Kung University, 2007
- 2005 K.T. Lee Research Award, National Cheng Kung University, 2006
- Lam Research Thesis Award (Master's Thesis), Advisor, 2006
- Distinguished Professor, National Cheng Kung University, 2006

Biography

Early Life and Education

Jen-Sue Chen was born and raised in Tainan, Taiwan. Tainan is a city located in the south of Taiwan that is rich in culture/history because it was the ancient capital of Taiwan. Jen-Sue received her early education in Tainan, where the sense of humanity and curiosity in science were rooted in her mind. She joined the Department of Materials Science and Engineering in National Tsing Hua University (in Hsinchu, about 200 km north of Tainan) after the university entrance examination in July 1985.

Originally, Jen-Sue was aiming for Electrical Engineering (EE) as her undergraduate major. In Taiwan, young people are encouraged to go to EE departments mainly because of the good job market in that area. However, she did not perform perfectly during the entrance examination so that she had to pursue her second choice, Materials Science. It turned out that Jen-Sue enjoyed quite a bit "looking into the materials." She finished her B.S. degree in 3 years because of her excellent academic performance. Afterward, she started to apply for M.S./Ph.D. programs at several institutions in USA. Propitiously, she received admission from Caltech with a full scholarship. In August 1989, Jen-Sue flew to California (the first flight in her life) to pursue her doctorate degree.

The days in Caltech were full of inspiration to Jen-Sue—filled with a different culture and different ways of thinking. The stimulation in many scientific aspects is certainly critical; however, she learned more about "communication." In Asian traditions, reading/listening is more important than asking/talking for young students. But, soon after she arrived at Caltech, Jen-Sue realized that "learning comes not only from reading/listening, but also from conversation/discussion." The countless discussions with her Ph.D. thesis advisor, Prof. Marc Nicolet, were pivotal in her growth from student to researcher. Jen-Sue defended her Ph.D. thesis on the investigations on contact metallization of beta silicon carbide in July 1994 and received the diploma from Caltech in June 1995.

Career History

After finishing her Ph.D. oral defense, Dr. Chen joined Integrated Device Technology (IDT) in San Jose, California, as an R&D process engineer. She stayed at IDT for 1 year—a year of experience in industry may not seem long, but it gave her a clear picture of semiconductor integrated circuit (IC) production. Upon returning to Taiwan in 1996, Dr. Chen resumed her academic career as an Associate Professor in Department of Materials Science and Engineering of National Cheng Kung University (NCKU). From her experience in IC industry, she kept her main research focus on materials for microelectronics but gradually switched her research interests from thin film material characterization to electrical device characteristics. Certainly, her ultimate goal has always

Jen-Sue Chen on the TGV train travelling from Lille to Paris, after attending E-MRS Spring Meeting in France, 2014.

Jen-Sue Chen presenting her seminar at the University of New South Wales, Sydney, Australia, in 2013 (at the invitation of Prof. Sammy L. I. Chan).

Prof. Jen-Sue Chen and Dean Wen-Teng Wu at the Red Square in Moscow, Russia when visiting Moscow State University, in 2009.

Jen-Sue delivering a mini-lecture of Basic Crystallography, 2014. (See insert for color version of figure.)

been on linking material characteristics to electrical performance.

Dr. Chen was promoted to full professor in 2002. Afterward, she started to get involved in academic administration work in her university. She had served as the Director of Project Management Division, Office of Research and Development (2003–2004), Associate Dean of College of Engineering (2004–2006), Chairperson of Department of Materials Science and Engineering (2006–2009), and Associate Vice President of Office for International Affairs (2011–2015). The administrative duties add a significant load on top of her research. However, serving as an academic administrator also brought her the opportunity to exchange thoughts with students and professors of diverse disciplines. For example, during her 4-year term of serving as the Associate Vice President for International Affairs, in addition to strengthening academic collaboration with renowned universities worldwide, she focused on connecting better with South/Southeast Asian universities for NCKU.

Dr. Chen is an active member of the Materials Research Society (MRS) and a member of Executive Committee of MRS-Taiwan.

3 Most Cited Publications

Title: Tantalum-based diffusion-barriers in Si/Cu VLSI metallizations
Author(s): Kolawa, E; Chen, JS; Reid, JS; et al.
Source: Journal of Applied Physics; volume: 70; issue: 3; pages: 1369–1373; published: August 1, 1991
Times Cited: 189 (from Web of Science)

Title: Bistable resistance switching of poly(N-vinylcarbazole) films for nonvolatile memory applications
Author(s): Lai, YS; Tu, CH; Kwong, DL; et al.
Source: Applied Physics Letters; volume: 87; issue: 12; article number: 122101; published: September 19, 2005
Times Cited: 123 (from Web of Science)

Title: Properties of reactively sputter-deposited Ta-N thin-films
Author(s): Sun, X; Kolawa, E; Chen, JS; et al.
Source: Thin Solid Films; volume: 236; issue: 1–2; pages: 347–351; published: December 15, 1993
Times Cited: 109 (from Web of Science)

ResearcherID (A-3298-2015)

Challenges

In August 2004, while I was on my first sabbatical in ETH Zürich, Switzerland, our newly inaugurated Dean of College of Engineering, Prof. Wen-Teng Wu, made an international phone call to me. He invited me to be one of his Associate Deans when I returned and simply said "let's try to do something for our university." Dean Wu is 20 years older than me and when he made the invitation, we did not know each other (since he had been at another university before taking the position). In my mind, all Deans and Associate Deans in the College of Engineering were male, and, uhm . . . pretty old. At that time, I was a woman under 40. Therefore, I was totally astonished when I received this phone call and request. After a few days of consideration, I accepted the position.

When I look back on it now, I was probably too young to measure the scale of the challenge. However, I took the position as a coordinator and tried to coordinate things as smoothly as possible. The lesson of "communication" learnt at Caltech helped me a lot then and still does today. As a matter of fact, I regard all my administrative positions as coordinators; all challenges may be resolved by appropriate communications. Certainly, these roles also require a lot of patience.

On being a woman in this field . . .

Like most of woman in science fields, I did not have many female classmates or colleagues throughout my career as a material scientist. Although always being in the minority, I hardly felt any difference for being a woman. Luckily, I got along with my male classmates and colleagues pretty well, I thought. However, over the past 5 years I started to notice the invited/plenary talks delivered by female material scientists. I also started to pay attention to the gender of journal editors. Well, I guess I simply got used to the "world of males" nicely. When some females step on the stage, I am more-or-less surprised, and want to find out if they are from another planet.

It sounds funny that I, a female material researcher, have such a silly thought. Nevertheless, it reflects the fact that "minority follows the majority and remains as the minority." I am happy for the ratio of female undergraduate students in our department (a typical engineering department in Taiwan) is about 30%. The best thing I can do is to encourage these young ladies to excel and persevere forward. Hopefully, soon, they will not be in the minority anymore.

Words of Wisdom

- Make the most of today.
- Work hard, then you will be a lucky person.
- Making the decision may be tough, but it could be even tougher if the decision has not been made.

Li-Chyong Chen

Distinguished Research Fellow and Director
Center for Condensed Matter Sciences
National Taiwan University, Taipei
Taiwan

e-mail: chenlc@ntu.edu.tw
telephone: 886-2-3366-5249

Li-Chyong Chen in 1983 at Harvard University during her graduate studies standing by a sputtering system nicknamed "the blue box" by the students. The blue box was problematic; accordingly, like many other students, she was trained to not be afraid of fighting against all odds. (See insert for color version of figure.)

Birthplace
Taipei, Taiwan
Born
March 12, 1959

Publication/Invention Record

>350 publications: h-index 50 (Web of Science)
14 patents, co-editor of 3 books/proceedings,
8 book chapters

Tags

❖ Academe
❖ Domicile: Taiwan
❖ Nationality: Taiwanese
❖ Asian
❖ Children: 2

Proudest Career Moment (to date)

After working in Taiwan for several years, I attended an international conference in San Diego. During the break immediately following my presentation, an attendee approached me and complimented my research work. Although this was not the first time that I

Successful Women Ceramic and Glass Scientists and Engineers: 100 Inspirational Profiles. Lynnette D. Madsen.
© 2016 The American Ceramic Society and John Wiley & Sons, Inc. Published 2016 by John Wiley & Sons, Inc.

received note of appreciation about a presentation, I felt the proudest at this time. When I first returned to Taiwan, I started a new lab and had not yet established a name for myself. Overall, significant investments were being made in R&D in Taiwan, but the research environment, and consequently, many of the accomplishments, still lagged behind United States. It was very encouraging to hear such a compliment from someone well established in the field.

In 2006, our work was published in *Nature Materials*, the first for an all-Taiwanese team. In this paper, we demonstrated a novel self-organized growth of Au nanopeapods of silica nanowires using a microreactor approach. Wavelength-dependent and reversible photoresponse behavior based on the hybrid nanowire ensemble devices were demonstrated, which may be ascribed to the surface plasmon resonance (SPR). By adjusting the process parameters, we obtained Au nanorods with different aspect ratios, which resulted in SPR peak shifts thereby causing the hybrid nanowire devices to switch with tuneable color. Our paper was subsequently featured in *Materials Today* and selected as a Fast Breaking Paper; only one paper is chosen bimonthly from the top 1% of highly cited papers.

In 2007, I received an Honorary Doctor degree from Linköping University in Sweden. This honor is primarily based on my efforts for enhancing joint research activities between my university and theirs. To date, I am the only Taiwanese to be given this honor in Sweden. The other technical faculty awardee in 2007 was Åke Svensson, the chief executive officer (CEO) of SAAB. One outcome was a joint workshop on nanoscience and technology among the universities in Taiwan, Linköping and SAAB in the following year; SAAB was celebrating their 70 years anniversary at this time. Based on my contact with Åke Svensson, all participants of this joint nano meeting were invited to an unforgettable Air Show as a finale of the workshop.

Academic Credentials

Ph.D. (1989) Applied Physics, Harvard University, Cambridge, Massachusetts, USA.
B.Sc. (1981) Physics, National Taiwan University, Taipei, Taiwan.

Research Expertise: nanomaterials, interfaces, plasma-based processes, optoelectronics, energy materials, sensors

Other Interests: enhancing diversity in science and engineering, volunteering in professional societies, promoting international collaboration

Key Accomplishments, Honors, Recognitions, and Awards

- Series Editor, Member of the Series Board: World Scientific Publishing Company (WSPC) Series in Nanoscience and Nanotechnology, 2015 to present
- Advisory Editor: Book Series on Energy and Sustainability, Cambridge University Press and Materials Research Society (MRS), 2013 to present
- Editorial Board: Advanced Materials Letters, VBRI Press, 2010 to present
- Distinguished Research Fellow, National Taiwan University, Taiwan, 2007 to present

- Editorial Board: International Journal of Molecular Engineering, Inderscience Publishers, 2007 to present
- Editorial Board: Critical Reviews in Solid State and Materials Sciences, Taylor and Francis, 2004 to present
- Outstanding Scholar Award, Foundation for the Advancement of Outstanding Scholarship, Taiwan, 2010–2015
- Editorial Board: Open Phys. Chem. Journal and Open Appl. Phys. Journal, Bentham Science Publishers, 2007–2014
- Acharya Vinova International Award in Materials Science and Technology, VBRI, India, 2013
- Outstanding Research Award, National Science Council, Taiwan, 2010–2013
- Editorial Board: Nanoscience and Nanotechnology Letters, American Scientific Publishers, 2010–2012
- Ho Chin-Tui Outstanding Scholar Award, Ho Chin-Tui Foundation, Taiwan, 2012
- Outstanding Scholar Research Project, National Science Council, Taiwan, 2008–2011
- Elected Fellow, MRS, USA, 2010
- Meeting Chair, 2009 Fall Meeting, MRS, USA, 2009
- Laureate of the Khwarizmi International Award, Iran, 2009
- International Federation of Inventors' Association Lady Prize, 2009
- Distinguished Visiting Research Fellow, Royal Academy of Engineering, UK, 2008
- Honorary Doctor, Linköping University, Sweden, 2007
- Outstanding Research Award, National Science Council, Taiwan, 2006
- Fellow, the Physical Society of Republic of China in Taiwan, 2006
- Associate Editor: Journal of Vacuum Science and Technology B: Microelectronics and Nanometer Structures, 2004–2006
- Research Achievement Award, National Taiwan University, Taiwan, 2004
- Academia Sinica Young Scholar Research Award, Taiwan, 2000
- Reinhold Rudenberg Memorial Prize for Graduate Student, Harvard, USA, 1989
- IBM Pre-doctoral Fellowship, USA, 1986–1988

Biography

Early Life and Education

Li-Chyong Chen was raised in Taipei, Taiwan by Taiwanese parents. Her ancestors came to Taiwan from China more than 200 years ago. She has four sisters and three brothers. Li-Chyong grew up in the suburb area of Taipei, received all her early education in the Taipei public school system and completed her undergraduate study in the Department of Physics from the National Taiwan University (NTU) (also located in Taipei). After receiving a B.Sc. degree, she remained in the same department at NTU as a full-time teaching assistant for 2 more years.

Then, Li-Chyong pursued her advanced study at Harvard University, Cambridge, MA, USA, with full financial support, that is, she had a fellowship for the first year and then research assistantships with fully waived tuition fees for the entire period while she was a graduate student. She was also awarded a Pre-doctoral Fellowship from IBM for the last 2 years of her Ph.D. studies.

Career History

After receiving her doctorate, Li-Chyong moved to Schenectady, New York. From 1989 to 1994, she held a position as a Materials Scientist at the Corporate Research and Development Center (CRD) for General Electric (GE). While working at GE, she undertook exploratory research of her own choice, and she was involved in a couple of industrial application-driven R&D activities, which are relevant to GE's core business, such as lighting and aircraft engines. For these applied research activities, ceramic material development is essential and joint efforts among the scientists from CRD and the R&D labs of the corresponding business unit were encouraged to ensure GE's position in leading-edge technologies.

In mid-1994, Li-Chyong joined the Center for Condensed Matter Sciences (CCMS), NTU, as an Associate Research Fellow. Since then, she has led the Advanced Materials Laboratory at CCMS. Li-Chyong was promoted to Research Fellow in 2000, honorably recognized as Distinguished Research Fellow in 2007, and was recently appointed (in 2012) as the Director of CCMS.

CCMS, the first university-level research center that was established in 1992 is a premier interdisciplinary research institute. The overall goal of CCMS is to be a leading research institute engaged in frontier research and technology development. It emphasizes the development of specialized technical expertise and the establishment of advanced facilities to promote the research and technological accomplish-

Li-Chyong Chen was appointed as the Director of Center for Condensed Matter Sciences (CCMS) at National Taiwan University in 2012. Credit: Director's Office of CCMS.

ments through collaborations with outstanding domestic and international research groups. Over the last decade, Li-Chyong has established a strong program in synthesizing low-dimensional nanomaterials and related hybrids based on carbon, Si, III-nitrides, and metal oxides. Her core competence areas are chemical and physical vapor deposition, in particular, those involving plasma or highly energized processes. Also, Li-Chyong has expertise in integrated micro-devices for the above-mentioned materials in optoelectronics, electronics, energy, and sensing applications. The entire research team in her group consists of more than 40 members (including post-doctoral scholars, graduate students, and research assistants) with various areas of expertise (e.g., physics,

Li-Chyong Chen (middle), Head of Advanced Materials Laboratory in Condensed Matter Sciences (CCMS) at National Taiwan University, in the lab with her graduate student Mr. Jay-Yang Lee (right) and research assistant Ms. Yesi (left). They are working with microwave plasma-enhanced chemical vapor deposition (MPECVD) system that has been a power horse in producing novel nanomaterials, including carbon nanotubes and graphene nanowalls.

chemistry, optoelectronics, electronics and electrical engineering, materials science, chemical engineering, and bio-photonics).

Li-Chyong is an active member of the USA Materials Research Society (MRS) and the American Vacuum Society (AVS), as well as a number of professional societies in Taiwan, including the Physics Society in Republic of China (PSROC), the Taiwan Association of Coating Technology (TACT), and the Society of Taiwan Women in Science and Technology (TWiST). She is recognized as a Fellow of MRS and PSROC.

Li-Chyong (middle in the last row) and her female peers at the first female physicists symposium, a two-day program: day one had an open-to-public forum at CCMS, followed by a second day program focused on brainstorming and networking in Wu-Lai, an aboriginal village located in the countryside of Taipei. Dr. Chen helped to establish the women in physics working group, which later became a committee in the Physics Society of the Republic of China (Taiwan).

She has also devoted a substantial time to professional activities as a volunteer, such as organizing international symposia and conferences, and serving various society committee and project review panels. Notably, Li-Chyong has served as one of the four Meeting Chairs for the MRS Fall Meeting in 2009, a Board Member of TACT, Vice President and Councillor of PSROC, and the Advisor of TWiST. As well, she was the founding President of the Taiwan Chapter of AVS and founding Chair of Committee on Women in Physics at PSROC.

3 Highly Cited Publications

Title: Improved broadband and quasi-omnidirectional anti-reflection properties with biomimetic silicon nanostructures
Author(s): Huang, Yi-Fan; Chattopadhyay, Surojit; Jen, Yi-Jun; et al.
Source: Nature Nanotechnology; volume: 2; issue: 12; pages: 770–774; published: December 2007
Times Cited: 472 (from Web of Science)

Title: Catalytic growth and characterization of gallium nitride nanowires
Author(s): Chen, C. C.; Yeh, C. C.; Chen, C. H.; et al.
Source: Journal of the American Chemical Society; volume: 123; issue: 12; pages: 2791–2798; published: 2001
Times Cited: 405 (from Web of Science)

Title: Heterostructures of ZnO-Zn coaxial nanocables and ZnO nanotubes
Author(s): Wu, J. J.; Liu, S. C.; Wu, C. T.; et al.
Source: Applied Physics Letters; volume: 81; issue: 7; pages: 1312–1314; published: 2002
Times Cited: 277 (from Web of Science)

ResearcherID (B-1705-2015)

Challenges

The challenges have been different in different stages of my career. First, I felt a lack of role models and support network; I know this situation is shared by many female scientists. Balancing my work and family is also quite challenging; in particular, it seemed like the time demands were excessive when the kids are young and I was new in my career; this effect is exacerbated because they are concurrent. Once my kids became grown and my career became more established, I enjoyed both a lot more. As a non-native English speaker, writing and presenting in English is a continuous challenge for me.

On being a woman in this field . . .

I have enjoyed being a woman, although there are challenges in being in the minority in this field. I consider myself a hard worker; probably others view me as a workaholic. However, I also wanted to get married and have a family; both these things came to pass during my graduate studies. I have no regret for wanting-it-all; it is possible that it slowed my progress a little, early in my career. Nevertheless, gradually I learned to manage my time to achieve a better work–family balance and quality of life. My husband and I also share our work—we co-established and jointly supervise the lab in Taiwan. His understanding and support in all aspects has been essential to my career.

    Perhaps being a minority in this field has been a blessing in disguise. Through the years, I had received more invitations, both for visiting and giving seminars in international research institutes, delivering invited talks or even keynotes at international conferences, and participating in international collaboration projects. I have enjoyed my work and tried my best to do it well, but I think I have slightly higher

The Chen family in 2008 before taking a boat trip along the River Cam near Cambridge University. Li-Chyong Chen received the Distinguished Visiting Research Fellowship from the Royal Academy of Engineering in UK; the accompanying fellowship supported her travels. (See insert for color version of figure.)

visibility than some of my male peers with similar qualification, which may in part be due to being a minority.

I also got involved in various gender and diversity matters, from participating in a working group on women in physics in 2001, to forming a formal committee of the local physics society (PSROC) in 2003 and more recently, establishing the Society of Taiwan Women in Science and Technology in 2012. Although organizing annual symposia and workshops is aimed at cultivating young brains, I have enjoyed the friendship and trust built over the years in these networks. The sisterhood with other women, who are peers, is significant for me and has been a positive influence. Foremost, I know that I am not alone.

Words of Wisdom

I do believe that teamwork with other women and/or with men is essential for almost any scientific endeavor. Teamwork helps to bring out the full potential of each individual.

Also, it is important that you do what you enjoy, and enjoy what you do.

Profile 14

Zhili Chen

Senior politician
People's Republic of China (PRC)

Zhili Chen at Penn State
University's Materials
Research Laboratory, ca.
1980. Photo courtesy of
Penn State University.

Birthplace

Xianyou County, Putian, Fujian Province, China

Born

November 1942

Publication/Invention Record

10 research publications
Author and chief editor of 12 books in the area of science, culture, and philosophy

Academic Credentials

M.S. (1968) Solid-State Physics, Shanghai Institute of Ceramics of Chinese Academy of Sciences, Shanghai, China.
B.S. (1964) Physics, Fudan University, Shanghai, China.

Research Expertise: ferroelectric and piezoelectric materials and devices

Other Interests: politics, education

Key Accomplishments, Honors, Recognitions, and Awards

• Vice Chairman of Standing Committee of the 11th National People's Congress, 2014 to present
• President, All-China Women's Federation, 2008–2013

Tags

❖ Administration and Leadership
❖ Government
❖ Domicile: China
❖ Nationality: Chinese
❖ Asian

Successful Women Ceramic and Glass Scientists and Engineers: 100 Inspirational Profiles. Lynnette D. Madsen.
© 2016 The American Ceramic Society and John Wiley & Sons, Inc. Published 2016 by John Wiley & Sons, Inc.

Secretary of State Hillary Clinton greets Zhili Chen, Vice Chairperson of the Standing Committee of the Congress of China at the State Department on March 28, 2012 in Washington, DC. The State Department hosted the delegation as China today celebrates the 53rd anniversary of the liberation of Tibet. Credit: T.J. Kirkpatrick/Getty Images, Inc. (See insert for color version of figure.)

- Vice Chairman, 11th Standing Committee of the National People's Congress (NPC), 2008–2013
- Soong Ching Ling Camphor Award, 2011
- Chairman, Children's Foundation of China, 2010
- Chief Editor, Great Chinese Encyclopedia of China, 2009
- State Councillor (in charge of education, culture, and sports), 2003–2008
- Governor of Beijing Olympic Village, 2008
- Vice Chairman of the Organization Commission of 29th Olympic Games, 2005
- Minister of Education for the People's Republic of China, 1988–2003
- Honorary degree of Doctor of Philosophy in Management on the

Zhili Chen, Vice Chairperson of the Standing Committee of the Congress of China, sits at a table with committee members for a meeting with U.S. Secretary of State Hillary Clinton at the State Department on March 28, 2012 in Washington, DC. The State Department hosted the delegation as China today celebrates the 53rd anniversary of the liberation of Tibet. Credit: T.J. Kirkpatrick/Getty Images, Inc.

Minister of Education of the People's Republic of China, Assumption University in Bangkok, Thailand, 2000

- Invited public lecture "China's Higher Education in the 21st Century," Penn State University, 2000
- Vice Minister of the State Education Commission, 1997–1998
- Secretary, Party Members' Group, 1997
- Deputy Secretary (Vice Mayor), CPC Shanghai Municipal Committee, 1988–1997
- Head, Propaganda Department of CPC Shanghai Municipal Committee, 1984–1988
- Secretary of the Science and Technology Sub-Committee of the CPC Shanghai Municipal Committee until 1984
- Deputy Secretary, Chinese Communist Party Committee, 1982
- Second prize for excellence by the Chinese Academy of Sciences, 1982
- Visiting Scholar, Penn State Materials Research Laboratory, 1980–1982

Zhili Chen, China's Minister of Education, was a visitor to Penn State University in 2000 where she, as well as members of her delegation, met with several officials, including President Graham B. Spanier. Photo credit: Greg Grieco.

Biography

Early Life and Education

Zhili Chen was born in Xianyou County, Putian, Fujian Province in the 1940s and her nationality is Han. She studied physics at Fudan University. Upon graduation, she pursued a postgraduate degree at Shanghai Institute of Ceramics of Chinese Academy of Sciences (SICCAS), also in Shanghai, conducting research in the area of solid-state physics.

Career History

Early in 1961, she joined the Communist Party of China (CPC). At the beginning of the Cultural Revolution, she was sent to work in an army farm (Danyang Lake Farm) for 2 years before returning to SICCAS. She continued with her research, first as an intern research fellow and then as an Associate Research Fellow. In 1982, she went abroad to Pennsylvania State University (USA) as a visiting scholar. After returning to China, she was elevated to Vice Party Chief, and thus transformed from a scholar to a CPC official.

Her political posts in Shanghai included Vice Secretary and later Secretary of the CPC Committee of Shanghai Science and Technology Commission, Director of the Propaganda Department of Shanghai, and Vice Secretary of CPC Shanghai Committee.

Zhili Chen at a U.S.–China meeting; Prof. L.E. Cross (with a beard) is to her right.

In 1997, she was transferred to the central government area and appointed as Vice Director and leader of the Party group of the National Education Commission. In 1998, she became the Minister of Education. In 2003, she was further elevated to the position of State Councillor, in charge of education, culture, and sports. In 2008, Chen was elected a Vice Chairman of the Standing Committee of National People's Congress. In the spring of 2008, she was appointed as Governor of Beijing Olympic Village. She served as an alternate member of 13th and 14th Central Committees of the Communist Party of China, and a full member of 15th, 16th, and 17th Central Committees.

3 Cited Publications

Title: Polarization and depolarization behavior of hot pressed lead lanthanum zirconate titanate ceramics
Author(s): Xi, Y; Chen, Z; Cross, LE
Source: Journal of Applied Physics; volume: 54; issue: 6; pages: 3399–3403; doi: 10.1063/1.332453; published: June 1983
Times Cited: 179 (from the American Institute of Physics (AIP))

Title: Depolarization behavior, and reversible pyroelectricity in lead scandium–tantalate ceramics under DC biases
Author(s): Chen, Z; Xia, Y; Cross, LE
Source: Ferroelectrics; volume: 49; issue: 1; published: 1983
Times Cited: 45 (from Google Scholar)

Title: Diffuse ferroelectric phase transition and cation order in the solid solution system $Pb(Sc_{1/2}Nb_{1/2})O_3:Pb(Sc_{1/2}Ta_{1/2})O_3$

Author(s): Chen, Z; Setter, N; Cross, LE
Source: Ferroelectrics; volume: 37; issue: 1; doi: 10.1080/00150198108223500; published: 1981
Times Cited: 20 (from Web of Science)

Sources used to create this profile

- Penn State University, http://www.psu.edu/ur/2000/chenzhili.html, accessed November 30, 2014.
- Wikipedia, http://en.wikipedia.org/wiki/Chen_Zhili, accessed November 30, 2014.
- All-China Women's Federation, http://www.womenofchina.cn/womenofchina/html1/special/15/2482-1.htm, accessed November 30, 2014.
- Yuwu Song, Biographical Dictionary of the People's Republic of China, accessed online November 30, 2014.
- Assumption University, http://www.library.au.edu/hall-of-fame/special-award5.html, accessed November 30, 2014.
- China Vitai, http://www.chinavitae.com/biography/Chen_Zhili/bio, accessed November 30, 2014.
- Soong Ching Ling Camphor Award, http://www.womenofchina.cn/womenofchina/html1/people/others/13/6332-1.htm, accessed November 30, 2014.

Profile 15

Uma Chowdhry

Senior Vice President and
Chief Science and Technology Officer,
Emeritus,
DuPont (retired December 2010)
Address:
104 Redwood Lane
Kennett Square
PA 19348
USA

e-mail: uma.chowdhry@yahoo.com

Uma Chowdhry at age 30.

Tags

❖ Administration and Leadership
❖ Industry
❖ Domicile: USA
❖ Nationality: American; Indian by birth
❖ Children: 0

Birthplace
Mumbai, India
Born
September 14, 1947

Publication/Invention Record

>30 publications: h-index 15

Proudest Career Moment (to date)

When I became the first woman to be appointed to the position of Chief Science and Technology Officer responsible for over 8000 scientists and engineers and over $1 billion of global R&D at DuPont—a Dow 30 multinational diversified chemicals and materials company, signifying that women "can" aspire to such positions.

Academic Credentials

Ph.D. (1976) Materials Science, Massachusetts Institute of Technology, Cambridge, Massachusetts, USA.
M.S. (1970) Engineering Science, California Institute of Technology, Pasadena, California, USA.
B.S. (1968) Physics, Mumbai University, Mumbai, India.

Successful Women Ceramic and Glass Scientists and Engineers: 100 Inspirational Profiles. Lynnette D. Madsen.
© 2016 The American Ceramic Society and John Wiley & Sons, Inc. Published 2016 by John Wiley & Sons, Inc.

Research Expertise: ceramic materials, catalysts, proton conductors, superconductors, and ceramic microelectronics packaging

Other Interests: Helping women be successful in science and engineering

Key Accomplishments, Honors, Recognitions, and Awards

- Awarded Caltech's Distinguished Alumnus Award, 2013
- Appointed to Governing Council for the National Academy of Engineering (NAE), 2012
- Appointed to Board of Directors, Baxter International, 2012
- Industrial Research Institute (IRI) Medal for DuPont Leadership, 2011
- American Chemical Society Award for Leadership in Chemical Research Management, 2011
- Appointed to MIT Corporate Visiting Committee for Sponsored Research, 2011
- Appointed to the Board of LORD corporation, 2010 (LORD is a privately held corporation; worldwide leader in adhesives and coatings, vibration and motion control, and magnetically responsive technologies)
- Appointed to Visiting Committee for Advanced Technologies, NIST, 2010
- Elected Member of the American Academy of Arts and Science, 2003
- Elected Member of NAE, 1996
- Fellow of the American Ceramic Society, 1989

Biography

Early Life and Education

Born and raised in Mumbai, India, Uma became interested in science during her high school years. Upon graduating, she studied Physics & Math at the University of Bombay in her home country of India, graduating with a B. Sc. (Hons) degree. After starting her graduate work at Caltech, she deviated from her initial plans to study nuclear physics, as she soon became convinced that the physics and chemistry of materials were fascinating to her. Prior to undertaking her doctoral studies at MIT, Uma got married to a colleague from Mumbai who was at the University of Michigan and worked at Ford Motor Company for a year in the early 1970s. Uma and her husband moved to Cambridge in 1972 to go to MIT and Harvard, respectively, and Uma began her doctoral program work at MIT.

Career History

Upon receiving her doctorate in materials science in 1976, Uma was offered a job by DuPont's corporate research labs in Wilmington, Delaware. Her first project was to find better battery materials, but soon thereafter she began to focus on ceramic catalyst materials for DuPont's chemical processes. Uma was offered many different

opportunities in both R&D and Business management and as she progressed through them she
became convinced that Technology management
was what she enjoyed most. When she was offered
the opportunity to head the entire global R&D
enterprise for DuPont, she was excited and humbled
at the same time. She focused the organization on
innovation leading to value creation through sustainability, globalization, and integrated science.
After a successful 33-year career at DuPont, she
retired as Senior Vice President and Chief Science
and Technology Officer.

A summary of her path follows:

1977, Research Scientist, Central Research

1981, Group Leader, Central Research

1983, Research Supervisor, Central Research

Uma Chowdhry in the laboratory
in 2007.

1985, Research Manager, Central Research

1988, Laboratory Director, Electronics

1990, Business Manager, Microcircuit Materials

1991, Quality Director, Electronics

1992, Laboratory Director, Jackson Laboratory, Chemicals

1993, R&D Director, Specialty Chemicals

1995, Business Director, Terathane Products

1997, Business Planning and Technology Director, Chemicals

1999, Director, DuPont Engineering Technology

2002, Vice President, DuPont Central Research and Development

2006, Senior Vice President and Chief Science and Technology Officer

3 Most Cited Publications

Title: A new high-temperature superconductor—$Bi_2Sr_{3-X}Ca_XCu_2O_{8+Y}$
Author(s): Subramanian, MA; Torardi, CC; Calabrese, JC; et al.
Source: Science; volume: 239; issue: 4843; pages: 1015–1017;
doi: 10.1126/science.239.4843.1015; published: February 26, 1988
Times Cited: 666 (from Web of Science)

Title: Crystal-structure of $Tl_2Ba_2Ca_2Cu_3O_{10}$, a 125-K superconductor
Author(s): Torardi, CC; Subramanian, MA; Calabrese, JC; et al.
Source: Science; volume: 240; issue: 4852; pages: 631–634;
doi: 10.1126/science.240.4852.631; published: April 29, 1988
Times Cited: 395 (from Web of Science)

Title: Structures of the superconducting oxides $Tl_2Ba_2CuO_6$ and $Bi_2Sr_2CuO_6$
Author(s): Torardi, CC; Subramanian, MA; Calabrese, JC; et al.
Source: Physical Review B; volume: 38; issue: 1; pages: 225–231;
doi: 10.1103/PhysRevB.38.225; published: July 1, 1988
Times Cited: 358 (from Web of Science)

Challenges

Leaving home and all the people I loved and cared about, to go to graduate school in a foreign land where I knew no one was a daunting challenge and now I often wonder where I found the courage to leave my comfort zone with $8 and embark upon an adventure into the unknown! Not only did I leave home, I left behind the man I was in love with and with whom I would continue to correspond daily through the two tough, academically challenging years at Caltech. I worked hard, sought help when I needed it from colleagues who were always generous with their time, and I learnt to love the process of learning new things every day! I became committed to a career in R&D in materials science while I was at Caltech where I discovered how influential good professors and colleagues can be. Today, I know that leaving one's comfort zone, taking prudent risks, working hard, seeking help when needed and dreaming courageously is important to discover oneself and to make one's life be all that it can be!

On being a woman in this field . . .

A recent image of Uma Chowdhry.

The early 1980s was an interesting time to be a woman in science and engineering in industry. There were very few of us around and so to build credibility and a track record of impactful results was very important in order to be given any position of responsibility. You had to prove you were capable of handling a new assignment no matter how many jobs you had progressed through. However, it was also a time when women were given many opportunities if they proved themselves capable of delivering results. I received many breaks because I was a woman but I also had to prove myself repeatedly to be taken seriously.

Words of Wisdom

- Women should discover their professional passion and then pursue their dreams with courage. It is human to make mistakes, but important to change course as often as needed even if it is risky, until you find what you love doing and giving it all you've got.
- Belief in oneself and having a purpose one is passionate about, are necessary for one's fulfillment.
- Having an internal compass with a set of core values and beliefs as a guide, cultivating a network of relationships, treating everyone with respect, and seeking help when needed are important attributes to be successful.

- It is critically important to learn to communicate effectively with both the spoken and written word.
- Building strong interpersonal relationships is key to doing an effective job.
- Enthusiasm for one's job that comes through when interacting with people is contagious and can energize the organization.
- Being authentic and engendering trust by demonstrating one's core values and principles in daily behaviors is very important.
- Helping other women succeed is what each of us has a responsibility to perform.

Profile 16

Deborah Duen Ling Chung

Composite Materials Research Laboratory
University at Buffalo, The State University
of New York
Buffalo, NY 14260-4400
USA

e-mail: ddlchung@buffalo.edu
telephone: (716) 645-3977

Deborah Chung as an
undergraduate student at
California Institute of
Technology (1972).

Birthplace
Hong Kong (British)
Born
September 12, 1952

Publication/Invention Record

>550 publications: h-index 55 (Web of Science)
18 patents, 14 books

Tags

- ❖ Academe
- ❖ Domicile: USA
- ❖ Nationality: American;
  Born British
- ❖ Asian (Chinese descent)
- ❖ Children: 0

Proudest Career Moment (to date)

There is nothing to be proud of, because all the recognitions received were due to God's grace and it was God who has been directing my life and my research. I guess I can be proud that God cares for me so much!

Academic Credentials

Ph.D. (1977) Materials Science, Massachusetts Institute of Technology, Cambridge, Massachusetts, USA.
S.M. (1975) Materials Science, Massachusetts Institute of Technology, Cambridge, Massachusetts, USA.
M.S. (1973) Engineering Science, California Institute of Technology (Caltech), Pasadena, California, USA.
B.S. (1973) Engineering and Applied Science, California Institute of Technology (Caltech), Pasadena, California, USA.

Research Expertise: materials science and engineering, particularly smart materials, concrete, thermal management, battery electrode materials, carbon fibers and nanofibers, composites processing and interfaces, metal–matrix composites, carbon–matrix composites, electronic packaging materials, activated carbon, clay, the removal of oil from water, and three-dimensional printing

Other Interests: integrating science and music, multidisciplinary research, practical applications, engineering material development, and scientific outreach to the public

Key Accomplishments, Honors, Recognitions, and Awards

- Recipient of the Niagara Mohawk Power Corp. Endowed Chair Professorship in Materials Research, University at Buffalo, State University of New York, 1991 to present
- Honorary Doctorate Degree, University of Alicante, Alicante, Spain, 2011
- Guest Professor, Tongji University, Shanghai, P.R. China, appointed in 2010
- Top Reviewer in 2008, an international award in relation to the journal Carbon, Elsevier, 2009
- Special Recognition Award, American Carbon Society, 2007
- Hsun Lee Award, jointly awarded by Institute of Metal Research (Chinese Academy of Sciences) and Shenyang National Laboratory for Materials Science, to recognize research accomplishment in materials science and technology, 2005

Deborah Chung receiving her honorary doctorate degree at University of Alicante, Spain (2011).

- Invited Professor, Tianjin University, Tianjin, P.R. China, appointed in 2005
- Visiting Professor, Jinan University, Jinan, P.R. China, appointed in 2005
- The Charles E. Pettinos Award, The American Carbon Society, 2004
- Chancellor's Award for Excellence in Scholarship and Creative Activities, Academic Year 2002–2003, The State University of New York
- Visiting Professor, Wuhan University of Technology, Wuhan, P.R. China, appointed in 2002
- Visiting Professor, Southeast University, Nanjing, P.R. China, appointed in 2002
- Visiting Professor, Beijing Technology and Business University, Beijing, P.R. China, appointed in 2002
- Outstanding Inventor, The State University of New York, 2002
- Fellow, American Carbon Society, conferred in 2001

- Honorary Professor, Shantou University, Shantou, Guangdong, P. R. China, appointed in 2000
- 24th Annual Inventor of the Year Award, Niagara Frontier Intellectual Property Law Association and the Technical Societies Council of the Niagara Frontier, 1999
- One of the top 10 inventors, 1998
- Fellow, ASM International, conferred in 1998
- Advisory Professor, Harbin Institute of Technology, Harbin, P.R. China, appointed in 1995
- Tau Beta Pi (New York Nu) "Teacher of the Year," 1992–1993

Deborah Chung speaking at Harbin Institute of Technology, China (2007). A student standing at the center of the room was asking a question.

- Recipient of the Ralph R. Teetor Educational Award, Society of Automotive Engineers, for being one of the top engineering educators in the United States, 1987
- Consultant to the Electro-Physics Section, NASA Lewis Research Center, Cleveland, OH, 1985
- Consultant to the International Advisory Panel and the Chinese Review Commission of the Chinese Ministry of Education, People's Republic of China, 1984
- Consultant to the General Technology Division, IBM, Endicott, NY, 1984
- Consultant to Semiconductor Research, Research and Development Center, Westinghouse Electric Corporation, Pittsburgh, PA, 1983
- Recipient of the American Institute of Mining, Metallurgical, and Petroleum Engineers Robert Lansing Hardy Gold Medal for the most promising metallurgist in the United States, 1980
- Recipient of the Ladd Award for one of the most promising engineering faculty members in Carnegie Mellon University, 1979
- Consultant to the Division of Materials Science, Research and Development Center, Westinghouse Electric Corporation, Pittsburgh, PA, 1978
- One of the four first woman graduates of California Institute of Technology, 1973
- Winner of Josephine de Karman Fellowship for graduate and senior undergraduate students of exceptional ability, 1972–1973

Biography

Early Life and Education

Deborah Chung was born and raised in Hong Kong when it was a British colony. Her initial education was also in Hong Kong at Ying Wa Girls' School and followed by King's College. In 1970 (at the age of 18), she moved on her own to the United States and subsequently received a B.S. degree in Engineering and Applied Science and an M.S. degree in Engineering Science from California Institute of Technology (Caltech) in 1973. She is one of the four first women to receive a B.S. degree from Caltech. She then completed S.M. and Ph.D. degrees in Materials Science at Massachusetts Institute of Technology (MIT) under the tutelage of Mildred S. Dresselhaus (who is also featured in this book).

Career History

Upon completion of her doctorate in 1977, she left MIT for an Assistant Professor of Metallurgical Engineering and Materials Science and Electrical Engineering at Carnegie Mellon University. In 1982, she was promoted to Associate Professor of Metallurgical Engineering and Materials Science.

She then moved in 1986 to the University at Buffalo, The State University of New York to become Professor of Mechanical and Aerospace Engineering. In 1991, she became the Niagara Mohawk Power Corp. Endowed Chair Professor of Materials Research.

She is best known for her invention of smart concrete in 1993. Smart concrete is concrete that is itself a sensor of strain and damage. This invention is featured in a film on NBC Learn (http://icue.nbcunifiles.com/icue/files/nbclearn/site/video/widget/NBC_Learn_Video_Widget2.swf?CUECARD_ID=62976). In addition, she is the inventor of "Exfoliated Graphite Fibers," "Carbon Fiber Reinforced Superconductor," and "Carbon Fiber Composites with Improved Fatigue Resistance," all of which were selected by the National Institute of Standards and Technology for funding through the Energy-Related Inventions Program of the Department of Energy.

She was a member of the Committee on Materials for High Density Electronic Packaging, National Materials Advisory Board, Commission on Engineering and Technical Systems, National Research Council, 1987–1990. Because of her extensive research on electronic packaging materials, she is the Editor-in-Chief of the Composite Materials Section of SpringerMaterials (2015–) and an Associate Editor of Journal of Electronic Materials (2008–). In addition, she is a member of the Honorary Editorial Advisory Board of Carbon (journal) (2001–) and is a member of the International Editorial Board of New Carbon Materials (journal) (1999–). In 1993, she served as the Conference Chairman, 21st Biennial Conference on Carbon, sponsored by American Carbon Society. In 1999–2006, she was a member of the Advisory Board of the American Carbon Society.

She is a dedicated teacher, having produced about 35 Ph.D. graduates, and received the "Teacher of the Year" award from Tau Beta Pi (New York Nu) in 1992–1993. The books that she has authored include "Functional Materials" (World Scientific, 2010) and "Composite Materials" (Springer, 2nd ed., 2010). Due to her interest in inspiring young people for science, she is the Editor of the book series "The Road to Scientific Success: Inspiring Life Stories of Prominent Researchers." Because of her experience as an international student and her ongoing interest in helping international students, she has served on the boards of International Students, Inc. and Ambassadors for Christ, Inc., and she speaks frequently to student groups.

Deborah Chung with her doctoral graduate student Yasuhiro Aoyagi (now Yasuhiro Yamada) in her laboratory in State University of New York (Buffalo) (2006). (See insert for color version of figure.)

Her talents extend beyond science. For example, she holds the L.R.S.M. diploma in piano performance from the Royal Schools of Music, UK (1971) and she took the second place finish at the Hong Kong Music Festival in the piano solo competition (Brahm's intermezzi) in 1970. She has been asked many times to speak on the intersection of science and music. In addition, she is a co-author of the book "Piloted to Serve," an autobiography of her mother, Rebecca Chan Chung (1920–2011), a nurse with the Flying Tigers, U.S. Army, and China National Aviation Corporation during World War II. Due to the wealth of history in this book, she also speaks on history to broad audiences and on TV.

Deborah Chung performing at the piano in Shatin Methodist Primary School, Hong Kong (2007).

3 Most Cited Publications

Title: Electromagnetic interference shielding effectiveness of carbon materials
Author(s): Chung, DDL
Source: Carbon; volume: 39; issue: 2; pages: 279–285; doi: 10.1016/S0008-6223(00)00184-6; published: 2001
Times Cited: 543 (from Web of Science)

Title: Exfoliation of graphite
Author(s): Chung, DDL
Source: Journal of Materials Science; volume: 22; issue: 12; pages: 4190–4198; doi: 10.1007/BF01132008; published: December 1987
Times Cited: 282 (from Web of Science)

Title: Thermally conducting aluminum nitride polymer-matrix composites
Author(s): Xu, Y; Chung, DDL; Mroz, C
Source: Composites, Part A; volume: 32; pages: 1749–1757; doi: 10.1016/S1359-835X(01)00023-9; published: 2001
Times Cited: 237 (from Web of Science)

Challenges

Scientific maturity took me decades to build up. After almost four decades of research, I still feel that my research and publications are getting better and better.

On being a woman in this field . . .

Engineering is a male-dominated profession, but women have done very well.

Words of Wisdom

Creativity is a key to success in scientific research.

Melanie Will Cole

Previously
Senior Research Physical Scientist
Leader of the Integrated Electromagnetic
Materials Research Group
U.S. Army Research Laboratory (ARL)
Weapons and Materials Research Directorate
(WMRD)
Bldg. 4600, Deer Creek Loop,
Attn: RDRL-WMM-E
Aberdeen Proving Ground
Aberdeen, MD 21005
USA

Melanie about 33 years old, at her first
job at ET& DL Ft Monmouth, using a
transmission electron microscope
(TEM) to study material interfaces.

e-mail: Melanie.willcole@gmail.com
telephone: (443) 528-7843

Tags

❖ Administration and Leadership
❖ Government
❖ Domicile: United States
❖ Nationality: American
❖ Caucasian
❖ Children: 1

Birthplace

Albuquerque, New Mexico, USA

Born

September 15, 1955

Publication/Invention Record

>125 referred journal publications and
>125 referred conference proceedings: h-index 25 (Web of Science)
Authored 5 invited book chapters; 10 U.S. patents,
2 invention disclosures pending, and 2 licensed patents

Pivotal Career Moment (to date)

Some of the most exciting moments in my career with the Army Research Lab occurred
at several points throughout my career path, whereby I have 5 times (1992, 1996, 2005,
2007, 2012) been the recipient of the Army Research and Development Achievement
(RDA) Award for my research innovations in materials physics. The RDA award is the
highest and most prestigious award for innovative science within the U.S. Army and is an
extremely competitive award since just over 1% receives the RDA award of the ~12,700

Successful Women Ceramic and Glass Scientists and Engineers: 100 Inspirational Profiles. Lynnette D. Madsen.
© 2016 The American Ceramic Society and John Wiley & Sons, Inc. Published 2016 by John Wiley & Sons, Inc.

scientists and engineers. It is awesome to do science, but it is exciting and quite humbling to be recognized by one's organization for innovation and creativity in science.

Outside Army, in 2008, I was very much honored to have been the recipient of the 2008 Society of Women Engineers (SWE) Lifetime Achievement Award for Lifetime Achievements and Sustained Contributions to the Field of Engineering. This award was a recognition of my pioneering research contributions, experimental creativity, and innovation in developing a fundamental understanding of the complex relationships between the structures, processing, and properties in thin film electronic materials. It also took me by surprise that many conference attendees, female scientists, and engineers asked me to autograph their conference books! I fondly remember my 14-year-old daughter watching me as I signed these books and her saying "mom, you're a science rock star"!

Melanie honored at the 2008 Society of Women Engineers (SWE) conference with the Lifetime Achievement Award, the highest honor given by the organization. Melanie holding the Steuben bowl given to her by Corning, and sharing the pride with her family: Alex, her daughter displaying the Achievement Award plaque, and her husband, Bob.

With all this aside, perhaps the proudest moment in my life was last year when my 19-year-old daughter Alex, a physics major at the University of Arizona, won three awards for best undergraduate research: (i) 1st place honors for her original research focused on mechanics of materials at the 2013 ARL Summer Student Symposium; (ii) 1st place award for her research in geophysics at the 2014 GeoDaze conference (celebration of geoscience research from the core to the clouds) held annually at the University of Arizona; and (iii) the "Al Weaver" award for best undergraduate physics research for her work focused on the electronic properties of graphene—University of AZ department of physics.

Academic Credentials

Ph.D. program—ABD/All But Dissertation (1985) Oceanography/Geochemistry, University of Rhode Island, Kingston, Rhode Island, USA.
M.S. (1980) Geo-Science/Geochemistry, Iowa State University, Ames, Iowa, USA.
B.S. (1977) Geology, University of Miami, Coral Gables, Florida, USA.

Research Expertise: complex oxides, semiconductor electronic materials, geochemistry, and geophysics

Other Interests: volunteerism within the technical community through symposia, workshop, and conference organization. Educating and mentoring others. Outreach to the public especially to underrepresented classes (namely, women) in science, technology, engineering, and mathematics (STEM) fields

Key Accomplishments, Honors, Recognitions, and Awards

- Editorial Board: Sensors Remote Sensing Section, 2007 to present
- Editorial Board: Integrated Ferroelectrics, 2004 to present
- Editorial Board: Ferroelectrics, 2004 to present
- Contributing Editor for Journal of the American Ceramic Society, 2002 to present
- Editorial Board: Journal Recent Patents on Materials Science, 2014
- "Distinguished Performance Award," annually (26 in total), 1988–2014
- Directors Research Initiative (DRI) Awards (11 in total): 1998, 1999, 2001, 2003, 2004, 2008, 2009, 2010, and 2011–2013
- Directors Strategic Initiatives (DSI) Awards (3 in total): 2011, 2012, 2013
- ARL Customer Service Awards (3 in total): 2012, 2013, 2014
- U.S. Army Research and Development Achievement (RDA) Awards (5 in total): 1992, 1996, 2005, 2007, 2012
- Materials Research Society (MRS) Spring Meeting Outstanding Paper Award, 2012
- US Army Small Business Innovation Research Program (SBIR) Achievement Award, 2010
- Elected "Fellow of the U.S. Army Research Laboratory," 2008
- Society of Women Engineers (SWE) Lifetime Achievement Award, *for her pioneering research contributions, experimental creativity, and innovation in developing a fundamental understanding of the complex relationships between the structures, processing, and properties in thin film electronic materials*, 2008
- Award/Recognition for Excellence in Directors Research Initiative (DRI) Research, 2008
- U.S. Patent (10 in total) awards: 2014, 2011, 2009, 2007, 2007, 2005, 2005, 2004, 2004, 2001
- Army Research Laboratory 1st Place (Gold) Mentor Award for Research Mentorship, 2008
- Research Academic Sabbatical Position in Applied Physics at California Institute of Technology, 2005–2007
- U.S. Army "Official Commendation" for innovative research in ferric oxide films, 2005
- U.S. Army Research Laboratory—Weapons and Materials Research Directorate "On-The Spot Award," 2001 and 2005 for excellent science in electronic thin film research
- U.S. Army Research Laboratory—Weapons and Materials Research Directorate "Official Commendation-Special Acts," 2000, 2001, and 2003 (contacts to SiC, and material designs to enable vibration-insensitive MEMS devices)
- U.S. Army Materiel Command "Certificate of Recognition," 1998, 2001, and 2002 for research innovations related to growth of low defect density GaN, Ni contacts to SiC, and methods for mitigation of film–substrate interdiffusion

- U.S. Army Materiel Command, U.S. Army Research Laboratory, Weapons and Materials Research Directorate "Official Commendation" for outstanding contributions to material science research, 2000
- Department of the Army "Official Commendation-Special Acts" and "Certificate of Award," for excellence in electronic materials research, 1998
- Army Science Conference: 1st Place Honor "Best Scientific Paper," 1992
- Department of the Army, Electronics Technology and Devices Laboratory (ET&DL), Electronic Devices Research Division, "Official Commendation" and "Certificate of Achievement," for distinguished performance at the 17th Army Science Conference, 1990
- Department of the Army Assistant Secretary of the Army for Research, Development, and Acquisition "Certificate of Outstanding Achievement" for best scientific paper at Army Science Conference, 1990
- Department of the Army, Electronics Technology and Devices Laboratory, Electronic Devices Research Division "Official Commendation" for science contributions toward understanding heterointerfaces in thin film materials, 1989

Biography

Early Life and Education

Melanie grew up in New Mexico in a region with many low-income earners of Native American or Mexican heritage that were poorly educated. Her father was a blue-collar worker and her mother an artist/teacher. Her high school had a limited repertoire of science classes, not offering anything beyond general biology. She was not particularly fond of biology, so she focused her efforts on math, a subject she seemed to have a natural talent for. Melanie never knew anyone who had attended college; however, she knew that a high school degree was not her educational endpoint, she wanted to attend college. One of her high school teachers told her that in order to go to college, she would have to take a test called the SAT. Melanie remembers the day she took the SAT test. She and ~50 other students from across the state took the SAT exam at the New Mexico state fair grounds on a warm Saturday morning sitting outside, she reminisces that working on picnic tables with such rough surfaces made it difficult to blacken the little circles on the answer sheet.

Attending college was not the norm in her neighborhood (or even the state). Her family's low income combined with the continual medical expenses for her father's life-long bout with cancer caused her family to struggle financially. Such financial difficulties made paying college tuition cost prohibitive for her parents. However, in spite of her family's financial status, with the aid of college scholarships, Melanie and her brother and sister were the first in their family to go to college. Her sister became an architect, her brother a medical doctor, and Melanie herself became a scientist. Melanie immediately loved college—she found the range of courses to be fantastic and fell in love with the concept that science is the key to understanding of the world around us. She began her undergraduate studies majoring in mathematics; however, second semester sophomore year she took a physical geology class and it was this class that forever changed her life. Switching her major in the middle of her 4-year degree program was challenging; she had to cram 32 credits of geoscience classes, along with the supporting physics and chemistry classes, into her last 2 years at college in order to graduate on time. Luckily, with a few 21 credit semesters and summer classes, everything worked out. While navigating her way through this heavy academic course load, Melanie found time to fall in love and marry her soulmate, Robert Cole. After receiving her B.S. in geology, Melanie obtained an M.S. in geochemistry from Iowa State University (ISU). While at ISU in the late 1970s, Melanie attended a seminar by Roger Larsen, a respected geophysical oceanographer; this seminar inspired her, she became intrigued with oceanographic science and how little was known and understood about plate tectonics of the ocean floor. Her fascination with this virtually unexplored area of science prompted her to study geological/geochemical oceanography with one of the world's most renowned scientists in the field, at the Graduate School of Oceanography University of Rhode Island (GSO-URI). After 4 years of intense graduate work, having passed the qualifying exams, completion of course work, oceanographic field work, and experimental research, due to an unexpected debilitating health issue, she was unable to complete the final requirements, i.e., her dissertation defense that would have resulted in the award of the Ph.D. degree. This is reflected in the 6, as opposed to 3, letters after her name, Ph.D. ABD (All But Dissertation).

Career History

In 1988, after several years of recovering from the physical/mental limitations associated with her health issues, she was hired as a research physical scientist at the Electronics, Technology and Devices Laboratory (ET&DL), Ft. Monmouth, NJ, which later became part of the Army Research Lab's (ARL) Sensors and Electron Devices Directorate (ARL-SEDD), where she held positions as Research Physical Scientist and Leader of the Electronics Materials Process Science Team. ARL is the corporate research laboratory for the U.S. Army. In 1997, Melanie joined ARL-WMRD, whereby she shifted her research focus from semiconductor materials research toward the development of complex oxide ceramic thin film materials for frequency agile microwave communications and RADAR systems. Although Melanie's formal education resides in geology, geochemistry, and oceanography, she was able to expand upon this expertise and became the Army's subject matter expert in electronic materials. Melanie has worked at ARL for the last 26 years and is a Fellow of the U.S. Army Research Lab (an honor bestowed on only 1% of the technical workforce) and serves as team leader of the Integrated Electromagnetic Materials Research team.

During her tenure as Principal Investigator and Team Leader, she was responsible for formulating major laboratory

Melanie in her current job at ARL where she uses a metal organic chemical vapor deposition (MOCVD) reactor that she codesigned and built to grow complex oxide ceramic thin films. (See insert for color version of figure.)

program objectives, directing and executing basic and applied research, establishing budgets and allocating resources, presenting/defending program results at high-level reviews, and writing proposals to secure funding and served as mentor to junior scientists and engineers within the organization. She also served as research advisor for numerous postdoctoral associates and graduate/undergraduate students. Throughout her entire technical career, Melanie has always remained extremely active in bench-level research, i.e., she has always followed a "hands-on laboratory-life style." Melanie Cole has built a team of skilled researchers (personally educating the "in-house" scientists and engineers, post-docs, and students in the area of thin film technology), established a state-of-the-art thin film growth and process science laboratory, acquired national and international recognition for her research in complex oxide materials physics, and established strong technical collaborations and funding relationships within the government, academia, national labs, and the commercial sector. Additionally, she held a sabbatical position (2005–2007) in applied physics at California Institute of Technology (Caltech), Pasadena, CA.

Melanie has been active in advancing the dissemination of scientific knowledge through various professional societies such as the International Materials Research Congress, American Ceramics Society, Materials Research Society, and Electrochemical Society, and has been a frequent invited speaker and/or organizer of such meetings. She holds a sustained record for technology transition and commercialization via Intellectual Property, Small Business Innovative Research (SBIR), Small Business Technology Transfer (STTR), and Manufacturing Technology (ManTech) programs.

3 Most Cited Publications

Title: Mg-doped $Ba_{0.6}Sr_{0.4}TiO_3$ thin films for tunable microwave applications
Author(s): Joshi, PC; Cole, MW
Source: Applied Physics Letters; volume: 77; issue: 2; pages: 289–291; published: July 10, 2000
Times Cited: 201 (from Web of Science)

Title: The influence of Mg doping on the materials properties of $Ba_{1-x}Sr_xTiO_3$ thin films for tunable device applications
Author(s): Cole, MW; Joshi, PC; Ervin, MH; et al.
Source: Thin Solid Films; volume: 374; issue: 1; pages: 34–41; published: October 3, 2000
Times Cited: 172 (from Web of Science)

Title: Low dielectric loss and enhanced tunability of $Ba_{0.6}Sr_{0.4}TiO_3$ based thin films via material compositional design and optimized film processing methods
Author(s): Cole, MW; Nothwang, WD; Hubbard, C; et al.
Source: Journal of Applied Physics, volume: 93; issue: 11; pages: 9218–9225; published: June 1, 2003
Times Cited: 165 (from Web of Science)

Challenges

There have been a boat load of challenges throughout my career, both professional and personal. My biggest professional challenge was changing technical fields (oceanography to electronic material physics) and trying to learn what everyone else already knew and was very proficient at in a very short time frame. Luckily, lots of reading, countless discussions with others, and not being afraid to ask "low-level" questions saved me! Not having female role models or mentors was also difficult. If I had had the good fortune of being mentored early on, I think I would have been much more successful, made fewer mistakes, and definitely would have felt more confident amongst my peers in the early to mid-stages of my career. However, having climbed the technical ladder as a "lone wolf" so to speak provided me the motivation, inspiration, and stamina to "never give up" regardless of the environment! Science itself is a challenge; always trying to learn new area of science is interesting, but can be difficult if one's technical background is deficient. I remember trying to learn the area of ferroelectric thin films when I had no idea what a ferroelectric/ceramic material was, no less how to grow, process, and make a device out of it. Even understanding the technical literature was challenging, it took me forever to understand phonons, loss mechanisms, and microwave measurement science. This is where attending technical conferences and spending hours upon hours talking with some of the best scientists in the field really benefited me. Although I think that my biggest challenge was to learn how to "Thrive" and not just survive in my career. I learned that to do this I had to figure out how to achieve balance between my personal and professional life, understand both the challenges and the opportunities, and accept that I could have it all, but probably not at the same time.

On being a woman in this field . . .

I would like to tell everyone that being women in the physical sciences is wonderful; however, this would not be true. What I can say is that the "science" is wonderful, in fact beautiful, stimulating and motivating, but the career path is challenging and all too often unfair. When I reflect on my career path in the physical sciences, I remember that in graduate school I was the only female grad student amongst my 108 male counterparts in geosciences and the only student who was married. At the time (nearly 30 years ago) it really did not register to me that being the only female in the class was odd, what actually bothered me most was that I was told that I could not use my hyphenated last name (my legal last name was a combination of my maiden and married name, i.e., Will-Cole), so the geosciences department took it upon themselves to officially list my name as Melanie Cole, hence the reason I now have both a professional (Cole) and personal/legal last name (Will-Cole). Also, 30 years ago being the sole female graduate student whose husband was a chemical-physics graduate student at the same university caused some issues; I was told that although my GPA was the highest of my class, it was determined by the geosciences department (notably, all male professors) that my stipend be half the amount as the male students and the reason for this was because I was married and my husband had a stipend, hence having a dual income lessened my needs for receiving a full-stipend. Unfortunately, this same statement echoed again when I accepted my first job (26 years ago); my starting salary was ~$10k lower than my equivalent male science and engineering counterparts hired at that time. When I questioned this salary discrepancy, I was told that the fact that my husband had a high-paying job at Bell Labs was factored into my salary equation (*take home message being*: "married female scientists did not require the same salary as either married/unmarried male scientists, Go Figure!"). So for me, being women in a male-dominated scientific career unfortunately caused my name to be changed, lowered my initial income, but on the bright side this inspired me to work harder to succeed in this field. I will comment that the name change is a still bit annoying, I often forget which name is listed on credit cards, on medical and school records, and with my acquaintances (both professional and social colleagues). As for the income issue, well after about 7 years of successful science, I was out-earning my male counterparts, so eventually everything worked positively.

Being a woman in science also had a large impact on my personal life. I married at 20 and with a husband also in science, we had the "two body science career issue" that required many compromises. Such compromises translated to living nearly 800 miles apart for a few years when I elected to study oceanography at URI and my husband was finishing his Ph.D. at ISU. I realize this is commonplace today; however, in the early 1980s, there were no cell phones and e-mail was not even a word yet—neither of us could afford landline phones, so letters and the yearly Christmas visits sufficed. Also, the job

Melanie sharing a special moment with her daughter Alex, a budding young physicist.

search for two professionals in science was quite challenging, but it worked out quite well, although we both make major career deviations from our educational disciplines. My husband, a theorist in statistical mechanics, became a network science engineer and I morphed from an oceanographer into a material physicist. We became technically agile long before it was fashionable, and somehow it worked out beautifully.

To avoid the "mommy track," I postponed having a family until my career was in a "safe state"; at this time I was 38, married for 18 years, and was a technical team leader. My daughter Alexandria was born in 1994, a time when Department of Defence (DOD) jobs did not have maternity or family leave; hence, sick leave and vacation time were the only mechanisms for childbirth recovery/bonding. I had saved 4 weeks of combined vacation and sick leave and after 6 months of being pregnant, I informed my management that I was going to have a baby and would be off work for 4 weeks. Unfortunately, I was informed that if I was off work for this length of time, I would lose my position as team leader, only to be moved down the technical ladder that had been so difficult to climb. Fortunately, my daughter was born on Martin Luther King Day (a federal holiday) and that same week New Jersey was stricken with the heavy snow that led to 3 canceled work days; thus, to make a full "work-free" week, I took a sick day and returned to work on the following Monday. Thus, the good fortune of my daughter's birth on a federal holiday, in concert with bad weather and 1 sick day, allowed me to keep my team leader position intact; I was exhausted but once again everything worked out.

My final comments on being a woman in the male-dominated physical sciences are that it is a wonderful career where one is always learning, discovering, and innovating, this is stimulating and motivating. Reflecting back on my career path: yes, it was difficult and not always fair, but I would do it all over again in a heartbeat! I cannot imagine doing anything else, science is one of the reasons I smile every day 365 days a year!

Words of Wisdom

- Network, Network, Network! Networking with other scientists, especially those who are technically your senior, will enable you to learn more, motivate you to think outside the box, and be more productive. Networking outside one's organization is important and rewarding. Networking within one's own organization is often inhibited by "internal competition" and hence may be less fruitful. I recommend reaching out as much as possible!

- Volunteer and Give Back! Yes, give your time and give back to the science community by organizing technical symposia, workshops, and conferences. Not only will you help the technical community, but you will also gain organization skills, easily meet others in your field, create collaboration opportunities, and most importantly others will know who you are! Giving back also involves mentoring others, it is rewarding, and teaching is the best way to learn!

- Collaborate, Collaborate, Collaborate! Science today is more different than in the past, today you cannot do everything by yourself, you cannot be "expert" and proficient at every part of your research topic, and time is so very limited. Collaboration yields a more comprehensive research product, which ultimately yields that journal paper that will be heavily cited. As an experimentalist, I found it extremely beneficial to collaborate with a theorist or modeling expert, a very complementary research collaboration.

- Read, Read, Read! Knowledge is power. Knowledge of the technical literature provides the basis for most innovations and discoveries. It also provides you facts and data that can be articulated in technical discussions with both peers and management, and hence allows you to stand out "technically" and be noticed.

- Don't be afraid to ask questions, this is the best way to learn.

- Document, Document, Document! Avidly documenting you results in archival journals gets your name "out there," it also builds your resume, which unlike your managers remains with you for life!

- You must take negatives and turn them into positives because bad things will inevitably happen. Life is not fair and it is difficult to change other individual's negative behaviors;

however, you can change the way you react to such behavior. Turn negative actions into positives, use them as motivators to be more proactive and productive, this will help you to "Thrive" in your science career!

- Never, Ever Give Up! Don't settle for "Surviving" in your science career, set your path to "Thrive"!
- Finally, you can have it all, but unfortunately you can't have it all at the same time.

Profile 18

Elizabeth C. Dickey

Professor
Department of Materials Science and Engineering
North Carolina State University
Raleigh, NC 27606
USA

e-mail: ecdickey@ncsu.edu
telephone: (919) 515-3920

Beth Dickey started her academic career at the University of Kentucky, where she founded the electron microscopy facility in the College of Engineering.

Birthplace
Lexington, KY, USA
Born
July 4, 1970

Publication/Invention Record
>150 publications: h-index 40 (Web of Science)

Tags
❖ Academe
❖ Domicile: USA
❖ Nationality: American
❖ Caucasian
❖ Children: 2

Proudest Career Moment (to date)

I gain a great deal of satisfaction seeing my students mature both scientifically and personally during the course of their Ph.D. studies. Ultimately, my proudest moments are when my students graduate and achieve success in their careers.

Academic Credentials

Ph.D. (1997) Materials Science and Engineering, Northwestern University, Evanston, Illinois, USA.
B.S. (1992) Materials Engineering, with High Distinction, University of Kentucky, Lexington, Kentucky, USA.

Successful Women Ceramic and Glass Scientists and Engineers: 100 Inspirational Profiles. Lynnette D. Madsen.
© 2016 The American Ceramic Society and John Wiley & Sons, Inc. Published 2016 by John Wiley & Sons, Inc.

Research Expertise: developing atomic-level structure/property relationships for internal interfaces; correlating atomic structure and chemistry with interfacial mechanical and electrical properties; understanding atomic mechanisms of electrochemically induced interfacial phase transformations in metal–ceramic and ceramic–ceramic composites; understanding and quantifying driving forces for interfacial segregation in ionic materials; applying advanced electron imaging and spectroscopy techniques to study statics and dynamics of novel nanomaterials

Other Interests: promoting a climate and infrastructure to advance scholarly creativity and engagement, and developing unique curricula and environments for educating undergraduate and graduate students at the forefront of materials science and engineering

Key Accomplishments, Honors, Recognitions, and Awards

- Microscopy and Microanalysis, Editor (Materials Applications), 2004–2010; Editorial Board 2011 to present
- Journal of the American Ceramic Society, Associate Editor, 2005 to present
- Journal of Nanoscience and Nanotechnology, Editorial Board, 2004 to present
- Board of Directors, American Ceramic Society, 2013–2015
- Guest Editor, Journal of the European Ceramic Society, 2013
- Northwestern University Early Career Achievement Award for Alumni in Materials Science and Engineering, 2013
- Materials Characterization; Editorial Board, 2009–2011; Associate Editor, 2011–2013
- Richard M. Fulrath Award, American Ceramic Society, 2012
- Fellow, American Ceramic Society, 2010
- John T. Ryan Faculty Fellow, Pennsylvania State University, 2005–2008
- Presidential Early Career Award for Scientists and Engineers (PECASE), 1999
- Outstanding Materials Engineering Professor, University of Kentucky, 1999 and 2001
- Oak Ridge National Laboratory's High Temperature Materials Laboratory Summer Faculty Fellow, 1997 and 1998
- Department of Energy, Basic Energy Sciences Shared Research Equipment (SHaRE) Junior Faculty Research Fellow, 1997

Beth Dickey became a Fellow of the American Ceramic Society in 2010. Here she is receiving the honor from Dr. Edwin Fuller, President of the American Ceramic Society.

- Materials Research Society Graduate Student Award Finalist, Fall 1996
- Oak Ridge National Laboratory's High Temperature Materials Laboratory (HTML) Graduate Student Fellow, 1995–1997
- Microscopy Society of America (MSA) Presidential Student Scholar, 1995

- National Science Foundation Graduate Research Fellow, 1993–1995
- Walter P. Murphy Fellow, Northwestern University, 1992–1993
- ALCOA Foundation Scholarship, University of Kentucky, 1988–1992

Biography

Early Life and Education

Beth was born in Lexington, Kentucky and spent most of her childhood and teenage years working with and riding horses. During her junior year of high school, she was introduced to the discipline of Materials Engineering through an outreach program of the University of Kentucky. The following year, which was during the high-temperature oxide superconductor craze of the late 1980s, she decided to base her physics-class research project on oxide superconductors. These experiences piqued her interest in materials engineering, which appeared to be a discipline in which she could develop her innate interests in science and math with practical engineering issues. She decided to pursue materials engineering in her undergraduate studies, leaving her passion with horses as a hobby instead of a career path. After graduating from the University of Kentucky with a B.S. in Materials Engineering, she then pursued her Ph.D. at Northwestern University under an NSF Graduate Research Fellowship.

Career History

Immediately after receiving her Ph.D., Beth returned to the University of Kentucky as an Assistant Professor in the Department of Chemical and Materials Engineering where she founded and directed the electron microscopy facility in the College of Engineering. As an Assistant Professor, she was awarded the Presidential Early Career Award for Scientists and Engineering. In 2001, she moved to Penn State University where she was later promoted to Associate and then Full Professor. She also served as an Associate Director of the Penn State Materials Research Institute, in which she directed the Materials Characterization Laboratory. In 2011, she moved to North Carolina State University where she is a Professor and Director of Graduate programs in the Materials Science and Engineering Department. She is also the Director of the Center for Dielectrics and Piezoelectrics, a National Science Foundation (NSF) Industry/University Collaborative Research Center (I/UCRC), jointly run with Penn State.

3 Most Cited Publications

Title: Load transfer and deformation mechanisms in carbon nanotube–polystyrene composites
Author(s): Qian, D; Dickey, EC; Andrews, R; et al.
Source: Applied Physics Letters; volume: 76; issue: 20; pages: 2868–2870; published: May 15, 2000
Times Cited: 1455 (from Web of Science)

Title: Titanium oxide nanotube arrays prepared by anodic oxidation
Author(s): Gong, D; Grimes, CA; Varghese, OK; et al.
Source: Journal of Materials Research; volume: 16; issue: 12; pages: 3331–3334; published: December 2001
Times Cited: 1247 (from Web of Science)

Title: Continuous production of aligned carbon nanotubes: a step closer to commercial realization
Author(s): Andrews, R; Jacques, D; Rao, AM; et al.
Source: Chemical Physics Letters; volume: 303; issue: 5–6; pages: 467–474; published: April 16, 1999
Times Cited: 704 (from Web of Science)

ResearcherID (A-3368-2011)

Challenges

A major career challenge has been learning to identify and focus my energy in a limited set of synergistic areas where I can make the biggest impact. This can be difficult to do in a profession where we are asked to contribute our time and knowledge in a myriad of ways. It is a constant battle for me to protect my time so that I can use it most wisely, which ultimately means saying "no" to otherwise worthwhile efforts.

Words of Wisdom

Every person and every career has ups and downs. Persevere through the downtimes and take full advantage of the ups, while giving credit to the people who helped you along the way.

Profile 19

Ulrike Diebold

Professor of Surface Science
Deputy Department Head
Institute of Applied Physics
Vienna University of Technology (TU Wien)
Wiedner Hauptrasse 8–130/134
1040 Vienna
Austria

e-mail: diebold@iap.tuwien.ac.at
telephone: +43-1-58801-13425

Dipl.Ing. Ulrike DIEBOLD
Vertragsassistentin

An official picture taken of
Ulrike Diebold while she
was a graduate student at
the institute that she is now
(co-)heading.

Birthplace
Kapfenberg, Austria
Born
December 12, 1961

Publication/Invention Record

>225 publications: h-index 51 (Web of Science)

Tags

❖ Academe
❖ Domicile: Austria
❖ Nationality: dual
   citizenship—Austrian
   and American
❖ Caucasian
❖ Children: 2

Proudest Career Moment (to date)

In 2013, I was awarded the "Wittgenstein-Prize." This is the highest science award in my home country, Austria. Only one award is given out for all areas of research—science, engineering, humanities, etc. The prize comes with an unrestricted research grant of 1.5 M Euros. One Saturday in June, I received a phone call from the president of the Austrian Science Fund that I would be awarded the prize the following Monday, and the most happy, proud, and hectic weekend of my entire professional career ensued. The prize has provided copious recognition, both amongst my peers in science and in the general public. And, of course, it is really nice to have available money to support my group over several years and pursue new lines of research.

Academic Credentials

Habilitation (1998) Experimental Physics, Technische Universität Wien (now TU Wien), Vienna, Austria.

Successful Women Ceramic and Glass Scientists and Engineers: 100 Inspirational Profiles. Lynnette D. Madsen.
© 2016 The American Ceramic Society and John Wiley & Sons, Inc. Published 2016 by John Wiley & Sons, Inc.

Ph.D. (1990) Physics, University of Technology (now TU Wien), Vienna, Austria. Diplom Ingenieur (equivalent to M.S.) (1986) Engineering Physics, University of Technology (now TU Wien), Vienna, Austria.

In 2013, Ulrike Diebold was awarded the Wittgenstein Prize, the highest science award in Austria. (U. Diebold with Christoph Kratky, the president of the Austrian Science Fund, Karlheinz Töchterle, Minister of Science, and Jan Ziolkowski, Harvard University). Credit: Austrian Science Fund.

Research Expertise: interdisciplinary research in surface science, physical chemistry, condensed matter physics, materials science, and nanoscience; investigating the atomic-scale geometric and electronic surface structure of pure and doped oxide materials, adsorption of gases and metals, correlating nanoscopic measurements with materials applications (in nanocatalysis, photocatalysis, gas sensing, (opto)electronics, and spintronics); growth of epitaxial thin films and supported nanoclusters; electrochemical surface science

Other Interests: teaching and mentoring

Key Accomplishments, Honors, Recognitions, and Awards

- Advisory Editorial Board for Surface Science Reports, 2003 to present
- Board of the Austrian "Chemisch-Physikalische Gesellschaft," 2014 to present
- Selection Committee L'Oreal Research Fellowships, Austria, 2014–2019
- Panel Member, European Research Council (ERC) Advanced Research Grants, 2008–2015
- Member, Senate of the Christian-Doppler Society, 2014–2017
- Elected to head the Surface, Interface, and Thin Film Division of the Austrian Physical Society, 2014–2016
- Advisory Editorial Board for the journal Surface Science, 2010–2016
- International Advisory Board, Advanced Materials Interfaces, 2013–2015
- Divisional Associate Editor (Materials Physics), Physical Review Letters, 2012–2017
- Erwin Schrödinger Memorial Lecture, Trinity College Dublin, Institute of Advanced Studies in Dublin, and Austrian Embassy in Ireland, 2015
- Debye Lecture, Utrecht University, The Netherlands, 2015
- Blaise Pascal Medal in Materials Science (awarded by the European Academy of Arts and Sciences), 2015
- Elected member of the German National Academy of Sciences Leopoldina, 2015
- 25th Brdička Memorial Lecture, Heyrovský Institute of Physical Chemistry, Prague, 2015
- Elected member of the European Academy of Sciences, 2015

- Local organization of the "Lise Meitner Lecture" of the German and Austrian Physical Society, 2013–2015
- Elected full member of the Austrian Academy of Sciences, 2014
- Eminent Visitor Award, Catalysis Society of South Africa (CATSA), 2014
- Wittgenstein-Prize (http://en.wikipedia.org/wiki/Wittgenstein-Preis), 2013
- Arthur W. Adamson Award for Distinguished Service in the Advancement of Surface Chemistry from the American Chemical Society, 2013
- Elected Corresponding Member, Austrian Academy of Sciences, 2012
- Shouheng Lecture, Zhejiang University of Technology, Hangzhou, China, 2012

Press conference where Ulrike Diebold was announced as the 2013 Wittgenstein Prize Winner. (U. Diebold with Christoph Kratky, the president of the Austrian Science Fund, Karlheinz Töchterle, Minister of Science, and Jan Ziolkowski, Harvard University).Credit: Austrian Science Fund.

- Guest Professor, Hubei University, Wuhan, China, 2012
- European Research Council (ERC), Advanced Research Grant, 2012
- Reader Panel, Nature, 2009–2010
- Advisory Editorial Board for Open Journal of Physical Chemistry, 2007–2010
- Guest Editor (together with T.M. Orlando), "Non-Thermal Processes on Surfaces. Dedicated to the Memory of Prof. Theodore E. Madey," Special Issue in Journal of Physics: Condensed Matter, 2009
- General Committee of the Physical Electronics Conference, 2006–2009
- Fellow, Research Center Dresden-Rossendorf, Germany, 2008
- Outstanding Researcher Award, Tulane's School of Science and Engineering, 2008
- Fellow, American Association for the Advancement of Science, 2007
- Surface, Interface and Atomic-Scale Science Editorial Board of Journal of Physics: Condensed Matter, 2006–2007
- Guest Editor (together with A. Selloni and C. Di Valentin), "Doping and Functionalization of Semiconducting Metal Oxides," Special Issue in Chemical Physics, 2006–2007
- American Physical Society, Selection Committee for the David Adler Lectureship Award, 2006–2007
- Elected member, CAMD (Center for Advanced Microstructures and Devices, LSU) User's Committee, 2005–2007
- Yahoo! Founder Chair in Science and Engineering, 2006
- Fellow, AVS—The Science and Technology Society, 2005
- Provost's Recognition Award for Research and Scholarly Achievement, 2005
- Tulane Liberal Arts and Sciences Faculty Research Award, 2004

- Fellow, American Physical Society, APS, 2004
- National Science Foundation, "Special Creativity Award," 2003
- Friedrich Wilhelm Bessel Research Prize from the Alexander von Humboldt Foundation, Germany, 2001
- Executive Committee, Surface Science Division, American Vacuum Society, 1998–2001
- NSF CAREER Award, 1997
- Oak Ridge Associated Universities, Junior Faculty Enhancement Award, 1995
- "Charlotte Bühler Fellowship" for Habilitation from the Austrian Science Foundation (not assumed), 1992
- Fellowships for "Especially Talented Students," University of Technology, Vienna, 1983, 1984, 1985

Biography

Early Life and Education

I spent much of my high school years reading books and skiing. When it was time to enter university, I found it hard to decide on a particular major. Too many things seemed interesting, and the choice overwhelming. After much agony I settled on engineering physics, as this left open many avenues yet promised solid job opportunities. My enthusiasm for physics really took off when I started working in the lab during my master's thesis. When I was offered a fellowship for graduate studies, I jumped at the opportunity. Although I had never dreamed of pursuing a career in research, this has become my calling.

Ulrike Diebold in her surface science lab. Credit: Aleksandra Pawloff.

Career History

I love to travel, and it slowly dawned on me that a Ph.D. in science opens up the world—you can work pretty much anywhere you wish. After receiving my Ph.D. from the TU in Vienna, I started a postdoc position at Rutgers University in New Jersey, USA. I was immersed in big research group, learned new experimental techniques, and obtained a wider scientific horizon. I had also my first encounter with oxide surfaces, which should become the love of my scientific life. During the week, we worked hard in the lab, and on weekends we became part of the bridge-and-tunnel-crowd that floods into Manhattan to party.

In 1993 I was offered a tenure-track faculty position at Tulane University in New Orleans. It was exhilarating and scary at the same time to define my own research program, learn how to teach and guide students, and set up and manage my own lab. I remember long and stressful hours of work, and much pressure and anxiety. It all paid off at the end, however. I am convinced that being a professor is the best job in the world.

After climbing up the academic ladder in the United States, we decided to return to Europe (to TU Wien). I very much enjoy the opportunities that come with running a bigger research program as well as being back home with family.

Ulrike Diebold with her family in Hangzhou, China.

3 Most Cited Publications

Title: The surface science of titanium dioxide
Author(s): Diebold, U
Source: Surface Science Reports; volume: 48; issue: 5–8; pages: 53–229; published: 2003
Times Cited: 3798 (from Web of Science)

Title: The surface and materials science of tin oxide
Author(s): Batzill, M; Diebold, U
Source: Progress in Surface Science; volume: 79; issue: 2–4; pages: 47–154; published: 2005; doi: 10.1016/j.progsurf.2005.09.002
Times Cited: 891 (from Web of Science)

Title: Epitaxial growth and properties of ferromagnetic co-doped TiO_2 anatase
Author(s): Chambers, SA; Thevuthasan, S; Farrow, RFC; et al.
Source: Applied Physics Letters; volume: 79; issue: 21; pages: 3467–3469; published: November 19, 2001
Times Cited: 334 (from Web of Science)

ResearcherID (A-3681-2010)

Challenges

The main challenge for me is time: why does a day have only 24 hours (of which I spent at least 8 sleeping and dreaming)? There are so many things I want to do—not only in science, but also with my family and friends. My life is very full as it is, yet I feel there are so many exciting things out there that would be worthwhile pursuing . . .

On being a woman in this field . . .

There are still relatively few women in physics. If you have the talent, stamina, and determination to see through your studies, you have an advantage: being one of the few successful females gives you a high visibility.

Words of Wisdom

I've seen too many talented women falter because of private issues. Pick your partner carefully. Life in science is hard. You will need all the support you can get. Does your partner celebrate your accomplishments more than his/her own? Is s/he a fair person? Would s/he take on the responsibilities of raising a family? Dual-science careers are particularly tricky, as it is not easy to get two jobs in research, especially when both are in the same field.

Mildred S. Dresselhaus

Institute Professor; Professor of Physics and Electrical
Engineering, Emerita
Massachusetts Institute of Technology (MIT)
77 Massachusetts Avenue, Bldg. 13-3005
Cambridge, MA 02139
USA

Mildred Dresselhaus at
MIT, ca. 1977–1978.
Credit: Georgia Litwack.

Birthplace
Brooklyn, New York, NY
Born
November 11, 1930

Publication/Invention Record
>1600 publications: h-index 122 (Web of Science)
8 books

Tags
❖ Academe
❖ Domicile: USA
❖ Nationality:
  American
❖ Caucasian
❖ Children: 4

Proudest Career Moment (to date)

It was perhaps my 80th birthday party, when 250 people, including family, former
students, friends, and collaborators worldwide, came to MIT to celebrate my birthday
and career with a scientific symposium and party. Close to that would be winning the
2012 Kavli Prize for nanoscience, and receiving the prize in Oslo from Fred Kavli and
King Harald. Winning the Kavli Prize has spurred me to new research interests. Some
special moments came with meeting various U.S. Presidents, starting with George Bush,
Sr., followed by most of them since then.

Academic Credentials

Ph.D. (1958) Physics, University of Chicago, IL, USA.
A.M. (1953) Physics, Radcliffe College, Cambridge, MA, USA.
Fulbright Fellow (1951–1952), Newnham College, Cambridge University, England,
United Kingdom.
A.B. (1951) Physics, Hunter College, New York City, NY, USA.

Successful Women Ceramic and Glass Scientists and Engineers: 100 Inspirational Profiles. Lynnette D. Madsen.
© 2016 The American Ceramic Society and John Wiley & Sons, Inc. Published 2016 by John Wiley & Sons, Inc.

Mildred Dresselhaus at Cornell University, 1960.

Research Expertise: electronic materials, particularly related to nanoscience and nanotechnology, with special regard to carbon-related materials; novel forms of carbon, including graphite, graphite intercalation compounds, fullerenes, carbon nanotubes, porous carbons, activated carbons, and carbon aerogels; as well as other nanostructures, such as bismuth nanowires, bismuth–antimony thin films, and the use of nanostructures in low-dimensional thermoelectricity

Other Interests: she is widely recognized for her considerable devotion to mentoring students, raising community awareness, and promoting progress on gender equity

Key Accomplishments, Honors, Recognitions, and Awards

- MIT Institute Professor, 1985–2007; emerita, 2008 to present
- 31 honorary doctorates worldwide, 1976 to present
- IEEE Medal of Honor, 2015
- Presidential Medal of Freedom, 2014
- Von Hippel Award, Materials Research Society, 2013
- Kavli Prize "for her pioneering contributions to the study of phonons, electron–phonon interactions, and thermal transport in nanostructures," 2012
- Co-recipient of the Enrico Fermi Award, 2012
- Acta Materialia Materials and Society Award, 2012
- ACS Award for Encouraging Women into Careers in the Chemical Sciences, 2010

- Oliver E. Buckley Condensed Matter Physics Prize, 2008
- Oersted Medal, 2008
- Chair of the governing board of the American Institute of Physics, 2003–2008
- UNESCO/L'Oreal, North American Woman Scientist of the Year, 2007
- Co-chaired a National Academy of Sciences Decadal Study on "Condensed Matter Materials Physics, CMMP2007"
- 11th Annual Heinz Award in the category of Technology, the Economy and Employment, 2005
- Karl T. Compton Medal for Leadership in Physics, American Institute of Physics, 2001
- Medal of Achievement in Carbon Science and Technology, American Carbon Society, 2001
- Director of the Office of Science at the U.S. Department of Energy, 2000–2001
- National Materials Advancement Award of the Federation of Materials Societies, 2000
- Weizmann Institute's Millennial Lifetime Achievement Award, 2000
- Harvard Overseer, 1997–2000
- Trustee, California Institute of Technology, 1993–2000
- Ioffe Institute, Russian Academy of Sciences, St. Petersburg, Russia, 2000
- Fellow of American Carbon Society, 1999
- American Physical Society: Dwight Nicholson Medal for Human Outreach, 1999
- Chairman of the Board, American Association for the Advancement of Science, 1998–1999
- Elected Board of Governors, Weizmann Institute, Rehovot, Israel, 1998
- Hall of Fame Award, Women in Technology International (WITI), 1998
- Award for Outstanding Professional Achievement, Hunter College, 1998
- President, American Association for the Advancement of Science, 1997–1998
- Sigri-Great Lakes Carbon Award, American Carbon Society, 1997
- Elected President of the American Association for the Advancement of Science, 1996
- Treasurer, National Academy of Sciences, 1992–1996
- Elected to Membership in the American Philosophical Society, 1995
- Elected as a Foreign Associate of the Engineering Academy of Japan, 1993
- Trustee, Rensselaer Polytechnic Institute, 1988–1992
- Achievement Award, New York Academy of Sciences, 1991
- National Medal of Science, 1990
- Elected to Council of National Academy of Sciences, 1987–1990
- Annual Achievement Award, Engineering Societies of New England, 1988
- Elected to Council of National Academy of Engineering, 1981–1987
- Elected to Board of Directors of the American Association for the Advancement of Science, 1985–1989
- MIT Killian Faculty Award, 1986

- Elected to membership in the National Academy of Sciences (Engineering Sciences Section), 1985
- Elected President of the American Physical Society, 1984
- MIT Abby Rockefeller Mauzé Professor of Electrical Engineering, 1973–1985
- Society of Women Engineers Achievement Award, 1977
- Corresponding Member, Brazilian Academy of Sciences, 1976
- Elected to membership in the U.S. National Academy of Engineering, 1974
- Elected to Fellow of the American Academy of Arts and Sciences, 1974
- Radcliffe College Alumni Medal, 1973
- Hunter College Hall of Fame Award, 1972
- American Physical Society Fellow, 1972
- NSF Postdoctoral Fellow, Cornell University, 1958–1960
- Bell Telephone Laboratory Fellow, 1956–1957
- Fulbright Fellowship, 1951–1952

Some activities where she served as trustee, on corporate boards, consulting, on visiting committees, editorships and publication boards, in professional society service, and guest lectureships are not included in the above list.

Biography

Adapted from her website (that is no longer available) that included the writings of Gene Dresselhaus.

Early Life and Education

She was born in Brooklyn, New York, in 1930 and was given the name Mildred Spiewak, named after her mother's mother, who died when her mother was about 5 years old. The Spiewak family's roots can be traced back to a small mostly Jewish village in southern Poland. When she was 4, her family moved to a Jewish ghetto in the Bronx where her parents took whatever unskilled jobs they could find. Their early years in New York were more difficult because of the Great Depression. Even though family resources were very limited, her mother managed to send money to European relatives what little she could— in the end, however, almost all family members remaining in Europe were killed in the Holocaust.

The family moved to the Bronx when she was 4 years old so that her older brother, Irving, could be close to a music school where he was given a scholarship to study the violin. But the teacher died soon after, and music lessons moved to Greenwich Village. As a very young child, Mildred accompanied her brother to his violin lessons and listened to him practicing his instrument, and soon she could sing the various pieces he was playing. This led to her starting violin lessons before she was 5 years old and before she started kindergarten. She learned to read music before she could read words in English. Throughout her elementary and junior high school years, she continued her music lessons as a scholarship student at the Greenwich House Music School (in Manhattan), which was nearly one hour away by subway from where she lived in the Bronx. Because of her considerable musical skills, she was also able during her childhood to make friends and move in more affluent circles outside of the ghetto where she lived. These contacts increased her aspiration level and exposed her to the cultural life of New York City.

Mildred attended the local public school, just two blocks from her apartment house, for grades K–6. For her junior high school (grades 7–9), Millie attended a public school only five blocks from her home. Her favorite subject was mathematics. She was a curious but shy student, and especially enjoyed the beginning of each semester when she received the books that were lent to her. This was an opportunity to read through the math book the first week to get a preview of what she would be learning. As she got older, science books for the semester were also lent to students, so she also read through those textbooks during the first week of class. Some of the memorable things she did in elementary school were given as special assignments. The first was a job to teach a "mentally retarded" boy to read. So at the age of 8 years, she had her first paying job, earning a total of 50 cents per week for three hours of hard work daily. When she was 11 years old, she had a non-paying job as an assistant to her sixth grade teacher, helping her with administrative work and other assignments.

Millie's aptitude and interest in science would also have taken her to Bronx Science, but girls were not allowed into the Bronx High School of Science at that time. Through her contacts she also learned about the possibility of attending a special high school for girls in New York City (Manhattan). Admittance to this high school was by examination, and only 80 girls from the whole city were accepted per semester. Since nobody from her junior high school had ever passed this entrance exam, she undertook an intensive self-study program of math and English exams given in past years for entrance into this school. Using this strategy, she managed to pass the examination, and she became a student at Hunter College High School. This event can be considered the "entry point/ junction" that led to her becoming a scientist.

The central focus of Hunter College High School was liberal arts, with emphasis on English, literature, history, Latin, modern languages (taught in the vernacular), current affairs, and speech. As it turned out in later life, a high level of exposure to the liberal arts provided excellent background for the large amount of writing necessary in a scientific career. The class on speech, which was required every semester, was the least interesting to her, but proved to be one of the most valuable in later life. It was here that she learned about public speaking, about improving her diction, projecting her voice, and also how to put herself to sleep at will for a short nap in the middle of the day (in an appropriate time and place for a nap). The emphasis on self-study provided skills at reading texts critically, and in identifying weaknesses or flaws in logical thinking. The subjects that interested Millie most as a high school student were her math and science classes. At this high school, there were generally two or three student teachers in the classroom who were valuable in helping students with self-study projects, in answering questions outside of classroom study, and in pointing students in new directions, by providing many one-on-one learning experiences. Another aspect of this high school that turned out to be valuable was the fact that the school had no cafeteria; therefore, students had to bring their own lunches with them. Lunch was eaten during "noon time club meetings" and she had opportunities to join many clubs, such as the Math Club, Orchestra, Public Affairs Club, etc. This is where students got to know one another, and where they developed leadership skills. It was through the high school clubs that she picked up her nickname "Millie" that has been subsequently used widely in place of her formal name of Mildred.

Hunter High channeled her into Hunter College with the goal of becoming an elementary school teacher. At that time, women typically chose one of three careers: K–12 teaching, nursing, and secretarial work. On the basis of her ranking on the New York State Regents exam, she was offered a scholarship to Cornell University. However, Cornell was far away from home and seemed to her like a wealthy man's school. So Millie decided to continue on to Hunter College where she felt very comfortable both socially and academically. Though she didn't know it at the time, this decision to attend Hunter College (which was relatively weak in the sciences) was decisive in launching her into a scientific career. The reason for this change in career objectives can be traced to one particular faculty member, Rosalyn Yalow, a future Nobel Laureate, who became Millie's advisor, mentor, and physics professor at Hunter College in her second year. Rosalyn followed Millie's career throughout her life and always seemed to be present whenever Millie was involved in either a minor talk or a more major event in the NYC area. This remained true for many years, as illustrated by Rosalyn's attendance in a wheelchair at an award ceremony for Millie at Hunter College in 1998. It is also fair to say that she chose her lifework in science because she loved it so much as a youngster, and this love and fascination for the unknown and for the discovery of new things, not yet known before, has remained with her throughout her life.

After graduating from Hunter College, Millie was awarded a Fulbright Fellowship, which allowed her to spend a year at Cambridge University in England. Her studies during this year focused on physics, and she was able to largely make up for the rather limited curriculum in physics that was offered at Hunter College.

Upon returning to the United States, Millie spent a year at Radcliffe College (now part of Harvard University) where she was awarded a master's degree. She then continued her graduate studies at the University of Chicago (where Enrico Fermi was one of her professors). The first year of her studies coincided with the very end of the Enrico Fermi era, and consequently her training in physics had a very strong flavor of the Fermi approach to research and teaching in physics. She met Gene F. Dresselhaus, now a well-known theorist, while doing her Ph.D. studies at the University of Chicago; they married in 1958.

Finally, she went to Cornell University where she worked as a postdoctoral fellow (with National Science Foundation support) while her newly acquired husband was on the faculty. While at Cornell, she worked on a model to explain the experimental results of her Ph.D. thesis. The times were such that women's careers were not taken very seriously by most men. For example, while at Cornell, Millie was told by the head of the laboratory with which she was affiliated that "a woman could never teach an engineering student." So with the expiration of the NSF postdoctoral fellowship in 1960 and the birth of a new baby, it was time to move on.

Career History

In the early 1960s, hiring of couples was rare. As they saw it, there were two choices—MIT and IBM. She was more interested in basic science and an academic career, so they chose MIT.

Lincoln Laboratory was affiliated with MIT and provided an excellent environment for a young person to develop a career. At Lincoln Laboratory, she was not able to continue with her superconductivity research, so she used this disappointment as an opportunity to branch into the study of magneto-optical effects in semiconductors and semimetals. The bias against superconductivity research came from her division leader, Benjamin Lax, who felt that the Bardeen–Cooper–Schrieffer theory had explained everything about superconductivity, and the field was now dead. Changing fields turned out to be the best thing that could have happened at this early time in her career in solid-state physics, and in her training of graduate students later on, she emphasized the importance for a young person to learn several research areas in the early career years.

Her earliest work at Lincoln Laboratory was on magneto-optics studies in semiconductors, where many others were also working at that time. Wanting to do something different from what others were doing, she started to study the electronic structure of graphite by magneto-optics, following a suggestion by her husband. She had little competition in this field because the magneto-optics experiment was considered difficult and the graphite electronic structure was at that time considered to be very complex. This situation was fortunate because the low general interest in this topic kept down the completion at a time when her next three children were born.

She supervised her first Ph.D. thesis in 1965 and now more than 60 people have earned their doctorates under her supervision. As a result of her research achievements during the years 1960–1967, she was invited to become a visiting professor in the Electrical Engineering Department at MIT under the auspices of the Abby Rockefeller Mauzé Fund for woman scholars. After one year as a visiting professor, she was appointed as a tenured full professor, the first female tenured professor in MIT's engineering school.

When she was appointed to a full faculty position and to the Abby Rockefeller Mauzé Chair in 1973, she had enough intellectual and financial independence to attempt high-risk experiments that federal funding agencies did not find worthy of support. This independence provided her with greater freedom and helped her on the path to becoming a resource person worldwide in the carbon science area, and giving her entry to many new fields of carbon science as they emerged. Along with this came many invitations to give invited talks and write review articles, books, and monographs, and the unofficial title "Queen of Carbon." When her visibility on the national scene was greatly enhanced (starting in 1974 with election to the National Academy of Engineering), she began receiving several recognitions; these have continued through the years.

In 1983, after she had been elected to become an upcoming president of the American Physical Society, she received a joint appointment between the Physics Department and the Department of Electrical Engineering and Computer Science. Her teaching focused on educating electrical engineering and materials science students about solid-state physics, group theory, and semiconductor physics, and in making this material accessible to such students. She held the Abby Rockefeller Mauzé Chair until 1985, when she was appointed to the position of Institute Professor. One indicator of her research success is that several physical theories include her name: the Hicks–Dresselhaus model for thermoelectric materials (1993), the SFDD ([Riichiro] Saito, [Mitsutaka] Fujita, [Gene] Dresselhaus, and [Mildred] Dresselhaus) model for carbon nanotubes (1992), and the Tang–Dresselhaus theory for bismuth–antimony thin films (2012).

She has made extensive research contributions and fundamental discoveries in condensed matter physics. One significant factor to her successful career is an exceptional skill in assessing a situation and deciding on exactly where she could fit in and make a significant contribution. Having made this decision, she gets in, and in a timely manner makes her contribution and then moves on to the next problem. It is likely that her early childhood experiences were instrumental in developing this intuitive strategy.

Mid-career, reflecting on her fortunate career development and recalling the training that she had received at Hunter College, Millie thought it was the time to return something to the pool from which she had been drawing all these years. To her, this "service to society" meant accepting administrative positions and service assignments in order to help other researchers and students do the science that was so important in her own life. One main emphasis of her leadership was to initiate new seed research programs, which were nurtured through funds from the Center for Materials Science and Engineering, which then grew into major programs. Another emphasis was in the grooming of young people for leadership positions, who could then become laboratory directors, so that she could go back to her other activities, after her term of office was over. Her ability to perform these service activities was possible in part due to two factors: their children had grown up and her husband joined the laboratory she worked in, which gave him freedom to work extensively with the Dresselhaus research group as well as on his own projects.

Throughout the years, Millie has been active in helping women in science and engineering in various ways. These efforts include mentoring women, examining how undergraduate admissions applications were handled at MIT, initiating a Women's Forum at MIT, serving on various studies and task forces, starting two initiatives for women in science and engineering while she was the Abby Rockefeller Mauzé Professor, serving on the Committee on the Status of Women in Physics of the American Physical Society, and chairing the National Research Council Committee on Women in Science and Engineering for three years (from 1990 to 1993). Millie has always taken great pride in the accomplishments of her students, including their work on broader issues in science and technology. She has also enjoyed working with graduate students, postdocs, and collaborators worldwide as science and technology have become more and more international.

Mildred Dresselhaus in the laboratory, 2011.
(See insert for color version of figure.)

In addition, Millie Dresselhaus has been blessed with five grandchildren—clearly making her one of the most famous grandmothers in science today.

3 Most Cited Publications

Title: Large area, few-layer graphene films on arbitrary substrates by chemical vapor deposition
Author(s): Reina, A; Jia, X; Ho, J; Nezich, D; Son, H; Bulovic, V; Dresselhaus, MS; Kong, J
Source: Nano Letters; volume: 9; issue: 1; pages: 30–35; doi: 10.1021/nl801827v; published: January 2009
Times Cited: 2294 (from Web of Science)

Title: Edge state in graphene ribbons: nanometer size effect and edge shape dependence
Author(s): Nakada, K; Fujita, M; Dresselhaus, G; Dresselhaus, MS
Source: Physical Review B; volume: 54; issue: 24; pages: 17954–17961; doi: 10.1103/PhysRevB.54.17954; published: December 15, 1996
Times Cited: 2061 (from Web of Science)

Title: Electronic structure of chiral graphene tubules
Author(s): Saito, R; Fujita, M; Dresselhaus, G; Dresselhaus, MS
Source: Applied Physics Letters; volume: 60; issue: 18; pages: 2204–2206; doi: 10.1063/1.107080; published: May 4, 1992
Times Cited: 1853 (from Web of Science)

Challenges

In retrospect, my greatest challenge came in my childhood, growing up in New York City "on the wrong side of the tracks." I had poor access to information about the possibilities for careers for women from either my home or school environments. This challenge was overcome by my musical education, which exposed me to education and advancement opportunities available to people "on the right side of the tracks." Once in Hunter High School, ability, dedication, and interest became the criteria for advancement and not family income or status.

A second challenge came when I was considering college education, and my school guidance advisors recommended a career for me as a school teacher, because it was the best available option for a person with no family funding. Going to Hunter College solved this problem, because the faculty members there pointed me along fruitful directions available to me, and provided me with tutoring jobs. Being told at Cornell that there was a limited future for me in science pointed me to give high priority to institutions emphasizing meritocracy, and this resulted in my coming to MIT in 1960, where actual accomplishment rather than what you looked like was what mattered.

My next challenge was in 1990 when the National Magnet Laboratory moved from Cambridge, Massachusetts, to Tallahassee, Florida. This necessitated a change in research focus away from high magnetic field research to studies on nanocarbons and nanothermoelectricity, two research areas that have since then grown dramatically in importance, more than high magnetic field research. The take-home message here is to learn how to use setbacks to reorient your personal goals, taking advantage of newly emerging opportunities. Learning new things and bringing a broad background and a high level of motivation allowed me to move forward successfully and to benefit from adversity.

On being a woman in this field . . .

Being a woman in this field has been both a challenge and an opportunity. Working in a research field where one is greatly outnumbered often leads to challenges, because the minority person is not expected to succeed so well in the given research field. If the research field and institution of employment emphasize meritocracy, then many of the barriers associated with being a minority are lowered, and evaluation emphasizes achievement rather than preconceived expectations. The advantage of being a woman in my field is that there are so few of us, especially in my age group, so my achievements are more noticed. This has in fact been a big advantage for me in my career. Furthermore, many women are less excited about their professional success and are happy with a successful family life in conformance with societal expectations. The take-home message here is to enjoy and gain satisfaction from what you are doing, both in your personal activities and in your professional activities. I believe that women can be happy and successful doing both at the same time.

Words of Wisdom

Words of wisdom that I give to women are that "they can do it also." I also say that careers in science and engineering are great careers for women, because the jobs are interesting, well paying, and appreciated by society. These are great career choices if the woman has an interest in such careers, the necessary talent to do the work, and a willingness to be flexible and adaptable to find ways to move forward when challenges present themselves.

Natalia Dubrovinskaia

Heisenberg Professor/Full Professor for Material
Physics and Technology at Extreme Conditions
Laboratory of Crystallography
University of Bayreuth
Universitätstraße 30
95440 Bayreuth, Germany

e-mail: natalia.dubrovinskaia@uni-bayreuth.d
telephone: +49 921 553880

Natalia Dubrovinskaia at the age of 34 at Uppsala University. Uppsala was for Natalia a place of great inspiration. Among the scientists at the historical photo hanged next to her office is Svante Arrhenius. Courtesy of the family.

Birthplace
Moscow, the Soviet Union (USSR)
Born
February 18, 1961

Publication/Invention Record
>125 publications: h-index 28 (Web of Science)
3 patents, 1 book

Tags
❖ Academe
❖ Domicile: Germany
❖ Nationality: Swedish
❖ Caucasian
❖ Children: 2

Memorable Career Moment (to date)
Awarding me Heisenberg Professorship by the German Research Foundation in 2011 and conferring on me *Doctor honoris causa* by Linköping University (Sweden) in 2014 have been the most memorable events of my professional life.

Academic Credentials
Umhabilitation (2008) Crystallography and Mineralogy, University of Heidelberg, Heidelberg, Germany.

Successful Women Ceramic and Glass Scientists and Engineers: 100 Inspirational Profiles. Lynnette D. Madsen.
© 2016 The American Ceramic Society and John Wiley & Sons, Inc. Published 2016 by John Wiley & Sons, Inc.

Habilitation (2007) Crystallography, University of Bayreuth, Bayreuth, Germany.
Ph.D. (1989) Crystallography and Crystal Physics, Lomonosov Moscow State University, Moscow, USSR.
M.Sc. (1983) Geochemistry, Lomonosov Moscow State University, Moscow, USSR.

Research Expertise: high-pressure and high-temperature (HPHT) crystallography, high-pressure synthesis of novel inorganic materials including nanocrystalline materials with advanced mechanical properties, research on superhard materials, crystal physics and crystal chemistry, properties of matter in nanocrystalline state, thermoelastic and transport properties of solids, experimental geochemistry and geophysics, superconductivity in composite materials, development of methods of HP generation using diamond anvil cell (DAC) and large-volume press (LVP) techniques, development of new scientific instruments for high and ultrahigh pressure and temperature generation, and application of the synchrotron radiation and neutron diffraction in materials science and solid-state physics

Other Interests: teaching and popularization of science

Key Accomplishments, Honors, Recognitions, and Awards

- Elected Secretary of the European High Pressure Research Group Committee, 2015 to present
- Editor-in-Chief of the International Journal of Materials and Chemistry, 2012 to present
- Heisenberg Professor for Material Physics and Technology at Extreme Conditions in the Excellence Program of the German Research Foundation (DFG), 2011 to present
- Elected Member of the European High Pressure Research Group Committee, 2011 to present
- *Doctor honoris causa* (*Dr.h.c.*) Linköping University, Linköping, Sweden, 2014
- Named Research Leader in materials progress within "2006 Scientific American 50", 2006
- Visiting Professor Fellow, Institut Galilée, Université Paris, France, 2005
- Award from the Ministry of Geology of the USSR for invention, 1989

Biography

Early Life and Education

Natalia Dubrovinskaia was raised in Moscow (USSR), attended a Soviet state comprehensive school, and graduated from the Moscow high school for advanced study of mathematics and physics No. 521. She still stays in contact and maintains the warmest relationship with her high school physics teacher, Nina I. Pavliuchenko.

After school, she studied geochemistry at the Geology Faculty of the Lomonosov, Moscow State University and later specialized in crystallography at the Department of Crystallography and Crystal Chemistry. It was an advantage to be taught by a number of the brightest contemporary Soviet crystallographers: academician N.V. Belov, G.N. Litvinskaya, Ju.G. Zagalskaya, and Ju.K. Egorov-Tismenko. Still in university, N. Dubrovinskaia got married and graduated from the University in 1983, not only with honors but also with a son of the age of 10 months.

Her research career started at the Central Research Institute of Geological Prospecting for Base and Precious Metals (TsNIGRI) of the Ministry of Geology of the USSR, where

Natalia Dubrovinskaia was employed at the X-ray laboratory. In parallel, she was working on a Ph.D. thesis in crystallography and crystal physics at the Moscow State University under the supervision of Prof. V.S. Urusov; she defended her thesis research in 1989.

Career History

1989 turned to be a landmark not only in Dubrovinskaia's personal career but also in the history of the whole country. Economic instability in the turbulent time of Gorbachev's *Perestroika* brought the young scientist to teaching in various Moscow high schools. Teaching in class for 36 hours per week gave her a salary, and an indispensable pedagogic experience. The fall of communism in the communist states of Central and Eastern Europe after the Revolutions of 1989 triggered the end of Gorbachev era marked by dissolution of the Soviet Union in 1991. The crumbling of the Russia's economic and political structures led to the collapse of many scientific institutes and a brain drain from Russia. In 1994, the family of four, two young academics with two kids (a daughter was born in 1993), moved to Sweden.

Natalia Dubrovinskaia at a beamline at the European Synchrotron Radiation Facility (ESRF) in 2013. Courtesy of the ESRF/Blascha Faust; photo by Blascha Faust. (See insert for color version of figure.)

At Uppsala University in Sweden, Dr. Dubrovinskaia worked in the Theoretical Geochemistry Group under the leadership of Prof. S.K. Saxena, where she learnt the high-pressure diamond anvil cell technique and gained her first experience in the application of the synchrotron radiation in high-pressure research using the brand new European Synchrotron Radiation Facility in Grenoble. In Uppsala she started working with carbon, diamond, and various ceramic materials in close collaboration with her husband Dr. Leonid Dubrovinsky.

In 2001 the development of the scientific couple's career led the family to Germany, where Dubrovinskaia first joined the Bavarian Research Institute of Experimental Geochemistry and Geophysics at the University of Bayreuth. In 2007 she passed habilitation in crystallography at the Physics Department, obtained the *venia legendi*, and became Privatdozent at the University of Bayreuth and later at the University of Heidelberg, where she worked from 2007 through 2011 as Research Staff. In 2011 Dubrovinskaia returned to the University of Bayreuth as Heisenberg Professor for Material Physics and

Prof. Dr. Natalia Dubrovinskaia and Prof. Dr. Leonid Dubrovinsky after conferring *Doctor honoris causa* (Dr.h.c.) in 2014, Linköping University, Linköping, Sweden. Photo by Siamen Dubravinski.

Technology at Extreme Conditions in the Excellence Program of the German Research Foundation (DFG) and Full Professor at the Laboratory of Crystallography.

3 Most Cited Publications

Title: Materials science: the hardest known oxide
Author(s): Dubrovinsky, LS; Dubrovinskaia, NA; Swamy, V; et al.
Source: Nature; volume: 410; issue: 6829; pages: 653–654; published: April 5, 2001
Times Cited: 206 (from Web of Science)

Title: Body-centered cubic iron–nickel alloy in Earth's core
Author(s): Dubrovinsky, L.; Dubrovinskaia, N.; Narygina, O.; et al.
Source: Science; volume: 316; issue: 5833; pages: 1880–1883; published: June 29, 2007
Times Cited: 109 (from Web of Science)

Title: Experimental and theoretical identification of a new high-pressure TiO_2 polymorph
Author(s): Dubrovinskaia, NA; Dubrovinsky, LS; Ahuja, R; et al.
Source: Physical Review Letters; volume: 87; issue: 27; article number: 275501; published: December 31; 2001
Times Cited: 101 (from Web of Science)

Challenges

Any circumstances can be considered either challenging or routine depending on the point of view from which they are judged. However, I think it is always a challenge to resist a common opinion or to act against the stream. In my career, such a challenge was to preserve our scientific couple (me plus my husband) working as a team. For some reason, people appreciate synergy of family members in the circus, but not in science. Not many would try to judge, whose contribution to the trick of an acrobatic couple is greater: It's clear that no couple, no trick. Synergy of a scientific couple may not be so evident for strangers, but it is extremely important, especially in such an interdisciplinary research field like ours. It is not easy to get two equally high positions at the same University—the academic system is not too friendly to such cases—but our example shows that it is possible; one just should not give up.

On being a woman in this field . . .

Well, once being in London with an invited talk at the International Meeting (Synthesis, Design and Function in New Materials Chemistry) sponsored in 2006 by the Royal Society of Chemistry, I was among other invited speakers gathering for an after-meeting banquet. I would probably not have paid attention that I was the only female scientist among male colleagues, unless I was asked by one of the ladies: "Whose wife are you?" Before I had time to answer, one of my colleagues clarified: "She is a scientist by herself." Funny!

As already said, I grew up in the Soviet society, where equality of genders was not only proclaimed but also realized: Women were represented in all professions, from cosmonauts to directors of factories. So, for me there was no question women should have a pathway to any profession in the same way a man does.

For me being a woman in a specific professional field means to have a chance to use in this field the advantages of a female character: patience, accuracy, thoroughness, and especially intuition. These properties, so characteristic for many women, are vital in experimental work.

Words of Wisdom

Like "all genius is simple," all wise is obvious. If we are talking about a key to success in any field of human activity, then the formula is simple and obvious: professionalism + persistence + resilience + broad erudition + perfection in very specific skills + self-confidence. However, some of your own life experience is necessary to understand and apply this formula.

On the way to your dream, keep balance: never sacrifice maternity/paternity and family to profession and vice versa. One never knows in which field his/her efforts will be most fruitful.

Bonnie Jeanne Dunbar

Director, STEM Center
Department of Mechanical Engineering
University of Houston
N207 Engineering Building 1
Houston, TX 77204-4006
USA
(Retired National Aeronautics and Space
Administration (NASA) Astronaut and Engineer)

e-mail: bjdunbar@uh.edu
telephone: (713) 743-4528

Bonnie J. Dunbar at the Rockwell
International Space Division,
1976–1978.

Birthplace
Sunnyside, Washington, USA
Born
March 3, 1949

Space Statistics

>50 days in space

Publication/Invention Record

6 publications: h-index 2 (Web of Science)

Tags

❖ Academe
❖ Government
❖ Domicile: USA
❖ Nationality: American
❖ Caucasian
❖ Children: 0

Academic Credentials

Ph.D. (1983) Mechanical/Biomedical Engineering, University of Houston, Houston,
Texas, USA.
M.S. (1975) Ceramic Engineering, University of Washington, Seattle, Washington, USA.
B.S. (1971) Ceramic Engineering, University of Washington, Seattle, Washington, USA.

Research Expertise: effects of spaceflight (microgravity) on the body and refractor
thermal protection systems for spacecraft

Successful Women Ceramic and Glass Scientists and Engineers: 100 Inspirational Profiles. Lynnette D. Madsen.
© 2016 The American Ceramic Society and John Wiley & Sons, Inc. Published 2016 by John Wiley & Sons, Inc.

Other Interests: education in science and engineering

Key Accomplishments, Honors, Recognitions, and Awards[1]

- United States Astronaut Hall of Fame, Class of 2013
- Fellow, American Institute of Aeronautics and Astronautics, 2006
- University of Washington Alumna Summa Laude Dignata, 2004
- Elected member, National Academy of Engineering, 2002
- Fellow, Royal Aeronautical Society
- Elected member, Royal Society of Edinburgh, 2001
- The American Ceramic Society (ACerS) James I. Mueller Award, 2000
- Inducted into the Women in Technology International Hall of Fame—one of five women in the world so honored annually, 2000
- Selected as one of the top 20 women in technology in Houston, Texas, 2000
- NASA Space Flight Medals, 1985, 1990, 1992, 1995, and 1998
- NASA Superior Accomplishment Award, 1997
- Member, National Science Foundation (NSF) Engineering Advisory Board, 1993–1996
- NASA Exceptional Achievement Medal, 1996
- NASA Outstanding Leadership Award, 1993
- Fellow of ACerS, 1993
- Design News Engineering Achievement Award, 1993
- The Institute of Electrical and Electronics Engineers (IEEE) Judith A. Resnik Award, 1993
- Society of Women Engineers Resnik Challenger Medal, 1993
- Museum of Flight Pathfinder Award, 1992
- American Association of Engineering Societies (AAES) National Engineering Award, 1992
- NASA Exceptional Service Award, 1991
- University of Houston Distinguished Engineering Alumna, 1991
- Materials Research Society President's Award, 1990
- ACerS Schwartzwalder P.A.C.E. Award, 1990
- University of Washington Engineering Alumni Achievement, 1989
- NASA Exceptional Service Medal, 1988
- General Jimmy Doolittle Fellow of the Aerospace Education Foundation, 1986
- Evergreen Safety Council Public Service in Space Award, 1986
- Kappa Delta Sorority, 1985
- ACerS Greaves-Walker Award, 1985

---

[1] Source: http://en.wikipedia.org/wiki/Bonnie_J._Dunbar.

- Rockwell International Engineer of the Year, 1978
- Graduated Cum Laude from the University of Washington, 1975

Biography[2]

Early Life and Education

Bonnie Jeanne Dunbar was raised in the Yakima Valley of Washington State on a cattle ranch homesteaded by her parents in 1948. She graduated in 1967 from a small high school in the Yakima Valley, Sunnyside High School, with 4 years of math and 3 years of science, which meant that she did not require any remedial math when entering college calculus and chemistry. She had a sense of what engineers and scientists "did" when she set out as the first in her family to attend college.

She is not easily discouraged: When she enrolled in ceramic engineering at the University of Washington in 1967, it was "not a field for women" in many people's eyes: "Only one professor tried to talk me out of it, all the rest were very supportive." (As well, many years later when she first applied to be an astronaut, she was turned down.) She was awarded Bachelor of Science and Master of Science degrees in ceramic engineering from the University of Washington in 1971 and 1975, respectively. As an undergraduate, she helped Ceramic Engineering Professor James I. Mueller develop the special ceramic tiles that NASA would use to coat the space shuttle to make it able to withstand re-entry into the atmosphere. Upon graduation in 1971, Dr. Dunbar worked for Boeing Computer Services for 2 years as a systems analyst. Subsequently from 1973 to 1975, she conducted research for her master's thesis in the field of mechanisms and kinetics of ionic diffusion in sodium beta-alumina. In 1975, she conducted research at Harwell Laboratories in Oxford, England, where her work focused on the wetting behavior of liquids on solid substrates.

She returned from England and accepted a Senior Research Engineer position with Rockwell International Space Division in Downey, California. Her responsibilities included developing equipment and processes for the manufacture of the space shuttle thermal protection system in Palmdale, California. She also represented Rockwell International as a member of the Dr. Kraft Ehricke evaluation committee on prospective space industrialization concepts. Dr. Dunbar subsequently completed her doctorate in 1983 at the University of Houston. Her multidisciplinary dissertation (materials science and physiology) involved evaluating the effects of simulated spaceflight on bone strength and fracture toughness.

Career History

She became a Certified Professional Engineer in Texas. Dr. Dunbar has served as an Adjunct Assistant Professor in Mechanical Engineering at the University of Houston.

---

[2] Sources: http://www.jsc.nasa.gov/Bios/htmlbios/dunbar.html, http://www.uh.edu/news-events/stories/2013/january/01242013DunbarSTEM.php, http://www.insidebainbridge.com/2013/05/07/an-interview-with-astronaut-bonnie-dunbar-who-will-walk-on-bainbridge-may-9/, and http://www.washington.edu/alumni/columns/june04/dunbar01.html.

Bonnie J. Dunbar, the Professor.

STS-71 mission specialist Bonnie J. Dunbar arrives at KSC's Shuttle Landing Facility from Johnson Space Center, Houston. (See insert for color version of figure.)

She learned to fly with the Rockwell Flying Club at the Orange County Airport in 1977. As a private pilot with over 200 hours in single-engine land aircraft, she has logged more than 1000 hours flying time in T-38 jets as co-pilot, and has over 100 hours as co-pilot in a Cessna Citation Jet. In addition, she logged more than 1000 hours as co-pilot in NASA T-38s as part of Spaceflight Readiness Training for the astronaut corps.

In 1978, she served as a payload officer/flight controller at the Lyndon B. Johnson Space Center. (It was also in 1978 that NASA opened the astronaut corps to women.) In 1979, she became a guidance and navigation officer/flight controller for the Skylab re-entry mission and was subsequently designated project officer/payload officer for the integration of several space shuttle payloads.

In May 1980, Dr. Dunbar was selected as an astronaut and completed her first year of "basic" training in August 1981.

By August, Dr. Dunbar had become a NASA astronaut. She is now a veteran of five spaceflights, and has logged more than 1208 hours (50 days) in space. She served as a mission specialist on STS 61-A in 1985, STS-32 in 1990, and STS-71 in 1995, and was the Payload Commander on STS-50 in 1992 and STS-89 in 1998. She was part of the first space shuttle mission to dock with the Russian space station *Mir*.

From 2005 to 2010, she served as President and Chief Executive Officer (CEO) of the Museum of Flight. Afterward, she consulted for several years. In 2013, she joined the University of Houston as faculty of the College of Engineering and to lead a new University STEM (science, technology, engineering, and math) Center. As the leader of this center, her main goal is to strengthen and support STEM-related educational programs for children of all ages—from kindergarteners to high school seniors.

3 Most Cited Publications

Title: Recent results and new hardware developments for protein crystal growth in microgravity
Author(s): Delucas, LJ; Long, MM; Moore, KM; et al.
Source: Journal of Crystal Growth; volume: 135; issue: 1–2; pages: 183–195; doi: 10.1016/0022-0248(94)90740-4; published: January 1994
Times Cited: 58 (from Web of Science)

Title: Effect of H_2O on Na^+ diffusivity in beta-alumina
Author(s): Dunbar, BJ; Sarian, S
Source: Solid State Communications; volume: 21; issue: 8; pages: 729–731; published: 1977
Times Cited: 3 (from Web of Science)

Title: Thermal protection design considerations for human-rated reusable space vehicles
Author(s): Dunbar, BJ; Korb, L
Book editor(s): Clark, DE; Folz, DC; McGee, TD
Source: Introduction to Ceramic Engineering Design; pages: 200–231; published: 2002
Times Cited: 2 (from Web of Science)

Bonnie J. Dunbar when she was President and
CEO of Museum of Flight.

Challenges

There were more opportunities than challenges. My biggest supporters have been primarily men and I wouldn't be in the career today if it were not for several significant male mentors, from my math and physics teachers to college professors.

Words of Wisdom

There is no substitute for being competent and knowing your discipline. Be professional, aspire to excellence, and believe the best in people.

Profile 23

Alicia Durán

Research Professor
Head of Research Group: Glass, Glass-Ceramics and
Sol-Gel Materials for a Sustainable Society (GlaSS)
Instituto de Cerámica y Vidrio
Consejo Superior de Investigaciones Científicas
(CSIC)
Madrid, Spain

e-mail: aduran@icv.csic.es
telephone: +34-91-7355840

Alicia Durán receiving the
ICG Gottardi Prize from the
President, Prof. Prindle,
1988.

Birthplace
Neuquén, Argentina
Born
August 27, 1950

Publications/Invention Record

>175 publications: h-index 32 (Web of Science)
10 patents, editor of 9 books, 9 book chapters

Tags

❖ Administration and
   Leadership
❖ Government
❖ Academe
❖ Domicile: Spain
❖ Nationality: Dual
   Argentinian/Spanish

Proudest Career Moment (to date)

One very important moment in my career occurred this year, when I received the
international prize RAICES (ROOTS) conceded by Argentinean Ministry of Science and
Technology to Argentinian scientists living abroad who maintained a strong and
continuous collaboration with research groups in the country. It was especially relevant
for me to receive a prize that recognized a long cooperation history with researchers of
my country of origin. Cooperation and international collaboration has been a constant
sign of identity during my entire professional career and it is a great satisfaction to know
that other people consider that we have done the things well.

Academic Credentials

Ph.D. (1984) Physics, Universidad Autónoma de Madrid, Madrid, Spain.
1st Degree (1974) Physics, Universidad Nacional de Córdoba, Córdoba, Argentina.

Successful Women Ceramic and Glass Scientists and Engineers: 100 Inspirational Profiles. Lynnette D. Madsen.
© 2016 The American Ceramic Society and John Wiley & Sons, Inc. Published 2016 by John Wiley & Sons, Inc.

Research Expertise: design, processing, and characterization of glasses, glass-ceramics, and sol-gel materials, going from the design to structural features, properties (e.g., optical, mechanical, chemical, thermal, and electrical), and applications

Other Interests: teaching; study and discussion of R&D policies in Spain and Europe

Key Accomplishments, Honors, Recognitions, and Awards

- Steering Committee, Management Board, and Treasurer of International Commission on Glass (ICG), 2002 to present
- Editorial Board of European Journal of Glass Science and Technology, 2002 to present
- Editorial Board of Journal of Sol-Gel Science and Technology, 2007 to present
- Associate Editor of International Journal of Applied Glass Science, 2009 to present
- ROOTS prize awarded by Argentinean Ministry of Science and Technology (to recognize Argentinian born scientists working abroad who have promoted ties that strengthen science and technology initiatives in Argentina), 2014
- Scientific Secretary and Editor of the Proceedings of International ICG Congress in Madrid, 1992
- International Commission on Glass (ICG) Prize Vittorio Gottardi, presented by the ICG to young researchers in the field of glass science and technology, 1998
- University Prize, Diploma of special mention, Universidad Nacional de Córdoba (Argentina), 1973

Alicia Durán receiving the prize RAICES (ROOTS) from the Argentinean Ministry of Science and Technology in November 2014.

Biography

Early Life and Education

I studied physics in Argentina at the Institute of Mathematics, Astronomy and Physics, obtaining the degree in 1974 with the specialty in materials science. In 1976, I moved to Spain with a Spanish fellowship for sons and grandsons of Spanish citizens to the Institute of Ceramic and Glass (CSIC), where I am presently working.

My Ph.D. thesis was focused on the study of color in glasses, from ionic color to ruby glasses produced by metal colloids. I obtained the Ph.D. degree in 1984 in the Universidad Autónoma de Madrid.

Career History

After obtaining a Ph.D., I devoted my career to broadening my horizons in glass research. I entered into the field of sol-gel processing, which was quite new at that time. I continued working in glasses and glass-ceramics produced by melting, adding new topics in glass coatings prepared by sol-gel. I continued with glass projects focused on glass industry topics (e.g., opal glass and coatings on flat glass) and studied glass sealing materials, in which we developed low-temperature seals for electronic devices, high chemical resistant sealing for molten carbonate fuel cells (MCFCs), and, in more recent years, glass-ceramic sealing for solid oxide fuel cells (SOFCs). Currently, we are focused on lithium phosphate glasses for solid electrolytes of Li batteries and transparent oxyfluoride glass-ceramics for photonic applications.

Alicia Durán in the lab, 2012.

In the field of sol-gel materials, we have developed coatings for different applications, from corrosion-resistant films for metals (stainless steel, aluminum, magnesium, etc.) to bioactive and protecting coatings for prosthesis alloys. In the last few years, we focused the work on mesoporous/mesostructure sol-gel coatings for photocatalytic application; these films also include bactericide and fungicide functionalities. All these research lines were financed through Spanish projects, UE projects, international projects, and industrial research projects.

During this career, I have visited in short stages (from 1 to 5 months) different international labs, from Rennes to Padova or Osaka Universities, and some Argentina research groups in Rosario, Mar del Plata, and Buenos Aires. International cooperation has been a constant in my career, collaborating with many R&D centers from Europe, Japan, and Latin America, and this explains the high percentage (>50%) of co-authored publications in the research group.

3 Most Cited Publications

Title: Structural considerations about SiO_2 glasses prepared by sol-gel
Author(s): Duran, A; Serna, C; Fornes, V; et al.
Source: Journal of Non-Crystalline Solids; volume: 82; issue: 1–3; pages: 69–77; published: June 1986
Times Cited: 171 (from Web of Science)

Title: Glass-forming ability, sinterability and thermal properties in the systems RO–BaO–SiO_2 (R = Mg, Zn)
Author(s): Lara, C; Pascual, MJ; Duran, A
Source: Journal of Non-Crystalline Solids; volume: 348; pages: 149–155; published: November 15, 2004
Times Cited: 85 (from Web of Science)

Title: Preparation and characterization of cerium doped silica sol-gel coatings on glass and aluminum substrates
Author(s): Pepe, A; Aparicio, M; Cere, S; et al.
Source: Journal of Non-Crystalline Solids; volume: 348; pages: 162–171; published: November 15, 2004
Times Cited: 82 (from Web of Science)

Challenges

The challenges of my research work were diverse and included not only the research topics in which I have worked, but also taking the time and energy to participate in different associations linked to my job. For example, I served as the Secretary of the Glass Section of the Spanish Society of Ceramic and Glass for more than 20 years to keep in touch with the glass industry and their research. In 1992, we organized the ICG International Congress for the first time in Spain, gathering more than 650 participants and 500 papers. Since that time, I have been fully interested in fostering international cooperation, in particular with Europeans and Latin Americans, because they provide significant opportunities in terms of training, projects, and personnel. I have participated in more than 10 European projects and several Latin American cooperation programs, thereby adding international visibility to the work developed in the GlaSS group. In addition, since 2002 I have served as the Treasurer of the International Commission on Glass; I am part of the Steering Committee, Management Board, and Council of the association; I participate in some of its Technical Committees; and I manage the Editorial Team. These efforts are critical to ensure the participation of Spanish industry and academia in the international glass world.

I have contributed to the R&D policy development in Spain by participating in the evaluation of national R&D programs as well as in the preliminary works of the new Science and Technology Law in 2011. I am convinced that it is necessary to devote a lot of time and effort to train both researchers and technicians—this will be our greatest inheritance.

On being a woman in this field . . .

Working for improvements in women's careers has been a key issue during my whole professional life. I was appointed to the Women and Science Commission of CSIC for 4 years where I represented the Material Science and Technology area of CSIC. Now, I participate in the Equality Commission of CSIC, which promotes equality between men and women in the research world.

The case of Spain is paradigmatic. First, the proportion of university female students in sciences exceeds 50% in many fields (though still a minority in engineering), and it has been proven that women get better grades than men. However, when they graduate and integrate into the working world of research, the percentage of women decreases dramatically as one moves up the ladder of responsibilities. There are no university female professors in several branches of Physics or Astronomy, nor in most Engineering, and surprisingly also in Obstetrics and Gynecology, or Pediatrics. Similarly, 60% of judges are women but only one has been appointed to the Supreme Court.

The total number of CSIC Research Professors increased from 13% to 23% in just a decade due only to a measure designed and implemented by the Women and Science Commission: to establish parity in numbers in all the juries for the fixed positions. On the other side, in Scientific Tenure positions women exceed 40% of the workforce.

There is still a long way to go to arrive at a real equality between women and men in the research world. Hopefully, a situation will be reached where it will be possible to work in science without renouncing one's personal life, i.e., to be at the same time mothers and researchers.

Words of Wisdom

These are words dedicated to young researchers, those who are beginning the careers and who face so many hard problems. It is certainly a difficult profession, but it is worth the fight/struggle to get a place in this world. Science is a very exciting career in which many things are waiting to be discovered. Arriving at the solution of a mathematic problem or an optical or electronic device always represents an irreplaceable feeling of surprise and happiness. But it is also worth knowing that it is a hard work, in which it is necessary to invest a lot of time searching for the solution that many times resists being found. A scientist has to study and think every day. This is the key issue to accept when we decide to enter the profession of scientists. If we love study, if we are happy to know more and more, this can be our profession. But it is also very important to think and know that science is not just the adventures of scientists, but a very strong and powerful tool to change society, to convert the world into a better place to live for all humanity.

Profile 24

Heike Ebendorff-Heidepriem

Associate Professor
Deputy Director of the Institute for Photonics
and Advanced Sensing (IPAS)
Associate Director of the Optofab Adelaide
Node of the Australian National Fabrication
Facility (ANFF)
ARC Centre of Excellence for Nanoscale
BioPhotonics (CNBP)
School of Physical Sciences
The University of Adelaide
Australia

e-mail: heike.ebendorff@adelaide.edu.au
telephone: +61 8 8313 1136

Heike Ebendorff-Heidepriem
in 2005.

Tags

❖ Administration and
  Leadership
❖ Academe
❖ Domicile: Australia
❖ Nationality: Australian
❖ Caucasian
❖ Children: 0

Birthplace
Jena, Germany
Born
January 2, 1966

Publication/Invention Record

>200 publications: h-index 25 (Web of Science)
7 patents

Pivotal and Proudest Career Moments (to date)

In 1985, at the end of my first undergraduate year, I undertook a summer project at the Otto Schott Institute for Glass Chemistry at the University of Jena. This experience was a pivotal moment as it sparked my passion for glass science and technology and was the start of my research career in this field.

In 2001, I was awarded my first international award, the Woldemar A. Weyl International Award by Pennsylvania State University. It is awarded once every 3 years to an outstanding young scientist working in glass research. I felt very honored to receive it.

Successful Women Ceramic and Glass Scientists and Engineers: 100 Inspirational Profiles. Lynnette D. Madsen.
© 2016 The American Ceramic Society and John Wiley & Sons, Inc. Published 2016 by John Wiley & Sons, Inc.

In 2013, I was awarded two Discovery Projects from the Australian Research Council (ARC). This was the first time that I was awarded funding as a lead investigator. This funding allowed me to undertake independent research, which was a critical transition in my career to becoming a research leader.

Academic Credentials

Ph.D. (1994) Chemistry, University of Jena, Jena, Thuringia, Germany.
Diploma (1989) Chemistry, University of Jena, Jena, Thuringia, Germany.

Research Expertise: development of fibers for a wide range of applications, glass science and technology, preform and fiber fabrication, fiber surface functionalization, nanotechnology

Other Interests: mentoring young researchers

Key Accomplishments, Honors, Recognitions, and Awards

- Member of the Scientific Advisory Council of the Leibniz Institute of Photonic Technology, 2014 to present
- Associate Director of the Optofab Adelaide Node of the Australian National Fabrication Facility (ANFF), 2014 to present
- Deputy Director of the Institute for Photonics and Advanced Sensing at The University of Adelaide, 2014 to present
- Leader of the Research Theme "Optical Materials and Structures" of the Institute for Photonics and Advanced Sensing at The University of Adelaide, 2009 to present
- Awarded over $12 million in competitive research grants, 2005 to present
- 16 plenary or invited talks, 2005–2014
- Member of the Organizing Committee of the International Conference on Nanoscience and Nanotechnology (ICONN), held in Adelaide, Australia, 2014
- Organized conference symposium of the 10th Pacific Rim Conference on Ceramic and Glass Technology, 2013
- Tutorial on "Glasses for infrared fibre applications," European Conference on Optical Communication (ECOC'2013), London, UK, 2013
- Associate Guest Editor of the International Journal of Applied Glass Science, 2012
- Co-organized conference sessions of the Annual Meeting of the American Ceramic Society in 2005, 9th Pacific Rim Conference on Ceramic and Glass Technology, 2010
- International Zwick Science Award, 2009
- Marie Curie Individual Fellowship of the European Union, 2001–2002
- Woldemar A. Weyl International Glass Science Award, 2001
- Habilitation Fellowship of the Land Thüringen, Germany, 1997–2000
- Award of the University of Jena, Germany, for an excellent Ph.D. thesis, 1995

Biography

Early Life and Education

Heike Ebendorff-Heidepriem was raised in Germany by German parents. She has a sister and a brother. Heike received her education from primary school to the University Diploma degree (approximately equivalent to M.Sc. degree) in the public school and university system of the former German Democratic Republic (Eastern Germany).

Heike received her undergraduate study in chemistry from the University of Jena in 1984–1989. She performed her Diploma degree research at the Otto Schott Institute for Glass Chemistry. She continued at this institute in undertaking her doctoral research, which commenced in the autumn of 1989, when the political system of Eastern Germany broke down.

Career History

After Heike was awarded a Ph.D. degree, she continued her research at the University of Jena on optical glasses until 2000. For the period of 2001–2004, she joined the Opto-electronics Research Centre at the University of Southampton (UK), where she worked on novel photo-sensitive glasses and soft glass micro-structured optical fibers with record optical nonlinearity.

In 2005, she accepted a position at The University of Adelaide in Australia. Currently, she is the Deputy Director and one of the leaders of the Optical Materials & Structures Research Theme at the Institute for Photonics and Advanced Sensing. She is also Senior Investigator at the ARC Centre of Excellence for Nano-scale BioPhotonics. Her research focuses on the development of glass, in particular mid-infrared, high-non-linearity, active, and nanocomposite glasses; glass, preform, and fiber fab-rication techniques; and surface func-tionalization of glass.

Heike Ebendorff-Heidepriem in the lab preparing a glass extrusion experiment in 2011. Credit: Jennie Groom.

3 Most Cited Publications

Title: Bismuth glass holey fibers with high nonlinearity
Author(s): Ebendorff-Heidepriem, H; Petropoulos, P; Asimakis, S; et al.
Source: Optics Express; volume: 12; issue: 21; pages: 5082–5087; published: October 18, 2004
Times Cited: 146 (from Web of Science)

Title: Highly nonlinear and anomalously dispersive lead silicate glass holey fibers
Author(s): Petropoulos, P; Ebendorff-Heidepriem, H; Finazzi, V; et al.
Source: Optics Express; volume: 11; issue: 26; pages: 3568–3573; published: December 29, 2003
Times Cited: 115 (from Web of Science)

Title: Mid-IR supercontinuum generation from nonsilica microstructured optical fibers
Author(s): Price, JHV; Monro, TM; Ebendorff-Heidepriem, H; et al.
Source: IEEE Journal of Selected Topics in Quantum Electronics; volume: 13; issue: 3; pages: 738–749; published: May–June 2007
Times Cited: 113 (from Web of Science)

Challenges

The single largest and ongoing challenge in my work life is to live up to leadership responsibilities, including giving non-scientific speeches. As with many things in life, "learning by doing it" has allowed me to successfully tackle this challenge.

On one hand, I struggle to begin the writing of papers and research proposals; on the other hand, I feel very rewarded when the writing and submission process is complete; this reward compensates for the uphill struggle at the beginning.

On being a woman in this field . . .

Until recently, all my supervisors and mentors have been strong women that made a stand in their fields. It is hard for me to say whether it just happened or whether I have subconsciously chosen to be mentored by women.

My research on optical glass has always been at the border of chemistry and physics. Transdisciplinary science is the way of the future and I am proud to be a leader in this area. From my experience in chemistry, the number of female and male students and young researchers is roughly balanced, whereas in physics women are clearly underrepresented. When doing experimental research in a physics environment, I never felt challenged being a woman, rather it made me proud to have the opportunity to share my skills.

Words of Wisdom

My research experiences have taught me the following lessons:
• Be clear about your motivation for undertaking a certain job or task. For example, I am passionate about doing research at a university. With this in mind, I have been able to

overcome the grand challenge of grant writing and dealing with deadlines on short notice and it helped me to persevere during tough times.

- Take advice from supervisors/mentors/friends, when it is "spot-on" even when it is painful. It is rare to have someone tell you the truth; use this opportunity to learn and improve.
- Be open to opportunities that arise. To a large degree, I achieved my success by seizing such opportunities.

Profile 25

Lina M. Echeverría

Innovation Leadership Consultant
Author
Previously:
Vice President
Science and Technology
Director
Exploratory Markets and Technologies
Corning Incorporated
USA

e-mail: contact@linaecheverria.com
telephone: (607) 936-3666

Close up of Lina Echeverría
in 1973.

Tags

❖ Management and
  Leadership
❖ Domicile: USA
❖ Nationality: Dual
  American and
  Colombian
❖ Caucasian Hispanic/
  Latina
❖ Children: 2

Birthplace
Medellín, Colombia
Born
December 26, 1949

Publication/Invention Record

23 publications (1974–1992)
7 patents (1990–1996), 1 book (2012)

Memorable Moments

We are at times unaware of the impact we may be having on others through our actions, so it is nice to have the perspective of time to help us find meaning. As I look back, opening the door to generations of women that followed my footsteps in entering the field of geological engineering at the Universidad Nacional de Colombia at Medellín is a memorable moment in my career, although I was just following my heart and pursuing my dream. It was only through the interviews by professor Pamela Murray in preparation for her book *Dreams of Development* that I first became aware of my impact. Today, a career once closed to them boasts a 50% participation of women at the undergraduate level.

Lina Echeverría doing field
geology in Colombia, 1973.

Successful Women Ceramic and Glass Scientists and Engineers: 100 Inspirational Profiles. Lynnette D. Madsen.
© 2016 The American Ceramic Society and John Wiley & Sons, Inc. Published 2016 by John Wiley & Sons, Inc.

Discovering and documenting the first ever reported occurrence of "young" (i.e., non-Archaean) komatiites (quenched high-temperature lavas), and doing so while carrying my first born on a backpack in a tropical rainforest is a second memorable moment for me. Once again, I was just following my heart and my scientific curiosity, but this time I was keenly aware of its significance to the scientific community.

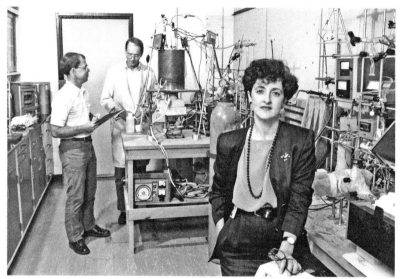

Lina Echeverría in her lab at Corning in 1990 when she was doing research in glass–ceramics.

Knowing that one of my inventions, a lithium disilicate glass–ceramic composition licensed to Ivoclar after being patented by Corning Incorporated, is providing high-quality, high-aesthetics dental restorations—only possible before with unsightly opaque metals—to people throughout the world also joins this list of memorables. It is of a different nature: It feels good to know that your invention is improving lives, and that makes it memorable.

Later in my career, having the guts and the savvy to create the space where people in my group could find their creative core and be themselves, regardless of their quirkiness or origin, and spread their wings, to create the cultures where creativity thrives and innovations are delivered is certainly one of my memorable "moments" (lasting years!). To have touched many and delivered so much together is certainly memorable.

A career is not only a series of professional accomplishments. It is an expression of the seamless coming-together of the many different facets of our being, blending to enable your performance. Two of my very meaningful memorable events are personal. One of them—a 20-year "moment"—is raising two wonderful children, who provided my career with inspiration and meaning for every step, particularly faltering moments. They have grown to be mature professionals, in biophysics and equine surgery, who enjoy life and do not take if for granted.

The second one is surviving a highly aggressive form of breast cancer and coming out of it with the wit to write a book about my passion for developing people, and to have it published. Hopefully, it will provide inspiration to others.

Academic Credentials

Post-Doctoral Fellow, 1978–1980, Carnegie Institution of Washington.
Ph.D. (1978) Geology, Stanford University, Stanford, CA, USA.
B.S. (1973) Geological Engineering, Universidad Nacional de Colombia at Medellín, Colombia.

Research Expertise: geology, ceramics, glass

Other Interests: creating cultures of creativity and innovation, developing people, enabling diversity

Key Accomplishments, Honors, Recognitions, and Awards

- Advisory Board, School of Earth Sciences, Stanford University, 1998–2014
- Board of Trustees, The Nature Conservancy, Western NY Chapter, 2009–2013
- Book author, *Idea Agent: Leadership that Liberates Creativity and Accelerates Innovation*, 2012
- Technologist of the Year Award, Career Communications Group, 2006
- Advisory Board, Department of Material Sciences, Cornell University, 2001–2004
- Board of Directors, National Physical Science Consortium, 1997–2004
- Women of Color Technology Award for Research Leadership, 1999
- Board of Directors, Meals on Wheels, Corning, NY, 1996–1999
- Board of Directors, Civic Music Association, Corning, NY, 1996–1999
- Advisory Board, Science and Discovery Center, Corning, NY, 1994–1996
- Chair, New York State Alliance for Girls and Women in Technology, Private Industry and Public Sector Subcommittee, 1992–1994

Biography

Early Life and Education

Born and raised in Colombia, Lina grew up as the third of six in a household that exposed the siblings to the creative world of artists and intellectuals and to the rigor and excellence of engineers. In 1969 she was the first woman to apply and be accepted into the geological engineering program at the Escuela de Minas at the Universidad Nacional de Colombia at Medellín, the most rigorous engineering program in her country's largest public university.

She graduated in 1973 and then worked at Geominas Ltd. as Junior Geologist staying in Medellín, Colombia. In 1974 Lina moved to California to pursue her doctorate in Geology at Stanford University under the guidance of Bob Coleman, where she was awarded a Ph.D. in 1978.

Credit: http://diversitygps.com/top-hispanics-in-stem-where-are-they-now-p192-115. htm. P. Murray, 1997, *Dreams of Development*, U. of Alabama Press

Career History

From 1978 to 1980, she was a Postdoctoral Research Fellow at the Carnegie Institution of Washington from where she moved on to become a Research Scientist at the Max Planck Institüt für Chemie in Mainz, Germany, until 1983. That year she joined Corning Incorporated where she held a variety of positions from research to management and leadership, with increasing responsibilities:

Lina Echeverría in 1986.

- 1983–1985 Senior Research Scientist, Ceramics Research
- 1985–1988 Senior Research Scientist, Environmental Products Development
- 1988–1991 Research Associate, Glass and Glass Ceramics Research
- 1991–1993 Supervisor, Glass Research
- 1993–1995 Product Line Manager, CorningWare™
- 1995–1999 Manager, Glass and Glass-Ceramics Research
- 1999–2001 Core Technology Director, Glass and Glass-Ceramics
- 2001–2003 Research Director, Inorganic Technologies
- 2004–2007 Division Vice President, Science and Technology, and Director, Corning European Technology Center
- 2007–2011 Vice President, Science and Technology, and Director, Exploratory Markets and Technologies.

From scientist to vice president, she helped drive new products at Corning Inc. that now

Lina Echeverría at Corning Laboratories, 1990.

underpin their technology-based economy, from faster optical fiber that powers the Internet to flat-panel glass used in everything from smart phones to LCD televisions.

Since her retirement from Corning in 2011, she acts as an Innovation Leadership Consultant and has authored the book *Idea Agent: Leadership That Liberates Creativity and Accelerates Innovation*, published by the American Management Association in 2012.

She is the mother of two children and lives in upstate New York with her husband, a research scientist. She is fluent in English, Spanish, and French.

3 Most Cited Publications

Title: Re-Os isotopic constraints on the origin of volcanic-rocks, Gorgona Island, Colombia: Os isotopic evidence for ancient heterogeneities in the mantle
Author(s): Walker, RJ; Echeverría, LM; Shirey, SB; et al.
Source: Contributions to Mineralogy and Petrology, volume: 107; issue: 2; pages: 150–162; doi: 10.1007/BF00310704; published: 1991
Times Cited: 95 (from Web of Science)

Title: Pyroclastic rocks: another manifestation of ultramafic volcanism on Gorgona Island, Colombia
Author(s): Echeverría, LM; Aitken, BG
Source: Contributions to Mineralogy and Petrology; volume: 92; issue: 4; pages: 428–436; doi: 10.1007/BF00374425; published: 1986
Times Cited: 44 (from Web of Science)

Title: Enstatite ceramics: a multicomponent system via sol–gel
Author(s): Echeverría, LM
Source: Journal of Non-Crystalline Solids; volume: 147; pages: 559–564; doi: 10.1016/S0022-3093(05)80676-3; published: October 1992
Times Cited: 8 (from Web of Science)

Challenges

As a child of the women's lib movement of the 1960s and 1970s, I wanted to have it all: to be a full player at work and at home. The challenge, as always, was one of prioritization of time and energy, and of choosing what to give up. Housekeeping work was an easy one for me to give up, personally. The work would still be done, I reckoned, and trusted that I would always find someone that would be reliable and trustworthy—and I did. The guilt of not doing it myself was never part of my burden. For decades I also gave up TV watching, as my husband and I used "flex-time" guidelines: He would do the early morning shift, getting the children up and out to school—and training them since an early age to pack their own lunches—while I woke up early to hit the gym before making it to work by 7 a.m. Early in my career I could be home mid-afternoon to be with the children and see them riding their first pair of cross-country skies and bikes. This meant going to bed early or risk being exhausted. We shared 50-50 cooking and cleaning duties, and reheating leftovers did not count as cooking!

On being a woman in this field . . .

Scientific and technology fields are still male-dominated worlds.
Though we are beyond the need to prove that women have the
brains and the commitment to be at par with men, the challenges of
making it to top leadership positions are still a reality for women.
So it is important that every woman be able to find her inner core,
the source of inspiration and strength. Every career is unique, but
every career needs to be fueled by that internal drive, the source of
passion that will make every barrier passable, every accomplishment momentous. Some of the principles that have guided me are
as follows:

A recent image of
Lina Echeverría.

- Be true to yourself. Have the clarity, the courage, and the inner
  strength to be the best authentic version of yourself—and not somebody else.
- Ignore the cultural encoding for girls to set aside your personal power, your drive, your
  energy, and to be nice, good, and accommodating. Forget about trying to be perfect and
  not offending. Be authentic instead!
- Be intuitive—learn to tune into that internal voice that guides you, even if you do not
  know why—and follow your heart.
- Create an active network. Of friends, of advisors, of mentors. Not people who will help
  you escalate in your professional life, but people who have the wisdom and the care
  that resonate with your own values. Go to them "in sickness and in health" and let them
  guide you.
- Balance your life. Your relationships, family, hobbies, and activities outside your work
  are the fuel that provides your professional engine with power. Make time to build and
  enjoy them.
- The three golden rules of a successful career are performance, performance, and
  performance. If you do what you love, you will be driven and you will give your best
  performance.
- Deliver through people's growth. Know the people in your group and understand their
  strengths and needs. Develop their lives and their careers.
- Your position does not define who you are. You are not above those who work for you,
  nor below your superiors. Your values are your core.
- Leadership is not an exercise in self-aggrandizement. Leadership *is* service. Leadership
  is an opportunity to serve others, to serve organizations. A commitment to self-
  awareness and self-acceptance is your wise guide.

Profile 26

Doreen D. Edwards

Dean, Inamori School of Engineering
Alfred University
1 Saxon Drive
Alfred, NY 14802
USA

e-mail: dedwards@alfred.edu
telephone: (607) 871-2422

Dr. Doreen Edwards when
she was 37 years old.

Tags

❖ Administration and
  Leadership
❖ Academe
❖ Domicile: USA
❖ Nationality: American
❖ Caucasian
❖ Children: 0

Birthplace
Lemmon, SD, USA
Born
January 28, 1964

Publication/Invention Record

>60 publications: h-index 13
2 patents

Proudest Career Moment (to date)

While I am proud of my personal accomplishments, I feel prouder when students who graduated years ago tell me that something I did, often something inconsequential in my landscape, was instrumental in their successes. These experiences have helped me realize that small words of encouragement delivered at the right time can have a profound effect on someone's life.

Academic Credentials

Ph.D. (1997) Materials Science and Engineering, Northwestern University, Evanston, Illinois, USA.
B.S. (1985) Chemistry, South Dakota School of Mines and Technology, Rapid City, South Dakota, USA.

Research Expertise: oxides for electronic, optical, energy, and environmental applications

Successful Women Ceramic and Glass Scientists and Engineers: 100 Inspirational Profiles. Lynnette D. Madsen.
© 2016 The American Ceramic Society and John Wiley & Sons, Inc. Published 2016 by John Wiley & Sons, Inc.

Other Interests: promoting diversity in science and engineering, leadership education for STEM students

Key Accomplishments, Honors, Recognitions, and Awards

• Chair, Publications Committee of The American Ceramic Society, 2012–2013
• Fellow, The American Ceramic Society, 2012
• Vice-Chair, Board of Directors of Alfred Technology Resources, Inc., 2012–2013
• Member, Board of Directors of Alfred Technology Resources, Inc., 2010–2013
• Member, Advisory Board, Department of Materials Science and Engineering, Northwestern University, 2011–2013
• Co-Chair, International Symposium on Defects, Transport, and Related Phenomena, 2009–2013
• Chair, Basic Science Division, MS&T Programming, The American Ceramic Society, 2009
• Chair, Phase Equilibria Program Subcommittee, The American Ceramic Society, 2006–2007 and 2007–2008
• Member, Organizing Committee for Solid State Ionics—17, 2007–2009
• National Science Foundation CAREER Award, 2001–2006
• Appointed Member, Phase Equilibria Program Subcommittee, The American Ceramic Society, 2004–2006
• Member, Basic Science Division Long-Range Planning Subcommittee, The American Ceramic Society, 2006–2009
• Chancellor's Award for Excellence in Teaching, SUNY, 2004
• McMahon Teaching Award, Alfred University, 2004
• Excellence in Teaching Award, Alfred University, 2002
• Alcoa Foundation Grant for Promoting Diversity in Ceramic Engineering/Materials Science, 1999–2001
• Engineering Education Scholars Workshop, National Science Foundation, 1998
• Appointed Member, Center for Advanced Ceramic Technology Internal Advisory Board, 2000–2013
• Advisor, Alfred University Student Chapter of the Materials Research Society, 2001–2008

Biography

Early Life and Education

Doreen Edwards was an only child raised on a ranch in South Dakota. She attended a one-room rural school through the eighth grade before enrolling in the high school at Lemmon, SD.

After earning a B.S. degree in chemistry from the South Dakota School of Mines and Technology in 1985, Doreen Edwards was a research scientist at Gould, Inc., near

Chicago, from 1985 to 1987. She then joined the Basic Industry Research Laboratory (a contract research lab operated by Northwestern University) and worked as a research scientist while attending graduate school part time. In 1994, she resigned from her job as a research scientist and enrolled full time in graduate school, where she served as both a research and a teaching assistant. She earned a Ph.D. degree in materials science and engineering from Northwestern University in 1997.

Dr. Doreen Edwards in her laboratory with a student.

Career History

She joined the Alfred University faculty in 1997 as an Assistant Professor of Materials Science and Engineering, and earned promotion to Associate Professor in 2003 and to Full Professor in 2007. Prof. Edwards had been Director of graduate engineering programs for the School from 2003 to 2007 and then served as the Associate Dean of the School of Engineering from 2007 to 2009. She was promoted to Dean of Engineering. In 2013, there were fewer than 35 female deans of engineering across the United States.

Dr. Doreen Edwards talking with students.

As Dean she holds active grants and continues to run a research program. To date, she has advised 6 Ph.D. students, including 3 current students; 11 master's students, including 2 current students; and more than 30 undergraduate thesis projects since 1998.

She is an active member of the American Ceramic Society.

Dr. Doreen Edwards in her laboratory with a student. (See insert for color version of figure.)

3 Most Cited Publications

Title: A new transparent conducting oxide in the Ga_2O_3–In_2O_3–SnO_2 system
Author(s): Edwards, DD; Mason, TO; Goutenoire, F; et al.

Source: Applied Physics Letters; volume: 70; issue: 13; pages: 1706–1708; doi: 10.1063/1.118676; published: March 31, 1997
Times Cited: 125 (from Web of Science)

Title: Phase relationships and physical properties of homologous compounds in the zinc oxide–indium oxide system
Author(s): Moriga, T; Edwards, DD; Mason, TO; et al.
Source: Journal of the American Ceramic Society; volume: 81; issue: 5; pages: 1310–1316; published: May 1998
Times Cited: 120 (from Web of Science)

Title: Point defects and electrical properties of Sn-doped In-based transparent conducting oxides
Author(s): Hwang, JH; Edwards, DD; Kammler, DR; et al.
Source: Solid State Ionics; volume: 129; issue: 1–4; pages: 135–144; doi: 10.1016/S0167-2738(99)00321-5; published: April 2000
Times Cited: 61 (from Web of Science)

Challenges

The greatest career challenge I have faced was my own attitude toward work and self-sufficiency. During my childhood, growing up in family of farmers and ranchers in rural South Dakota, I had plenty of role models that helped shape my character but few that could provide informed career advice. I developed many of the traits that served as a foundation for a successful career, but did not always understood how to apply these to advance in my career. As an example, I was raised with a strong work ethic, but in a culture that placed a great deal of value on physical labor compared to intellectual and creative activities. It was only when I critically examined this value system that I could envision myself in a job outside the laboratory. Because I was raised in a culture where self-reliance and stoicism was valued, I often had difficulties with seeking mentors and advocating for myself.

On being a woman in this field . . .

As a young woman, I would have argued that gender was not an issue in my career development. As I became aware of the gender-and-career studies conducted over the past couple of decades, I have had to re-evaluate my thoughts on this subject. Today, I am ambivalent about how to frame the issue of gender in the science and engineering fields. From a personal viewpoint, I try not to focus on these issues because I think it is self-defeating. From a more detached viewpoint, I think it's important to shed light on these issues to minimize their effect on the upcoming generations. I continue to be disheartened by the fact that I am often the only woman, or one of only a few women, in many of the technical and business meetings I attend. When hearing about the troubles that some young women are having in their careers, I often struggle with how to give advice that encourages self-determination and technical excellence while acknowledging the realities of gender bias.

From my own experience, I know that my sense of self (positive and negative) has had a far bigger impact on my career advancement than external gender-related biases (good and bad). Personally knowing women who succeeded in the field helped me see the possibilities for my own advancement. Their influence was particularly strong when I viewed the woman as friend or peer, rather than as an idealized role model. One of the most important things we can do as women in the field is to be an approachable role model for the younger generation.

Words of Wisdom

We need to make our own definitions of success. We follow our own paths during our short time on Earth. If we're fortunate, this path will have many interesting twists and turns and will bring us joy along the way. While it's beneficial to set goals and to take the necessary steps to move toward them, it's equally important to take the time to discover our passions and to allow ourselves the flexibility to take advantage of opportunities as they arise.

Mari-Ann Einarsrud

Professor
Department of Materials Science and Engineering
Norwegian University of Science and Technology
Sem Sælands vei 12
7491 Trondheim
Norway

e-mail: mari-ann.einarsrud@ntnu.no
telephone: +47 73 59 40 02

Mari-Ann
Einarsrud about
35 years old.

Tags

- ❖ Academe
- ❖ Domicile: Norway
- ❖ Nationality: Norwegian
- ❖ Caucasian
- ❖ Children: 1

Birthplace

Hof i Solør, Norway

Born

November 26, 1960

Publication/Invention Record

>175 publications: h-index 30 (Web of Science)
2 patents

Proudest Career Moment (to date)

Among my proudest moments is the establishment of the spin-off company CerPoTech together with colleagues (Professors Tor Grande and Kjell Wiik) because this creates values for the society and new workplaces are established. I am also very proud each time my students defend their thesis and are awarded the Ph.D. degree.

Mari-Ann Einarsrud together with colleagues (from left, Professor Kjell Wiik, Professor Tor Grande, and engineer Rune T. Barland) during the establishment of the spray pyrolysis pilot-scale laboratory. Photo credit: Julian Tolchard.

Successful Women Ceramic and Glass Scientists and Engineers: 100 Inspirational Profiles. Lynnette D. Madsen.
© 2016 The American Ceramic Society and John Wiley & Sons, Inc. Published 2016 by John Wiley & Sons, Inc.

Academic Credentials

Ph.D. ~ Dr.ing. (1987) Inorganic Chemistry, Norwegian Institute of Technology (NTH), Trondheim, Norway.
M.Sc. ~ Siv.ing. (1983) Inorganic Chemistry, NTH, Trondheim, Norway.

Research Expertise: ceramic processing and advanced synthesis of inorganic and hybrid materials, ionic or mixed conducting oxide ceramics for energy technology (solid oxide fuel cells and membranes), ferroic oxide materials, thermoelectric oxide materials, and nanostructured materials

Other Interests: industrial innovation

Key Accomplishments, Honors, Recognitions, and Awards

- Examination committee member for doctoral student (Jill Sundberg), Uppsala University, 2014
- Faculty opponent for doctoral defense (Mohsin Saleemi), Royal Institute of Technology (KTH), 2014
- Examination committee member for licentiate degree (Neda Keshavarzi), Stockholm University, 2013
- Examination committee member for doctoral student (Seikh M.H. Rahman), Chalmers University, 2013
- Examination committee member for doctoral student (Mari Næss), NTNU, 2013
- Examination committee member for licentiate degree (Samrand Shafeie), Stockholm University, 2011
- Examination committee member for doctoral student (Geir Watterud), NTNU, 2011
- Examination committee member for doctoral student (Øystein Dahl), NTNU, 2010
- Member of committee to summarize chemistry evaluation in Norway, The Norwegian Research Council, 2009–2010
- Elected member, The Norwegian Academy of Science and Letters (DNVA), 2009
- Elected member, The Royal Norwegian Society of Sciences and Letters (DKNVS), 2009
- Examination committee member for doctoral student (Crina Silvia Ilea), University of Bergen, 2007
- Member of committee to elaborate nanotechnology in Norway, The Norwegian Research Council, 2006
- Examination committee member for doctoral student (Hedda Vikan), NTNU, 2005
- Examination committee member for doctoral student (Jon Arvid Lie), NTNU, 2005
- Member of international committee to elaborate nanotechnology at NTNU, 2002
- Examination committee member for doctoral student (Åsa Ekstrand), Uppsala University, 2002

- Member of the board for the national PROSBIO research program, The Norwegian Research Council, 2001–2005
- Member of Council for Strategic Programs, The Norwegian Research Council, 2000–2002
- Elected member, The Norwegian Academy of Technological Sciences (NTVA), 1999
- Examination committee member for doctoral student (Oluf Böckman), NTNU, 1999
- Examination committee member for doctoral student (Mika Jokinen), Åbo Akademi, 1998
- Member of the Executive Board of NTNU, 1996–2001
- Member of Program for Research in Materials Technology, The Norwegian Research Council, 1995–1999
- Head and Deputy Head of Department of Inorganic Chemistry, NTH, 1990–1996
- Science and Technology Agency (STA) fellowship, Japan, 1992
- Chair for the Equal Opportunities Committee, NTH, 1990–1991
- Esso Research Price for Fundamental Research for best Ph.D. thesis, NTH, 1987
- NTH Ph.D. Scholarship, 1984–1985

Biography

Early Life and Education

Mari-Ann Einarsrud was raised on a farm in Norway. She has one brother and two sisters. She received her early education in local schools in Grue and Åsnes counties and was very determined to continue for higher education in natural science after graduating from high school. She moved to Trondheim and started on the engineering program in chemistry at the Norwegian Institute of Technology (NTH). After 4½ years of study, she received her engineering degree (equivalent to M.Sc.) in inorganic chemistry. She was then hired as a Scientific Assistant at the Department of Inorganic Chemistry at NTH to teach freshmen students in chemistry laboratory work. At the same time, she also started on her Ph.D. studies in inorganic chemistry. After 1 year as Scientific Assistant, she was awarded the NTH Ph.D. Scholarship and completed her Ph.D. degree after just 2 years, at an age of 26. Her Ph.D. thesis was both an experimental and a theoretical study on the structure of molten salts.

Career History

After receiving her doctorate in 1987, she was employed as an Associate Professor at NTH in Trondheim and started teaching structural chemistry. At that time, she was determined to change fields and was awarded a sabbatical leave (academic year 1987/1988) from NTH to go to the Ceramics Processing Laboratory at MIT in the United States to study ceramic engineering. After returning to NTH, she started to build her own group in ceramic engineering and has together with colleagues built the Inorganic Materials and Ceramics Research Group. She developed a graduate course in ceramic engineering that she has been teaching annually since then. In 1992, she was awarded a Science and Technology Agency Fellowship to go to Japan and was working on SiAlON materials at National Institute for Research in Inorganic Materials (now NIMS) in Tsukuba for about 8 months.

After returning to Trondheim, she continued her work on ceramic materials (focused on ceramic processing and synthesis of materials) and she was appointed as a Full

Professor of Ceramic Engineering in 1997 at Norwegian University of Science and Technology, NTNU (formed by merging NTH with Norwegian College of General Sciences in 1996). When NTNU was established, she was elected as an Executive Board member of the university and served for two periods of 4 years each. She also started to teach a freshman course in inorganic chemistry for engineering students and has taught this class annually since then. In the academic year 2002/2003, she was awarded a sabbatical leave and spent 4.5 months at the National Institute for Advanced Industrial Science and

Mari-Ann Einarsrud in her office at NTNU in 2005. Photo credit: Kjell Røkke.

Technology, Tsukuba, Japan working on oxide ceramics for energy technology. During most of 2007, she was on maternity leave after giving birth to her son in December 2006. In 2007, she established the spin-off company CerPoTech together with colleagues based on their developments in ceramic powder synthesis by spray pyrolysis. During the years at NTH and NTNU, she has served several committees and has been instrumental for development of several national strategies in materials science and nanotechnology and also played a key role in the establishment of NTNU NanoLab. Einarsrud has also been instrumental for the development of bilateral research collaboration between Norway and Japan through KIFEE (Kyoto International Forum for Energy and Environment). During the academic year 2013/2014, she was on sabbatical leave to Department of Materials and Environmental Chemistry, Stockholm University, Sweden working on spark plasma sintering of ceramics.

3 Most Cited Publications

Title: On the thermodynamic stability of $BiFeO_3$
Author(s): Selbach, SM; Einarsrud, M-A; Grande, T
Source: Chemistry of Materials; volume: 21; issue: 1; pages: 169–173; published: January 13, 2009
Times Cited: 117 (from Web of Science)

Title: Size-dependent properties of multiferroic $BiFeO_3$ nanoparticles
Author(s): Selbach, SM; Tybell, T; Einarsrud, M-A; Grande, T
Source: Chemistry of Materials; volume: 19; issue: 26; pages: 6478–6484; published: December 25, 2007
Times Cited: 116 (from Web of Science)

Title: Combustion synthesis and characterization of nanocrystalline CeO_2-based powders
Author(s): Mokkelbost, T; Kaus, I; Grande, T; Einarsrud, M-A
Source: Chemistry of Materials; volume: 16; issue: 25; pages: 5489–5494; published: December 14, 2004
Times Cited: 100 (from Web of Science)

ResearcherID (I-5085-2014)

Challenges

What has been a challenge over time since I am involved in engineering study programs with many students is to find enough time for research when the teaching load has been quite heavy. However, teaching students is a fantastic job as you can see how young students develop and learn during their studies.

On being a woman in this field . . .

Mari-Ann Einarsrud in the Spark Plasma Sintering Laboratory at Department of Material and Environmental Chemistry at Stockholm University, Sweden, where she was on sabbatical leave (2014). Photo credit: Jon Herman Rismoen.

Already as a very young girl, I made up my mind that I wanted to be a scientist and work in chemistry-related field. This destiny was not obvious for me as I was raised on a farm in the rural part of Norway and no one in my family or neighborhood had an academic education. My dedication to become a scientist started when my father was subscribing to a magazine called "Farmers Magazine" and I found a picture in one of the issues showing a chemist working in a laboratory. I was only 6 years old at that time, but I was sure that "this is what I want to do!" I kept the picture under my pillow in the bed for a long time and my dedication to become a scientist grew. Since I saw that picture (even when I, at a young age, did not know what it meant) followed the dream it gave me. Later, when in my professional life I have met challenges (as a woman), I do not think I have looked upon them as real hindrances because my dedication to become a researcher has been so strong.

Being raised on a farm also means that you learn to work hard as there is always more work than there is workforce available and it did not matter if you were a girl or a boy—everyone could do the same kind of work. From there I learned I could be as good as a man and I could do the same kind of work. I think this has been very helpful throughout my career.

In my early career there were not many female professors, later on to increase the number of the underrepresented gender in any field; regulations were introduced in such a way that it should be at least 40% of each gender in committees (promotion committees, boards, etc.). However, a small fraction of female scientists gave us a challenge because a rather large fraction of our time could be occupied doing non-meriting work. I partly solved this by choosing the most influencing committees, but it also gave me more non-meriting work than my male colleagues. Nowadays, there are more female professors in most engineering fields as well, showing that politics can make changes.

I have throughout my career not felt so many challenges except the scientific ones, probably because I have been working in Norway most of the time. Norway has an excellent social security system and is one of the countries in the world with the most well-developed equal opportunities systems. I became mother at a late age and with the social security system in Norway my husband and I were given 54 weeks of paid leave that we shared. After ending the period of maternity leave, my son was guaranteed a child day care that we got very close to our home.

Words of Wisdom

Follow your dreams and believe that you can achieve what you want to do.

Mari-Ann Einarsrud in 2008. Photo credit: Ole Morten Melgård.

Profile 28

Katherine T. Faber

Simon Ramo Professor of Materials Science
California Institute of Technology
MC 138-78
Pasadena, CA 91125
USA

e-mail: ktfaber@caltech.edu
telephone: (626) 395-4448

Katherine Faber as a
new professor at
Northwestern University.
Credit: Northwestern
University.

Tags

❖ Academe
❖ Domicile: USA
❖ Nationality: American
❖ Caucasian
❖ Children: 2

Publication/Invention Record

>150 publications: h-index 24 (Web of Science)
Edited 3 books

Proudest Career Moment (to date)

My greatest achievements are my students! I enjoy working with them, having them challenge me, and watching them grow to be my colleagues with successful careers of their own. A recent accomplishment, which has also been particularly rewarding, is a program that I, with Dr. Francesca Casadio, established between Northwestern University and the Art Institute of Chicago in conservation science. We are using our engineering and science know-how to study and contribute to preservation strategies for objects of cultural heritage from one of the great museums of the world. Little did I imagine as a student that an engineering degree would allow me to work on priceless pieces of art! (Modified from http://www.engineergirl.org/Engineers/interviews/7080.aspx.)

Academic Credentials

Ph.D. (1982) Materials Science, University of California, Berkeley, California, USA.
M.S. (1978) Ceramic Science, The Pennsylvania State University, University Park, Pennsylvania, USA.
B.S. (1975) Ceramic Engineering, Alfred University, Alfred, New York, USA.

Successful Women Ceramic and Glass Scientists and Engineers: 100 Inspirational Profiles. Lynnette D. Madsen.
© 2016 The American Ceramic Society and John Wiley & Sons, Inc. Published 2016 by John Wiley & Sons, Inc.

Research Expertise: fracture of brittle materials; ceramics and ceramic composites for structural applications; cellular ceramics and processing methods for porous materials; thermal and environmental barrier coatings for engine components needed for enhanced efficiency in the next generation of engines such as microturbines and industrial gas turbines; cultural heritage science, with emphasis on porcelains and jades

Other Interests: working with professional societies to build community and advance the discipline; developing career development and leadership programs for female faculty in STEM disciplines

Key Accomplishments, Honors, Recognitions, and Awards

* John Jeppson Award, American Ceramic Society, 2015
* American Academy of Arts and Sciences, 2014
* Distinguished Life Member, American Ceramic Society, 2013
* Toledo Glass and Ceramics Award, Michigan/Northwest Ohio Section of the American Ceramic Society, 2012
* National Science Foundation American Competitiveness and Innovation Fellow and Creativity Extension Award, 2010
* The American Ceramic Society, President, 2006–2007
* Scientific Advisory Committee of the Advanced Photon Source at Argonne National Lab, 2005–2007
* Northwestern University Women's Center 20th Anniversary Faculty Award, 2007
* The John F. McMahon Memorial Lecture Award, Alfred University, 2006
* Featured in Changing Our World, True Stories of Women Engineers by S.E. Hatch, ASCE Press, 2006
* Charles L. Hosler Alumni Scholar Medal, College of Earth and Mineral Sciences, The Pennsylvania State University, 2004
* ISI Highly Cited Author in Materials, 2003

Katherine Faber (seated, third from left) as President of the American Ceramic Society joins other officers and directors for 2006 Annual ACerS Awards Banquet. Credit: American Ceramic Society.

* Fellow, ASM International, 2003
* University Materials Council, Chair, 2001–2002
* NSF Creativity Extension Award, 2001
* NAE Gallery of Women Engineers: http://www.engineergirl.org/CMS/WomenEngineers/2985/5949.aspx, 2000
* YWCA Achievement Award for Education, 1997

- World Academy of Ceramics, 1996
- Centennial Fellow, College of Earth and Mineral Sciences, The Pennsylvania State University, 1996
- National Science Foundation Faculty Award for Women in Science and Engineering, 1991–1996
- Society of Women Engineers Distinguished Engineering Educator Award, 1995
- Department of Materials Science and Engineering Teacher of the Year, Northwestern University, 1992
- Fellow, American Ceramic Society, 1990
- Ceramic Education Council, President, 1989–1990
- McCormick School of Engineering and Applied Science, Faculty Teaching Honor Roll, Northwestern University, 1990
- Society for Automotive Engineers Ralph R. Teetor Educational Award, 1989
- National Science Foundation Presidential Young Investigator Award, 1984–1989
- Alumni Fellow, Alfred University, 1988
- American Society for Engineering Education, North Central Section, AT&T Foundation Award, 1986
- Selected for and participated in the first class of the Defense Science Study Group, 1985–1988
- IBM Faculty Development Award, 1984–1986
- Charles E. MacQuigg Award for Outstanding Teaching, College of Engineering, The Ohio State University, 1985

Biography

Early Life and Education

Katherine Faber was the fifth of five children of Robert and Agnes Faber in Buffalo, New York. Her father, who had to cut short his aeronautical engineering studies at the University of Detroit due to the Great Depression, instilled in his daughter a love for science and technology. He gladly drove her to the Buffalo Museum of Science many Saturday mornings for classes in astronomy and provided her with a telescope for her personal stargazing. By high school she had developed an interest in chemistry, and ultimately chose to attend the New York State College of Ceramics at Alfred University. There she studied ceramic engineering, after having seen how satisfying engineering problem solving could be compared to organic chemistry.

Her graduate work began at the Pennsylvania State University where she earned a Master of Science in Ceramic Science with Guy Rindone on phase separation in glasses. Ready for an academic break, she worked for the Carborundum Co. in Niagara Falls, New York as a development engineer. Here she discovered her favorite ceramic—silicon carbide—one of the hardest materials known and one so versatile that it is used for structural, wear, electronic, thermal, and nuclear applications. She has continued to include silicon carbide in her research portfolio over her 30+-year career.

Realizing that she was not suited for a long-term career in industry, she pursued a Ph.D. in materials science and engineering at the University of California, Berkeley working under the tutelage of Anthony G. Evans. It was during this period that she realized that she aspired to an academic career, and was most interested in problems having to do with fracture and material design of fracture-resistant ceramics.

Career History

Katherine's first academic position was at Ohio State University. She was Assistant, then Associate Professor of Ceramic Engineering through 1987, at which point she joined the materials science and engineering faculty at Northwestern University. In addition to her faculty duties, she has served a 5-year term as Associate Dean for Graduate Studies and Research in the McCormick School of Engineering and Applied Science where she oversaw more than $25M in faculty research. Subsequently, she completed a 5-year term as department chair of one of the top three materials departments nationwide. During her term as department chair, he served as chair of the University Materials Council (2001–2002), a group composed of department chairs, directors, and leaders of academic materials programs in U.S. and Canadian universities.

Katherine Faber in the classroom at Northwestern University in a course on mechanical behavior of solids. Credit: Northwestern University.

In 2004, she established a university-wide collaborative program in conservation science with the Art Institute of Chicago with support from the Andrew W. Mellon Foundation. She co-founded and co-directs the Northwestern University–Art Institute of Chicago Center for Scientific Studies in the Arts, the first center of its kind, which provides a hub where individuals from other museums or cultural institutions can make use of Center facilities and expertise to study their collections. In July 2014, she joined the faculty of the California Institute of Technology.

Katherine Faber addresses 2006 American Ceramic Society Business Meeting as its new president.

She is an active member of the American Ceramic Society, in which she served as its president in 2006–2007, ASM International, American Society for Engineering Education, Sigma Xi, Society of Women Engineers, and the American Association for the Advancement of Science.

3 Most Cited Publications

Title: Crack deflection processes. 1. Theory
Author(s): Faber, KT; Evans, AG
Source: Acta Metallurgica; volume: 31; issue: 4; pages: 565–576; published: 1983
Times Cited: 903 (Web of Science)

Title: Crack-growth resistance of microcracking brittle materials
Author(s): Evans, AG; Faber, KT
Source: Journal of the American Ceramic Society; volume: 67; issue: 4; pages: 255–260; published: 1984
Times Cited: 327 (Web of Science)

Title: Crack deflection processes. 2. Experiment
Author(s): Faber, KT; Evans, AG
Source: Acta Metallurgica; volume: 31; issue: 4; pages: 577–584; published: 1983
Times Cited: 323 (Web of Science)

ResearcherID (B-6741-2009)

Challenges

One of the challenges, yet thrills, of an academic career is that you are always being tested. With each lecture given, each proposal written, and each paper submitted, I am pushed to do my best. Strange as this seems, it is both exhausting and invigorating.

On being a woman in this field . . .

Perhaps the greatest challenge of being a woman in academia has been how to balance my career with raising a family. I have had to make compromises on both fronts. When my children were small, I would do engineering-related activities in their classrooms. Such activities allowed me to kill two birds with one stone—educate children about science and engineering AND spend some special time with my children. As my children grew, and when not in school, they accompanied my husband and me on our professional travels to meetings and conferences around the world. We all gained from these trips in mixing vacation with our careers. (Modified from http://www.engineergirl.org/Engineers/interviews/7080.aspx.)

Words of Wisdom

- Find a professional home where your colleagues are a source of inspiration and collaboration. This makes for a much richer career.

- My mentor and dear friend Julia Weertman once said to me, "You have to walk on both sides of the street at once." By this she meant that there will be occasions when you are denied opportunities because of your gender, yet there will be other occasions when being a female will prove to be advantageous. Both of these situations can be unpleasant and even infuriating, but it's good to learn how best to deal with both.
- No matter how much career planning you do, it is important to consider unplanned opportunities. Take advantage of the serendipitous.

Profile 29

Monica Ferraris

Professor
Department of Applied Science and
Technology (DISAT)
Politecnico di Torino
24, corso Duca degli Abruzzi
10129 Torino
Italy

Monica Ferraris in Washington, DC in
1989, at her first conference in the
United States.

e-mail: monica.ferraris@polito.it
telephone: (+39) 011 090 4687

Tags

❖ Academia
❖ Domicile: Italy
❖ Nationality: Italian
❖ Caucasian
❖ Children: 2

Birthplace
Casale Monferrato (AL), Italy
Born
September 27, 1961

Publication/Invention Record

>150 publications: h-index 17 (Web of Science)
14 patents, 13 book chapters

Memorable Career Moment (to date)

At the very beginning of my career (1987), I was invited as a visiting researcher at
Rennes University in France to work on the synthesis of fluoride glasses for ultra-low-
loss optical fibers within the group of Professor J. Lucas. He was the world expert in this
field and I felt extremely honored to work with him and his group. I enjoyed each and
every moment of my stay in Rennes; I still have good friends from there. Thanks to this
experience and other very happy work experiences abroad, I encourage my students and
young colleagues to go abroad for a while . . .

Another memorable moment was in 2007, during our annual group meeting at
Politecnico di Torino. We had been working on silver nanocluster doped silica thin films
for photonics for years and the results had been quite disappointing. Some colleagues had

Successful Women Ceramic and Glass Scientists and Engineers: 100 Inspirational Profiles. Lynnette D. Madsen.
© 2016 The American Ceramic Society and John Wiley & Sons, Inc. Published 2016 by John Wiley & Sons, Inc.

Monica Ferraris with her colleague and friend Andrea celebrating the success of torsion tests (May 2013). (See insert for color version of figure.)

just shown their results on antibacterial effects of silver ions . . . and then the idea suddenly popped out: "why don't we test if these silver nanocluster doped silica thin films are antibacterial?" They were. We patented this concept in 2008. In addition, we submitted two projects to work on these new antibacterial films and they were both funded for a total of 3.5 million euro!

More recently, I have been contacted by two companies in a time span of 3 days . . . both willing to make what I have been studying for 15 years! Two completely different subjects (joining of dissimilar materials and products for vitrified wastes); both equally challenging! When a company is interested in using what I have thought and done just on a laboratory scale . . . well, it is pure joy!

Finally, I started by chance an activity on joining of materials for nuclear plants in 1992. Now the joining materials we developed are among the very few able to work in a nuclear environment and the way we proposed to test their mechanical strength is going to be adopted by ASTM (formerly known as the American Society for Testing and Materials). I still remember many key points—the moment when I was told that joined materials resisted after neutron irradiation, and when we experimentally validated their theoretically predicted mechanical strength.

Last but not least, I had the pleasure of obtaining the above with several colleagues that are also good friends. I cannot ask for more in my professional life.

Academic Credentials

M.S. (1985) Chemistry (summa cum laude), University of Torino, Turin, Italy.
B.Sc. (1980) Maturita' Scientifica, Liceo Scientifico N.S.I. Palli, Casale Monferrato, Alessandria, Italy.

Research Expertise: glasses, ceramics, and composites for energy production, photonics, biomedical applications, and waste management

Other Interests: problem solving for companies—I love when colleagues from companies ask me to solve a materials-related issue and I'm able to help!

Key Accomplishments, Honors, Recognitions, and Awards

- Co-Editor-in-Chief, International Journal of Applied Ceramic Society, since 2015
- Associate Editor, International Journal of Applied Ceramic Society, 2014 to present

- Academician of the World Academy of Ceramics, elected 2014
- "Global Star Award," American Ceramic Society, Daytona Beach, FL, USA, 2011

Biography

Early Life and Education

I did primary and secondary schools in a small town between Torino and Milano (Casale Monferrato, Italy), and then moved to Torino to start my studies at the university. I have always loved science and my first choice was biology. My idea was to work in a lab and discover new viruses or bacteria, but a friend of mine, 4 years older than me, told me that there wasn't a good school of biology at Torino University and it would have been better for me to study chemistry. I will never know whether he was right or not, but I followed his advice and I've never regretted it!

I fell in love with solid-state chemistry and spectroscopy from the very beginning. I received a M.S. in Chemistry in 1985 from the University of Torino, Italy, summa cum laude. I decided to start Physics immediately after, but I was offered a position at the Italian Telecom Research Centre (CSELT-Torino) and I accepted it. A Ph.D. was mostly unknown in Italy at that point.

Career History

In 1985, I joined the Italian Telecom Research Centre (CSELT-Torino) to work on electron paramagnetic resonance spectroscopy on silica fibers, something I'd done during my master's thesis. Then I was asked to work on the synthesis of fluoride glasses for ultra-low-loss optical fibers, which I did for 5 years. The discovery of doped silica fiber-based optical amplifiers made this activity less interesting and some disagreements with my boss made me decide to quit CSELT. It was a big and painful step in my career.

In 1990, I moved to Fiat Research Centre (CRF-Torino) where I worked on scanning electron microscopy of metal and ceramic matrix composites for automotive applications, with a tight cooperation with Politecnico di Torino. It was an interesting experience, and I kept spending 1–2 days per week at CRF also after quitting it in 1992. We still have research projects together today.

I joined Politecnico di Torino as a researcher in 1991. My boss told me: "You have expertise on glasses, bring it here." This is how I started again a new career and research subject. Since his research group was active on metal matrix composite materials, I thought about starting new research on metal-reinforced glass matrix composites. I've found some papers on bioactive glasses, so I started preparing titanium particle-reinforced bioactive glass matrix composites for bone substitution!

Politecnico di Torino is a very stimulating place and I enjoyed from the very beginning the freedom of working on subjects of my choice. I've started several other research projects (joining, vitrification of wastes, fiber optics again, etc.). The main issue was the lack of funds, a big problem I've never ever faced before at CSELT or CRF! My experimental research slowly became a search for funding.

I've got my Associate Professorship in 2000 and Full Professorship in 2005 (Professor of Materials Science and Technology); my beloved boss, Professor Pietro Appendino, retired in 2008 and I've been group leader since then.

I enjoy gardening, skiing, tennis, horse riding, sailing, jogging, and cooking in my spare time.

The EU Researchers' Night 2013, with all flags of nationalities represented in Monica Ferraris' group.

3 Most Cited Publications

Title: Oxidation protective multilayer coatings for carbon–carbon composites
Author(s): Smeacetto, F; Salvo, M; Ferraris, M
Source: Carbon; volume: 40; issue: 4; pages: 583–587; article number: PII S0008-6223 (01)00151-8; published: 2002
Times Cited: 99 (from Web of Science)

Title: Multilayer coating with self-sealing properties for carbon–carbon composites
Author(s): Smeacetto, F; Ferraris, M; Salvo, M
Source: Carbon; volume: 41; issue: 11; pages: 2105–2111; published: 2003
Times Cited: 70 (from Web of Science)

Title: Glass–ceramic seal to join Crofer 22 APU alloy to YSZ ceramic in planar SOFCs
Author(s): Smeacetto, F; Salvo, M; Ferraris, M; et al.
Source: Journal of the European Ceramic Society; volume: 28; issue: 1; pages: 61–68; published: 2008
Times Cited: 68 (from Web of Science)

Challenges

My everyday challenges are so many . . . first of all to provide enough funds to pay for postdocs' salaries and research equipment . . . ; if you mean a more scientific challenge, well, I would like to find a reliable method to measure the shear strength of joined ceramics.

On being a woman in this field . . .

When I started at CSELT, I was one of two women out of 80 people in that division. The other woman was the secretary. Once, I answered the phone in my boss's office and the man on the phone apologized for having dialed the wrong number. When I told him the number was the right one and my boss was just outside, he answered: "Impossible, there are no women working with him!"

I cannot say I've been discriminated during my career just because I'm a woman. I've felt more discriminated because of not belonging to "power" groups.

For sure, the worst period was when I had my two children. I've been lucky enough to have two very easy pregnancies and I worked in the lab or office until the last day for both. However, I remember with sadness when they were sick and I had to stay home with them. Or worse, leaving and knowing they were sick at home without me.

Monica Ferraris giving a lecture in 1993, a few weeks before her daughter was born.

Words of Wisdom

I borrowed this sentence from S. Sandberg. "Better done than perfect.." I agree with it 100%.

Profile 30

Catherine (Cathy) Patricia Foley

Chief, Materials Science and Engineering
Commonwealth Scientific and Industrial Research
Organisation (CSIRO)
Bradfield Road
West Lindfield, NSW 2070
Australia

e-mail: Cathy.Foley@csiro.au
telephone: +61 2 9413 7413

Tags

❖ Administration and
   Leadership
❖ Government
❖ Domicile: Australia
❖ Caucasian
❖ Children: 6 (including 3
   stepchildren)

Birthplace
New South Wales (NSW), Australia
Born
October 10, 1957

Publication/Invention Record

>75 publications: h-index 18 (Web of Science)
1 patent, editor of 1 book, 4 book chapters

Academic Credentials

Ph.D. (1984) Physics, Macquarie University, Sydney, NSW, Australia.
B.Sc. (1980) Physics, Macquarie University, Sydney, NSW, Australia.

Research Expertise: superconductivity, nanotechnology, nitride semiconductors, magnetic materials, combinatorial and high-throughout techniques, and developing materials informatics

Other Interests: communicating science to the public, science education, encouraging women in science careers, nuclear disarmament

Key Accomplishments, Honors, Recognitions, and Awards

• Rapid Communications Editor of Superconductor Science and Technology, 2008 to
  present
• Editorial Board Member of the Institute of Physics (IOP) journal Superconducting
  Science and Technology, UK, 2003 to present

Successful Women Ceramic and Glass Scientists and Engineers: 100 Inspirational Profiles. Lynnette D. Madsen.
© 2016 The American Ceramic Society and John Wiley & Sons, Inc. Published 2016 by John Wiley & Sons, Inc.

- Australasian Institute of Mining and Metallurgy (AUSIMM) Mineral Industry Operating Technique Award (MIOTA) for LANDTEM^{TM}, which accurately measures the magnetic (B) field, rather than the rate of decay of induced eddy currents used by conventional coil systems, 2011
- Telstra National and NSW Business Women of Year—Innovation Award, 2009
- President of the Australian Institute of Physics, 2007–2009
- CSIRO Divisional Awards: 2002, 2003, 2004, and 2008
- Institute of Electrical and Electronics Engineers (IEEE) "Distinguished Lecturer" Award, 2007–2008
- CSIRO Medal for Research Achievement, for LANDTEM^{TM}, 2007
- Macquarie University Alumni Award for Distinguished Service, 2007
- Vice President of the Australian Institute of Physics, 2004–2006
- Elected Fellow of the Australian Institute of Physics, 2004
- National Treasurer of the Australian Institute of Physics, 2001–2004
- A Public Service Medal in the Australia Day Honours for service to physics and promotion of women in science, 2003
- Eureka Prize for the Promotion of Science, 2003
- Elected Fellow of the UK Institute of Physics, 2000
- International Union for Vacuum Science, Technique and Applications (IUVSTA) Welch Foundation Scholarship, 1984
- Commonwealth Postgraduate Research Award, 1981–1984
- CSIRO Vacation Scholarship, 1980
- Plessey Prize for Proficiency in Materials Science, 1980
- NSW Education Department University Scholarship, 1977–1980
- NSW Department of Education High School Scholarship, 1975

Biography

Early Life and Education

Catherine Patricia Foley was born in Sydney in the NSW region of Australia in the late 1950s. Early influences were her mother, an accountant, who died when Cathy was 9 years old, her elder brothers and sister, and her father who worked as an architect.

She attended St. Anthony's Primary School and Santa Sabina College for secondary school. For her tertiary education, she attended Macquarie University earning a Diploma of Education for High School Physics in 1979. Initially, she had thought about being a school teacher, since only really smart people were scientists and she did not know any. However, she loved nature and was naturally gifted at science at school. So, she continued at Macquarie University and was awarded a Bachelor of Science majoring in physics with first-class honors in 1980. Moreover, when the opportunity arose, she enrolled as a doctoral student in physics—she thought she could make a difference. While writing up her dissertation, she spent half a year on a scholarship as a Research

Fellow in the Department of Electrical Engineering at Oregon State University in the United States. She was awarded a Ph.D. in 1984.

Along the way, Cathy realized that she was dyslexic; she still struggles today to recognize words. However, she capitalized on her strengths—mathematics and laboratory work—and did not let her disability deter her from pursuing her interests.

Career History

Upon graduation, Dr. Foley joined the CSIRO Division of Materials Science as a National Research Fellow. In subsequent years, she received a series of promotions: in 1991 to Senior Research Scientist, 1996 to Principal Research Scientist, 2000 to Senior Principal Research Scientist, and 2008 to Chief Research Scientist.

Her research expertise spans superconducting materials and electronics and their applications. She has held various project responsibilities:

1995–1996: Acting Project Leader, Superconducting Sensors and Technology Project

1996–2009: Project Leader, Superconducting Devices and Applications Project

2003–2004: Stream Leader, Applied Quantum Systems

2005–2007: Theme Leader, Nanoscale Manufacturing

2007: Research Program Leader

Then, she served as Coordinator, Advanced Materials Transformational Capability Platform (2008–2011) and Acting Deputy Chief-Research (in 2009). In 2011, she was appointed as Chief of the Division of Materials Science and Engineering.

She was a science correspondent for the ABC 2BL radio evening program (from 1988 to 1993) and since 1989 has been an Honorary Research Associate of the School of Physics, University of Sydney.

3 Most Cited Publications

Title: Optical bandgap of indium nitride
Author(s): Tansley, TL; Foley, CP
Source: Journal of Applied Physics; volume: 59; issue: 9; pages: 3241–3244; published: May 1, 1986
Times Cited: 493 (from Web of Science)

Title: Electron mobility in indium nitride
Author(s): Tansley, TL; Foley, CP
Source: Electronics Letters; volume: 20; issue: 25–26; pages: 1066–1068; published: 1984
Times Cited: 90 (from Web of Science)

Title: Pseudopotential band structure of indium nitride
Author(s): Foley, CP; Tansley, TL
Source: Physical Review B; volume: 33; issue: 2; pages: 1430–1433; published: January 15, 1986
Times Cited: 69 (from Web of Science)

Written sources consulted in preparing this profile

http://www.csiropedia.csiro.au/display/CSIROpedia/Foley,+Catherine+Patricia, accessed January 31, 2015.

http://www.csiro.au/Organisation-Structure/Divisions/CMSE/CathyFoley/Distinctions-and-awards.aspx, accessed January 31, 2015.

http://www.womensagenda.com.au/talking-about/editor-s-agenda/six-kids-later-and-managing-a-team-of-870-cathy-foleys-mission-to-get-more-women-into-leadership/201307112516#.VM1QVi6UJ-8, accessed January 31, 2015.

http://www.abc.net.au/science/articles/2011/11/30/3380173.htm, accessed January 31, 2015.

https://theconversation.com/profiles/cathy-foley-1066, accessed January 31, 2015.

Disclaimer

Profile 31

Liesl Folks

Dean of Engineering
University at Buffalo (UB)
208 Davis Hall
Buffalo, NY 14260
USA

e-mail: lfolks@buffalo.edu
telephone: (716) 645-2771

Liesl Folks during her early
days at IBM when she was
about 33 years old.

Birthplace
Australia
Born
August 1, 1967

Publication/Invention Record

>50 publications: h-index 21 (Web of Science)
14 patents

Tags
❖ Administration and
  Leadership
❖ Academe
❖ Domicile: USA
❖ Nationality: American
❖ Caucasian
❖ Children: 2

Proudest Career Moment (to date)[1]

Crafting the 2015 Diversity Initiative of the Engineering Deans Council of the American Society for Engineering Education, with deans from across the country. In that effort, engineering deans from more than 100 universities in North America committed their schools to building more diverse and inclusive programs, with the objective of improving the culture of engineering for women and minorities in the broader workforce. "While gains have been made in the participation of Hispanics, African Americans, women and other underrepresented groups in engineering, significant progress is still needed to reach all segments of our increasingly diverse society." "We must promote engineering education to those historically underrepresented, provide an experience that's equitable and inclusive, and improve the broader engineering culture to fully engage future generations."

[1] Source: http://www.buffalo.edu/news/releases/2015/08/010.html#sthash.RBuao1Su.dpuf, released August 2015, accessed August 2015.

Successful Women Ceramic and Glass Scientists and Engineers: 100 Inspirational Profiles. Lynnette D. Madsen.
© 2016 The American Ceramic Society and John Wiley & Sons, Inc. Published 2016 by John Wiley & Sons, Inc.

Academic Credentials

Ph.D. (1994) Physics, University of Western Australia, Perth, Australia.
M.B.A. (2004), Cornell University, Ithaca, New York, USA.
B.S. (1989) Physics, University of Western Australia, Perth, Australia.

Research Expertise: thin films, nanotechnology, physics, metrology, magnetics, magnetic materials

Other Interests: mentoring

Key Accomplishments, Honors, Recognitions, and Awards

- President of the IEEE Magnetics Society, 2013–2014
- National Academy of Sciences panel for the Triennial Review of the National Nanotechnology Initiative, 2012–2013
- AVS Excellence in Leadership recognition, 2012
- President-Elect of the IEEE Magnetics Society, 2011–2012

Biography[2]

Early Life and Education

Liesl Folks, a native of Australia, earned a B.Sc. in Physics with Honours (in 1989) and Doctorate in Physics (in 1994) both from the University of Western Australia (UWA). A decade later (in 2004), she obtained an M.B.A. with Distinction from Cornell University.

Career History

After completing her doctorate, she stayed at UWA, first as a Research Associate, Special Research Center for Advanced Mineral & Materials Processing (1994–1995), and then as a Research Fellow, Faculty of Science (1995–1997), where she enjoyed generous external grant funding.

In 1997, she joined the IBM Almaden Research Center as a Senior Postdoctoral Scientist and then as a Research Staff Member (1999–2003). When the IBM storage division was sold in 2003, she became Research Staff Member at the Hitachi GST San Jose Research Center (2003–2008). In 2008, she was promoted to Manager, Media Advanced Technologies (2008 to 2012), where she oversaw and managed technology transfer from Research to Advanced Development for Recording Media for all hard disk drive (HDD) products and all platforms.

Early in 2013, she became Dean of Engineering and Applied Sciences at the University at Buffalo. Dr. Folks' experience in academia includes roles as a teacher, researcher, mentor, and advisor. She has taught undergraduate and graduate students,

---

[2] Sources included: http://www.buffalo.edu/news/releases/2012/10/13761.html, released Oct. 2012; accessed Jan. 2015.

supervised postdoctoral researchers, student researchers, and interns, and served as a dissertation advisor to doctoral students. She has had many university collaborations that include serving as an advisor to Cornell University's Center for Nanoscale Systems and collaborating with scientists at Oxford University, UC Santa Barbara, Ohio State, University of Colorado, and the Rochester Institute of Technology.

She is active with her professional community. For example, she has been an organizer of several academic conferences, including the 2018 International Conference on Magnetism, which she is helping to bring back to the United States for the first time in 30 years. She served on a committee for the National Academy of Sciences that conducted the congressionally mandated Triennial Review of the National Nanotechnology Initiative. In addition, she regularly reviewed for the National Science Foundation, writing proposal reviews and serving on-site visit teams covering many U.S. research centers in her field of expertise. She has been a member of AVS, IEEE Magnetics Society, the American Institute of Physics, and the Australian Institute of Physics. She has an extensive network of professional connections and has been an invited lecturer at many universities.

When Liesl is not working, she enjoys the outdoors—hiking, mountaineering, snowboarding, swimming, skiing, and cycling. In addition, she is an avid reader of both fiction and non-fiction and likes to cook. Also, she spends time "messing about" with her children, building with Lego, or playing games.

Liesl Folks enjoying a Snow Day in Buffalo, 2014. (See insert for color version of figure.)

3 Most Cited Publications

Title: Monodisperse FePt nanoparticles and ferromagnetic FePt nanocrystal superlattices
Author(s): Sun, SH; Murray, CB; Weller, D; Folks, L; Moser, A
Source: Science; volume: 287; issue: 5460; pages: 1989–1992; doi: 10.1126/science .287.5460.1989; published: March 17, 2000
Times Cited: 3942 (from Web of Science)

Title: High K-u materials approach to 100 Gbits/in^2
Author(s): Weller, D; Moser, A; Folks, L; Best, ME
Source: IEEE Transactions on Magnetics; volume: 36; issue: 1; pages: 10–15; doi: 10.1109/20.824418; part: Part 1; published: January 2000
Times Cited: 927 (from Web of Science)

Title: Epitaxial growth and properties of ferromagnetic co-doped TiO_2 anatase
Author(s): Chambers, SA; Thevuthasan, S; Farrow, RFC; Marks, RF; Thiele, JU; Folks, L; Samanti, MG; Kellock, AJ; Ruzycki, N; Ederer, DL; Diebold, U
Source: Applied Physics Letters; volume: 79; issue: 21; pages: 3467–3469; doi: 10.1063/1.1420434; published: November 19, 2001
Times Cited: 330 (from Web of Science)

Challenges

I was the first woman to serve as President-Elect of the Institute of Electrical and Electronics Engineers' Magnetics Society; don't let breaking new ground deter you.

On being a woman in this field . . .

The President of the University at Buffalo noted that even though engineering is a traditionally male-dominated field, women such as Liesl Folks continue to break down barriers. Dr. Folks joins many other women who serve as engineering deans at institutes of higher learning in both the United States and Canada (e.g., Stanford, Alfred, Harvard, Yale, Purdue, Texas A&M, Florida, Toronto, and Waterloo).

Words of Wisdom[3]

"Say yes to opportunities when they present themselves!" I am often surprised at the opportunities that people say "no" to, small and large, and how it limits their career trajectories. Even small opportunities can introduce us to new people and new perspectives, and these can so easily turn into the next big opportunity, especially if we have made a positive impression.

Liesl Folks at work, at the University at Buffalo. Credit: Onion Studio.

---

[3] Taken from https://www.avs.org/AVS/files/58/5869681b-7231-4288-a033-84fda9a2963e.pdf, accessed January 2015.

Profile 32

Katharine G. Frase

Vice President and Chief Technology Officer
IBM Global Public Sector
IBM
294 Route 100
Somers, NY 10589
USA

Katharine Frase
in 1999.

e-mail: frase@us.ibm.com
telephone: (914) 766-2347

Tags

❖ Administration and Leadership
❖ Industry
❖ Domicile: USA
❖ Nationality: American
❖ Caucasian
❖ Children: 4

Birthplace
Washington, DC, USA
Born
December 19, 1957

Publication/Invention Record

15 publications: h-index 8
2 patents

Proudest Career Moment (to date)

Wow, this is a hard question, because I have done so many different things in my career . . . early on, it was exciting to be part of the High Tc Superconductor mania . . . then later when I worked in process development, it was probably seeing my process transition to high-volume manufacturing . . . then later it was transforming a small business inside IBM, bringing value to the customers as well as more revenue to my company . . . and now? I am very proud that I have been part of proving that using data and analytics can help us run physical systems (water, traffic, energy) more effectively.

Academic Credentials

Ph.D. (1983) Materials Science and Engineering, University of Pennsylvania, USA.
A.B. (1979) Chemistry, Bryn Mawr College, Bryn Mawr, PA, USA.

Successful Women Ceramic and Glass Scientists and Engineers: 100 Inspirational Profiles. Lynnette D. Madsen.
© 2016 The American Ceramic Society and John Wiley & Sons, Inc. Published 2016 by John Wiley & Sons, Inc.

Research Expertise: materials engineering, high-temperature superconductors, mechanical properties, composites

Other Interests: communicating clearly and adding value

Key Accomplishments, Honors, Recognitions, and Awards

- Contributor to several NRC studies on materials, innovation, supply chain, and technology topics, 2004 to present
- Board of Assessment for the Army Research Laboratory, 2008–2011
- Chair of the National Materials Advisory Board, 2005–2009
- NRC Review Panel for NIST Laboratory of Materials Science, 2001–2005
- Elected Member National Academy of Engineering for engineering contributions, including the use of lead-free materials, to the development of electronic packaging materials and processes, 2006
- Elected to the IBM Academy of Technology, 2000

Biography

Early Life and Education

Katharine Frase received her first degree in chemistry from Bryn Mawr College in 1979. She then moved just slightly south to Philadelphia and was awarded a doctorate in materials science and engineering from the University of Pennsylvania 4 years later. She studied at the University of Uppsala, Sweden as part of her doctorate research. Her thesis topic was the structure and electrochemical properties of solid-state ionic conducting materials for batteries.

Career History

She has made her career at IBM. She began her career in IBM Research, focused on composite materials and high-temperature superconducting materials. She subsequently spent over a decade in IBM Microelectronics, in roles that included process development, quality and reliability, product development, and vendor management across IBM's portfolio of ceramic and organic chip carriers, assemblies, and test. She has led the technical and business strategy for IBM's software business and completed corporate assignments on technology assessment and strategy. She has served as Vice President of Industry Solutions Research, working across IBM Research on behalf of their clients, to create transformational industry-focused solutions, including the application of IBM Watson technologies to business applications and the realization of Smarter Planet solutions.

Katharine Frase in 2011.

Since March 2013, Dr. Frase has been Vice President and Chief Technology Officer of IBM Public Sector at International Business Machines Corporation. According to her Bloomberg Businessweek profile, she provides "thought leadership for IBM and its customers on innovation and strategic transformation specific to government, education, life sciences, health care and cities, driving the creation of new solutions."

In 2006, she was elected as a member of the (U.S.) National Academy of Engineering. She is a member of the IBM Academy of Technology and sits on numerous external committees and boards.

3 Most Cited Publications

Title: Phase compatibilities in the system Y_2O_3-BaO-CuO
Author(s): Frase, KG; Clarke, DR
Source: Advanced Ceramic Materials; volume: 2; issue: 3B; pages: 295–302; published: July 1987
Times Cited: 125 (from Web of Science)

Title: Phase compatibilities in the system Y_2O_3-BaO-CaO at 950°C
Author(s): Frase, KG; Liniger, EG; Clarke, DR
Source: Journal of the American Ceramic Society; volume: 70; issue: 9; pages: C204–C205; published: September 1987
Times Cited: 98 (from Web of Science)

Title: Environmental and solvent effects on yttrium barium cuprate ($Y_1BA_2CU_3O_x$)
Author(s): Frase, KG; Liniger, EG; Clarke, DR
Source: Advanced Ceramic Materials; volume: 2; issue: 3B; pages: 698–700; published: July 1987
Times Cited: 33 (from Web of Science)

Challenges

I think the biggest challenge for any of us is to be clear on what it is we want to do. Once we decide that the "doing" is just a matter of setting a plan in place and executing. Well, it's not quite that simple, we have to be realistic in our goals . . . but truly the hardest part is deciding what we want!

On being a woman in this field . . .

When I began my career, there were quite a few women in the lower ranks, but very few in the upper ranks, which made it hard to see them as role models: There were so few of them that it was hard to separate their individual personalities from the behaviors that made them successful. There were also essentially no women in management who had kids . . . the women managers all went "back to the bench" once they had children. I chose to challenge that, to prove that it could be done, in case anyone after me wanted to try it. Certainly, the situation is much better now, a much better understanding of work life balance and the different styles that women bring to the workplace.

Refuse to behave "like one of the guys."

Words of Wisdom

My Ph.D. advisor said: It is important to do good science, but it is even more important to be able to communicate about it . . . and I think that is really true. To be able to explain to a variety of audiences why our technical work "matters" is really an important skill, and can help guide us to the work that will have the most societal impact as well. Also, the ability to present the facts using vocabulary that anybody can understand forces people to take you seriously.

Profile 33

María Verónica Ganduglia-Pirovano

Scientific Researcher
Instituto de Catálisis y Petroleoquímica/Institute of
Catalysis and Petrochemistry
Consejo Superior de Investigaciones Científicas (CSIC)/
The Spanish National Research Council
Marie Curie 2
28149 Madrid
Spain

e-mail: vgp@icp.csic.es
telephone: +34-91-585-4631

Maria Verónica Ganduglia-Pirovano on the day she received her doctorate degree, 1989.

Birthplace
Buenos Aires, Argentina
Born
November 17, 1962

Publication/Invention Record
>60 publications: h-index 29 (Web of Science)

Tags
❖ Government
❖ Domicile: Spain
❖ Nationality: German and Argentinian
❖ Caucasian
❖ Children: 2

Important/Memorable Career Moment (to date)

In 1989 obtaining my Ph.D. from the University of Stuttgart and Max-Planck-Institut für Festkörperforschung in Stuttgart, Germany, was a very important personal moment in my career. The moment occurred 3½ years after I had relocated from Argentina to Germany, and by the time of my Ph.D. defense, I had one daughter (2½ years old) and a second one on its way. Prof. Peter Fulde was a great scientific father figure; the German *Doktorvater* seems to fit him perfectly.

A very memorable moment was my first invited talk to a large international conference, namely, the 13th International Conference on Vacuum Ultraviolet Radiation Physics (VUV-XIII) in 2001, Trieste, Italy. Many internationally renowned scientists in the field of experimental and theoretical surface science attended that conference. The feeling of belonging to a scientific community that emerged at that time is still strong; belonging uncertainty may contribute to the underrepresentation of women in science fields.

Successful Women Ceramic and Glass Scientists and Engineers: 100 Inspirational Profiles. Lynnette D. Madsen.
© 2016 The American Ceramic Society and John Wiley & Sons, Inc. Published 2016 by John Wiley & Sons, Inc.

My most memorable moment as a professor (so far) came when I got my first students' evaluation of teaching after my Quantum Chemistry course for chemists at the Free University in Berlin in 2007. Physical Chemistry and Quantum Chemistry courses have a reputation of being difficult subjects for many students in chemistry. I was truly honored to receive so much positive feedback from my students.

Academic Credentials

Habilitation (2010) Theoretical Chemistry, Humboldt University, Berlin, Germany.
Ph.D. (1989) Physics, University of Stuttgart and Max-Planck-Institut für Festkörper-forschung, Stuttgart, Germany.
M.Sc. (1985) Physics, Balseiro Institute of the National Atomic Energy Commission, National University of Cuyo, S.C. de Bariloche, Argentina.

Research Expertise: modeling and computer simulations of materials, surface science and heterogeneous catalysis, functional (nano)materials, adsorption and oxide formation on metal surfaces, supported transition metal oxide aggregates for oxidative dehydrogenation reactions, ceria-supported metal catalysts for hydrogen production

Other Interests: networking and gender equality activities

Key Accomplishments, Honors, Recognitions, and Awards

- Management committee member of the European Cooperation in Science and Technology (COST) Action CM1104—*Reducible oxide chemistry, structure*, 2012 to present
- Workgroup leader within the COST Action CM1104—*Reducible oxide chemistry, structure*, 2012 to present
- >80 invited talks at conferences, universities, and research centers, 1993–2014
- Supervision of Postdoctoral Associates at the Institute of Catalysis and Petrochemistry, CSIC, Madrid: Javier Carrasco, 2010–2013; Gustavo E. Murgida, 2010; David López-Durán, 2012–2014; Pablo G. Lustemberg, 2015
- Project leader within the Marie Curie Career Integration Grant-FP7-PEOPLE-2011-CIG: *Water-gas shift reaction on metal-oxide nanocatalysts for hydrogen production*, 2012–2013
- European Research Council referee of proposals submitted under the Seventh Framework Programme (FP7) *Ideas* Specific Programme, ERC-2013-CoG, 2013
- Examination committee member for doctoral student (José Javier Plata Ramos; *Ceria for all seasons*), Universidad de Sevilla, Sevilla, Spain, 2013
- Project leader within the EULANEST—European–Latin American Network for Science and Technology project: *Cerium-based catalysts for the purification of hydrogen from renewable sources*, 2010–2012
- Member of the CECAM (Centre Européen de Calcul Atomique et Moléculaire)–ZCAM (Zaragoza Scientific Center for Advanced Modeling) Workshop organizing

committee: *Understanding structure and functions of reducible oxide systems: a challenge for theory and experiment*, Zaragoza, Spain, 2011

- Member of the organizing committee of the workshop in honor of Víctor Hugo Ponce: *Advances in atomic collisions and particle–surface interactions*, S.C. de Bariloche, Río Negro, Argentina, 2011
- Co-supervision of doctoral students at the Humboldt University Berlin: Veronika Brázdová, *Plane-wave density functional calculations on transition metal oxides*, 2000–2005; Tanya Kumanova Todorova, *Periodic density functional study on supported vanadium oxides*, 2003–2007; Gang Feng, *Theoretical investigation of the mechanism of hydrogen-transfer reaction over gamma-alumina supported metal catalysts*, 2009–2010
- Project co-leader within the German Science Foundation (DFG)–Sonderforschungsbereich (SFB 546) Collaborative Research Centre: *Structure, dynamics, and reactivity of aggregates of transition metal oxides*, 2000–2009
- Co-supervision of Postdoctoral Associates at the Humboldt University Berlin: Juarez L.F. da Silva, 2004–2006; Veronika Brázdová, 2006; Yoshiki Shimodaira, 2007–2008; Cristina Popa, 2008–2009
- Docent in Quantum Chemistry, Free University Berlin, Berlin, Germany, 2007
- Docent in Physical Chemistry of Surfaces and Heterogeneous Catalysis, Humboldt University Berlin, Berlin, Germany, 2003–2006
- Examination committee member for doctoral student (Javier Carrasco Rodríguez; *Estructura electrónica y propiedades de defectos puntuales en óxidos binarios y ternarios*), Universidad de Barcelona, Barcelona, Spain, 2006
- Examination committee member for doctoral student (Øyvind Borck; *Adsorption of methanol, phenol, and methylamine on* α-Al_2O_3 *and* Cr_2O_3 *surfaces*), Norwegian University of Science and Technology, Trondheim, Norway, 2006
- Examination committee member for doctoral student (Svetla D. Chakarova Käck, *Towards first-principles understanding of biomolecular adsorption*), Chalmers University of Technology, Gothenburg, Sweden, 2006
- Examination committee member for Habilitation (Dr. Thorten Klüner), Humboldt University Berlin, Berlin, Germany, 2004
- Member of the local organizing committee of the International Symposium of the Collaborative Research Centre 546, *Transition metal oxides—clusters, surfaces and solids—structure, dynamics and reactivity*, Berlin-Schmöckwitz, 2004
- Examination committee member for doctoral student (David Domínguez Ariza; *Estudio teórico de estados excitados en superficies y sólidos*), Universidad de Barcelona, Barcelona, Spain, 2003
- Examination committee member for doctoral student (Anders Hellmann; *Electron transfer and molecular dynamics at metal surfaces*), Chalmers University of Technology, Gothenburg, Sweden, 2003
- Member of the CECAM (Centre Européen de Calcul Atomique et Moléculaire) Workshop organizing committee: *Discussion meeting on catalysis from first principles*, Lyon, France, 2000

- Habilitation Stipendium of the DFG, 1999–2001
- Ph.D. Stipendium of the Max-Planck Society, 1986–1989

Biography

Early Life and Education

María Verónica Ganduglia-Pirovano was raised in Buenos Aires by her parents; her father is a medical doctor, who had specialized in Public Health Medicine, and her mother has a M.A. in Philosophy. She has three sisters. M. Verónica received her early and higher education in the Buenos Aires private catholic school system; she attended a girls-only school and she was motivated by science and math.

After about 2 years of undergraduate studies at the private Instituto Tecnológico de Buenos Aires (ITBA), she continued her undergraduate studies in physics at the Instituto Balseiro of the National Atomic Energy Commission in S.C. de Bariloche, Argentina for an additional 3 years. This institution is one of the most prestigious research centers in Latin America. Students are admitted after a tough examination where between 30 and 40 are selected out of around 100–150 applicants. They are all given a full scholarship that allows them to be devoted full time to their studies. Professors are all active researchers at the Bariloche Atomic Center. Early on she understood that she liked theoretical modeling and computer simulations in physics. She was awarded her Lic. (M.Sc.) in Physics in 1985. She published her first scientific paper in The Journal of Chemical Physics, together with her advisor Víctor Hugo Ponce—a great teacher and mentor.

Soon after obtaining her M.Sc. degree, she got married to a university classmate and moved to Stuttgart, Germany, because her husband had obtained a Max-Planck Ph.D. Stipendium. Three months later, she not only had started her doctoral research

at the Max-Planck-Institut für Festkörperforschung in Stuttgart under the supervision of Prof. Peter Fulde, but also was pregnant with their first child. These were not easy times, in particular because there was no family help available in Germany. However, by sharing duties and having stable childcare, both she and her husband completed their doctorates in 1989. Their second child was born a few months after her graduation.

Maria Verónica Ganduglia-Pirovano on the day she received her doctorate degree with (from right to left) her Ph.D. advisor, Prof. Peter Fulde, Prof. Eberhard Umbach, Prof. Max Wagner, and daughter Daniela, 1989.

Career History

After receiving their doctorates, they moved to New Jersey, USA. They came to this decision by once again following the path of the one who had the best career opportunity at the time—her husband became a post-doctoral fellow at AT&T Bell Labs. It is fair to say that the agreement was to look for jobs in cities where there would be more than one university or research center nearby. She soon started as a postdoctoral fellow at the Exxon Research and Engineering Labs in Annandale, NJ. The postdoctoral work, under the supervision of Prof. Morrel H. Cohen, on surface reactivity theory and modeling and computer simulations of catalytic metal surfaces paved the way for her future career in the field of the fundamental aspects of the physical chemistry of surfaces and heterogeneous catalysis.

Back in Germany, from 1994 to 1995, she held a Research Associate contract position at the Max-Planck-Institut für Physik komplexer Systeme, Branch Stuttgart, Stuttgart, Germany and at the Fritz-Haber-Institute Berlin, Berlin, Germany, working in

Maria Verónica Ganduglia-Pirovano at the library of the Quantum Chemistry Group at the Humboldt University in Berlin, 2001.

the groups of Prof. Peter Fulde and Matthias Scheffler, respectively. In 1996, she spent about a year as a Research Associate at the Center for Atomic-Scale Materials Physics, Lyngby, Denmark, in the group of Prof. Jens K. Nørskov. During that year, her husband took on the primary day-to-day management of the family and she commuted between Copenhagen and Berlin. Returning to Berlin meant missing fewer things that were important for the kids. From late 1996 until Spring 1999, she was a Research Associate at the Theory Department at the Fritz-Haber-Institute. This period was followed by a DFG Habilitation Stipendium until late 2001.

Maria Verónica Ganduglia-Pirovano on the day she obtained her Habilitation in Theoretical Chemistry at the Humboldt University in Berlin with (from left to right) Profs. Joachim Sauer, Helmut Winter, and Christian Limberg, 2010.

Dr. Ganduglia-Pirovano joined the Quantum Chemistry Group led by Prof. Joachim Sauer at the Humboldt University in Berlin as a Research and Teaching Assistant late in 2001 until mid-2009. It was a very productive period in her career. She was co-leader within the DFG–Sonderforschungsbereich (SFB 546) Collaborative Research Centre: *Structure, dynamics, and reactivity of aggregates of transition metal oxides*, conducted research, supervised students and postdocs, and regularly taught Physical Chemistry courses at various levels. She obtained her Habilitation in Theoretical Chemistry in 2010.

Since then, Dr. Ganduglia-Pirovano has become a Scientific Researcher at the Institute of Catalysis and Petrochemistry of the Spanish Council for Scientific Research (CSIC), Madrid, Spain, where she is the leader of Modeling for Theoretical Catalysis Group. The Spanish National Research Council (CSIC) is the largest public institution dedicated to research in Spain and the third largest in Europe. It belongs to the Spanish Ministry of Economy and Competitiveness through the Secretary of State for Research, Development and Innovation. CSIC is not academia in the strictest sense; research toward a Ph.D. is carried out, but the degree is given by a university nearby.

3 Most Cited Publications

Title: Oxygen vacancies in transition metal and rare earth oxides: current state of understanding and remaining challenges
Author(s): Ganduglia-Pirovano, MV; Hofmann, A; Sauer, J
Source: Surface Science Reports; volume: 62; issue: 6; pages: 219–270; published: June 30, 2007
Times Cited: 400 (from Web of Science)

Title: Hybrid functionals applied to rare-earth oxides: the example of ceria
Author(s): Da Silva, JLF; Ganduglia-Pirovano, MV; Sauer, J; et al.
Source: Physical Review B; volume: 75; issue: 4; article number: 045121; published: January 2007
Times Cited: 230 (from Web of Science)

Title: Density-functional calculations of the structure of near-surface oxygen vacancies and electron localization on $CeO_2(111)$
Author(s): Ganduglia-Pirovano, MV; Da Silva, JLF; Sauer, J
Source: Physical Review Letters; volume: 102; issue: 2; article number: 026101; published: January 16, 2009
Times Cited: 197 (from Web of Science)

Challenges

Moving from Argentina to Germany, learning the German language, pursuing a Ph.D. degree, and being a mother—all in about 4 years—were extremely challenging. Pursuing a career in science and having a family has not been entirely straightforward, but the unconditional family support, a healthy dose of stubborn determination, and the support of my colleagues, in particular Profs. Peter Fulde, Morrel H. Cohen, Matthias Scheffler, Jens K. Nørskov, and Joachim Sauer, have helped me overcome many obstacles.

I know by now that countries vary greatly in their organization of research and the procedures for higher level appointments. I did not really plan my career tailored to the "rules" of any specific country and did not seriously consider the importance of thoughtful mentoring along the way, but now I know that mentoring and career planning do matter. If I would do it again, I would reconsider these aspects.

On being a woman in this field . . .

Being a woman in this field is surely different from being a man, because we are in essence a minority group in science and engineering. The image of science and scientists seems to be predominantly male and one may risk not be taken too seriously if one looks too feminine, particularly at early stages of the career.

I am not conscious of having received promotion delays specifically because of being a woman, but I understand that formal and informal networking is vital as tools for integration in science and research as well as for the promotion of scientific careers, and that men and women networking styles are different. It is likely that old-boy networks lack inclusion and transparency.

I never considered not having kids. Early on I have decided for motherhood and for pursuing a career in science with the understanding that these are two very rewarding and demanding occupations. Hence, I have learned to be multitasking and to divide available time between the two activities. At times, one feels that one is

Maria Verónica Ganduglia-Pirovano in 2005.

neither being a great mother nor becoming a successful scientist. Today, I am happy of having had juggled the demands of a career and motherhood—certainly not a challenge for the faint-hearted. I have two resourceful and independent daughters and I am certain of having made the best out of the opportunities created by myself or given to me within the restrictions imposed by my own personal decisions and choices. I am proud of both my scientific achievements and my daughters.

Words of Wisdom

- Be determined.
- Choose your mentors wisely. Mentors can recommend you as a speaker at a meeting, as a lead scientist on a project, or as an author of a review for a prestigious journal.
- Plan your career tailored to the "rules" of the specific country you may want to live in.
- Be active, publish, present your work at conferences, and network.
- A supporting partner who shares your aims and ambitions as well as family duties is crucial for mothers in science; reliable childcare is also most important.
- Make sure that your family understands that your work is important not only to you but also to them—a frustrated mother is not good for anyone.
- Don't use family responsibilities as an excuse for not getting a job done.

Profile 34

Elsa M. Garmire

Sydney E. Junkins Professor of
Engineering Sciences
Dartmouth College
Thayer School of Engineering at
Dartmouth
14 Engineering Drive
Hanover, NH 03755
USA

e-mail: elsa.garmire@dartmouth.edu
telephone: (603) 646-3154

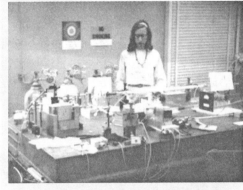

Garmire in her lab, c. 1970.

Dr. Elsa Garmire as a postdoc in the Caltech
Laboratory of Amnon Yariv, about 1970.

Birthplace
Buffalo, NY, USA

Born
November 9, 1939

Publication/Invention Record

>200 publications: h-index 38 (Web of Science)
14 patents, editor of 7 books, 14 book chapters

Tags

❖ Academe
❖ Domicile: USA
❖ Nationality: American
❖ Caucasian
❖ Children: 2

Proudest Career Moment (to date)

In 1995 I was awarded the highest award of the Society of Women Engineers, their
Achievement Award, and presented a talk at their annual meeting. The speech I gave was
my proudest moment, as both my mother and my spouse were there to share it with me. It
happened to be my spouse's birthday and the entire room sang happy birthday to him.
I felt close to and accepted by the society of achieving women engineers. In other
circumstances, I always felt like an outsider.

Academic Credentials

Ph.D. (1965) Physics, Massachusetts Institute of Technology, Cambridge, MA, USA.
A.B. (1961) Physics, Radcliffe College, Harvard University, Cambridge, MA, USA.

Successful Women Ceramic and Glass Scientists and Engineers: 100 Inspirational Profiles. Lynnette D. Madsen.
© 2016 The American Ceramic Society and John Wiley & Sons, Inc. Published 2016 by John Wiley & Sons, Inc.

Research Expertise: glass; lithium niobate; nonlinear optics; integrated optics and semiconductor photonic devices; lasers; electro-optics; fiber optics

Other Interests: global economy and related science policy, technology literacy, standards for technology education in the K-12 arena, equality, explaining career options for scientists and engineers

Key Accomplishments, Honors, Recognitions, and Awards

- Councillor, National Academy of Engineering, 2000–2008
- National Associate of the National Academies for pro bono service to the National Research Council, 2005
- USA Lasers and Optics Delegation to South Africa, People-to-People, 2005
- Fellow, Society of Women Engineers, 1997
- Councillor at Large, American Physical Society, 1995–1997
- American Academy of Arts and Sciences, elected member, 1996
- Dartmouth College, Honorary Phi Beta Kappa, 1996
- Fulbright Senior Lecturer (to Australia), 1995
- Visiting Professor, Sydney University, Australia, 1994–1995
- President (also Vice-President and President-elect), Optical Society of America, 1990–1994
- Society of Women Engineers Achievement Award, 1994 (their highest award)
- University of Southern California University Research Award, 1994 (their highest award)
- Fellow, American Physical Society, 1994
- Visiting Professor, Telebras and University of Sao Paolo, Brazil, 1992
- National Academy of Engineering, elected member for her contributions to nonlinear optics and optoelectronics and leadership in education, 1989
- US Delegate to Bi-National USSR Symposium on Laser Optics, 1989
- World Bank Technical Expert to Harbin Institute of Technology, PRC, 1989
- PRC Visiting Scientist Exchange Program Fellowship, awarded by NAS, 1989
- Visiting Professor at Southeast University, Nanjing, PRC, 1989
- Visiting Professor at Macquarie University, Sydney, Australia, 1988–1989
- Visiting Professor, National Defense Academy, Japan, 1987
- Councillor, IEEE Lasers and Electro-Optics Society, 1984–1987
- Radcliffe College, Distinguished Alumna Award, 1986
- Visiting Professor, Nippon Telephone and Telegraph, Japan, 1986
- Director-at-Large, Optical Society of America, 1983–1986
- Visiting Consulting Expert to Taiwanese National Government, Taipei, 1985
- US Delegation to China, IEEE Quantum Electronics Study Group, 1982
- Fellow, Optical Society of America, 1981

- Fellow, Institute of Electronic and Electrical Engineers, 1980
- U.S. Delegate to the International Commission on Optics, 1980

Frequently, Dr. Garmire consulted for many companies on technical matters, and served on advisory panels for the U.S. Government.

Biography[1]

Early Life and Education

Elsa Garmire received her first degree from Harvard University via Radcliffe (the women's arm of Harvard) in 1961 and her Ph.D. from MIT in 1965, both in physics. She was the first from her high school to be accepted to Harvard. In those days, there were only 300 women per class at Radcliffe and 1000 men per class at Harvard. All her classes were joint with Harvard and they took the same exams and were graded on the same curve. But they lived in a separate quadrangle and had separate extracurricular activities. And, they were clearly products of the 1950s. There were three girls in her class who majored in physics. At MIT, it seemed that she was the only girl (that's what they were called in those days) in physics graduate school. Much later she found out that there was one other, but they never shared classes. Along the way she made some good choices, e.g., she asked to work with Charles Townes; later he received the Nobel Prize for his introduction of the field of quantum electronics and introduction of the maser.

Career History

Dr. Garmire completed postdoctoral work at Caltech working with Professor Amnon Yariv, where she contributed to early research on mode-locking to produce ultrashort laser pulses. Their interests shifted to a new field: integrated optics. They demonstrated optical waveguide components in gallium arsenide, developed to monolithically integrate with lasers and detectors into optical waveguide circuits.

After Caltech, Dr. Garmire spent a year in industry working on integrated optics at Standard Telecommunications Laboratory in England, and at Thomson

Elsa Garmire giving a lecture at Thayer School of Engineering, Dartmouth College, in 2008.

---

[1] Credit: http://engineering.dartmouth.edu/~d34863r/biography.pdf

CSF in France. Then, Dr. Garmire joined the University of Southern California (USC) in 1974, where she progressed as a faculty member with final positions as the William Hogue Professor of Electrical Engineering and director of the Center for Laser Studies. At USC, she continued research on integrated optics in other materials: glass and lithium niobate. She held a joint appointment at The Aerospace Corporation, where she researched new geometries for semiconductor lasers. In 1978, she entered yet another new field, optical bistability, which held promise for optical computers. Material limitations inspired research directed toward improving optical nonlinearities using specially constructed layered materials to improve nonlinear optical properties in semiconductors.

Elsa Garmire as Dean at Thayer School of Engineering, Dartmouth College, in 1995.

While at USC, Dr. Garmire started their optics program and taught optics at both the undergraduate and graduate levels. While in California, she started the first commercial public laser light show, "Laserium," and also built one of the first semiconductor epitaxial growth facilities in a university.

In 1995, she joined Dartmouth College where she was named Sydney E. Junkins Professor at the Thayer School of Engineering. Her shift to engineering was intentional—initially she worked in Physics and then Applied Physics, but through the years she saw that engineers had broader opportunities to impact the world (and often higher salaries). Dr. Garmire investigated optical devices used in a variety of applications: optical sensors, image identification, and fiber optics systems. These devices included semiconductors grown in ultrathin layers, called multiple quantum wells, which enhanced optical performance. Their sensitivity involves the transport of carriers—electrons and holes—across the device. Dr. Garmire's research has improved the understanding of the mechanisms for carrier transport, enabling more optimal designs. Recently closed down, her general-purpose laser laboratory provided equipment for both research and student classes. Dr. Garmire has been a leading spokesperson for developing technological literacy for everyone. She has taught courses at Dartmouth on this subject and participated through the National Academy of Engineering (NAE) in national efforts to develop standards in technological literacy for K-12 education. She chaired the NAE's Committee on Assessing Technological Literacy.

Dr. Garmire was elected a member of the NAE in 1989 and served for two terms as NAE Councillor. She has been on the National Research Council (NRC) Report Review Committee and continues serving as monitor to NRC reports. As a result of this pro bono work, she has been named a National Academy Associate. She served for 6 years on the National Academies Committee on Science, Engineering, and Public Policy, which organized the workshop and report "Rising above the Gathering Storm" that inspired Congress to focus on the need for better STEM education. Among her previous NAE contributions are service on the following committees: Nominating Committee, Membership, Electronics Engineering, Charles Stark Draper Prize, Technology Education Standards; and the Academic Advisory Board. She has served on the following NRC study committees: STEM Higher Education, Active Electro-optic Sensing, Optical Science and Engineering, Research-Doctorate Programs, and Potential Applications of Concentrated Solar Photons.

Dr. Garmire is a member of the American Academy of Arts and Sciences, and a fellow of the Institute of Electrical and Electronics Engineers, the American Physical Society, and the Optical Society of America, of which she was at one time president, and has served on the boards of three other professional societies. In 1994, she received the Society of Women Engineers Achievement Award. She has been a Fulbright senior lecturer and a visiting faculty member in Japan, Australia, Germany, and China. Dr. Garmire has graduated 30 Ph.D. and 15 M.S. students. She has been a consultant in lasers and optics for over 20 companies.

Elsa Garmire discussing section activities at a Society of Women Engineers event at University of Southern California in 1981.

3 Most Cited Publications

Title: Self-trapping of optical beams
Author(s): Chiao, RY; Garmire, E; Townes, CH
Source: Physical Review Letters; volume 13; issue: 15; pages 479–481; published: 1964
Times Cited: 1195 (from Web of Science)

Title: Theory of bistability in non-linear distributed feedback structures
Author(s): Winful, HG; Marburger, JH; Garmire, E
Source: Applied Physics Letters; volume 35; issue: 5, pages 379–381; published: 1979
Times Cited: 362 (from Web of Science)

Title: Coherently driven molecular vibrations and light modulation
Author(s): Garmire, E; Townes, CH; Pandarese, F
Source: Physical Review Letters; volume: 11; issue: 4; pages 160–163; published: 1963
Times Cited: 192 (from Web of Science)

Challenges

Science and engineering are changing very rapidly and producing more and more impact on humans, the environment, societal, cultural, and political values. This is why we need women in there, to connect the research being done with its value to society and to identify possible negative consequences.

On being a woman in this field . . .

As long as women are in the minority, there will be the stereotype threat. You have to acknowledge this and make it a part of your approach to science/engineering. As long as you are in the minority, whatever that minority is, if you want to play in the majority culture, you will have to play by their rules. This means you often have to stretch yourself to do things the way that the majority does, rather than to accept the stereotype and do what they expect you to do. For example, in group meetings you may have to speak up rather than wait to be called on. In general, you will probably have to be better than the average of the majority. That's just the way it is; they won't accept you otherwise. Fortunately, this is not difficult if you follow your dreams and do not let yourself fall into the stereotype boxes that the majority tries to put minorities in. There is plenty of proof that women can succeed at the highest levels, but this doesn't happen without focused effort on the part of women. It is valuable to read a report of the National Academy Press called "Beyond Bias and Barriers: Fulfilling the Potential of Women in Academic Science and Engineering," published in 2007.[2]

Words of Wisdom[3]

I have found that many factors play into career success: dreams, acculturation, competition, stick-to-it-iveness, luck overcoming bias, etc. While at one level I knew since 6th grade that I wanted to be a scientist and dreamt of replacing Marie Curie, at another level, as a product of the 1950s, I wanted to be the world's most perfect mom. I found it challenging to reconcile these two. My innate competitiveness (as a second daughter) and my ability to stick to it when the going got tough (which I attribute to years of piano training) kept me focused. I had good luck at the beginning, but also I reached out for it and worked to the best of my ability, I was able to obtain the highest quality education available and whenever I had a choice, I sought excellence over all else.

I believe these factors play into the life of every successful career woman. Professionals have come to understand that expertise takes 10,000 hours to develop. It will take deep, inward dreams to get through 10,000 hours of problem-solving when you begin as a novice. It helps to be competitive, at least with yourself, yet to keep in mind that the culture in which you find yourself will invariably affect your success in competition. You may not make it to the top of the pyramid, but its fun to be well up there, where the view is broad and international. This climb may require deviations from initial expectations. It requires the ability and willingness to hang on when the going gets tough. Understanding the role of luck helps you accept what the world throws at you— both the good and the bad. You don't do it all yourself. If you're lucky, life will be a positive experience and you'll be glad that you made the climb. For the record, I married before graduate school, had two children while a post-doc, divorced and eventually married again; my two daughters are happily married, with their own children. While neither discovered an aptitude for science or engineering, both have been fulfilled by balancing family with their own life pattern (at present one works part-time for a nonprofit and the other is a full-time homemaker). My passion for them has balanced my passion for physics and engineering. Along with my spouse, they are the light of my life.

---

[2] http://www.nap.edu/catalog/11741/beyond-bias-and-barriers-fulfilling-the-potential-of-women-in.

[3] Credit: "Blazing the Trail: Essays by Leading Women in Science," edited by Emma Ideal and Rhiannon Meharchand (http://www.amazon.com/Blazing-Trail-Essays-Leading-Science/dp/1482709430/).

Profile 35

Rosario A. Gerhardt

Professor and Goizueta Foundation
Endowed Faculty Chair
Georgia Institute of Technology
Materials Science and Engineering
Atlanta
Georgia 30332
USA

e-mail: rosario.gerhardt@mse.gatech.edu
telephone: (404) 894-6886

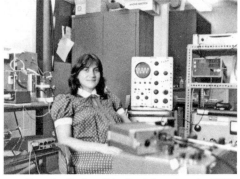
Rosario Gerhardt in the laboratory in 1982.

Birthplace
Lima, Peru

Born
May 20, 1953

Publication/Invention Record

>200 publications: h-index: 21 (Web of Science)
3 patents issued, 10 disclosures, 5 published books,
7 book chapters

Tags

❖ Academe
❖ Domicile: USA
❖ Nationality: American
❖ Hispanic
❖ Children: 2

Proudest Career Moment (to date)

The first time that I met a certain scientist at a conference who came to see my single author poster and stated, "I never expected a lady!" (This was only possible before the Internet became commonplace).

Academic Credentials

Eng.Sc.D. (1983) Metallurgy and Materials Science, Columbia University, New York City, New York, USA.
M.S. (1979) Metallurgy and Materials Science, Columbia University, New York City, New York, USA.
B.A. (1978) Mathematics, Carroll College, Helena, Montana, USA. Combined Plan Program with Columbia University.
B.A. (1976) English and French, Carroll College, Helena, Montana, USA.

Successful Women Ceramic and Glass Scientists and Engineers: 100 Inspirational Profiles. Lynnette D. Madsen.
© 2016 The American Ceramic Society and John Wiley & Sons, Inc. Published 2016 by John Wiley & Sons, Inc.

Research Expertise: ceramic materials, determining structure–property–processing relationships, polymer and ceramic composites, synthesis and assembly of nanoparticles, dielectric insulators, ionic conductors and ceramic superconductors, electrical and microstructural characterization using impedance and dielectric spectroscopy, resistivity measurements, structural characterization via microscopic, optical, and scattering techniques.

Other Interests: diversity/broadening participation, life/career balance

Key Accomplishments, Honors, Recognitions, and Awards

- National Academies Defense Materials Manufacturing and Infrastructure Workshops Planning, 2012 to present
- Faculty Advisor for the Student Chapter of Materials Research Society (MRS) at Georgia Tech, 2010 to present
- Faculty Advisor for the Student Chapter of ACerS at Georgia Tech, 1994 to present
- Member, Keramos, 1991 to present
- Member Sigma Xi, 1982 to present
- Mentor and Organizer, Future Faculty Workshops, 2012 to present
- Author of a textbook on Impedance and Dielectric Spectroscopy, est. 2015
- International Association of Advanced Materials Medal - 2015, Stockholm, Sweden
- Member of the National Research Council Associateship Panel Review Program, 2002–2013
- Editor of a book on Properties and Applications of Silicon Carbide, 2011
- Together with one of her students: a fellowship award from The American Society for Nondestructive Testing, Inc. (ASNT), 2011
- Visiting Scientist, Center for Nanophase Materials Sciences, Oak Ridge National laboratory, 2007–2008
- Author of chapter for Encyclopedia of Condensed Matter Physics, 2005
- Keynote Speaker at the International Conference on Electrochemical Impedance Spectroscopy, 2004
- Main Editor of three books on Electrically Based Microstructural Characterization, 2002, 1998, and 1996
- Executive officer of the Electronics Division of the American Ceramic Society, 1996–2001
- Chair of Electronics Division of the American Ceramic Society, 2000–2001
- Fellow of the American Ceramic Society (ACerS), 1998
- NASA/ASEE Faculty Fellow at Marshall Space Flight Center, Summer 1995
- Eminent Scholar Member, Tau Beta Pi, 1994
- NSF Career Advancement Award, 1993
- NASA Faculty Award for Research, 1993
- Participant in numerous women and minority leadership workshops

- Co-organizer of numerous symposia at ACerS, MRS, and other societies
- Served as research advisor to more than 40 graduate students

Biography

Early Life and Education

Rosario Gerhardt was born and raised in Lima, Peru by her German father and Peruvian mother. She has two older siblings, a sister, and a now deceased brother. She attended Rosa de Santa Maria, a highly regarded girls' high school in Lima until she was chosen to participate in the American Field Service Program (now known as AFS International) during the 1969–1970 academic year in Iowa City High school. She still maintains close ties with her American host family from then and considers them a second family.

After returning to Lima, she taught English at the ICPNA (U.S.–Peru Binational Center) and then started her college education at San Marcos University in Lima, where she studied geography and geology from 1971 to 1973. She came back to the United States in August 1973 with a scholarship from the Catholic Church to attend Carroll College in Helena, Montana, where she was a language major with a math minor. She received her first Bachelor's degree in English and French in 1976.

Her love of science and math resulted in her staying on at Carroll for another year to take all of the necessary pre-engineering courses before transferring to Columbia University in the fall of 1977. There she completed the Combined Plan Program (often called dual degree program), and received a second B.A. from Carroll College in Mathematics in 1978 (while she was working for her Master's degree in Metallurgy and Materials Science at Columbia University). She was supported by a Henry Krumb Fellowship at Columbia University during the 1977–1978 academic year and graduate research assistantships from 1978 to 1983 facilitated by her graduate advisor, Prof. A.S. Nowick. She was awarded her M.S. in Metallurgy and Materials Science from Columbia University in New York City in 1979 and her Doctor of Engineering Science in 1983.

Her doctoral research was focused on understanding the effect of changing the dopant size on the ionic conductivity, and defect association behavior in cerium dioxide materials. It was during the development of her Ph.D. thesis that her interest in identifying trends between microstructural features and electrical response in materials was awakened. While she was able to carry out all of the electrical characterization experiments in the laboratory of her Ph.D. advisor, she had to develop a collaboration with the Center for Microanalysis at the University of Illinois in Urbana–Champaign (UIUC) in order to conduct one of the first studies that related the microstructure of these hard ceramic materials to their electrical response and especially the grain boundary behavior. She had to learn how to prepare the difficult to grind and polish cerium dioxide samples and make them electron transparent way before dimplers were invented. After many months of elbow grease, she succeeded and sent her <100 µm thick slices to UIUC for ion milling (which took 3–4 days each!) She then traveled to UIUC in 1982 to acquire the STEM images, and associated spectra. Her transmission electron microscopy images on CeO_2 doped with various trivalent ions were first presented at an American Ceramic society meeting in 1982 and then at a Department of Energy investigators meeting in 1983, but were not published in a peer reviewed journal until 1986.

Career History

Upon receiving her doctorate, Dr. Gerhardt stayed on at Columbia as a post-doctoral researcher for a year while she searched for other employment. The year she graduated was a difficult year for getting a position due to the breakup of the giant telecommunications company Bell Laboratories. This was compounded further by being geograph-

ically limited due to family responsibilities (she got married and had her first child during graduate school). She was offered the opportunity to join the NSF-funded Center for Ceramics Research at Rutgers-The State University of New Jersey late in 1984 as a post-doctoral research associate to work with Dr. John B. Wachtman, Jr. This position, together with her graduate thesis research on solid elec-

Rosario Gerhardt with her two children in 1991. (See insert for color version of figure.)

trolytes for high-temperature fuel cells, cemented her interest and background in the ceramics field. She became active in the American Ceramic Society and has been a regular attendee of the annual meetings culminating in her becoming an executive officer of the Electronics division of ACerS from 1996 to 2001 and the chair of that division in 2000–2001.

Subsequent to her 2-year stint as a post-doctoral researcher, she was promoted to Assistant Research Professor at the Center for Ceramics Research at Rutgers University from 1986 to1990. During this time, she supervised the research of two doctoral students and numerous undergraduate researchers. Her research during this time was mostly funded by the industrial consortium at the Center for Ceramics Research at Rutgers University and was related to low dielectric constant materials, porosity characterization, and fabrication and characterization of high-temperature ceramic superconductors. She conducted some paid research for General Electric on using dielectric measurements to monitor porosity in thermal barrier coatings and assisted in the acquisition of a high-resolution transmission electron microscope using Keck Foundation and State of New Jersey funds. In addition, during this time Rosario carried out engineering consulting in the area of dielectric properties for companies such as Foster-Miller, Inc.; Norton Ceramics; National Starch and Chemical, and Union Carbide. She also assisted Bell Communications Research with the installation of ionic conductivity measurement equipment and Hercules Inc. with superconductor testing and gave overview talks at the Hewlett Packard Semiconductor Test Symposium and Exposition sponsored by IEEE.

Dr. Gerhardt joined the faculty of the School of Materials Science and Engineering at the Georgia Institute of Technology (commonly known as Georgia Tech) as an untenured Associate Professor in January 1991. She conducted paid research for Delco Electronics, W.R. Grace, and Coors Electronic Package Company. In 1993, she was awarded an NSF Career Advancement Award (CAA, a 1-year award that was a predecessor of the current CAREER program) and also a NASA Faculty Award for Research (FAR) and an NSF grant for monitoring sintering using small angle scattering techniques, which are especially sensitive to porosity. These grants allowed her to make some inroads into establishing her research program at Georgia Tech. She was granted tenure in 1997 after having worked as an ASEE/NASA Faculty Fellow at the NASA Marshall Space Flight Center in Huntsville, Alabama during summer 1995. During this time, many of her students were also awarded NSF and NASA fellowships.

Prior to being promoted to full professor in 2001, Prof. Gerhardt had shifted her interest back to trying to use electrical measurements as a way to monitor the microstructure of materials at all length scales and revisited her earlier work on silicon carbide whisker—mullite composites that she had done 10 years earlier and finally published a complete paper on it in 2001 together with some work done by one of her doctoral students on other similar materials. Her biggest hurdle was obtaining the funds to buy more advanced equipment that could more easily be interfaced with newer computers and that could be used to collect and analyze data more quickly. Seemingly, any other researchers who mentioned that they wanted to use impedance spectroscopy, even when they had no experience with it quickly received funds. Her big breakthrough came when she was awarded an NSF grant in fall 2000 which finally permitted her to purchase the equipment that has allowed her group to make the progress that she has

Rosario Gerhardt
in 2008.

been able to make in the last decade or so. A separate grant from the US Department of Energy (DOE) in 2003, which was renewed in 2007, made it possible to continue that momentum. With assistance from DOE, she was able to spend the 2007–2008 academic year as a Visiting Professor at the Center for Nanomaterials Science (CNMS) at Oak Ridge National Laboratory (ORNL) learning how to use the atomic force microscope and buy a machine of her own.

While her students have not taken to using atomic force microscopy (AFM) at the level she would wish they would, this is understandable as this is a very tedious technique when coupled with frequency measurements. Nevertheless, she is confident that the initial work she started at ORNL (most of it is still unpublished) will carry her for the rest of her career. She has been fortunate to continue getting additional NSF grants (after persistent multiple submissions) for conducting work on percolating ceramic composites, which have been instrumental in letting her group make the tremendous headway that they have made in recent years. During the last few years, her group has also forayed into the field of transparent oxides and printed electronics. Most of that work has been funded by the Institute for Paper Science and Technology and continued collaboration with researchers at ORNL, the Nanotechnology Center at Georgia Tech, Novacentrix and additional funds from NEI Corp and Exide Technologies. Much of the research work in the last 15 years has led to the awarding of three patents and several other patent disclosures. Her students have continued to be of the highest caliber and are being recognized with a variety of fellowships. Dr. Gerhardt has also continued carrying out engineering consulting, primarily in the dielectric property measurement arena with AlSiMag, Inc., Advanced Refractory Technologies, Microcoating Technologies (now called nGMat), NEETRAC, and GTRI as well as long-time collaborator Advanced Composite Materials LLC.

In 2011, she also became the Executive Director for Research and Institute Collaborations in the Office of Vice President for Institute Diversity at Georgia Tech on a part-time basis. In this role, her goal is to assist the Vice President in identifying opportunities to involve diverse populations in research at all levels and facilitate collaborations across different colleges and schools at the institute. So far she has been involved in projects at the faculty, student, and staff levels.

Prof. Gerhardt was chosen as the Goizueta Foundation Faculty Endowed Chair at the Georgia Institute of Technology in 2015 in recognition of her research, teaching and service to the Institute.

3 Most Cited Publications

Title: Impedance and dielectric-spectroscopy revisited: Distinguishing localized relaxation from long-range conductivity
Author(s): Gerhardt, R
Source: Journal of Physics and Chemistry of Solids; volume: 55; issue: 12; pages: 1491–1506; doi: 10.1016/0022-3697(94)90575-4; published: December 1994
Times Cited: 357(from Web of Science)

Title: Grain-boundary effect in ceria doped with trivalent cations: 1. Electrical measurements
Author(s): Gerhardt, R; Nowick, AS
Source: Journal Of The American Ceramic Society; volume: 69; issue: 9; pages: 641–646; doi: 10.1111/j.1151-2916.1986.tb07464.x; Published: September 1986
Times Cited: 217(from Web of Science)

Title: Ionic-conductivity of CeO_2 with trivalent dopants of different ionic-radii
Author(s): Gerhardtanderson, R; Nowick, AS
Source: Solid State Ionics; volume: 5; issue: October; pages: 547–550; published: 1981
Times Cited: 159 (from Web of Science)

Challenges

One challenge is getting recognition for her work and contributions. It is very lonely to be so far ahead of your peers that it prevents you from getting the resources, assistance and cooperation needed to truly make an impact.

On being a woman in this field . . .

It definitely makes a difference. Being a woman means that your contributions will often be ignored and sometimes ridiculed. And even when your contributions are recognized, they are hardly ever accepted as a stand-alone effort. Every time a woman is recognized, everyone scrambles to find out what 'amazing' male has had an impact on her scientific acumen rather than giving the woman her proper due. If you don't believe me, you don't have to go too far to remember that Rosalind Franklin was excluded from being recognized for her contributions to understanding the structure of DNA. We may never find out for sure what role, if any, Mileva Maric may have played on Albert Einstein's prolific writings in 1905. While things are definitely getting better and more and more women are beginning to be recognized for their contributions; unfortunately, sometimes the recognition goes to the undeserving, further compounding the problem. This may not necessarily be a gender issue, but nevertheless it does occur.

Words of Wisdom

- Work hard, play hard, and love a lot!
- Find what makes you happy and go for it.
- It is not necessary to look like a man to be a good scientist and an engineer.
- On the other hand, stay away from suggestive or revealing clothing in the workplace.
- Be respectful of others ideas, give credit to everyone for their contributions, however small.
- Refrain from only quoting work related to the material at hand! Use the literature from wide-ranging fields, but be sure to reference all papers that planted an idea in your mind.
- Don't be afraid to stand for what you believe in; it may take years before the messages start to sink in but keep trying.
- Never give up! Just when you are ready to throw in the towel, a door will open if you look hard enough.

Dagmar Gerthsen

Dagmar Gerthsen at age 30.

Professor of Electron Microscopy
Head of the Central Facility for Electron Microscopy
Laboratory for Electron Microscopy
Karlsruhe Institute of Technology (KIT)
Engesserstr. 7
76131 Karlsruhe
Germany

e-mail: gerthsen@kit.edu
telephone: +49 721 608-4 3200

Birthplace

Aachen, Germany

Born

January 23, 1958

Publication/Invention Record

>350 publications: h-index 40 (Web of Science)

Tags

❖ Academe
❖ Domicile: Germany
❖ Nationality: German
❖ Caucasian
❖ Children: 0

Proudest Career Moment (to date)

There have been some occasions and it is difficult to decide which was most eminent. One of them was certainly being offered a full professorship at the age of 34. In addition, I find teaching rewarding and always enjoy if members of my group are particularly successful by winning prizes or achieving good professional positions.

Academic Credentials

Ph.D. (1986) Physics, University of Göttingen, Germany.
Diploma (1981) Physics, University of Göttingen, Germany.

Research Expertise: electron microscopy (application and methodological developments), materials for solid oxide fuel cells and gas separation membranes, ZnO nanorods/layers, niobium oxides, carbon fiber/carbon matrix composites, diamond and diamond-like carbon, mechanisms of epitaxial growth, semiconductor quantum dots and quantum wells, nanoparticles and clusters, organic solar cells, interactions of nanoparticles with biological cells

Successful Women Ceramic and Glass Scientists and Engineers: 100 Inspirational Profiles. Lynnette D. Madsen.
© 2016 The American Ceramic Society and John Wiley & Sons, Inc. Published 2016 by John Wiley & Sons, Inc.

Key Accomplishments, Honors, Recognitions, and Awards

- Member of the Senate Evaluation Committee of the Leibniz Association (an umbrella organization for 86 research institutions that addresses scientific issues of importance to society as a whole), 2009 to present
- Member of the Board of Directors of the University of Bielefeld, 2003–2007
- Spokesperson of the Collaborative Research Center SFB551 "Carbon from the gas phase: elementary reactions, structures, materials," 2003–2007
- Elected "Fachgutachterin" reviewer of the German Research Foundation (DFG) for solid-state physics, 2000–2004
- Equal opportunity commissioner of the Physics Department of the University of Karlsruhe, 1997–2004
- Board member of the German Society for Electron Microscopy, 2000–2003

Biography

Early Life and Education

Dagmar Gerthsen completed her first degree in physics in 1981 after studying physics at the University of Karlsruhe and the University of Göttingen. Her Diploma thesis entitled "Demixture of particle suspensions in a capillary duct" was completed at the Max-Planck-Institute for Flow Research in Göttingen.

She then served as a Research Associate at the Aeronautics Department of the California Institute of Technology with Prof. Dr. H.W. Liepmann in 1981 and 1982 working on second sound shock waves in superfluid helium.

In 1983, she embarked on doctoral studies at the Institute for Metal Physics at the University of Göttingen with Prof. Dr. P. Haasen. In 1986, she defended her thesis, "Electrical properties of plastically deformed GaAs," and was awarded a Ph.D.

Career History

Upon graduation, she took a position as a Research Scientist at the Institute for Metal Physics at the University of Göttingen. Later that year, she moved to Xerox Palo Alto Research Center as a Postdoctoral Associate where she started electron microscopy.

In 1988, she returned to Germany to serve as the Group Leader at the Department of Solid State Research, Institute for Microstructure Research at the Research Center Jülich.

Late in 1993, she accepted an offer for the position of Professor of Electron Microscopy and Head of the Central Facility for Electron Microscopy, then the University of Karlsruhe (and since renamed to Karlsruhe Institute of Technology, KIT).

A long-time research interest of Prof. Gerthsen has been electron microscopy where her emphasis has been on the development of techniques for quantification of image information, as well as working on the experimental and theoretical developments in the field of phase-plate transmission electron microscopy. Applying a broad range of electron microscopic techniques in materials science, she is interested in structure–property relationships in semiconductors, metals, ceramics, and composite materials. A long-time focus has been structural properties and chemical composition of semi-conductor nanostructures and their correlation with optical properties. More recently, she became interested in the structural and chemical properties of nanoparticles that she studies in context with catalysis or toxicological effects in cells.

Her interests include sports (bike riding), travel to places with spectacular natural beauty, and animal watching.

3 Most Cited Publications

Title: Ultra-large-scale directed assembly of single-walled carbon nanotube devices
Author(s): Vijayaraghavan, A; Blatt, S; Weissenberger, D; et al.
Source: Nano Letters; volume: 7; issue: 6; pages: 1556–1560; doi: 10.1021/nl0703727; published: June 2007
Times Cited: 169 (from Web of Science)

Title: Gain studies of (Cd, Zn)Se quantum islands in a ZnSe matrix
Author(s): Strassburg, M; Kutzer, V; Pohl, UW; et al.
Source: Applied Physics Letters; volume: 72; issue: 8; pages: 942–944; doi: 10.1063/1.120880; published: February 23, 1998
Times Cited: 136 (from Web of Science)

Although not readily apparent, Dagmar Gerthsen is overjoyed about the almost complete installation of their new aberration-corrected transmission electron microscope in 2009. Photo credit: Andrea Fabry. (See insert for color version of figure.)

Title: Digital analysis of high resolution transmission electron microscopy lattice images
Author(s): Rosenauer, A; Kaiser, S; Reisinger, T; et al.
Source: Optik; volume: 102; issue: 2; pages: 63–69; published: April 1996
Times Cited: 121 (from Web of Science)
ResearcherID (I-4448-2012)

Challenges

Demands on young scientists (women and men) have increased during the past 20 years with respect to scientific output, development of teaching and organizational skills, and being active in scientific outreach and academic self-administration. Nevertheless, it is still more difficult for women bringing family life in line with professional demands and maintaining interests apart from work. Increasingly, I see young couples, both with excellent education, who share tasks more evenly and successfully manage to pursue their career because both compromise. In addition, in Germany, the substantial lack of highly qualified people may help to improve the situation here because employers increasingly try to attract and keep female scientists and engineers. However, progress is slow and maintaining a reasonable work–life balance remains a challenge.

On being a woman in this field . . .

Despite all efforts, the fraction of female students and scientists is still small in most engineering sciences and physics in Germany. Increased attention is often paid to the few women in the field, which can be an advantage because excellent achievements will typically get quite a bit of attention. On the other hand, the opposite applies as well. Performance far from perfect will be "well" remembered by the community. Cases of obvious discrimination are meanwhile rare and political pressure has led to an increased awareness to actively promote women in academia and industry. Today more subtle discriminations prevail. This may start by being rudely interrupted in discussions and end with not being seriously considered for high-level grants or prizes.

Nevertheless, the situation has improved with increasing number of female scientists in top positions. In 1993, we were two females among about 300 professors at the University of Karlsruhe. A fraction close to 10% is now achieved, which is, of course, still not satisfactory at all. With increasing number of female colleagues, cooperation with women becomes more frequent. This is enjoyable because the atmosphere is often more open and relaxed.

Words of Wisdom

Maybe in 20 years . . .

Profile 37

Dana G. Goski

Director of Research & Technology
Allied Mineral Products, Inc.
2700 Scioto Parkway
Columbus, OH 43221
USA

e-mail: dgg@alliedmin.com
telephone: (614) 876-0244

Dana Goski with a winning student
poster entry, circa 1995.

Birthplace
Nova Scotia, Canada

Publication/Invention Record

9 publications
6 patents, co-editor of 1 book

Most Memorable Moments (to date)

Tags

❖ Administration and Leadership
❖ Industry
❖ Domicile: USA
❖ Nationality: Canadian, US
   Permanent Resident
❖ Caucasian
❖ Children: 1

A recent memorable moment was being with my research & corporate team, as we broke ground on a new research building expansion at corporate headquarters in Columbus, OH in 2014.

A collage of great memories were created while working with industry colleagues to develop the successful event of UNITECR 2013, coordinating the technical program for the global refractories industry, and development a Wiley publication of the subsequent proceedings. The team involved was an inspiring group of individuals.

Academic Credentials

Ph.D. (1997) Applied Science, Dalhousie University, Halifax, Nova Scotia, Canada.
M.Sc. (1992) Chemistry, Dalhousie University, Halifax, Nova Scotia, Canada.
B.Sc. (1989) Chemistry, Dalhousie University, Halifax, Nova Scotia, Canada.

Successful Women Ceramic and Glass Scientists and Engineers: 100 Inspirational Profiles. Lynnette D. Madsen.
© 2016 The American Ceramic Society and John Wiley & Sons, Inc. Published 2016 by John Wiley & Sons, Inc.

Research Expertise: refractory minerals and additives, monolithic ceramics systems, high temperature composite materials, fiber reinforcement of ceramics, refractories for metal and carbon induction processes, biomass and waste to energy refractory, silica-based refractory applications, precast refractory practices

Other Interests: science and engineering educational outreach activities, increasing diversity in science and engineering, encouraging interest in materials and ceramic engineering fields

Key Accomplishments, Honors, Recognitions, and Awards

- The Refractories Institute, Scholarship Committee: 2010 to present
- The Unified International Technical Conference on Refractories (UNITECR) North American Executive Committee: 2008 to present
- Chair ACerS Meetings Committee 2014–2015, member 2011 to present

Dana Goski receiving a token of appreciation for organizing the technical program for UNITECR 2013, credit: The American Ceramic Society.

- ACerS Refractory Ceramics Division Allen Award Selection Committee Member: 2009 to present
- Fellow of the American Ceramic Society (ACerS), elected 2015
- ACerS Nominating Committee: 2012–2015
- Board of Trustees Advisor, The Edward Orton Jr. Ceramic Foundation, 2014
- University of Chicago, Leadership Insight, 2014
- Completed The Ohio State University, Alber Enterprise Center, Problem Solving & Management Development certificate program, 2014
- The American Ceramic Society (ACerS) Meetings Sub-committee on Technology and Manufacturing member: 2011–2014
- ACerS Ceramic Leadership Summit, Advisory Group member, 2014
- The Unified International Technical Conference on Refractories (UNITECR) Technical Program Chair, UNITECR, 2013
- Co-chair, Refractory Innovations and Novel Applications in Iron and Steel Manufacture Symposium, Materials Science and Technology Conference, 2008
- ACerS Refractory Ceramics Division: secretary 2004–2005, vice-chair 2005–2006, and chair 2006–2007

Dana Goski receiving her Ph.D. at Dalhousie University in 1997.

- ACerS Membership Committee: member 2002–2005, chair 2005–2006
- ACerS Jeppson Award Committee: member 2003–2005 and chair 2005–2006
- ACerS Central Ohio Section: officer chain 1999–2005, chair 2001–2002
- Co-Chair 45th Annual Symposium on Refractories, St. Louis Section & Refractory Ceramics Division Meeting, 2009
- Best Student Paper in Materials Physics, 6th Canadian Materials Science Conference, 1994

Biography

Early Life and Education

Dana Goski was raised in Canada. She received her early education crossing the country at Royal Canadian Air Force Base schools, her father being a mechanical engineer by education, and a navigation and communications officer. She finished high school in the Nova Scotia public school system, and continued with her college education at Dalhousie University in Nova Scotia.

Sport and coaching were an early influence. Exposure to coaches through the National Coaching Certification Program (NCCP), Sport Canada, Synchro Canada, and Sport Nova Scotia gave her the opportunity to work with world-class leadership and learn from mentors in various sports before she left college. She was a certified NCCP coach and course conductor for Synchro Canada, a Canada Winter Games athlete, manager and coach over the years. Upon reflection, these were wonderful learning opportunities to improve from mistakes, create strategies, manage budgets, develop leadership skills, and be inspired (as well as to inspire others).

After completing her undergraduate degree in chemistry with an honors thesis in nuclear analytical chemistry under Dr. Amares Chatt, she joined the physical chemistry graduate group of Dr. Jan Kwak (then chair of the chemistry department). Her master's thesis was developed as a joint project with the National Research Council (NRC) of Canada, specifically Dr. Kristoff J. Konzstowicz, of their ceramics group, who was nearby campus at that time. Her work focused on colloidal processing and surface chemistry of alumina and zirconia systems. In retrospect, the most formative event during her masters was attending and presenting a poster at the American Ceramic Society traditional Cocoa Beach meeting. The breadth of ceramic materials topics at the meeting, the opportunity to network with people with similar interests (remember this was just before the internet was available to the public), and a chance to watch the space shuttle launch while sitting on the beach of the meeting hotel all left a lingering impression.

Dana was encouraged to enter a doctoral program with Dr. Bill Caley, who at that time was the associate dean of engineering at the Technical University of Nova Scotia (TUNS). TUNS was an independent institution at the time, and the Department of Mining and Metallurgical Engineering was a great choice for Dana. It provided a different view to the broad reaching chemistry of materials. "During my Ph.D., I felt like I was encouraged by a team of mentors. That might be because we used equipment and resources in various departments, but there were no visible research boundaries among those departments. They shared their time, talent, and equipment. I felt like I was a product of not only the Mining and Metallurgical Engineering department, but also Chemical and Food Science Engineering. It was a great experience." Today, TUNS is part of Dalhousie University and became the Faculty of Engineering with a restructured materials engineering program, and an Institute for Research in Materials. After her Ph. D. laboratory work on natural mineral applications in engineered ceramics was wrapped up, and she started drafting her thesis, which she finished after taking a position in the United States.

Career History

When Dana and her husband Keith Souchereau, relocated to Columbus, Ohio in 1996, Dana joined Allied Mineral Products to support the refractory business for radioactive metal melting consolidation processes as a senior research engineer. Simultaneously, she was the technical consultant for their American Precast Refractories operation, located at that time in Gary, Indiana; she traveled there regularly to help optimize their cast refractory processes. At the time, Allied was about a 150 employee monolithic refractory business with international sales and no wholly owned corporate offshore production. They were an employee stock ownership company (known as ESOP company) and while that was unique in itself, she was also impressed with the leadership and research team . . . and also the warmer weather in comparison to Halifax.

Dana Goski at the ground breaking ceremony for Allied Mineral Products research and technology building expansion May 2014, left to right Jon R. Tabor (President), Doug Doza (Sr. Exec. VP) and Dana, credit: Madeline Palefsky.

The next decade was spent studying local and global raw materials, iron and steel refractories, light metals melting contact materials, biomass refractory specialization, regional product diversification, corporate development projects, and working with a second research laboratory and team in China. Being part of a sales-driven organization has made for a variety of research and professional experiences. "Some of the most remarkable experiences I have had were working with customers to solve complex materials challenges."

In 2014, Allied Mineral Products, Inc. has more than 700 employees with wholly owned production facilities in China (2), Europe, South Africa, United States (2), India and a joint venture in Brazil. Today, Dana focuses on innovation, strategic planning activities and research management.

3 Most Cited Publications

Title: Reaction sintering of kyanite and alumina to form mullite composites
Author(s): Goski, DG; Caley, WF
Source: Canadian Metallurgical Quarterly; volume: 38; issue: 2; pages: 119–126; published: April 1999
Times Cited: 9 (from Web of Science)

Title: Determination of trace-elements in acid-rain by reversed-phase extraction chromatography and neutron-activation
Author(s): Rao, RR; Goski, DG; Chatt, A
Source: Journal of Radioanalytical and Nuclear Chemistry; volume: 161; issue: 1; pages: 89–99; published: August 1992
Times Cited: 8 (from Web of Science)

Title: Lime-alumina-silica processing incorporating minerals
Author(s): Bryden, RH; Goski, DG; Caley, WF
Source: Journal of the European Ceramic Society; volume: 19; issue: 8; pages: 1599–1604; published: 1999
Times Cited: 4 (from Web of Science)

Challenges

On the personal/professional front, an earlier life challenge was managing a family with a small child. Having flexibility in my work structure, and a spouse with scheduling flexibility, certainly made life easier. A different type of challenge I experienced was defining a leadership style that is adaptable in an industry where one can find >50-year-old technology co-existing in the same plant as start of the art processes.

On being a woman in this field . . .

The field of refractory ceramics and composite materials is sparse with women; however, I connected with women who were customers, competitors, vendors, and in academia and they gave me an opportunity to network, learn, mentor, and encourage. I had very positive graduate school experiences and academic mentors. I was given a wide range of technical application opportunities in industry, and a family style support atmosphere within my company and my leadership team. My experiences were probably not typical, and may have more to do with the fact that I was working as an

Dana Goski with her husband Keith Souchereau and son Reid in 2012, credit: Dana Goski.

employee owner in an employee owned company. Have I had some Sara Sandberg-like experiences (as described in Sandberg's co-authored book: *Lean In: Women, Work, and the Will to Lead*)? Those stories are for sharing over a Canadian beer. The fact is that companies that have diversity within their leadership, as well as ranks, are statistically more profitable and innovative.

Words of Wisdom

Focus on what you want to achieve. Do reach out to network. Engage people who are experts in their area. Remember to create opportunities for other people to develop into experts.

Profile 38

Laura H. Greene

Chief Scientist
National High Magnetic Field Laboratory
Florida State University, University of Florida,
Los Alamos National Laboratory
and
Francis Eppes Professor of Physics
Florida State University
1800 E. Paul Dirac Drive
Tallahassee, FL 32310
USA

e-mails: lhgreene@magnet.fsu.edu;
lhgreene@illinois.edu
telephone: (850) 644-0311

Birthplace
Cleveland, Ohio, USA
Born
June 12, 1952

Laura Greene in the laboratory
in 1996.

Tags
❖ Academe
❖ Domicile: USA
❖ Nationality: American
❖ Caucasian
❖ Children: 9

Publication/Invention Record

>200 publications: h-index 48 (Web of Science)

Proudest Career Moments (to date)

Beyond my physics accomplishments, which I am so very proud of—amazing to understand something for a time that no one else knows about—I will get more general here. From the platform of being a professor, I am able to effect science beyond my own research. For example, I impact the research community through my engagement with professional societies, and I try to reach a broader audience, including young scientists in developing countries, with my outreach and public engagement endeavors.

Academic Credentials

Ph.D. (1984) Physics, Cornell University, Ithaca, New York, USA.
M.S. (1980) Experimental Physics, Cornell University, Ithaca, New York, USA.
M.S. (1978) Physics, The Ohio State University, Columbus, Ohio, USA.
B.S. (1974) Physics, The Ohio State University, Columbus, Ohio, USA.

Successful Women Ceramic and Glass Scientists and Engineers: 100 Inspirational Profiles. Lynnette D. Madsen.
© 2016 The American Ceramic Society and John Wiley & Sons, Inc. Published 2016 by John Wiley & Sons, Inc.

Research Expertise: experimental condensed matter physics investigating strongly correlated electron systems with a focus on elucidating the mechanisms of high-temperature superconductivity and developing methods for predictive design of new superconductors

Other Interests: working toward broadening our diversity in science, both in people and in subjects, as this enhances our resources for exploration and discovery in fundamental physics; ethics in science

- Plenary at International Physics Young Ambassador Symposium (middle and high school students from 29 countries), 2006, Taipei
- U.S. Delegate to 2nd IUPAP (The International Union of Pure and Applied Physics) Conference on Women in Physics, 2005, Rio de Janeiro
- Speaker at American Physical Society (APS)-04 and -08 Committee on the Status of Women in Physics (CSWP) Symposia
- Mentor for APS Committee on Minorities
- Maria Goeppert-Mayer Award used to lecture at minority institutions
- Dozens of women/minority students and postdocs in her lab. Mentor through research, teaching (large high-profile courses), and informally. Works with Committee on the Advancement of Women Chemists (COACh) International to increase the number and career success of women and all young scientists in developing nations
- In APS, a founding member of the Committee on Informing the Public, co-founder of the new Forum on Outreach and Engaging the Public (FOEP), and served on Maria Goeppert-Mayer and Bouchet Award Committees for women and minorities

Web Public Relations Outreach and Engagement:

- APS-TV interview for the Kavli Lecture, March 6, 2014, Denver, CO: https://www.youtube.com/watch?v=WWmtzgv102Y
- Physics World Interview, April 14, 2011, "Life after the cuprates": http://physicsworld.com/cws/article/multimedia/45686
- National Academy of Sciences Interview, 2009: http://www.nasonline.org/news-and-multimedia/podcasts/interviews/laura-greene.html
- The "Year of Science 2009" interview: http://www.yearofscience2009.org/themes_physics_technology/meet-scientists/
- APS Physics Central Interview, 2008: http://www.youtube.com/watch?v=ptswilP4yi0
- Women in Technology International (WITI) profiled in 2001: http://www.witi.com/center/witimuseum/womeninsciencet/2001/060901.shtml

Key Accomplishments, Honors, Recognitions, and Awards

- Associate Director, Center for Emergent Superconductivity, 2009 to present
- Elected Vice President, American Physical Society (APS), 2015; President Elect, 2016; President, 2017; and Past President, 2018

- Distinguished Visiting Professor, Institute for Basic Sciences Center for Correlated Electron Systems (IBS CCES), Seoul National University, Seoul, South Korea, 2015–2018
- University of Maryland Carr Lectureship, 2014
- John S. Guggenheim Foundation Fellowship, 2009–2010
- Editor-in-Chief, Reports on Progress in Physics, Institute of Physics Publishing, Bristol, UK, 2007–2010
- Visiting Fellow Commoner, Trinity College, Cambridge University, UK, Lent Term, 2010
- Center for Advanced Study Professor of Physics, University of Illinois, elected 2009
- Visiting Professor, University of California at Irvine, Fall 2009
- Center for Advanced Study Professor of Physics, University of Illinois, elected 2009
- Fellow, Institute of Physics, "FinsP," UK, elected 2007
- Brookhaven National Laboratory "Condensed Matter Sciences Distinguished Lecturer," 2007
- Tufts University "Kathryn McCarthy Lecturer," 2007
- Center for Advanced Study Research Associate, University of Illinois Urbana-Champaign, 2006–2007
- Member, National Academy of Science, elected 2006
- Phi Beta Kappa Visiting Scholar, 2005–2006
- The Ohio State University Department of Physics "1st Distinguished Alumnus Lecturer," 2005
- Fellow, Phi-Kappa-Phi Honor Society, elected 2001
- Chosen as New Student Convocation Speaker for the ~6000 incoming freshman class at the University of Illinois, Urbana-Champaign, 2001
- Center for Advanced Study Resident Associate, University of Illinois Urbana-Champaign, 2000–2001
- Swanlund Endowed Chair, University of Illinois at Urbana-Champaign, named 2000
- E.O. Lawrence Award for Materials Research, Department of Energy, 1999
- Fellow, American Academy of Arts and Sciences, elected 1997
- APS Centennial Speaker, chosen to give a series of colloquia commemorating the first 100 years of the American Physical Society, selected 1997
- Kalamazoo College, Department of Physics "Jennifer Mills Lecturer," 1997
- Center for Advanced Study, Beckman Associate, University of Illinois Urbana-Champaign, 1996–1997
- Fellow, The American Association for the Advancement of Science, elected 1996
- National Science Foundation MAPP Workshop Committee. Neal Lane and other NSF officials assembled this committee in 1995 to answer questions put forth by Senate
- Superconductivity Review, Gordon and Breach Science Publishers, 1992–1995
- Maria Goeppert-Mayer Award of the APS, 1994

- Fellow, APS, elected 1993
- Beckman Award from the University of Illinois Campus Research Board, 1993
- Bellcore, Red Bank, NJ: Award of Excellence, 1989
- Hazel S. Brown Scholarship Award, The Ohio State University, 1974

Numerous national and international appointments/elected positions/committees.

Biography

Early Life and Education

From her earliest memories, she had a keen interest in all things science; throughout her life, she was also absorbed in the music arena.

At college, all her attention turned to science. Although she was met with resistance in the science or engineering programs she approached, the Physics Department at The Ohio State University welcomed her, and she threw herself into the field. She received her bachelor and master's degrees there, and in 1984 received a doctorate in physics from Cornell University investigating the linear and nonlinear far-infrared properties of materials. She then joined Bell Laboratories, and then Bellcore, where she researched thin-film growth and tunneling of metallic multilayers, heavy fermions, superconductor–semiconductor hybrid structures, and high-temperature superconductors.

Credits: http://www.youtube.com/watch?v=ptswilP4yi0; http://physics.illinois.edu/people/profile.asp?lhgreene; http://www.nasonline.org/news-and-multimedia/podcasts/interviews/laura-greene.html.

Career History

In 1992, she joined the senior physics faculty at the University of Illinois at Urbana-Champaign. Her research continues in experimental condensed matter physics, investigating strongly correlated electron systems and focusing primarily on revealing the mechanisms of unconventional superconductivity by planar tunneling and point-contact electron spectroscopies. Her research also involves growing novel materials and developing methods for predictive design of new families of superconductors. She is recognized for her work on superconductor/semiconductor proximity effects, elucidating the physical properties of the pure and doped high-temperature superconductors, the discovery of broken time-reversal symmetry in high-temperature superconductors, and spectroscopic studies of the electronic structure in heavy-fermion metals.

Her service to science includes presently being Vice President of the APS (President in 2017), Chair of the Division of Materials Physics (DMP) of the APS, Board of Directors of the American Association for the Advancement of Science (AAAS), and Chair of the Board of Governors for the International Institute for Complex and Adaptive Matter (I2CAM). She is serving her second 6-year term on the IUPAP and is Vice Chair of the Commission on the Structure and Dynamics of Condensed Matter (C10). In the APS, she has served on Council, Executive Board, Committee on Committees, was a founding member of the Committee on Informing the Public, co-founder of the new

Forum on Outreach and Engaging the Public (FOEP), and served nominating, fellowship, and multiple prize committees. She has completed two terms on the NAS Board on Physics and Astronomy (BPA), and served for 13 years on the Basic Energy Sciences Advisory Committee (BESAC) for the DoE. Greene has been a visiting scientist in Orsay, UCI, and Cambridge, UK,

Laura Greene's COACh class in Medan, Indonesia, June 2014.

and has co-chaired and advised numerous international conferences. She works with COACh International to increase the number and career success of women and all young scientists in developing nations. Her various editorial positions include Reports on the Progress in Physics (Editor-in-Chief), Philosophical Transactions A, and Current Opinions in Solid State & Materials Science (COSSMS).

Greene is a member of the National Academy of Sciences, and a Fellow of the American Academy of Arts and Sciences, Institute of Physics (UK), American Association for the Advancement of Science, and the APS. She has been a Guggenheim Fellow, and received the E.O. Lawrence Award for Materials Research, the Maria Goeppert-Mayer Award, and the Award of Excellence from Bellcore.

She is married with two children and seven stepchildren, and is a breast cancer survivor.

Additional credit: http://www.nasonline.org/news-and-multimedia/podcasts/interviews/laura-greene.html.

A recent photo of Laura Greene, 2014.

3 Most Cited Publications

Title: Structural and physical properties of the metal (M) substituted $YBa_2Cu_{3-x}M_xO_{7-y}$ perovskite
Author(s): Tarascon, JM; Barboux, P; Miceli, PF; et al.
Source: Physical Review B; volume: 37; issue: 13; pages: 7458–7469; published: May 1, 1988
Times Cited: 776 (from Web of Science)

Title: Crystal substructure and physical properties of the superconducting phase $Bi_4(Sr, Ca)_6Cu_4O_{16+x}$
Author(s): Tarascon, JM; Lepage, Y; Barboux, P; et al.

Source: Physical Review B; volume: 37; issue: 16; pages: 9382–9389; published: June 1, 1988
Times Cited: 624 (from Web of Science)

Title: Oxygen and rare-earth doping of the 90-K superconducting perovskite $YBa_2Cu_3O_{7-x}$
Author(s): Tarascon, JM; McKinnon, WR; Greene, LH; et al.
Source: Physical Review B; volume: 36; issue: 1; pages: 226–234; published: July 1, 1987
Times Cited: 560 (from Web of Science)

Challenges

A key challenge is balancing it all. I used to say my life was a "semi-controlled crash"; I have since removed the word "controlled." But it is worth every second—the passion for the science gets you through more than you ever thought.

On being a woman in this field . . .

Being a minority in a field always poses its challenges. In physics, the sociality of the field is basically to "weed out" all but the "most talented" students (and "most talented" is not well defined), and there is the explicit or implicit belief that "if you have not made your great discovery (such as a Nobel Prize) by your early 20s, you are not worthy." This approach is counter to what many women view and accept as a positive challenge. Furthermore, there are many "little things" that weigh on you. For example, as a woman in physics you often get a very polite question, "why would you major in physics?" Nothing negative is meant by that—usually it is just friendly curiosity, but you get that often and the men usually do not. I have heard this as being compared to running a marathon with an extra few ounces of weight on each ankle—no big deal for the first 15 miles, but after that and beyond, it becomes a race breaker. Having more women in physics would positively alter the sociality of the field and public's perception and would certainly help alleviate these challenges.

Words of Wisdom

Some of the "Greene's rules":
a. In times of stress: Keep a sense of humor and throw money at it.
b. Keep in contact with your girlfriends.
c. It is more important to look confident than to be confident.
d. For travel: Never save money on shoes, phones, or laptops.
e. EVERYTHING is an opportunity. Stay aware and don't miss even one. You never know which one is the key.

Clare P. Grey

Geoffrey Moorhouse Gibson Professor of Chemistry
Department of Chemistry
University of Cambridge
Lensfield Road
Cambridge CB2 1EW
UK
and
Adjunct Professor
Department of Chemistry
Stony Brook University
669 Nicolls Road
Stony Brook, NY 11794-3400
USA

e-mail: cpg27@cam.ac.uk
telephone: +44 1223 336509

Dr. Clare Grey, ca. 2002.

Tags

❖ Academe
❖ Domicile: USA and UK
❖ Nationality: British
❖ Caucasian
❖ Children: 0

Birthplace
Middlesbrough, UK
Born
March 17, 1965

Publication/Invention Record

>300 publications: h-index 53

Academic Credentials

D.Phil. (1991) Chemistry, University of Oxford, Oxford, United Kingdom.
B.A. (1987) Chemistry, University of Oxford, Oxford, United Kingdom.

Research Expertise: solid-state materials, nuclear magnetic resonance (NMR), batteries, supercapacitors, fuel cells, zeolites

Other Interests: outreach to the public. For example, Dr. Grey has served as a panelist for the BBC World Service "The future of renewable energy" for the program "The Forum" (http://www.bbc.co.uk/programmes/p01g94yj) in 2013.

Successful Women Ceramic and Glass Scientists and Engineers: 100 Inspirational Profiles. Lynnette D. Madsen.
© 2016 The American Ceramic Society and John Wiley & Sons, Inc. Published 2016 by John Wiley & Sons, Inc.

In addition, she was a member of the Emerging Technologies Panel for the World Economic Forum (2012–2014). She also helped write the 2013 report "Energy Harnessing: New Solutions for Sustainability and Growing Demand" (http://www.weforum.org/reports/energy-harnessing-new-solutions-sustainability-and-growing-demand).

Key Accomplishments, Honors, Recognitions, and Awards

- Arfvedson Schlenk Award that honors outstanding scientific and technical achievements in the field of lithium chemistry and is given by the Gesellschaft Deutscher Chemiker (GDCh), 2015
- Davy Medal from the Royal Society, 2014
- Honorary Ph.D. Degree from Lancaster University in England, 2013
- Laukien Award from the Experimental NMR Conference for innovative applications of solid-state NMR to energy storage systems, especially lithium ion batteries, 2013

Prof. Clare Grey receives the Laukien Prize from Richard Ernst in 2013.

- Research Award, International Battery Association, 2013
- Honorary Ph.D. Degree "Docteur Honoris Causa" from the University of Orleans in France, 2012
- Kavli Medal from Royal Society, 1st recipient of this medal, 2011
- Fellow of the Royal Society, 2011
- Fellow of the International Society of Magnetic Resonance, 2011
- Royal Society of Chemistry John Jeyes Award, 2010
- AMPERE Award, 2010
- Department of Energy Award to fund the Northeastern Center for Chemical Energy Storage (NECCES), 2009
- Vaughan Lecturer in connection with the Rocky Mountain Conference on Magnetic Resonance, 2008
- New York State Foundation for Science, Technology and Innovation (NYSTAR) Award, 2008

Clare Grey receives an honorary doctorate from the University of Orleans, 2012.

- Research Award of the Battery Division, Electrochemical Society, 2007
- National Science Foundation (NSF) Professional Opportunities for Women in Research and Education (POWRE) Award, 2000
- Camille and Henry Dreyfus Foundation Teacher–Scholar, 1998
- Alfred P. Sloan Research Fellowship, 1998
- Cottrell Scholar, 1997
- DuPont Young Investigator Award, 1997
- NSF National Young Investigator Award, 1994
- Royal Society Postdoctoral Fellow, University of Nijmegen, The Netherlands, 1991–1992
- Junior Research Fellow, Balliol College, University of Oxford, 1990

Biography

Early Life and Education

Clare Grey was born in the UK but lived in The Netherlands and Belgium for 5 and 7 years, respectively, before finishing the last 4 years of high school in the UK (in St. Albans). She received a B.A. (in 1987) and a D.Phil. (in 1991) from the University of Oxford. Afterward, she spent a year at the University of Nijmegen, as a Royal Society Postdoctoral Fellow (1991) in the laboratory of Prof. W.S. Veeman, and 2 years as a Visiting Scientist at DuPont Central Research and Development in Wilmington, Delaware (1992–1993).

Career History

Dr. Grey joined the faculty at Stony Brook University in 1994; she was promoted to Associate Professor in 1997 and Full Professor in 2001. She has made forays to France as a Visiting Professor at the Université Louis Pasteur, Strasbourg (in 2000) and at the Université de Picardie, Jules Verne, Amiens (in 2007 and 2008, respectively).

Since 2002, she has served as the Associate Director, National Science Foundation Center for Environmental Molecular Science at Stony Brook University. In 2009, she became the Director of the Northeastern Center for Chemical Energy Storage, a Department of Energy Frontier Center.

Prof. Clare Grey in her office, 2011.

She lived a double life from 2009 to 2015 also serving first as the Head of the Inorganic Sector, and then the Head of Materials Chemistry, and as the Geoffrey Moorhouse Gibson Professor in Chemistry at the University of Cambridge in the UK.

She resigned the Stony Brook post in 2015 remaining as an Adjunct Professor at Stony Brook.

Her research interests include the use of solid-state NMR and diffraction methods to investigate structure and dynamics in materials for energy storage and conversion and environmental chemistry. In the lithium ion battery area, she uses lithium NMR to determine local structure and mechanisms for intercalation and deintercalation of Li^+ in a wide range of electrode materials including layered materials, spinels, and phosphates. She developed a systematic understanding of the causes of the large NMR shifts that are often seen in these systems, and showed how this information could be used to obtain local structure and oxidation state changes. Her recent interests include the development of *in situ* NMR methods to investigate batteries and supercapacitors. She continues to work on the development and application of novel NMR methodology to investigate complex and disordered materials.

Prof. Clare Grey (in brown near the middle) with her research group, 2011.

3 Most Cited Publications

Title: Electrodes with high power and high capacity for rechargeable lithium batteries
Author(s): Kang, KS; Meng, YS; Breger, J; et al.
Source: Science; volume: 311; issue: 5763; pages: 977–980; doi: 10.1126/science.1122152; published: February 17, 2006
Times Cited: 684 (from Web of Science)

Title: Selective oxidation of methane to synthesis gas using transition metal catalysts
Author(s): Ashcroft, AT; Cheetham, AK; Foord, JS; et al.
Source: Nature; volume: 344; issue: 6264; pages: 319–321; published: March 22, 1990
Times Cited: 504 (from Web of Science)

Title: Determination of the quadrupole coupling constant of the invisible aluminum spins in zeolite HY with H-1/Al-27 TRAPDOR NMR
Author(s): Grey, CP; Vega, AJ
Source: Journal of the American Chemical Society; volume: 117; issue: 31; pages: 8232–8242; doi: 10.1021/ja00136a022; published: August 9, 1995
Times Cited: 235 (from Web of Science)

Challenges

A major challenge involves juggling research, teaching, and administration, while maintaining time to think.

Words of Wisdom

Nothing compares to the freedom of academia and the joys of working in an area that one is deeply committed to.

Sources used to create this profile

- CV, http://www.ch.cam.ac.uk/sites/ch/files/CPG_short_CV_web_2014.pdf, accessed December 13, 2014.
- http://www.rakcam.com/en/internationalworkshops/invitedspeakers/invitedspeakersprofiles/09-05-06/Clare_P_Grey.aspx, accessed February 16, 2015.
- http://www.chem.stonybrook.edu/awards/, accessed February 16, 2015.
- Laukien Prize details, http://www.enc-conference.org/LaukienRecipients/PastRecipients/tabid/73/Default.aspx, accessed February 16, 2015.

Profile 40

Sossina M. Haile

Walter P. Murphy Professor of Materials Science
and Engineering
Professor of Applied Physics
McCormick School of Engineering
and Applied Science
Northwestern University
Evanston, IL 60208
and
Carl F. Braun Professor of Materials Science
and of Chemical Engineering (on leave)
California Institute of Technology
Pasadena, CA 91125
USA

e-mail: sossina.haile@northwestern.edu
telephone: (847) 491-3197

Sossina Haile,
1994. Credit:
University of
Washington PR
office.

Tags

❖ Academe
❖ Domicile: USA
❖ Nationality: American
❖ African American
❖ Children: 2

Birthplace

Addis Ababa, Ethiopia

Born

July 28, 1966

Publication/Invention Record

>150 publications: h-index 39

Proudest Career Moment (to date)

Transitioning our fundamental scientific advances in superprotonic solid acids to a viable fuel cell technology.

Academic Credentials

Ph.D. (1992) Materials Science and Engineering (Ceramic Science), Massachusetts Institute of Technology, Cambridge, Massachusetts, USA.
M.S. (1988) Materials Science and Engineering, University of California, Berkeley, California, USA.
B.S. (1986) Materials Science and Engineering, Massachusetts Institute of Technology, Cambridge, Massachusetts, USA.

Successful Women Ceramic and Glass Scientists and Engineers: 100 Inspirational Profiles. Lynnette D. Madsen.
© 2016 The American Ceramic Society and John Wiley & Sons, Inc. Published 2016 by John Wiley & Sons, Inc.

Research Expertise: materials chemistry, fuel cells, solid-state ionics, high-temperature electrochemistry, electroceramics, materials synthesis; thermoelectrics, crystallography

Other Interests: inclusive mentoring, education, and outreach across many age groups and educational levels

Key Accomplishments, Honors, Recognitions, and Awards

- World Academy of Ceramics Laureate for the outstanding contribution opening new horizons in electrochemical ceramic research as well as for the success in fabricating and demonstrating innovative solid electrolyte fuel cells based on these achievements, 2012
- Member of the National Materials Advisory Board, a committee serving the National Academies of Sciences and of Engineering, 2005–2011
- Outstanding Women in Science Lecturer, 2010
- Chemical Pioneers Award of the Chemical Heritage Foundation, 2010
- National Science Foundation (NSF) American Competitiveness and Innovation Fellow for her timely and transformative research in the energy field and her dedication to inclusive mentoring, education, and outreach across many levels, 2008
- Named as one of 12 people to watch by Newsweek Magazine in its 2008 end-of-the-year issue
- J.B. Wagner Award of the High Temperature Materials Division of the Electro-chemical Society, 2001
- Principal editor for the Journal of Materials Research, 1997–2001
- Coble Award from the American Ceramics Society, 2000
- Minerals, Metals and Materials Society (TMS) Robert Lansing Hardy Award, 1997
- NSF National Young Investigator Award, 1994–1999
- Humboldt Fellowship, 1992–1993
- Fulbright Fellowship, 1991–1992
- AT&T Cooperative Research Fellowship, 1986–1992

Biography

Early Life and Education

After soldiers arrested and nearly killed her father, Sossina Haile (then Sossina Getat-chew) left Ethiopia with her family during the coup in the mid-1970s. They first settled in rural Minnesota where she continued with her primary and secondary education. Upon completion of high school, she proceeded to university and received a B.S. from the Massachusetts Institute of Technology (MIT), an M.S. from the University of California, Berkeley, and then returned to MIT to complete her Ph.D. in 1992. As part of her studies, she spent 2 years at the Max Planck Institute for Solid State Research in Stuttgart, Germany, first as a Fulbright Fellow and then as a Humboldt Fellow.

Career History

Upon completion of her postdoctoral studies, Dr. Haile spent 3 years as an assistant professor at the University of Washington, Seattle. She then moved due south to join the California Institute of Technology (Caltech) faculty in 1996. She subsequently received a series of promotions to full professorship in 2006 and was named the Carl F. Braun Professor of Materials Science and of Chemical Engineering in 2012. (She was in the first pair of women to receive chair professorships in engineering at Caltech.) In 2014, she began her transition to Northwestern University, where she will serve as the Walter P. Murphy Professor of Materials Science and Engineering.

Sossina Haile in the lab with graduate student Rob Usiskin, 2013. Credit: Caltech EAS Communications office.

Her research in solid-state ionics and electrochemistry has been supported from several sources, including NSF, the Army Research Office, the Stanford Global Climate and Energy Program (GCEP), the Moore Foundation (through its support of the Caltech Center for Sustainable Energy Research (CCSER)), the Defense Advanced Research Projects Agency (DARPA), the Office of Naval Research, the California Energy Commission, the Department of Energy (DOE), Advanced Research Projects Agency—Energy (ARPA-e), the Powell Foundation, and the Kirsch Foundation. Industrial support has been provided by eSolar, SAFCell Inc. (a start-up out of her laboratory), LiOx, General Motors, EPRI (formerly Electric Power Research Institute), HRL (formerly Hughes Research Labs), and Honeywell (now General Electric).

She is the mother of two children.

3 Most Cited Publications

Title: A high-performance cathode for the next generation of solid-oxide fuel cells
Author(s): Shao, ZP; Haile, SM
Source: Nature; volume: 431; issue: 7005; pages: 170–173; doi: 10.1038/nature02863; published: September 9, 2004
Times Cited: 1424 (from Web of Science)

Title: Fuel cell materials and components
Author(s): Haile, SM
Source: Acta Materialia; volume: 51; issue: 19; pages: 5981–6000; doi: 10.1016/j.actamat.2003.08.004; published: November 25, 2003
Times Cited: 539 (from Web of Science)

Title: Solid acids as fuel cell electrolytes
Author(s): Haile, SM; Boysen, DA; Chisholm, CRI; et al.
Source: Nature; volume: 410; issue: 6831; pages: 910–913; doi: 10.1038/35073536; published: April 19, 2001
Times Cited: 512 (from Web of Science)

Challenges

As scientific research becomes more complex and interdisciplinary, it is increasingly important to be part of a team. Being included in those teams has sometimes been a challenge.

On being a woman in this field . . .

The stagnation in the fraction of women that enters the field is heartbreaking. I am grateful for the women and the people of color who have gone before me and opened the way, but I wonder how I can transfer that progress to the next generation so that we remain on (or return to) an upward trajectory.

Words of Wisdom

Your self-worth is ultimately much more important than how others see you. Maintain your own internal high standards, and even if the rest doesn't follow, you'll take pride in what you see in the mirror.

Kersti Hermansson

Professor of Inorganic Chemistry
The Ångström Laboratory
Uppsala University
Box 538
SE-75121 Uppsala
Sweden

e-mail: kersti@kemi.uu.se
telephone: +46 18 4713767

Recent photo of Kersti Hermansson.

Birthplace
Västerås, Sweden

Tags
- ❖ Academe
- ❖ Domicile: Sweden
- ❖ Nationality: Swedish
- ❖ Caucasian

Publication/Invention Record

>200 refereed publications:
h-index 36 (Web of Science)

Proudest Career Moments (to date)

I am proud when a Ph.D. student of mine demonstrates scientific maturity and depth during her/his Ph.D. dissertation defense. A doctoral defense is a tough event in Sweden since the external examiner called an "opponent" discusses the thesis work in depth with the candidate for several hours and the audience is large with the supervisor, examination committee, colleagues and other students, and family members; it is usually a very revealing exercise.

On a more personal note, it was fun (and quite the honor) to be elected into The Royal Swedish Academy of Sciences.

Academic Credentials

Ph.D. (1984) Chemistry, Uppsala University, Uppsala, Sweden.

Research Expertise: development of multiscale models to bring large-scale computer simulations closer to the complex dynamical systems of the real world; development of computational vibrational spectroscopy methodology; applications such as catalysis, nanochemistry, metal oxide nanoparticles, hydration phenomena, water and aqueous ionic solutions; fundamental chemical bonding and intermolecular interaction

Successful Women Ceramic and Glass Scientists and Engineers: 100 Inspirational Profiles. Lynnette D. Madsen.
© 2016 The American Ceramic Society and John Wiley & Sons, Inc. Published 2016 by John Wiley & Sons, Inc.

Other Interests: promotion of the standing and quality of theoretical calculations in materials research; promotion of the standing and quality of theoretical calculations in the undergraduate education in chemistry

Key Accomplishments, Honors, Recognitions, and Awards

- Member of the European Commission (EC) Operative Management Board of the newly launched European Materials Modelling Council under Horizon 2020, 2014 to present

- One of the PIs of the Swedish national strategic research program in e-science, *eSSENCE*, 2010 to present; program coordinator, 2010–2011

- Elected deputy trustee of The Nobel Foundation, 2010 to present

- Appointed by the Norwegian Ministry of Education as member of the Norwegian national committee for academic tenure in chemistry, 2012–2019

- Commissions of trust at a national level and elected member of national boards: for the Swedish Research Council (VR), the Swedish National Infrastructure for Computing (SNIC), 2012–2016

Kersti Hermansson (as part of a small delegation from Uppsala University) discussing with professors from Al-Baha University about the development of scientific staff exchanges between Al-Baha University and Uppsala University, on site in Saudi Arabia, 2011. Credit: Dr. Peter Sundin, Director of the International Science Programmes (ISP), Uppsala University.

- Member of the Management Committee of the EU-COST Action CM1104 (Reducible oxide chemistry, structure and function), 2012–2016

- Elected foreign member of the Macedonian Academy of Sciences and Arts, 2015

- Elected board member of the strategic Norwegian e-Science Centre for Theoretical and Computational Chemistry (CTCC) in Oslo and Tromsö (Norwegian Research Council), 2010–2014

- Appointed Chairman and rapporteur of the session "Future for modelling: materials and industrial process" at the LET'S (Leading Enabling Technologies for Societal Challenges) in Bologna in connection with the Italian "Presidency of the Council of the European Union," 2014

- Invited as rapporteur for the EC–US workshop (San Francisco) on future EC–US cooperation in materials modeling, 2014

- Elected by the School of Biotechnology and Chemistry as their candidate for one of approximately 10 Guest Professorships announced worldwide by KTH and subsequently selected as an awardee, 2008, and was Guest Professor in the Group of Theoretical Chemistry and Biology, 2008–2013

- Served on boards of several of the national supercomputer centers (NSC, PDC, UPPMAX) and for the national organization for the Swedish University Computer Network (SUNET), ca. 2002–2012
- Elected Honorary Professor at the Department of Ion Physics and Applied Physics, Innsbruck University, 2009
- Elected member of The Royal Swedish Academy of Sciences, which decides on the Nobel prizes in the sciences, 2007
- The "Norblad-Ekstrand" medal in gold from the Swedish Chemical Society, 2003
- Elected member of The Royal Society of Science, the oldest academy in Sweden, 2002
- The "Oskarspriset" from Uppsala University, 1988
- The "Letterstedska priset" from the Swedish Royal Academy of Sciences (KVA), 1987

Biography

Early Life and Education

I was born and raised in Sweden. I received my basic education through the Swedish public school system. After finishing my high school education in Västerås, I attended 1 year at Bryn Mawr College (BMC) in the United States on a scholarship (granted by BMC). After returning to Sweden, I began university studies at Uppsala University focusing on mathematics, chemistry, and physics for 3 years and received my M.Sc. degree after a final year at the teachers' training college in Uppsala.

As a doctoral student at Uppsala University, I started out as a crystallographer specializing in X-ray and neutron diffraction to help characterize the features of intra- and intermolecular bonding in crystals via their electron distribution, which can give information about the nature of covalent bonding, molecular polarization, electron transfer, and so on. This expertise was then combined with quantum mechanical calculations that were the start of my computational chemistry career. The crystals I started with contained crystal water, which marked the start of my interest in water.

Career History

Here is a quick summary of my career path:

- Postdoctoral stay (1984–1986) at the IBM Corporation in Kingston, NY with Professor Enrico Clementi.
- Initial faculty appointment in inorganic chemistry (1986), Uppsala University, Uppsala, Sweden.
- Full Professor of Inorganic Chemistry (2000), Uppsala University.
- During 2008–2013, I was also part-time Guest Professor at The Royal Institute of Technology (KTH), Stockholm, Sweden.

Kersti Hermansson giving a lecture about computational chemistry in connection with a visit by a small delegation from Uppsala University to Al-Baha University in Saudi Arabia with the overall purpose to discuss staff exchanges between the universities and the development of the Master's program curricula at Al-Baha University.

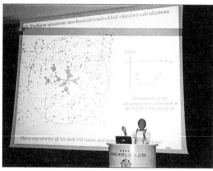

Kersti Hermansson lecturing at the 29th International Conference on Solution Chemistry in Slovenia in August 2005. (See insert for color version of figure.)

On a scholarship from the Swedish Research Council, I spent a little more than 2 years in the dynamic computational laboratory of Dr. Enrico Clementi and Dr. Gina Corongiu at IBM in Kingston, NY, where I learnt much about the tricks of the trade of large-scale computation, molecular dynamics simulations, force-field development, and parallel programming, the latter from one of the masters in the field, Dr. Clemens Roothaan (University of Chicago, "Papa Clem"). Many from our tight group of 20 postdoctoral fellows in Clementi's laboratory have since made significant impacts on the development of theoretical chemistry worldwide and several are still close friends of mine.

When I returned to Uppsala University as a faculty member, I started the rather painstaking (as every junior researcher knows) struggle of building a new, yet sustainable, research line of my own. Continued financial support from the Swedish Research Council, and perhaps the fact that I was the only theoretical practitioner in an otherwise experimental department, helped me to chisel out my research profile and role and build a strong research group in the field of condensed matter chemistry modeling. Even today, there are only a few research groups with a focus on *materials chemistry* modeling in Sweden, and even fewer groups involved in (multiscale) model development in materials chemistry. In view of all the new insight that this research branch provides, it is clearly an area that needs further reinforcement.

For a 5-year period, I was a part-time Guest Professor at KTH, which gave me new insights about a range of problems at the intersection of chemistry, physics, and biology.

If I shall make my own subjective list of my key accomplishments, they are (i) the scientific structure-property relations and models that I have discovered in my research and problems that can be explained through my research, (ii) the smart young scholars whom I have hopefully helped to become even more accomplished, and (iii) the establishment of an accomplished research group in condensed matter theoretical chemistry.

3 Most Cited Publications

Title: Blue-shifting hydrogen bonds
Author(s): Hermansson, K
Source: Journal of Physical Chemistry A; volume: 106; issue: 18; pages: 4695–4702; published: May 9, 2002
Times Cited: 312 (from Web of Science)

Title: Atomic and electronic structure of unreduced and reduced CeO_2 surfaces: a first-principles study
Author(s): Yang, ZX; Woo, TK; Baudin, M; Hermansson, K
Source: Journal of Chemical Physics; volume: 120; issue: 16; pages: 7741–7749; published: April 22, 2004
Times Cited: 198 (from Web of Science)

Title: Hydration of the calcium ion. An EXAFS, large-angle X-ray scattering, and molecular dynamics simulation study
Author(s): Jalilehvand, F; Spångberg, D; Lindqvist-Reis, P; Hermansson, K; Persson, I; Sandström, M
Source: Journal of the American Chemical Society; volume: 123; issue: 3; pages: 431–441; published: January 24, 2001
Times Cited: 169 (from Web of Science)

Challenges

Moving abroad for a couple of years for my postdoctoral research was among the most palpable challenges for me. I reckon that moving abroad for any significant time to a new research environment poses a challenge to anyone. How well can you manage on your own? It is an exciting test of your competences and an opportunity to grow scientifically and develop your networking capabilities.

Another challenge, from my own experience, is the painstaking endeavor of changing research fields or adding some totally new and significant area to your current expertise. I have accepted this challenge a few times and although in the long run it is rewarding, the path toward it is sometimes perilous.

In my current research, I am concerned with the following overriding challenge: to push the boundaries of electronic and atomistic materials modeling closer to realistic applications. The main barrier to achieving this goal is the lack of accurate and efficient *interaction models* that manage to describe the physics and chemistry of complex compounds, and the chemical and physical processes that take place on their surfaces and interfaces (catalysis, adhesion, electrode processes, and so on). A large and conscious effort to prioritize such method and model development efforts (and the associated software development) is needed both within the scientific community and by grant-giving bodies.

For the latter, there should be a methodology to attach a high merit (and support) to the painstaking efforts of method and model development, which are often not directly reflected in a high publication rate, but which are very beneficial to the scientific community. This recognition is particularly crucial for junior researchers in the beginning of their careers. Personally, I try to accept this challenge in my research, and more broadly with my involvement in the new European Materials Modelling Council, participation in national and European grant evaluations and research assessments, and organization of thematic workshops, meetings, etc.

On being a woman in this field . . .

There are few professionally established women in the field of materials modeling in Europe. On the whole, I have had positive experiences from being a woman in this (still very male-dominated) field and I know that I have been, and am being, listened to by other scientists and decision-makers (both women and men). Our community generally seems to work quite well collectively in moving the understanding in this research field ahead.

Words of Wisdom

- *Theoretical ceramics research—powers and opportunities.* Computer simulation methods for materials are powerful techniques as they are able to provide the very much needed structural and mechanistic detail! Rightly used they deepen our "understanding", give rise to new ideas, and allow us to test, predict, and play with tailored (computer) experiments under very controlled conditions. This ability is tremendously valuable, particularly in tandem with laboratory experiments.

- *Not reality (yet?).* With the powers and opportunities of simulations comes a responsibility: simulations today can display a tantalizing detail that may be mistaken for reality. It is important to remember (not the least for the practitioners in the field) that simulations just reflect the properties of the man-made models employed. With better models, the *predictive power* of simulations will grow, as will their impact on the experimental sciences and (hopefully) on industrial procedures.

- *Novelty and substance?* More surface and less substance may be a good target in the design of clever and competitive ceramics, but the opposite is hopefully true for us scientists (and engineers) who are supposed to do the design job. Erroneous and carelessly obtained results are detrimental to the scientific community. So is poor referencing (and it is unethical as well). Doctoral students need encouragement (from their busy supervisors) to check and double-check every number that will be published and to check that every reference, recent or old, that ought to be included is in fact included.

Profile 42

Huey Hoon Hng

Associate Professor
and Associate Chair (Academic)
School of Materials Science and Engineering
Nanyang Technological University
Nanyang Avenue, Singapore 639798

e-mail: ashhhng@ntu.edu.sg
telephone: +65 6790 4140

Huey Hoon Hng at work in her office with students' thank you cards and photos in the background.

Birthplace
Singapore

Born
1971

Publication/Invention Record

>200 publications: h-index 41 (Web of Science)
1 book

Tags

❖ Administration and Leadership
❖ Academe
❖ Domicile: Singapore
❖ Nationality: Singaporean
❖ Asian
❖ Children: 1

Proudest Career Moment (to date)

In 2008, I received the Nanyang Teaching Excellence Award, which "recognizes the dedication and achievements of faculty members who displayed excellent teaching practices and enriched the learning experiences of NTU undergraduates." This award has reflected on my passion to teach and inspire students to learn.

Academic Credentials

Postgraduate Diploma of Teaching in Higher Education (2001), National Institute of Education, Singapore.
Ph.D. (1999) Materials Science and Metallurgy, University of Cambridge, Cambridge, United Kingdom.
M.Sc. (1995) Materials Science, National University of Singapore, Singapore.
B.Sc. (1994) Materials Science, National University of Singapore, Singapore.

Successful Women Ceramic and Glass Scientists and Engineers: 100 Inspirational Profiles. Lynnette D. Madsen.
© 2016 The American Ceramic Society and John Wiley & Sons, Inc. Published 2016 by John Wiley & Sons, Inc.

Research Expertise: thermoelectric materials, energy storage materials, energetic materials, nanostructured materials, materials characterization (processing–microstructure–property relationship)

Other Interests: teaching

Key Accomplishments, Honors, Recognitions, and Awards

- Editorial Board Member for Journal of Technology, 2011 to present
- Editorial Board Member for Applied Sciences, 2010 to present
- Science and Engineering Research Council (SERC) Technical Review Panel (TRP), Chemicals, Materials and Energy Cluster, co-Chairperson, 2014–2016
- A∗STAR review committee, 2010–2011
- Jury member for L'Oreal Singapore for Women in Science National Fellowship, 2010 and 2011
- Cheng Jialiang (Prof. Hng's doctoral student) won the Graduate Student Award at the E-MRS, Strasbourg, France, in recognition of the outstanding paper contributed to Symposium D, 2009
- Nanyang Teaching Excellence Award, 2008
- Council member for Microscopy Society of Singapore, 2000–2001
- Nanyang Technological University Overseas Scholarship for pursuing the PhD degree at University of Cambridge, 1996–1999
- National University of Singapore Research Scholarship for pursuing the M.Sc. degree at National University of Singapore, 1994–1995
- Sakura Bank Scholarship, 1993–1994

Prof. Guaning Su, President of Nanyang Technological University, presenting the Nanyang Teaching Excellence Award to Prof. Huey Hoon Hng in 2008. Courtesy of Nanyang Technological University.

- Rachel Meyer Book Prize for being the best woman student, 1993
- Glaxo Gold Medal for being the best student in Chemistry A completing the Final Examination for the degree of Bachelor of Science, 1993
- Singapore National Academy Science Award for outstanding performance in the Bachelor of Science degree examination of the National University of Singapore in Chemistry, 1993

Biography

Early Life and Education

Huey Hoon Hng was raised in a typical Singaporean family with her elder sister and younger brother. She received her early education in the Singaporean public school system, and continued to pursue her tertiary education at the National University of

Singapore (NUS) majoring in Chemistry and Materials Science. After graduating with her first degree, she obtained a M.Sc. degree through her research with the support of a National University of Singapore Research Scholarship. While working on her M.Sc. degree, she managed to secure financial support to fund her Ph.D. studies at the University of Cambridge through the Nanyang Technological University Overseas Scholarship. She was awarded a Ph.D. degree in 1999. In 2001, while working in Nanyang Technological University, she also finished a part-time programme to obtain a Postgraduate Diploma of Teaching in Higher Education from the National Institute of Education, Singapore.

Career History

After receiving her doctorate, Huey Hoon came back to Singapore in late 1999 to serve her scholarship bond with Nanyang Technological University as an Assistant Professor. She has been with the School of Materials Science ever since. In 2008, she was granted tenure and promoted to Associate Professor. Over the years, she had assumed various senior positions in the school. First, she was the Assistant Chair (Undergraduates) from 2000 to 2004 and then reappointed again from 2008 to 2009, taking care of undergraduate student matters. She also completed a 1-year stint as the Head of Division of Materials Science from 2009 to 2010. Since then, Prof. Hng has served as Associate Chair (Academic) for the School. As the Associate Chair (Academic), she is responsible for the school's materials engineering bachelor degree programme. She also holds the post of Director, Facility for Analysis, Characterization, Testing, and Simulation (FACTS), which is a university shared facility.

As a faculty member, Huey Hoon continued her research in understanding the processing–microstructure–property relationships of materials. Her current research interest is focused on the synthesis of inorganic materials using chemical and mechanical processing techniques. The materials of interest are thermoelectric materials for power generation and energetic materials including metal alloys, intermetallics, and functional ceramics. She has secured major research projects amounting to more than S\$2.5 million.

3 Most Cited Publications

Title: Nanostructured metal oxide-based materials as advanced anodes for lithium-ion batteries
Author(s): Wu, HB; Chen, JS; Hng, HH; et al.
Source: Nanoscale; volume: 4; issue: 8; pages: 2526–2542; published: 2012
Times Cited: 301 (from Web of Science)

Title: Photoluminescence study of ZnO films prepared by thermal oxidation of Zn metallic films in air
Author(s): Wang, YG; Lau, SP; Lee, HW; et al.
Source: Journal of Applied Physics; volume: 94; issue: 1; pages: 354–358; published: July 1, 2003
Times Cited: 264 (from Web of Science)

Title: Formation of Fe_2O_3 microboxes with hierarchical shell structures from metal-organic frameworks and their lithium storage properties Author(s): Zhang, L; Wu, HB; Madhavi, S; et al. Source: Journal of the American Chemical Society; volume: 134; issue: 42; pages: 17388–17391; published: October 24, 2012 Times Cited: 214 (from Web of Science)

ResearcherID (A-2246-2011)

Challenges

Having to excel in both research and teaching, while completing administration duties is the main challenge that I have faced in my career. Striking a balance in life taking care of family and career is a challenging task.

On being a woman in this field . . .

As a woman, I am in the minority in the science and engineering field, and there are times when I (and all other minorities) need to make greater effort to be seen or heard. Having said that, there are still a few advantages and rewards being the fairer sex!

Words of Wisdom

Stay positive and be proactive to make things work!

Profile 43

Susan N. Houde-Walter

CEO and Co-founder
LaserMax, Inc.
Rochester, NY
USA

e-mail: shw@lasermax.com
telephone: (800) 527-3703

Susan Houde-Walter with newborn daughter in the lab where Susan measured gradient-index profiles in glass as they formed. (Her baby rests on the vacuum chamber designed and built by Susan for that purpose, also known as the "Tuna Can"). Photo credit: Photograph was taken with a gradient-index objective made by fellow graduate student Brian Caldwell.

Tags

❖ Administration and Leadership
❖ Industry
❖ Domicile: USA
❖ Nationality: American
❖ Caucasian
❖ Children: 1

Birthplace
New York, NY

Publication/Invention Record

>50 publications
6 patents

Proudest Career Moment (to date)

I am proud of my former students; I like to hear about their successes. I am also very proud of my daughter; it gives me great joy when we discuss common interests in science and engineering.

Successful Women Ceramic and Glass Scientists and Engineers: 100 Inspirational Profiles. Lynnette D. Madsen.
© 2016 The American Ceramic Society and John Wiley & Sons, Inc. Published 2016 by John Wiley & Sons, Inc.

Academic Credentials

Ph.D. (1987) Optics, University of Rochester, Rochester, NY, USA.
M.S. (1983) Optics, University of Rochester, Rochester, NY, USA.
B.A. (1976) Liberal Arts, Sarah Lawrence College, Bronxville, NY, USA.

Research Expertise: laser and optical materials, optoelectronic design; seminal X-ray absorption fine structure studies of molecular structure in multicomponent glasses (conducted at four major synchrotrons in the USA and the UK); theoretical studies of leaky waveguiding in optoelectronic devices; design/fab/demo of waveguide lasers, high-gain materials, superlattice disordering in Group III–V semiconductors, quantum cascade lasers (QCLs)

Other Interests: encouraging young people to develop their careers in science and engineering, entrepreneurship

Key Accomplishments, Honors, Recognitions, and Awards

- U.S. Air Force Science Advisory Board, 2013 to present
- Technical Advisor, Force Science Research Center, Minnesota State University, 2005 to present
- U.S. Army Science Board, 2006–2015
- Adjunct Professor, College of Optical Sciences, University of Arizona, 2005 to present
- Commander's Award for Public Service, Department of the Army, 2014
- Joint Civilian Orientation Conference—one of 43 business and community leaders selected for Pentagon program to visit troops and review military assets
- President, Optical Society of America (OSA); elected to the Presidential chain (4-year duty, President in 2005); OSA is an international association of optical physicists and engineers, with rosters that include 33 Nobel Laureates, the inventors of the laser, fiber optics, holography, and other advanced optical technologies. Susan's development of international leadership, her monthly essays in the award-winning "Optics and Photonics News" membership magazine, her live interviews of the 2005 Nobel Prize winners in Physics (all OSA members), executive team development, and unprecedented fundraising success characterized her year as President
- Elected Fellow, American Ceramic Society, 2000
- Elected Fellow, Optical Society of America, 1996
- Elected to the Board of Directors, Optical Society of America, 1994–1996

Biography

Early Life and Education

Well, this was a long time ago! I was brought up as an artist (painting, printmaking, and sculpture). For some reason, I was exempted from math and science in both high school and college, and was instead given my own private art studios where I spent nearly every waking hour. However, during my senior year in college I took a bus in New York City and I saw someone wearing a hologram belt buckle. I was fascinated with the ephemeral imagery and decided to use holography as an art medium. I scraped my pennies together and eventually bought a laser. I let other holographers use my laser in exchange for holographic studio time in New York City, where I lived at the time. They talked about "electromagnetic radiation" and "diffraction." I had never heard of these concepts before, but figured I'd better find out about them. In between running the bookstore at the Museum of Holography in Manhattan and working the (unpaid) night shift at the Smithsonian Institution's pulsed holography laboratory at Brookhaven National Labs on Long Island, I took unmatriculated classes at the Courant Institute of Mathematical Sciences at New York University (NYU), where I discovered that I loved mathematics. However, I was also very hungry for optics, and NYU didn't have much of that at the time, so I moved to Rochester, NY so that I could take courses at the famous Institute of Optics at the University of Rochester.

Once at Rochester, I crammed all of the technical courses required for a bachelor's degree into three semesters, including two summer courses at MIT. (Although MIT didn't have much optics at the time, either.) All of the courses were out of their normal sequence, so it was a bit of a mash-up, but I got through it, and was invited to apply to the Master's program in Optics at the University of Rochester. After my first semester in graduate school there, I was invited to apply to the Doctoral program. I had already started a Master's thesis, so I ended up doing both a Master's and a Doctoral thesis, with a marriage (an elopement!) and a newborn baby thrown in at the end.

Career History

Once done with my doctoral work, I was invited to stay on as regular faculty (as you can see, this optics business is a slippery slope). For the first 80 years of my department's history, I was the first and only regular faculty member who also happened to be female.

The funny thing about my scientific work is that no matter what I was asked to work on, somehow I or my students and I would stumble upon some key fundamental issue that had not been previously considered and we would dive into that, only to emerge once we had uncovered some hitherto unknown secret. I suppose it is my nature.

I was open to my students' personal interests and goals, so I bent my own research efforts around their interests to the degree that I could. It was intellectually stimulating to me; we were simultaneously doing

Undergraduate students in Susan Houde-Walter's Geometrical Optics class. Each student had an "optics kit" of plastic lenses and a ruler to build simple optical systems alongside the lecture. Here, a student shows off his own art photography using a slide projector built in the lecture hall. Photo credit: Betsy Benedict.

new methods in complex variable theory, developing new thermodynamical theory of ion transport in disordered systems, optical spectroscopy of a couple of rare earth ions, X-ray spectroscopy of multicomponent glasses, optimization theory, and probing the energy levels of impurities and defects in Group III–V semiconductor superlattices. This and more went on for years. One day in our weekly group meetings I noticed that my students were having difficulty communicating about their work to each other. Even their math language was divergent. I think in retrospect that having a more tightly themed research group might have worked better for them. However, I had a grand old time; it was great fun for me.

I also enjoyed teaching. Rochester is rich in the optics industry, so I frequently took the students on tours, and the optics companies made them very welcome. We did a lot of demos and hands on in class at both the undergraduate and graduate (doctoral) levels.

My husband and I started a business while I was Assistant Professor. It really was my husband's passion; I was much more interested in basic research. Nonetheless, through marriage, I was also wedded to this company. In the beginning, I wrote business plans and did technical firefighting; more on that below. The company is called LaserMax.

The main professional society for academics in optics is the Optical Society of America (somewhat of a misnomer because it is a distinctly international organization). I felt it was important to give back and volunteer for the OSA. I organized meetings, etc., and was eventually elected to their Board of Directors, and ultimately to the Presidential chain (a 4+-year commitment).

Dr. Houde-Walter at LaserMax. Photo credit: LaserMax.

In the meantime, my academic department was going through leadership changes. I was exhausted from continually teaching new courses (see the "tough as nails" part in "Challenges" section below) and had never taken a sabbatical. Through negotiation, I was ultimately granted a 2-year unpaid leave of absence. I was not mobile (with both a company and family), so I opted to work for LaserMax on a temporary basis. I had active research grants and research students; I could more easily keep that going by staying local also.

During the 2 years at the company, it became apparent that my husband was not well and that I would be needed there longer and in the top leadership position. I continued to run the company, returning to teaching (but not committee work), and of course the OSA work. At the same time, my father also became gravely ill. It all came to a head during my year as OSA President, when I was shuttling around several continents for professional meetings, and variously running my husband and father to their respective hospitals. My husband became very ill and my father passed away. It was clear that I would have to stay with the company and give up my beloved science for good, so that is what I did. Since I was elected to the OSA Presidential chain, I felt that I owed it to the membership to serve out the 4+-year term, so I did that, too.

Shortly after my father passed, I was introduced to the U.S. military community. I learned from the soldiers and their families how to stay resilient and maintain a sense of humor. Through all this, I have realized that every day with my husband is a cherished gift. I now use my scientific background to the best of my ability in service to my country.

3 Most Cited Publications

Title: Cation microsegregation and ionic mobility in mixed alkali glasses
Author(s): Vessal, B; Greaves, GN; Marten, PT; Chadwick, AV; Mole, R; Houde-Walter, S
Source: Nature; volume: 356; pages: 504–506; published: April 1992
Times Cited: 112 (from Web of Science)

Title: Sodium and silver environments and ion-exchange processes in silicate and aluminosilicate glasses
Author(s): Houde-Walter, SN; Inman, JM; Dent, AJ; Greaves, GN
Source: Journal of Physical Chemistry; volume: 97; issue: 37; pages: 9330–9336; published: September 1993
Times Cited: 105 (from Web of Science)

Title: Yb/Er-codoped and Yb-doped waveguide lasers in phosphate glass
Author(s): Veasey, DL; Funk, DS; Peters, PM; Sanford NA; Obarski, GE; Fontaine, N; Young, M; Peskin, AP; Liu, W-C; Houde-Walter, SN; Hayden, JS
Source: Journal of Non-Crystalline Solids; volume: 263; issue: 1–4; pages: 369–381; published: March 2000
Times Cited: 103 (from Web of Science)

Challenges

Keeping an upbeat attitude is critical; also keeping focused on priorities. As a Professor, I was targeted for a disproportionate amount of committee work, presumably because people wanted "a woman" on their committee. Know what is standard and take no more. I was also something of a curiosity to my (all-male) colleagues, who wanted to see how much I could take ("she's tough as nails!"). Keep a sense of humor and don't take it personally. Science is a human endeavor, and the spoils come only to the winner, whether fairly or not.

Finally, science requires sustained concentration. You must protect your ability to concentrate if you are to deliver meaningful results to society.

On being a woman in this field . . .

As just mentioned, uninterrupted focus and mental acuity are critical in scientific work, but these can be compromised if you are also raising children and/or caring for aging parents. Still, if you're going to have kids, someone has to care for them. It doesn't have to be you personally every single second, but children need love and safety, and you must provide a reliable situation for them. Everyone works out their own solution.

Words of Wisdom

Do not hyphenate your last name! Hyphens cause all sorts of mischief in databases, from citation indices to airline tickets. I did not anticipate that long years ago. If there's any decision I would change in retrospect, this one is it.

Profile 44

Ellen Ivers-Tiffée

Prof. Dr.-Ing.
Karlsruhe Institute of Technology
Institute for Applied Materials—Materials for Electrical
and Electronic
Engineering (IAM-WET), Adenauerring 20 B
76131 Karlsruhe
Germany

Prof. Ellen Ivers-
Tiffée.

e-mail: ellen.ivers@kit.edu
telephone: +49 721 608 47490

Birthplace
Frankfurt/Main, Germany
Born
September 24, 1951

Tags

❖ Academe
❖ Domicile: Germany
❖ Nationality: German
❖ Caucasian
❖ Children: 1

Publication/Invention Record

>275 publications: h-index 34 (Web of Science)
10 patents, co-author/author of 12 books/book chapters

Academic Credentials

Ph.D. (Dr.-Ing.) (1980) Materials Science, Friedrich-Alexander-Universität, Erlangen,
Germany.
M.Sc. (Dipl.-Min.) (1975) Mineralogy/Crystallography, Philipps-Universität, Marburg,
Germany.

Research Expertise: functional ceramics for the energy sector (fuel cells, batteries, gas
permeation membranes) with a focus on the characterization of electrical and electrochemical reactions and transport processes, development of nanoscaled functional layers,
methods of model-based materials development and model-based *in-situ* diagnosis and
lifetime prediction

Other Interests: international collaborations

Successful Women Ceramic and Glass Scientists and Engineers: 100 Inspirational Profiles. Lynnette D. Madsen.
© 2016 The American Ceramic Society and John Wiley & Sons, Inc. Published 2016 by John Wiley & Sons, Inc.

Key Accomplishments, Honors, Recognitions, and Awards

- Fellow of The Electrochemical Society (ECS), since 2015
- Member of the Senate of the German Research Foundation ("Deutsche Forschungs-gemeinschaft", DFG), since 2014
- Member of the Board of Trustees, "Friedrich und Elisabeth Boysen-Stiftung", since 2014
- Member of the Council for Innovation and Competitiveness ("Rat für Innovation und Wettbewerbsfähigkeit") of the Baden-Württemberg State Ministry of Finance and Economy, Germany, since 2014
- Member of the Review Board of the German Science Foundation (DFG) ("Synthesis and Properties of Functional Materials"), 2012–2014
- Member of the International Board of Advisors, European Fuel Cell Forum, since 2011
- Working group member of the "World Premier International Research Center (WPI) Initiative" of the Japan Society for the Promotion of Science (JSPS), 2010 to present
- Member of the Board of Trustees of the Institut für Mikroelektronik Stuttgart/Germany (IMS CHIPS), 2008 to present
- Elected Member of acatech—the German Academy of Science and Engineering, 2007 to present
- Member of the Senate of the Karlsruhe Institute of Technology (KIT) (until 2009: Universität Karlsruhe (TH)), 2006–2011
- Elected Member of the German Academy of Sciences Leopoldina (Engineering Sciences), 2005 to present
- Member of the Scientific Advisory Board of the Fraunhofer Institute for Silicate Research (FhG-ISC) Würzburg/Germany, 2004 to present
- Member of the Coordination Board of the DFG-Center for Functional Nanostructures (CFN), Karlsruhe Institute of Technology (KIT) (until 2009: Universität Karlsruhe (TH)), 2001–2013
- Member of the Innovation Think-Tank of the State of Baden-Württemberg, 2007–2010
- Elected Member/Academician of the "World Academy of Ceramics," Faenza/Italy, 2009
- Member of the Board of Directors of the International Society for Solid State Ionics, 2003–2007
- Member of the Senate of the Helmholtz Association of the National Research Centers (HGF), Bonn/Germany, 2001–2007
- Member of the Search Committee "Technical Sciences" of the Körber European Science Award, 2004–2006
- Member of the Scientific Advisory Board of the Leibniz Institute for Solid State and Materials Research (IFW) Dresden/Germany, 1999–2006

- Organizer and Chairman of the Scientific Committee of the "International Conference on Solid State Ionics (SSI-15)" in Baden-Baden/Germany, in cooperation with Prof. Dr. Joachim Maier, Max Planck Institute for Solid State Research, Stuttgart/Germany, 2005
- Member of the Perspective Committee of Forschungszentrum Jülich/Germany, 2004
- Member of the Selection Committee for the State Research Prize of Baden-Württemberg, 1999–2004
- Member of the Research Council of the State of Baden-Württemberg, 1999–2002
- Vice-President of Universität Karlsruhe (TH), 1998–2002
- Liaison Officer for the Foundation of German Business ("Stiftung der deutschen Wirtschaft"), Klaus Murmann Fellowship Programme, 1996–2002
- Visiting Professor at Oita University, Japan, 2000
- Member of the Selection Committee of the "Margarethe von Wrangell" Habilitation Program for Female Scientists, Baden-Württemberg, 1997–2000
- Member of the Board of Trustees of the Academy for Technology Assessment in the State of Baden-Württemberg, 1997–2000
- Member of the Review Board BMBF-NMT, Project Management Jülich (PTJ), Germany, 1996–1998
- Member of the Senate of Universität Karlsruhe (TH), 1996–1998
- Swiss Society for Thermal Analysis and Calorimetry (STK): Award for "Applied Chemical Thermodynamics," 1996
- Head of Technical Committee on "Continuing Education" within the German Ceramic Society (DKG), 1986–1996

Biography

Early Life and Education

During her undergraduate years (1970–1975), she studied mineralogy/crystallography at Philipps-Universität in Marburg. Immediately following the completion of her undergraduate studies, she moved to Friedrich-Alexander-Universität in Erlangen where she completed her doctorate thesis in materials science (1980).

Career History

Upon completion of her doctorate, she took a position in research at Siemens AG, Corporate Research and Technology, Center of Applied Materials Research. She focused on research & development of functional ceramics for electronic devices (capacitors, nonlinear resistors) and electrochemical devices (sensors and fuel cells). She became the project leader of various national (BMBF: BEO, MATFO) and European research projects (EU: JOULE, BRITE-EURAM).

In addition, from 1991 to 1994, she took a teaching assignment in the Department of Geosciences at Philipps-Universität in Marburg.

Prof. Ivers-Tiffée with one of her doctoral students, Michael Schönleber, in 2015.

Since 1996, she has been a full professor for Electrical Engineering and Information Technology and also the Head of the Institute for Applied Materials—Materials for Electrical and Electronic Engineering (IWE), Karlsruhe Institute of Technology (KIT) (until 2009: Universität Karlsruhe (TH)).

She is the author or co-author of a dozen books and book contributions, including the German textbook, Werkstoffe der Elektrotechnik (2007).

She is married to Dr.-Ing. Thomas Ivers-Tiffée and they have one child, Anna, who was born in 1985. She is fluent in both English and German.

3 Most Cited Publications

Title: Materials and technologies for SOFC-components
Author(s): Ivers-Tiffee, E; Weber, A; Herbstritt, D
Source: Journal of the European Ceramic Society; volume: 21; issue: 10–11; pages: 1805–1811; doi: 10.1016/S0955-2219(01)00120-0
Published: 2001
Times Cited: 235 (from Web of Science)

Title: Materials and concepts for solid oxide fuel cells (SOFCs) in stationary and mobile applications
Author(s): Weber, A; Ivers-Tiffee, E
Source: Journal of Power Sources; volume: 127; issue: 1–2; pages: 273–283; doi: 10.1016/j.jpowsour.2003.09.024
Published: March 10, 2004
Times Cited: 205 (from Web of Science)

Title: Evaluation and modeling of the cell resistance in anode-supported solid oxide fuel cells
Author(s): Leonide, A; Sonn, V; Weber, A; et al.
Source: Journal of the Electrochemical Society; volume: 155; issue: 1; pages: B36–B41; published: 2008
Times Cited: 169 (from Web of Science)

Profile 45

Carol M. Jantzen
(aka Carol M. Fredericks and Carol M. Fredericks Jantzen)

Consulting Scientist
Environmental & Chemical Process Technology
Savannah River National Laboratory
Bldg. 773-A
Aiken, SC 29808
USA

e-mail: carol.jantzen@srnl.doe.gov
telephone: (803) 725-2374

Carol Jantzen in her postdoctoral position at University of Aberdeen in Scotland grinding simulated nuclear waste samples to homogenize them. Credit: Department of Chemistry, University of Aberdeen.

Birthplace
Brooklyn, NY, USA
Born
January 14, 1946

Publication/Invention Record

>300 publications: h-index 13
(Web of Science)
11 patents, editor of 3 books

Tags

❖ Government
❖ Domicile: USA
❖ Nationality: American
❖ Caucasian
❖ Children: 0

Proudest Career Moments[1] (to date)

I am most proud of the fact that in the early 1990s I developed the process control models that have run the world's largest high-level waste (HLW) vitrification facility safely and accurately for the last 20 years (1994–2015). The process models ensure that the waste that is highly variable in composition gets mixed with the right amount of glass forming

---

[1] Adapted from K.L. Woodward, Profiles in ceramics: Carol M. Jantzen—a woman ahead of her time, *Am. Ceram. Soc. Bull.*, 78(7): 50–56 (July 1999).

Successful Women Ceramic and Glass Scientists and Engineers: 100 Inspirational Profiles. Lynnette D. Madsen.
© 2016 The American Ceramic Society and John Wiley & Sons, Inc. Published 2016 by John Wiley & Sons, Inc.

frit, and (i) does not crystallize in the melter, (ii) has the right physical properties to melt and pour out of the melter into a $10' \times 2'$ steel canister, and (iii) ensures a waste form durability that will be acceptable to a national geologic repository when one is available. These process models have run the vitrification facility, known as the Defense Waste Process Facility (DWPF), at 95% confidence for the last 20 years (1994–2015). This has proven that the nuclear fuel cycle can be safely closed. While the DWPF processes HLW that was generated during the Cold War between the United States and Russia, the fact that HLW can be safely processed into a solid waste form for ultimate geologic disposal serves as a model for future clean energy production by nuclear power plants. For the development of this process control system, the first of a kind implemented anywhere in the world, I received a Ph.D. Honoris Causa in May 2014 from Queens College (QC) for using my knowledge in geochemistry learned at QC to develop the DWPF process models. The HLW process models also demonstrate that ceramics (glass) can be used for environmental cleanup of many types of radioactive and hazardous wastes.

One of my other proudest achievements was that I received the George Westinghouse Corporate Gold Award of Excellence three times as this competition involved all of the Westinghouse companies and subsidiaries in the United States: three is the most Gold awards ever received by one person.

I am most proud of the 1996 Gold Award as it was presented jointly to both, my husband Dr. John B. Pickett and myself—we were the first married couple to achieve this distinction.

Academic Credentials

Ph.D. (1978) Materials Science & Engineering, State University of New York at Stony Brook, Stony Brook, NY, USA.
M.S. (1970) Geochemistry, Queens College of the City University of New York, NY, USA.
B.S. (1967) Geology, Queens College of the City University of New York, NY, USA.

Research Expertise: glass chemistry; physical property determination; thermodynamic and structural process models; glass leaching and dissolution mechanisms; geochemistry related to nuclear, mixed, and hazardous waste disposal; rare earth glasses for sensors; ceramic, zeolite, and mineral waste forms; development of ASTM standards for the nuclear industry, many of which have been adopted worldwide

Other Interests: making the world a cleaner and safer place, geoscience education

Key Accomplishments, Honors, Recognitions, and Awards

- Associate Editor, Journal of the American Ceramic Society, 1989 to present
- National Academy of Science Board Nuclear and Radiation Studies Board (NRSB) formally known as the Board on Radioactive Waste Management, 2012–2015
- Ph.D. Honoris Causa, Doctorate of Science, Queens College of the City University of New York, NY, 2014

- U.S. Expert on waste forms at an International Atomic Energy Agency (IAEA) workshop on waste forms in Buenos Aires, Argentina, 2012
- SRNL Donald A. Orth Award for Technical Excellence, 2010
- National Academy of Science Committee on Waste Forms Technology and Performance, 2009–2010
- Honorary Professorship, Department of Inorganic Chemistry, University of Aberdeen, Scotland, 2005–2010
- American Ceramic Society Distinguished Life Award, 2008
- Wendell Weart Lifetime Achievement Award in Waste Management (sponsored by Waste Management Symposium and Sandia National Laboratory), 2008
- Savannah River Site Career Achievement Award for Technical Achievement, 2006
- Alfred Victor Bleininger Award (ACerS Pittsburgh Section) for excellence and lifetime achievements in the field of ceramics, 2005
- D.T. Rankin Award, American Ceramic Society for outstanding contributions to the Nuclear and Environmental Technology Division, 2003
- State University of New York at Stony Brook, Distinguished Alumna Award, 2000
- Alfred University Scholes Lecture Award in Industrial Ceramics Engineering, 2000
- National Research Council of the National Academy of Sciences Commission on Physical Sciences, Mathematics and Astronomy (CPSMA), 1997–2000
- "Famous American Women" Nationwide Awardee, American Society Mechanical Engineers (ASME) Section at City University of New York, NY, 1999
- Queens College Alumni Star Awardee, Queens College, NY, 1998
- The American Ceramic Society, President, 1996–1997
- Distinguished Scientist Award, Citizens for Nuclear Technology Awareness (CNTA), Aiken, SC, 1997
- George Westinghouse Corporate Gold Awards of Excellence (highest corporate award achievable nationwide), 1992, 1993, 1996
- George Westinghouse Bronze Signature Awards of Excellence (second highest corporate award achievable nationwide), 1989, 1991,1996
- Fellow of the American Ceramic Society (ACerS), 1990
- Winner of American Ceramic Society Nuclear Division Best Paper Award, 1981
- NATO Grant to examine phase transformations of high-temperature oxides and glasses by small-angle neutron scattering at Kernforschunanlanger (KFA), Julich, West Germany, June 1975
- Research-Teaching Assistantship, Department of Materials Science, State of New York at Stony Brook (research on nucleation and growth of manganese nodules, phase transformations and decomposition of oxide, silicate and metallic systems; instruction of X-ray diffraction, crystallography, and phase equilibria), 1972–1976
- Research-Teaching Assistantship, Department of Earth and Space Science, State University of New York at Stony Brook (research in theoretical petrology and phase equilibria; instruction of master's-level courses), 1970–1971

- Research Assistantship, Department of Geochemistry, The Pennsylvania State University (PSU) (thermodynamics and hydrothermal geochemistry), 1969–1970
- Teaching and Field Assistantships, Queens College of the City University, 1969 of New York Department of Geology (research of Canadian Shield on Sigma Xi Grant-In-Aid of Research), 1967

Biography

Early Life and Education[1]

Carol recalls a strong affinity in geological materials that was developed through her father's business. The entire family collected rocks and minerals to be used in

educational kits for the Hayden Planetarium, the New York Museum of Natural History gift shops, high schools in New York City, and colleges and universities around the United States. Other rocks and minerals were made into polished gemstones and sold. Some of her fondest memories are of family excursions throughout the country collecting minerals and rocks for the business.

She also recalls a determination on the part of her father that she should become a scientist. Her grades were excellent and she was accepted at Queens College, which (at the time) was a tuition-free university with stiff competition for entrance.

While a love for science was instilled in her, a passion for earth sciences was even more deeply ingrained into Carol by others. One such role model was her earth science teacher in high school, Irving Horowitz, a man she maintains contact with even today. Another inspiration was Martin Groffman, a general science teacher in junior high school. Both teachers delivered good content, but more importantly they taught her to think and analyze.

Carol at 1.5 years old where her dad is determined that she becomes a geologist (or some other kind of scientist).

During the first 4 years at Queens College, Carol majored in geology and took a minor in education since early on her goal was to be a general science or earth science teacher at the junior or high school level. In 1967, a required 6-week geology field course took her to Puerto Rico to study the geology of that terrain. That was when she met her first husband who was her mapping partner. Rather than venture immediately into a teaching career, she decided to pursue her master's degree in geology. While pursuing her master's, she had a teaching assistantship to offset her expenses. She enjoyed teaching college students in comparison to teaching high school students. At that point, she decided to pursue a doctorate in order to teach at the university level.

She began her doctoral studies at PSU in 1969, while simultaneously completing her masters' dissertation. However, her husband's job took him to New York City, so she returned to New York and enrolled at the State University of New York at Stony Brook. She once again received a research assistantship; but the university geology curriculum, a branch of geology called igneous petrology (the study of how molten lavas form) in which Carol was interested, did not offer much in the way of career. One faculty member at PSU (Arnulf Muan) had a doctoral degree in materials science and engineering, but was the chair of the geochemistry department. Also, her favorite class in the geology department (now the School of Earth and Environmental Sciences) at Queens College was phase equilibria. The class was taught by David Speidel, a geochemist, who had learned phase equilibria from Arnulf Muan. She wondered if going from geochemistry to materials (the reverse of what Muan had done) could successfully work for her: "So I went into the material sciences and engineering program at Stony Brook." She discovered a ceramics area of study—glass chemistry—that was the closest thing to molten magma. Once her doctoral research was complete in 1976, she searched for a job but there was a minirecession in the United States and even postdoctoral positions in the United States were difficult to obtain.

Career History

She secured a postdoctoral fellowship from the Department of Inorganic Chemistry at the University of Aberdeen in Scotland from 1976 to 1978. At Aberdeen University, she worked for Dr. Fred Glasser who had obtained his Ph.D. from PSU. PSU was the U.S. center for cement research and the University of Aberdeen was the Scottish center for cement research. There were many visits made by PSU professors to Aberdeen through the years while Jantzen was there.

Jantzen's task was to stabilize high-level radioactive wastes from the United States and United Kingdom in cement, i.e., to make and test cementitious waste forms. While she was now in Chemistry Department, the research heavily relied on phase equilibria, thermodynamics, and crystal chemistry, i.e., all the things that she had learned about in geology at QC and in materials science at Stony Brook. She and Glasser continued to collaborate until 1984 and published many papers based on the postdoctoral research she carried out.

Toward the latter part of her postdoctoral fellowship in Aberdeen, PSU entered into a joint R&D program with what is now Rockwell International Science Center (RISC) in Thousand Oaks, CA, to formulate and test ceramic waste forms. One of the PSU professors whom Jantzen had met in Aberdeen suggested her for the job and she got a telephone call in Aberdeen one day from her future Rockwell boss. Jantzen worked at RISC for about 2 years (1979–1981) and made frequent visits to PSU for the collaborative R&D. She and the ceramic waste form team formulated high Al_2O_3-containing waste forms and her first paper won the ACerS Nuclear Division Best paper award. The Rockwell–PSU program was managed by Savannah River Laboratory (SRL)

American Ceramic Society Annual Meeting, 1997.

who was interested in downselecting either a glass waste form or a ceramic waste form for the processing of HLW dependent on the results of the R&D program.

She left Rockwell in late 1981 and took a job with Bechtel, Inc. in San Francisco. She did not feel challenged enough at Bechtel and she wanted to relocate to the east coast as her parents were getting older. She was hired at SRL (now the Savannah River National Laboratory (SRNL)) in 1982. The job entailed processing of HLW vitrified (glass) waste forms, which was her field of specialty from undergraduate and graduate school. In 1982, through a downselection process, glass was chosen as the HLW waste form. Subsequently, the first U.S.-based HLW vitrification facility was built at the Savannah River Site; her process models have been used successfully since its completion (in 1994, i.e., for the past 20^+ years) to stabilize 4,380,000 gallons of HLW into 15,200,000 lbs. of borosilicate glass. The DWPF began processing non-radioactive waste in April 1994 and radioactive waste in April 1996. Jantzen had an integral part in the statistical process control system, in the materials of selection, waste form acceptance, and interactions with various potential repositories of varying geology. The job was very challenging as operating a nuclear waste glass processing facility using statistical process control had never been achieved before.

She served as the first woman president of the American Ceramic Society from 1996 to 1997, in the society's 98th year. Two men followed her, so she was the only woman president in the first century! Her spouse was a member of the Nuclear & Environmental Technology Division of ACerS and likely the only spouse to have taken an ACerS Short Course on Powder Processing.[2] At this time, the ACerS Past President's Wives Luncheon was changed to the Past President's Spouses Luncheon. Thereafter there

---

[2] Excerpt from C.M. Jantzen and J.B. Pickett, On being the first woman president of ACerS, *Am. Ceram. Soc. Bull.*, 76(5): 7 (May 1997).

were several other women (e.g., Kathryn Logan, 2003–2004; Katherine Faber, 2006–2007; Marina Pascucci, 2010–2011; and Kathleen Cerqua-Richardson, 2014–2015) . . . they all know each other, of course.

Her position at the SRNL is the "highest technical level one can go"; it took her about 30 years to achieve that level. As you probably can figure out—she is the first and only woman to make this technical level in the R&D arena since the laboratory was built in 1951. Through the years, SRNL had a few other men who were level 40s, but at the time of this writing, they all retired, so now she is now the only one!

Carol Jantzen, 2014.
Credit: SRNL
photography.

She received a doctorate of Science Honoris Causa from Queens College of CUNY in May 2014. The rank of honorary doctorate is an honorific title granted in order to distinguish a learned person whose knowledge and wisdom were considered exemplary, but also, in return, to pride the university for having recognized such an outstanding person. The day of the presentation was the annual commencement ceremony at Queens College, CUNY, and Jantzen also gave the commencement speech ensuring to highlight that global warming must be stopped and one way to achieve this, perhaps the only way, is with clean energy and that includes nuclear power. The United States must conquer its fear of nuclear power. France produces 75% of its power through nuclear plants and the United Kingdom also generates over 50% of their power through nuclear plants. France, Russia, and the United States have excess weapons-grade plutonium from the Cold War that can be made into clean energy. In France and Russia, they consider their plutonium as their "national treasure"—as they can create cheap electricity for their nation's people. In the United States, we are contemplating burying our plutonium as nuclear waste.

Carol Jantzen washing a 250 lb feldspar crystal in April 2014 before donating it to the Ruth Patrick Science Education Center (RPSEC).

She is an active member of the Mineralogical Society of America

(1968 to present), the American Ceramic Society (1970 to present), the Materials Research Society (1979 to present), the National Institute of Ceramic Engineers (Member 1982 to present), ASTM Committee C26 on Nuclear Fuel Cycle (1987 to present), and a SC state registered geologist (December 1987–June 2014). She was an Adjunct Professor in the Department of Ceramic Engineering, Clemson University, SC, from 1989 to1999.

In April 2014, while still working full time at SRNL, Carol decided to become involved in science education again. She donated her museum quality rock and mineral collection and funding to the Ruth Patrick Science Education Center (RPSEC) in Aiken, SC. The RPSEC has its own telescope and planetarium and it seemed the right venue for the mineral collection that my dad and I amassed during his lifetime. "I wanted to make sure that the collection went somewhere where it could be kept intact, have longevity and be enjoyed by the community," Jantzen said. The specimens will become the Fredericks Mineral Gallery at the RPSEC that she hopes will generate "a love of geology and earth science" and make people appreciate "how natural minerals form in wondrous shapes and sizes."

The first specimen donated was the largest she had, a 250 lb interpenetrating twin feldspar crystal that is so large that it can be touched, stroked, and hugged, so children can run their hands along the crystal faces and consider how neat it is that it wasn't cut that way, but it grew that way. Jantzen said it is vital for scientists and researchers to share their knowledge with the next generation. "If the current scientists, researchers and collectors do not reach out to the next generation and beyond, we will be a country without scientists. The time to interest a child in science is early. The earlier the interest, the more certain one can be that the boy or girl will carry that interest into adulthood."

3 Most Cited Publications

Title: High-alumina tailored nuclear waste ceramics
Author(s): Morgan, PED; Clarke, DR; Jantzen, CM; et al.
Source: Journal of the American Ceramic Society; volume: 64; issue: 5; pages: 249–258; published: 1981
Times Cited: 74 (from Web of Science)

Title: Thermodynamic model of natural, medieval and nuclear waste glass durability
Author(s): Jantzen, CM; Plodinec, MJ
Source: Journal of Non-Crystalline Solids; volume: 67; issue: 1–3; pages: 207–223; published: 1984
Times Cited: 72 (from Web of Science)

Title: Durable glass for thousands of years
Author(s): Jantzen, CM; Brown, KG; Pickett, JB
Source: International Journal of Applied Glass Science; volume: 1; issue: 1; special issue: SI; pages: 38–62; published: March 2010
Times Cited: 34 (from Web of Science)

Challenges

Being a woman in science in the early 1960s was not considered "fashionable" and so there were few women in many of my science classes. Most notably, in my mining engineering class, when I was sent to a mine on a field trip with the rest of the class, I was not allowed in the mine as it was "bad luck" for the miners. Consequently, I stayed in the parking lot with the bus driver.

I had tried to get into the Colorado School of Mines in the early 1960s to study geology, but they did not accept female students at that time. Queens College accepted women, but women were not allowed into class with trousers on, not even what were referred to as "snow pants" when it was snowing heavily in New York.

The Materials Science department at Stony Brook had very few women, but interestingly, in Europe when she was there for her postdoctoral fellowship, there were more women in chemistry and other science disciplines than in the United States.

I have struggled with the balance between career and home life . . . , but I note that challenge is gender neutral.

On being a woman in this field . . .

When I was hired into SRL/SRNL, I was one of three women Ph.D.'s in a building of ~1000 scientists (B.S., M.S., and Ph.D.'s). I was often the only woman in a given meeting whether it was a meeting of 10 or 30 persons. About 2% of the audience at technical conferences were women even in the late 1970s and early 1980s. While science in general was male dominated, ceramic science and nuclear science were almost totally male dominated in the 1980s through the late 1990s.

One of the secrets to my career was to "never refuse a task I was assigned" even if it was the most difficult task my boss doled out to anyone in his group. He soon realized that I would take on jobs that others would not . . . and I would be successful at completing them to both his surprise and sometimes even my own surprise. He would come into my office and say "I have a rare opportunity for you" and I knew that we were going where no one had dared to go before. I was always appreciated for what I achieved technically and gender was not a part of those achievements.

Words of Wisdom[3]

I "solve seemingly unsolvable problems." And not only do I solve them, but I also make the solution look "simple." In the research teams that I lead, I also mentor my team to "solve the seemingly unsolvable" and it is not always easy.

My success in solving the unsolvable stems from an early inspirational book I read, actually it was read to me the first few times by my mother . . . "The Little Engine That Could." If you have never read this book, it does not matter whether you are in your 20s or 60s, it is inspirational. It is about a long train that must be pulled over a high mountain. Other larger trains (the Shiny New Engine or the Big Strong Engine) refused the job for various reasons. The request is sent to a small engine that agrees to try. The engine

[3] Excerpt from Queens College Commencement Speech, May 29, 2014.

succeeds in pulling the train over the mountain by chanting "I-think-I-can, I-think-I-can" even as it goes slower and slower up the mountain. But the Little Engine overcomes a seemingly impossible task. On the "downside" of the mountain, the little engine says "I-knew-I-could."

The "Little Engine That Could" is one of the greatest tales of motivation and the power of positive thinking ever told. It is the main theme of many of our favorite superhero movies such as "Raiders of the Lost Ark." It is also a story of initiative and risk. Most people try to avoid risk. But for those who take the initiative, there is a chance of much greater gains, of growing, of doing something you didn't think you could do before. All the experienced engines in the story merely found an acceptable way of failing. It was only the small engine who was willing to put *her* own neck on the line to progress.

My story is a story of hardwork, positive thinking to tackle the unknown, motivation, and optimism—never giving up. And I never give up because I never say "I can't"—I only know how to say "I'll try" or "I-think-I-can."

Sylvia Marian Johnson

Chief Materials Technologist
Entry Systems and Technology Division
National Aeronautics and Space Administration (NASA)
Ames Research Center
M/S 234-1
Moffett Field, CA 94035-1000
USA

e-mail: sylvia.m.johnson@nasa.gov
telephone: (650) 604-2646

A graduation picture of Sylvia from 1977 at the University of New South Wales. She said: "Oh, and we had to wear white. At least no gloves!"

Birthplace
Sydney, Australia

Born
August 29, 1954

Publication/Invention Record

21 publications: h-index 11 (Web of Science)
6 patents; editor of 2 books

Proudest Career Moment (to date)

Giving the Mueller Award Lecture.

Tags

❖ Administration and Leadership
❖ Government
❖ Domicile: USA
❖ Nationality: born Australian, now American
❖ Caucasian
❖ Children: 2

Successful Women Ceramic and Glass Scientists and Engineers: 100 Inspirational Profiles. Lynnette D. Madsen.
© 2016 The American Ceramic Society and John Wiley & Sons, Inc. Published 2016 by John Wiley & Sons, Inc.

Academic Credentials

Ph.D. (1983) Materials Science and Engineering (Ceramics), University of California, Berkeley, California, USA.
M.S. (1979) Materials Science and Engineering (Ceramics), University of California, Berkeley, California, USA.
B.Sc. (1976) Ceramic Engineering, University of New South Wales, Sydney, Australia.

Research Expertise: ceramic and composite processing and characterization, ultra-high-temperature ceramics

Other Interests: understanding how complex technical issues interrelate with policy, business, and industry issues; encouraging girls to pursue engineering and science careers

Key Accomplishments, Honors, Recognitions, and Awards

- Technical Committee Chair, International Ceramic Federation, 2013 to present
- Edward Orton, Jr. Award and Lecture, The American Ceramic Society (ACerS), 2015
- Women@NASA, 2015 (http://women.nasa.gov/sylvia-johnson/)
- U.S. Representative to the International Ceramic Federation, 2009–2015
- Elected to the World Academy of Ceramics, 2014
- James I. Mueller Award, The American Ceramic Society, 2011
- Evaluation Board, Materials Science and Technology at the Sandia National Laboratories, 2008–2010
- Member, National Research Council Committee on Materials Needs and R&D Strategy for Future Military Aerospace Propulsion Systems, 2008–2010
- Member, National Research Council Committee on Assessing Corrosion Education, 2007–2008
- Lifetime National Associate of the National Research Council, 2008
- Advisory Committee, Second International Congress on Ceramics in Verona, 2008
- Counselor of the San Francisco Bay Area section of The American Ceramic Society, 1998–2005
- Chair, Pacific Rim 5 International Conference on Ceramic and Glass Technology, 2005
- Member, National Research Council Panel for Materials Science and Engineering, 2000–2005
- Elected Director, The American Ceramic Society, 2002–2005
- Chair, National Research Council Committee for Materials and Society: From Research to Manufacturing, 2002–2003
- Member, National Research Council National Materials and Manufacturing Board, 1997–2002
- Member, National Research Council Committee on Materials Technologies for Process Industries, 1999–2001
- Vice President, The American Ceramic Society, 1996–1997

- Member, National Research Council Committee on Materials Science Research for the International Space Station, 1997
- Member, National Research Council Committee on Microwave Processing of Materials: An Emerging Industrial Technology, 1992–1994
- Fellow, The American Ceramic Society, 1992

Biography

Early Life and Education

For as long as I can remember, I always wanted to be a chemist. I grew up in Sydney, Australia and attended Willoughby Girls High School, where I first became interested in my profession after reading an article that talked about ceramic engineering. There are events in a young person's life that make a profound impression. For me, one of these events was the Apollo 11 landing on the moon on July 20, 1969. I can still recall listening to the radio commentary (there were no television sets at school in those days). The mission was considered important enough that we were sent home, so that we could watch it on the television.

Sylvia in 1982 at the University of California, Berkeley doing the happy dance upon graduation. (See insert for color version of figure.)

Coupled with my interest in chemistry and all science was my love of reading, which drove me to read anything I could get my hands on. Later on, when I embarked on a professional career, I was determined to show that women could do whatever they set out to accomplish. One of my favorite quotes is: "If at first you don't succeed, try, try, and try again." I also like "Pick yourself up, dust yourself off, and start all over again."

I attended the University of New South Wales in Australia, and graduated in 1976 with a bachelor's degree in ceramic engineering (1st class honors). I then moved to the United States, and came to the University of California in Berkeley, where I got my master's degree in 1979 and my Ph.D. in 1983, both in materials science.

Credit: http://quest.nasa.gov/people/bios/aero/johnsons.html.

A graduation picture of Sylvia from 1982 at the University of California, Berkeley.

Career History

Dr. Johnson worked at SRI International in Menlo Park, California for 18 years primarily in the area of ceramic and composite processing and characterization. She gained expertise in the mechanical behavior of ceramics, application of preceramic polymers,

Dr. Johnson delivering a talk at the Electronic Materials and Applications 2014 meeting. Credit: E. DeGuire, The American Ceramic Society.

sintering kinetics and phase equilibria, ceramic forming techniques, joining of ceramics, ceramic processing, powder synthesis, composite fabrication, degradation behavior of ceramics, and synthesis and processing of high-temperature superconductors. During this period, she also did business consulting for materials companies. When she left SRI, she was the Director of Ceramic and Chemical Product Development.

She joined NASA Ames in 2000 as Chief of the Thermal Protection Materials and Systems Branch. Her goal as Branch Chief was to make our Thermal Protection Systems Branch a leader in the field, and to pursue technical excellence in everything they did. During this time, she developed extensive materials processing and evaluation capabilities for NASA Ames. Her own research focused on ultrahigh-temperature ceramics and composites.

Since late 2009 she has been the Chief Materials Technologist in the Entry Systems and Technology Division at NASA Ames. She holds six U.S. patents, has edited 2 books, and published 50 articles. Dr. Johnson is a lifetime member of the National Associates Program (http://www.nationalacademies.org/includes/nationalassociates.pdf), a program that recognizes individuals for outstanding contributions to the work of the National Research Council and Institute of Medicine programs.

She is married and lives with her husband (James W. Evans); they have two children. Reading is still one of her favorite pastimes. She also enjoys traveling, hiking, cooking, and sailing.

3 Most Cited Publications

Title: Strength and creep behavior of rare-earth disilicate–silicon nitride ceramics
Author(s): Cinibulk, MK; Thomas, G; Johnson, SM
Source: Journal of the American Ceramic Society; volume: 75; issue: 8; pages: 2050–2055; doi: 10.1111/j.1151-2916.1992.tb04464.x; published: August 1992
Times Cited: 122 (from Web of Science)

Title: Grain-boundary-phase crystallization and strength of silicon nitride sintered with a YSiAlON glass
Author(s): Cinibulk, MK; Thomas, G; Johnson, SM
Source: Journal of the American Ceramic Society; volume: 73; issue: 6; pages: 1606–1612; doi: 10.1111/j.1151-2916.1990.tb09803.x; published: June 1990
Times Cited: 112 (from Web of Science)

Title: Oxidation behavior of rare-earth disilicate–silicon nitride ceramics
Author(s): Cinibulk, MK; Thomas, G; Johnson, SM
Source: Journal of the American Ceramic Society; volume: 75; issue: 8; pages: 2044–2049; doi: 10.1111/j.1151-2916.1992.tb04463.x; published: August 1992
Times Cited: 101 (from Web of Science)

Challenges

As I write this, I realize that my reluctance to write about myself is really related to being a woman brought up in a different time and place. Politeness and modesty are commendable, but the problem of being sufficiently assertive in the work environment continues to be a challenge. Being able to speak up and not be talked over is difficult when one's natural instinct is to defer. As the years have gone by, I look back at times when I lacked both the self-confidence and the necessary assertiveness to speak up on an issue, and wonder what the outcomes could have been. I have become more assertive and less easily vanquished by a loud voice with time and experience. I now find it quite easy to point out when another person is being ignored!

Other challenges were family related, in trying to juggle two careers and two travel schedules, and commuting a long distance to work (mostly in a time before cell phones were common). We worked out an operating plan but there were always times when no one was available. Our relatives all lived in other countries, and thus neighbors and friends became the backup plan. It did teach me to plan and plan again!

On being a woman in this field . . .

I started in this field when it was very uncommon for women to be in engineering or applied science, and I did so in a country that held conservative views on women. I attended an all-girls high school as was common, and so lacked some of the background for engineering (engineering drawing was a nightmare!). Some stubborn streak in me does not take well to being told no or being dismissed as just a girl. So I fought back and did better than the guys.

It was always challenging to overcome the sexism and harassment that was more overt when I was younger. A sense of humor and forethought on how to avoid situations works well. I always wanted a family and have been very lucky to have two wonderful children. There were a couple of years I do not remember all that clearly, but were really very productive. I did choose a non-academic career but mostly because I would be a lousy teacher. I have made career choices that allowed me some balance between work and family. One always wonders about a path not taken, but I always come back to my family. A big challenge in the younger years was balancing and coordinating travel with my husband and childcare. I did make sure that I had a network of friends whom I could call for emergency assistance, which although rarely used made life less stressful.

So there are challenges, and they can be stressful, but you can get a good story out of them! I would not have it any other way.

Words of Wisdom

- My advice to students is that you get a really good education, both in depth and in breadth, remembering that as a part of a team you depend on others for results; therefore, you must continue to cultivate and develop the people side of your profession.
- You never know what might come in handy, so take advantage of all your opportunities.
- Favorite quotes:
  - "If at first you don't succeed, try, try, and try again."
  - "Pick yourself up, dust yourself off, and start all over again."
- Laughing is often the best strategy.

Profile 47

Linda E. Jones

Provost and V.P. for Academic Affairs
Western New England University
Springfield, MA 01119
USA

e-mail: linda.jones@wne.edu
telephone: (413) 782-1223

Linda Jones when she first started
working after college (in 1987).

Birthplace
Philadelphia, PA, USA
Born
September 12, 1958

Publication/Invention Record

37 publications: h-index 9 (Web of Science)
1 patent, edited 1 book

Tags

❖ Administration and Leadership
❖ Academe
❖ Domicile: USA
❖ Nationality: American
❖ Caucasian
❖ Children: 0

Proudest Career Moment (to date)

My work to establish the Picker Engineering Program at Smith College was an element of my career that I hold most dear. It represents a conscious decision to actively work while teaching and now as an administrator to create opportunities for others to learn and grow and specifically to help women and underrepresented individuals to achieve careers in the sciences and engineering.

Academic Credentials

Ph.D. (1987) Fuel Science, The Pennsylvania State University, University Park, State College, Pennsylvania, USA.
M.S. (1984) Fuel Science, The Pennsylvania State University, University Park, State College, Pennsylvania, USA.
B.S. (1980) Chemistry, Mary Washington College, Fredericksburg, Virginia, USA.

Research Expertise: synthesis, structure and property relationships of carbons and carbides used in energy applications, high-temperature decomposition behavior of carbon and ceramic materials including glass

Other Interests: addressing women's role and advancement in engineering, improving engineering education

Successful Women Ceramic and Glass Scientists and Engineers: 100 Inspirational Profiles. Lynnette D. Madsen.
© 2016 The American Ceramic Society and John Wiley & Sons, Inc. Published 2016 by John Wiley & Sons, Inc.

Key Accomplishments, Honors, Recognitions, and Awards

- The American Ceramic Society's Sub-Committee on John Jeppson Award, 2012–2015
- The American Ceramic Society Bulletin's Editorial Advisory Board, 2010–2014
- The American Ceramic Society's Outstanding Educator Award, 2013
- American Carbon Society's Executive Board, 2006–2012
- Sigma Xi—President of the Smith College Chapter, 2008–2010
- American Society of Engineering Educators Engineering Deans Council, 2005–2010
- Alfred University John F. McMahon Award and delivered annual McMahon Lecture, 2008
- Fellow of the American Ceramic Society—Nuclear and Environmental Division, 2008
- International Carbon Societies meeting opening keynote lecture, 2006
- Joseph Kruson Trust Fund Award for Excellence in Teaching, 2004 and 1994
- State University of New York Research Foundation's Award for Excellence in Research and Scholarship, 2003
- President of the Ceramic Education Council, 1999–2003
- McMahon Award for Excellence in Teaching, 2001 and 1998
- Ruth Berger Ginsberg Award for Excellence in Teaching, 2000
- Chancellor's Award for Excellence in Teaching in the State University of New York, 1999
- American Carbon Society Griffin Lectureship in recognition of her research work with carbon and carbon-composite materials, 1997
- American Carbon Society's Mrozowski Award for best research paper, 1999
- President of the Western New York Chapter of The American Ceramic Society, 1994–1997
- The American Ceramic Society's Glass Division Education Secretary, 1993–1997

Biography

Early Life and Education

Her father was a Naval Aviator, which exposed her very early in life to jets and the excitement of flight. Her early interest was in high-temperature chemistry and predictive thermochemistry. She began her early career in the aerospace industry, designing and fabricating rocket propellants. This experience led her to pursue both her M.S. and Ph.D. at Penn State in the field of fuel science. Her research involved the study of carbon fiber-reinforced carbon composite materials that house and control highly energetic fuels.

Career History

After completing her doctorate, she worked in the field of hypervelocity flight and the design and fabrication of lightweight oxidation-resistant carbon and ceramic matrix

Linda Jones with her class of students after they finished their final design clinic presentation.

composites. She returned to Penn State University joining the Center for Advanced Materials where she was engaged in NASA's High-Temperature Program and the study of the creep behavior of sapphire fiber. Her research interests were tempered only by her desire to return to academia to teach.

In 1991, she took an Assistant Professor position at Alfred University in the Department of Ceramic Engineering. She rose through the ranks becoming Full Professor. She served as the Chair of the Materials Science and Engineering Department in 2003. Throughout her time at Alfred University, she became increasingly interested in innovative and effective pedagogy particularly as it influences the engagement of women and underrepresented individuals in the study of engineering.

In 2005, she joined the faculty of Smith College as the Rosemary Bradford Hewlett '40 Professor of Engineering and Director of the Picker Engineering Program. The program is the nation's only accredited engineering program at an all-women's institution. The position was perfectly suited to her interests in science, technology, engineering, and math (STEM) education and desire to increase diversity in STEM disciplines. She advocated for a learner-centered engineering education integrated in a liberal arts educational experience. She took this new program through the process of building the curriculum, faculty, and staff. She shepherded an unranked undergraduate program to a national ranking and Accreditation Board for Engineering and Technology (ABET) accreditation.

Linda Jones, Director of the Picker
Engineering Program, in the Smith
College Museum of Art. The glass
bench is a piece by Jason Berg, 2003.

In 2010, she returned to Alfred University where she was responsible for the New York State College of Ceramics at Alfred University that is a statutory college of the State University of New York. The College is comprised of the School of Art and Design, the Inamori School of Engineering, Scholes Library, and the Schein-Joseph Museum (and its associated collections). This is an institution that she values because of its unique place in the ceramic arts and ceramic engineering science worlds. The curriculum juxtaposes the arts, sciences, and engineering and affords students the opportunity to work in all fields of study. Her work in the College of Ceramics was motivated by her belief that creativity leads to innovation and access to technology and new media informs the arts. Educating both artists and engineers in one educational community using hands-on learning as a primary educational tool shapes students uniquely. In addition to her management responsibilities, she is very proud of her role as the advisor for all students seeking a double degree in art and engineering.

In June 2014, Dr. Jones was appointed Provost and Vice President for Academic Affairs at Western New England University where she continues to work on building programs and programming that advances interdisciplinary study and affords students access to STEM disciplines.

3 Most Cited Publications

Title: Influence of boron on carbon fiber microstructure, physical properties, and oxidation behavior
Author(s): Jones, LE; Thrower, PA
Source: Carbon; volume: 29; issue: 2; pages: 251–269; published: 1991
Times Cited: 112 (from Web of Science)

Title: The effect of boron on carbon fiber microstructure and reactivity
Author(s): Jones, LE; Thrower, PA
Source: Journal de Chimie Physique et de Physico-Chimie Biologique; volume: 84; issue: 11–12; pages: 1431–1438; published: November–December 1987
Times Cited: 53 (from Web of Science)

Title: The formation and oxidation of BC_3, a new graphite-like material
Author(s): Fecko, DL; Jones, LE; Thrower, PA
Source: Carbon; volume: 31; issue: 4; pages: 637–644; published: 1993
Times Cited: 41 (from Web of Science)

Linda Jones in the lab by her CVD equipment.

Gretchen Kalonji

Independent Expert: International Collaborations in Research, Education, and Economic Development
and
Former Assistant Director-General for Natural Sciences
United Nations Educational, Scientific and Cultural Organization (UNESCO)
Paris
France

Gretchen Kalonji during her time at the University of California.

e-mail: gretchen.kalonji@gmail.com

Birthplace
Chicago, Illinois, USA
Born
April 13, 1953

Publication/Invention Record

ca. 75 publications: h-index 14 (Web of Science)
1 patent

Tags

❖ Administration and Leadership
❖ Domicile: USA
❖ Nationality: American
❖ Caucasian
❖ Children: 3

Proudest Career Moment (to date)

I would cite my first General Conference, as Assistant Director-General for Natural Sciences for UNESCO, in November 2011. It was tremendously satisfying to be successful in bringing the then 194 member states of UNESCO to enthusiastic consensus on a new vision for science for the organization.

Academic Credentials

Ph.D. (1982) Materials Science and Engineering, Massachusetts Institute of Technology, Cambridge, MA, USA.
B.Sc. (1980) Materials Science and Engineering, Massachusetts Institute of Technology, Cambridge, MA, USA.

Research Expertise: symmetry constraints on the structure and properties of crystalline defects, phase transformations, microstructural evolution, atomistic computer simulation techniques, rapid solidification of ceramics

Successful Women Ceramic and Glass Scientists and Engineers: 100 Inspirational Profiles. Lynnette D. Madsen.
© 2016 The American Ceramic Society and John Wiley & Sons, Inc. Published 2016 by John Wiley & Sons, Inc.

Other Interests: internationalization of research and education, higher education transformation, promoting equity and access for women and other underrepresented groups, science and engineering for sustainable development, intergovernmental scientific processes, science, technology, and innovation policy

Key Accomplishments, Honors, Recognitions, and Awards

Dr. Kalonji's work, in materials science, in educational transformation, and on new models for international scientific collaboration, has been recognized by numerous awards and honors, including

- Leadership Award from the International Network for Engineering Education and Research, 2003
- National Science Foundation (NSF) Director's Award for Distinguished Teaching Scholars, 2001
- George E. Westinghouse Award from the American Society for Engineering Education, 1994
- NSF's Presidential Young Investigator Award (Materials Research), 1984
- Visiting faculty appointments at numerous universities and institutes around the world, including the Max Planck Institute (Germany), the University of Paris (France), Tohoku University (Japan), Sichuan University (China), and Tsinghua University (China)
- More than 200 invited lectures in 43 countries around the world

Kalonji has played leadership roles on numerous national and international advisory boards and committees. A partial list is provided below:

- Member, External Advisory Board, Arthur C. Clark Center for the Human Imagination, UC San Diego, 2014 to present
- Coordinating Group GenderInSITE: Gender in Science, Innovation, Technology and Engineering, 2013 to present
- Member, Scientific Advisory Council, International Union for the Conservation of Nature (IUCN), 2011 to present
- Steering Committee, Abdus Salam International Center for Theoretical Physics (ICTP), Trieste, Italy, 2010–2014
- Steering Committee, The World Academy of Science (TWAS), Trieste, Italy, 2010–2014
- Steering Committee, The World Science Forum, Budapest, 2011, and the World Science Forum, Rio de Janeiro, 2013
- Governing Board, International Science, Technology and Innovation Center for South–South Cooperation, Kuala Lumpur, Malaysia, 2011–2014
- Governing Board, International Sustainable Development Research Center, Moscow, Russia, 2011–2014

- Governing Board, International Center on Space Technologies for Natural and Cultural Heritage, Beijing, 2011–2014
- Representative of UNESCO on the "Transition Team" of "Future Earth" and on the "Science and Technology Alliance for Global Sustainability," 2011–2013
- Organizing Committee, "First Africa Forum on STI for Youth Employment, Human Capital Development and Inclusive Growth," a Ministerial Conference co-organized by UNESCO, the African Union, the African Development Bank, and the Association for the Development of Education in Africa, Nairobi, 2012
- Chair, Advisory Committee for International Science and Engineering, National Science Foundation, member from 2006; Chair from 2007 to 2010
- Governing Board, Indo–US Science and Technology Forum, 2006–2009
- Executive Committee, Indo–US Collaborations for Engineering Education, 2007–2010
- Chair, International Scientific Advisory Committee, Centre for Interdisciplinary Science, Indian Institute of Science (a new initiative presented to the Sir Ratan Tata Trust), 2007–2010
- Board of Directors, US–China Green Energy Council, 2007–2010
- Board of Directors, International Forum of Public Universities, 2008–2010
- Steering Committee, Canada–California Strategic Innovation Partnerships, 2005–2010
- Advisory Council on US–China Science Policy Dialogues, 2002–2006
- Sigma Xi International Committee, 2006–2010
- Senior Advisory Committee, UC School of Global Health, 2007–2010
- Board of Directors, Tohoku University Global Promotion Center, 2007–2010
- Steering Committee, NSF project on "Assuring a Globally Engaged Workforce," 2005–2007
- Steering Committee, Indian Institute of Science Global Conference: Celebrating Leadership in Science, Technology and Innovation, 2007
- Member of the International Reference Group for the Tertiary Education Advisory Commission (TEAC), a body appointed to advise the government on the future of higher education in New Zealand, 2000–2005
- Steering Committee, Pacific Rim Experiences for Undergraduates: PRIME, UC San Diego
- Advisory Council on International Science, AAAS, 2001–2004
- Advisory Board to the Education and Human Resources Directorate of the NSF, 2002–2004
- Chair, Committee of Visitors, Programs for Minorities and Minority Serving Institutions, National Science Foundation, 2004
- Chair, Washington State–Sichuan Province Friendship Association, 2000–2004
- Advisory Board, NSF-sponsored Rural Systemic Reform Initiative for the Navajo Nation, 1998–2003

- Task Force, Knowledge Exchange Through Learning Partnerships (KELP), a southern Africa initiative of USAID, 1999–2003
- Advisory Board, the Integrated Teaching and Learning Laboratory, University of Colorado, 1995–2000
- Editorial Board, The International Journal of Engineering Education, 1995–2000
- Organizing Committee, International Workshop on Creativity and International Working Ability, Osaka, 1998
- Co-Director for the NSF-sponsored Engineering Coalition of Schools for Excellence in Education and Leadership (ECSEL), 1994–1997
- Chair, Initiative on Diversity in the Professoriate, MIT, 1995–1997
- Advisory Board for the NSF-sponsored project "A Slice of Life: An Introductory Biology Course," Northern Arizona University, 1993–1996
- Co-organizer, "Teaching Invention" Workshop of the National Collegiate Inventors and Innovators Alliance (NCIIA), Hampshire College, 1996
- Organizing Committee, International Symposium on Engineering Education and Evaluation, Osaka, Japan, November 1995
- Advisory Board for the "Technology for Science" Project, Technical Education Research Centers (TERC), Cambridge, MA, 1992–1995
- Advisory Board for the Global Laboratory Project, TERC, Cambridge, MA, 1992–1995
- Advisory Board for the SUCCEED Coalition's Center for Professional Success, 1993–1995
- Co-Director in the United States for the Computer Science and Electronics Program at the Solomon Mahlangu Freedom College, Mazimbu, Tanzania (ANC School), 1987–1991

Biography

Early Life and Education

The child of journalists, Kalonji grew up largely outside the United States, in India, Hong Kong, Thailand, and Kenya.

She earned two degrees at Massachusetts Institute of Technology (MIT) in short succession—a bachelor's degree in 1980 and then a doctorate in 1982. With this background, she stepped immediately into a faculty position.

Career History

Kalonji began her academic career at the MIT, where she served as Assistant and Associate Professor in the Department of Materials Science and Engineering, from 1982 to 1990. At MIT, her research foci expanded to include efforts in rapid solidification of ceramics and in atomistic simulations of interfaces, in addition to the work on applications of symmetry theory to the structure and properties of crystalline defects that had been the focus of her doctoral work. Her primary teaching responsibilities were the graduate level core thermodynamics course and the undergraduate course in chemical physics of materials, though she also created a Freshman Seminar in Kiswahili.

While at MIT, Kalonji was very active with the anti-apartheid movement in the greater Boston area. Working with the academic and high-tech communities, and with scientific leadership of the African National Congress in exile, she was instrumental in the design of a "cradle to grave" computer science and electronics program for the Solomon Mahlangu Freedom College (SOMAFCO) in Tanzania, an institution set up to serve the needs of the South African exile community. She served as Co-Director in the United States for this program from 1987 to 1990, and through this work began a lifetime commitment to innovations in science and engineering education. It was also at SOMAFCO that Kalonji first met her two adopted sons, Daniel Lungu and Mbombo Maleka, who subsequently joined Kalonji and her son, Hussein Charles Kalonji, in Seattle where all three boys graduated from high school. All three subsequently completed their university educations in the United States.

In 1990, she joined the University of Washington (UW) in Seattle as the Kyocera Professor of Materials Science. (Kyocera Corporation is a multinational electronics and ceramics manufacturer that is headquartered in Kyoto, Japan. It trades on the New York Stock Exchange.) She was the first woman at UW to hold an endowed chair.

From 1990 to 1997, Kalonji served in multiple leadership roles in the NSF-sponsored Engineering Coalition of Schools for Excellence in Education and Leadership (ECSEL), including serving as Co-Director with the Dean Emeritus of Howard University, M. Lucius Walker. ECSEL was a coalition of seven universities, head-quartered at Howard, which received $29.5 million in funding from the NSF over a 10-year period, and which had as its key theme increasing participation in engineering by groups historically underrepresented, through the integration of multidisciplinary, team-based, design projects throughout the curriculum.

At the UW, Kalonji developed a strong professional focus on new models of international collaboration in research and education. She led a campus-wide effort to integrate collaborative international research activities into curricular pathways of students, across the disciplines and from freshman to doctoral level. In this initiative, entitled *UW Worldwide*, teams of faculty and students, starting with first-year under-graduates, were brought together across national boundaries to work collaboratively on common pressing practical challenges, such as water quality, biodiversity, forest ecology, "eco-materials," arts and graphic design, public health, urban planning, and the quality of education itself. The *UW Worldwide* program was honored with multiple grants and awards, particularly in the United States, Japan, and China, where it was designated by the Chinese Ministry of Education as a "national key model in under-graduate education reform." It was largely on the basis of this work that Kalonji was honored with the NSF's Distinguished Teaching Scholar Award, in 2001.

In March 2005, Kalonji joined the University of California (UC) Office of the President, where she served in the newly created position of Director of International Strategy Development for the 10-campus UC system until 2009, and as Director of Systemwide Research Development from 2009 to 2010. In her role as Director of International Strategy Development, Kalonji worked with faculty and administrators from throughout the 10 campuses of the UC, and with partners around the world, to

Gretchen Kalonji at the Indian Science Congress, Bhubaneshwar, January 2012. (See insert for color version of figure.)

catalyze pilot projects that put forward new models of targeting UC's extraordinary research capacity to the solution of pressing problems facing both California and partner regions. Central to the new approaches was the more effective integration of collaborative, multidisciplinary research addressing common "grand challenges" into the curriculum, at both undergraduate and postgraduate levels. Pilot projects were launched with China, India, Canada, Mexico, and Chile. With China, Kalonji was the principal architect of the "10+10," a program that brought together the 10 UC campuses with 10 of the premier research universities in China on a collaborative research and educational agenda. Similar large-scale "binational" efforts included the *Canada–California Strategic Innovation Partnerships (CCSIP)*, the *UC–India Initiative*, and the *Chile–California Program on Human Capital Development*. Each of these initiatives garnered significant support at the national level from governmental, academic, and industrial partners. In addition to these regionally focused initiatives, Kalonji was also active in conceptualization and planning for the UC systemwide Global Health Sciences Initiative, which has a major strategic focus in East Africa.

In her role as Director of Systemwide Research Development, from March 2009 to March 2010, Kalonji's primary role was to assemble external review teams to help the UC assess priorities for its portfolio of internally funded "Multi-Campus Research Programs and Initiatives" (MRPIs). This effort constituted the most fundamental reassessment of UC's research priorities in decades. Multicampus research proposals were solicited in six areas, including Emerging Sciences and Technology; Arts and

Humanities; International and Area Studies; Social Sciences; Biological and Health Sciences; and "Critical California Issues." As a result of this competition, 37 proposals were selected for a total of $68 million in funding from the UC. In parallel to her roles in the UC Office of the President, from 2006 until the end of 2013, Kalonji held an appointment as Professor in the Department of Electrical Engineering at the University of California, Santa Cruz.

In July 2010, Kalonji joined the United Nations Educational, Scientific and Cultural Organization (UNESCO) as the Assistant Director-General for Natural Sciences, a position she held until the end of February 2014. Reporting to UNESCO's Director-General, Kalonji was responsible for strengthening the efforts of the organization in all aspects of the science and engineering enterprise, and for working together with the rest of UNESCO's senior leadership team in the design and implementation of interdisciplinary programs focusing on poverty eradication and sustainable development. At UNESCO, Kalonji was responsible for managing a large and complex network of individuals and institutions in the fields of science and engineering. These included several large-scale intergovernmental science programs, including the International Hydrological Program (IHP), the Man in the Biosphere Program (MAB), the International Geosciences Program (IGCP), and the International Basic Sciences Program (IBSP). Several other major international entities also reported to Kalonji, including the International Center for Theoretical Physics (ICTP), in Trieste; TWAS, also in Trieste; and the UNESCO Institute for Hydrological Education (UNESCO-IHE), in Delft. In addition, she was responsible for catalyzing and coordinating the activities of approximately 30 UNESCO-affiliated research centers around the world in science and engineering, the so-called UNESCO "category 2" institutions, and 300+ "UNESCO Chairs" in science and engineering. The total biennial regular program budget for the natural sciences at UNESCO was approximately $62 million, with an additional amount of approximately $180 million raised from external sources. Kalonji managed a staff of approximately 175 individuals, located not only in UNESCO headquarters but also in multiple field offices around the world. UNESCO's two global priorities are Africa and gender equality, and Kalonji benefited from strong networks worldwide in both of these domains.

Kalonji's primary accomplishments during her tenure at UNESCO include (1) raising the profile of UNESCO science in the international community, as evidenced, for example, by the recent designation of UNESCO by UN Secretary General Ban Ki-moon as the host of the Secretariat of his new "Scientific Advisory Board"; (2) strengthening ties with other important organizations in international science and engineering, for example, the International Council for Science (ICSU), the World Federation of Engineering Organizations (WFEO), and CERN, the European Organization for Nuclear Research; (3) building UNESCO's participation in new initiatives, including *Future Earth* and IPBES (the Intergovernmental Platform on Biodiversity and Ecosystem Services); (4) significantly enhancing collaborations with the private sector and with professional societies, including new projects with Airbus, Intel, Roche, IEEE, ASME, Microsoft, and Nature Publishing Group, among multiple others; (5) greatly enhancing UNESCO's roles in science and engineering for disaster risk reduction and mitigation, including notable efforts in the Pakistan floods in 2010 and in the drought in the Horn of

Africa in 2011–2013, among other regions; (6) undertaking a comprehensive restructuring of the natural sciences sector, emphasizing several strategic cross-cutting themes, including natural disaster risk reduction, science education, engineering, and biodiversity; (7) launching of a new "UNESCO Engineering Initiative," focusing on more effectively targeting the efforts of the engineering community, including students, to the challenges of sustainable development; and (8) playing important leadership roles in bringing the centrality of science and technology to the "post-2015" development agenda, including leadership roles in the Rio+20 process.

Kalonji is currently working as an independent consultant on international collaborations in research, education, and economic development.

3 Most Cited Publications

Title: The cooling rate dependence of cation distributions in $CoFe_2O_4$
Author(s): DeGuire, MR; O'Handley, RC; Kalonji, G
Source: Journal of Applied Physics; volume: 65; issue: 8; pages: 3167–3172; doi: 10.1063/1.342667; published: April 15, 1989
Times Cited: 57 (from Web of Science)

Title: Excess energy of grain-boundary trijunctions: an atomistic simulation study
Author(s): Srinivasan, SG; Cahn, JW; Jonsson, H; et al.
Source: Acta Materialia; volume: 47; issue: 9; pages: 2821–2829; doi: 10.1016/S1359-6454(99)00120-2; published: July 9, 1999
Times Cited: 48 (from Web of Science)

Title: A grain boundary phase transition studied by molecular dynamics
Author(s): Deymier, P; Taiwo, A; Kalonji, G
Source: Acta Metallurgica; volume: 35; issue: 11; pages: 2719–2730; published: November 1987
Times Cited: 32 (from Web of Science)

Challenges

My primary professional challenge has been finding a balance between my personal materials science research interests and my desire to contribute more broadly to strengthening science, technology, and innovation ecosystems in service to society. This is an ongoing struggle.

On being a woman in this field . . .

Materials science and engineering is a field that has in general been friendlier to women than a number of other technical disciplines. And, over the years, I have seen a very significant improvement in the climate for women on average. Nevertheless, there is, I believe, tremendous institutional variation in commitment to equity and access. While logically men and women should bear equal responsibility to assure that our institutions are fair to all, in reality a disproportionate share of the burden tends to be taken up by women, and particularly by senior women. Especially in the early years of my career, at MIT, I expended an enormous amount of my personal energy addressing issues of equity for women and minority faculty members. I by no means regret that, and believe that I most certainly contributed to the strengthening of the institution as a consequence. Nevertheless, it represented a major diversion of my personal energy that could have been spent on more enjoyable and professionally rewarding activities.

Words of Wisdom

I would like to focus what advice I offer primarily to those who are interested in combining a career in science and engineering with efforts in service to society more broadly. It seems to me there are three elements that are critically important. The first is the choice of field/ subfield in which to work; some topics are more amenable than others for such a combined focus, for example, some of the aspects of addressing concrete sustainable development challenges. Second, the choice of an institutional home is *vitally* important, particularly for junior faculty members. One needs to search out institutions that truly value "service" contributions broadly construed, that have a history and culture of interdisciplinary collaboration, and that promote links between academia and other sectors of society. Finally, young researchers, starting even at the stage of undergraduate and graduate studies, should try to design their educational and research activities to enable multiple subsequent career pathways. Most science and engineering professors are most comfortable in guiding their students to follow their own pathways, so it is important to seek out advice from others,

Gretchen Kalonji with a subset of her grandchildren, including the second Gretchen Kalonji (standing in the front), Seattle, October 2012. (See insert for color version of figure.)

who might be able to shed light for you on the pros and cons of different professional possibilities, including in industry, government, civil society, entrepreneurship, or even in intergovernmental agencies. There are really an enormous number of ways to contribute to society with a scientific background, and getting as much knowledge as possible to be able to craft sound decisions is vitally important.

Finally, and this is advice I hope would be useful for everyone: Do not feel that you have to put off your family for the sake of your career. Of course, it is a challenge to have small children when you are still in school, or in the early stages of a tenure track position. And for some, waiting might be the most appropriate thing to do. But it is *not* necessary. I had my son Hussein before I started university, and in addition to the joy and satisfaction of being with him, I believe his presence had a stabilizing effect on me, and in fact helped me get through the undergraduate and doctoral programs at MIT in record time. Again, each person is different, but it makes me sad to know that there are still many women who think they have to sacrifice the opportunity to have children for the sake of their professional advancement.

Maarit Johanna Karppinen
(née Viljavuori)

Professor of Inorganic Chemistry
Deputy Head of Department
Department of Chemistry
School of Chemical Technology
Aalto University
Finland

e-mail: maarit.karppinen@aalto.fi
telephone: +358 50 384 1726

Maarit Karppinen at the age of 46 years when she returned back to Finland.

Birthplace
Helsinki, Finland

Born
April 17, 1959

Publication/Invention Record
>350 publications: h-index 32 (Web of Science)
7 patents; 1 textbook; 10 book chapters

Tags
❖ Administration and Leadership
❖ Academe
❖ Domicile: Finland
❖ Nationality: Finnish
❖ Caucasian
❖ Children: 1

Proudest Career Moment (to date)

In 2008, I was nominated for the prestigious Academy Professor position in Finland for the five-year period, 2009–2013. Academy of Finland funds 44 five-year Academy Professorships, placed in fields that cover everything from culture, society, and health to biosciences, natural sciences, and engineering. Hence, each Academy Professor can be considered as an honorable representative of his/her own field in Finland. Academy Professorship had been my secret dream for a long time, but I never was self-confident enough to anticipate that I would be selected.

Successful Women Ceramic and Glass Scientists and Engineers: 100 Inspirational Profiles. Lynnette D. Madsen.
© 2016 The American Ceramic Society and John Wiley & Sons, Inc. Published 2016 by John Wiley & Sons, Inc.

Academic Credentials

D.Tech. (1993) Chemistry, Helsinki University of Technology, Espoo, Finland.
Lic.Tech. (1990) Chemistry, Helsinki University of Technology, Espoo, Finland.
M.Sc.Tech. (1987) Chemistry, Helsinki University of Technology, Espoo, Finland.

Research Expertise: functional oxide materials (both bulk and thin films) for the next-generation energy and nanotechnologies, including material categories such as high-T_c superconductors, thermoelectrics, multiferroics, half-metals for spintronics, ionic conductors for fuel cells, batteries, oxygen storage, etc.; nanocomposite materials where inorganics are combined down to molecular-level precision with organic molecules, polymers, biomaterials, nanotubes, and graphene sheets for novel thin-film structures and 3D architectures

Other Interests: tutoring younger scientists and participating in various domestic and international research assessment/evaluation panels/committees

Key Accomplishments, Honors, Recognitions, and Awards

- Member of Editorial Board: Advances in Materials Science and Engineering, 2014 to present
- European Research Council (ERC) Advanced Grant, 2014–2018
- Chair of the Chemistry Panel: Publication Forum Project (UNIFI), 2010–2016
- Finnish representative of experts: COST-CMST, 2010–2014
- Head of Department of Chemistry, Aalto University, 2008–2013
- Listed in AcademiaNet (profiles of leading women scientists, by invitation only), 2013
- Knight, First Class, of the Order of the White Rose of Finland, 2011
- Elected member of Finnish Academy of Science and Letters, 2010
- Vice Chair of External Expert Panel: Research Assessment of University of Gothenburg (Department of Chemistry), 2010
- Member of journal Editorial Advisory Board: Kemia (Finnish Chemistry Magazine), 2006–2008
- Research Prize of Materials & Structures Laboratory, Tokyo Institute of Technology, Japan, 2006
- Prize of Doctoral Thesis of Finnish Academy of Sciences (best doctoral thesis published in 1993–1994 in Finland), 1995
- Research Prize of International Superconductivity Technology Center (ISTEC, Japan), 1993
- External expert evaluator for professorships: 16 times (in Finland, Sweden, Norway, Estonia)
- Chairman/member of Selection Committee for Professorships: 12 times (at Aalto University)
- External evaluator of major grants: 12 times (in Sweden, Switzerland, Germany, European Research Council, COST)

- External opponent for doctoral theses: 17 times (in Finland, Japan, Norway, Sweden)
- Invited/keynote/plenary speaker at major international conferences: 57 times (in Japan, USA, China, UK, Germany, Finland, Australia, Taiwan, Spain, Lithuania, UK, France, Switzerland, Italy, Korea, India, Russia, Sweden, Estonia, Mexico)
- Session Chairman in international conferences: more than 50 times
- Main/major organizer of international conferences: 23 times (in Japan, Finland, Spain)
- Editor of conference proceedings: 9 times
- Invited lectures at universities and graduate schools: more than 70 times (in Japan, USA, UK, China, Taiwan, Canada, Australia, France, Spain, Sweden, Estonia, Lithuania)
- Tens of external grants (in total around 10 million euros)

Biography

Early Life and Education

Maarit Karppinen was born in Helsinki, Finland, where she also received her basic education up to the matriculation examination in 1978. In the same way as her parents, brother, and sister, she selected university studies in natural sciences and technology. She entered Helsinki University of Technology in 1978, and completed her master's degree in 1987, licentiate degree in 1990, and doctoral degree in 1993, all in inorganic chemistry.

She was 33 years old when she received her Ph.D. degree. Her university studies had been decelerated several times, due to her marriage in 1979, her career in basketball at the highest national level up until 1985 including a one-year visit to Sweden in 1983–1984, the birth of her daughter in 1984, her regular teaching duties since 1985, and her one-year research stay in Japan in 1991–1992.

Career History

After receiving her Ph.D. degree in 1993, Maarit Karppinen continued her academic pursuit until 2001 at Helsinki University of Technology where she held various

Maarit Karppinen (upper right corner) with her research group at
Tokyo Institute of Technology.

positions (Senior Assistant, Lecturer, Docent, Academy Fellow, Acting Associate Professor, and Acting Professor). Meanwhile, for the period of 1995–1996 she was appointed as a Visiting Associate Professor at Tokyo Institute of Technology (Tokyo Tech) in Japan. Although she returned to Finland, she kept in touch with her Japanese students providing them with much appreciated supervision. In 2001, she was appointed as a regular Associate Professor at the Materials & Structures Laboratory of Tokyo Tech.

In 2006, Maarit Karppinen was nominated (by invitation) as a Full Professor in Inorganic Chemistry at her alma mater, Helsinki University of Technology. In 2008, she was additionally appointed as the Head of the Department of Chemistry for the next six years. Meanwhile, she was nominated as Academy Professor for 2009–2013, and in 2013 she was the first chemist in Finland to receive a European Research Council (ERC) Advanced Grant; the ERC funded her research on atomic/molecular layer-by-layer (ALD/MLD) deposited hybrid inorganic–organic thin-film materials.

Maarit Karppinen at the age of 50 as a newly nominated Academy Professor.

3 Most Cited Publications

Title: Unconventional magnetic transition and transport behavior in $Na_{0.75}CoO_2$
Author(s): Motohashi, T; Ueda, R; Naujalis, E; et al.
Source: Physical Review B; volume: 67; issue: 6; doi: 10.1103/PhysRevB.67.064406; published: February 1, 2003
Times Cited: 164 (from Web of Science)

Title: Evidence for valence fluctuation of Fe in Sr_2FeMoO_{6-w} double perovskite
Author(s): Linden, J; Yamamoto, T; Karppinen, M; et al.
Source: Applied Physics Letters; volume: 76; issue: 20; pages: 2925–2927; published: May 15, 2000
Times Cited: 149 (from Web of Science)

Title: Oxygen nonstoichiometry in $YBaCo_4O_{7+d}$: large low-temperature oxygen absorption/desorption capability
Author(s): Karppinen, M; Yamauchi, H; Otani, S; et al.
Source: Chemistry of Materials; volume: 18; issue: 2; pages: 490–494; doi: 10.1021/cm0523081; published: January 24, 2006
Times Cited: 87 (from Web of Science)

Challenges

There have been continuous challenges throughout my career, or at least that is what I felt when I was younger. Now, looking back at my early career, it appears as if I had planned everything very well.

When I was young, I was extremely shy and my self-confidence was low: as a student, public speaking was something I thought I could not overcome. I was not sure that I could someday get a master's degree as it required a final presentation, to get a doctoral degree required a public defense, and so on. Then to be able to work in academia I would have to teach in front of students, speak to conference audiences, lead my research group, be in leadership positions at the university, and so on. Despite all my fears, at the same time, I always knew that there could not be any other career option for me than being a professor someday.

My family has lived in many foreign countries and very often not all of us in the same place at the same time; we are all Finnish, but I have worked in Japan for 10 years, my husband is a basketball coach—a profession that commonly and regularly changes cities and countries, and my daughter has studied in Japan, Hong Kong, the Netherlands, and the UK. These visits abroad with various durations have been required from all of us, but at the same time have been the source of joy and a kind of proudness as well.

On being a woman in this field . . .

I have always enjoyed of being a woman, both in my private life and in my academic career; I do not know if men think in the same way, but I wish they would be as happy in being a man, as I am in being a woman. I could not be happier in any other role than being a mother for my daughter, although sometimes I am mother from a geographical distance.

When I was appointed to my regular Chair at Tokyo Institute of Technology in 2001, there were 51 other applicants, all Japanese men. At Tokyo Tech, I was the only foreign professor from the Western world. I was also—at that time—the only female professor in science and engineering. The funny thing was that initially there were not even toilets for ladies in the building where I was working! I was told that I was not the only female professor; there was another woman at Tokyo Tech, as an Associate Professor in Economics, but I have never met her. As a Finnish basketball player, I was also taller than 95% of my colleagues in Japan. Hence, I was an alien in an otherwise unbelievably

Maarit Karppinen with her colleagues in Japan.

homogeneous society. I mostly enjoyed this aspect, in particular the fact that all the comparisons to the colleagues were essentially impossible. I am sure that the situation would again be different for a Finnish male professor, a Japanese female professor, or just an unexceptionally tall person in Japan.

In Finland being a woman in science or technology is not that exceptional. Of course, even in my present environment most of my colleagues are men, but it is very rarely I even notice that. I have never felt that I have either suffered from or benefited of being a woman. Occasionally it is discussed, even in Finland, whether a certain portion of grants or positions should be guaranteed for female applicants in different academic selections. I would definitely not like to be selected to any position just because I am a woman; even if that had not been the reason for the selection, there would always be somebody who would think so later. Finally, I should acknowledge my husband who always let me go ahead, even when our daughter was small.

Words of Wisdom

- You can only love this job, otherwise you cannot do it.
- If you love your work, you do not feel like you are working even though you do not do anything else but work.
- Do not plan your career too seriously, the best things tend to happen by accident; you just need to be open-minded and ready to accept the good things.
- Be persistent in your work if you believe in it.
- Praise your closest colleagues and other people around you; from outside you are seen and judged as a member of that group.
- Be optimistic and motivated; everybody likes to collaborate with such a shiny researcher.
- Every successful researcher should have at least one unique competence, whether it is big or small; try to find yours and develop it further.

Kazumi Kato

Prime Senior Research Scientist
Research Group Leader and Deputy Director
National Institute of Advanced Industrial Science and
Technology (AIST)
Inorganic Functional Materials Research Institute
2266-98 Anagahora, Shimoshidami
Moriyama
Nagoya, Aichi 463–8560
Japan

e-mail: kzm.kato@aist.go.jp
telephone: +81-29-861-2000

Tags

❖ Administration and
  Leadership
❖ Academe
❖ Domicile: Japan
❖ Nationality: Japanese
❖ Asian

Publication/Invention Record

>150 publications: h-index 20 (Web of Science)
30 patents, editor of 3 books, 3 book chapters

Academic Credentials

Ph.D. (1989) Applied Chemistry, Nagoya University, Nagoya, Japan.
M.S. (1986) Applied Chemistry, Nagoya University, Nagoya, Japan.
B.S. (1984) Applied Chemistry, Nagoya University, Nagoya, Japan.

Research Expertise: solution chemistry of thin film oxides, including ferroelectrics, superconductors, and porous oxides

Key Accomplishments, Honors, Recognitions, and Awards

• 2nd Kodate Kashiko Award, The Japan Society of Applied Physics, 2011
• 3rd "Monodukuri" Award, Ministry of Economy, Trade and Industry, 2009
• 61st Academic Award, The Ceramic Society of Japan, 2007
• 35th Academic Award, The Tokai Kagaku Kogyo Kai, 2000
• 59th Award for Outstanding Patents, Science and Technology Agency, 2000

Successful Women Ceramic and Glass Scientists and Engineers: 100 Inspirational Profiles. Lynnette D. Madsen.
© 2016 The American Ceramic Society and John Wiley & Sons, Inc. Published 2016 by John Wiley & Sons, Inc.

Biography

Early Life and Education

She received her B.S. (1984), M.S. (1986), and Ph.D. (1989) all in applied chemistry from Nagoya University, Japan.

Career History

She worked as a Researcher at the Toyama Prefectural Industrial Technology Center (from 1989 to 1994), and as a Researcher (1994–1995), and then as a Chief Researcher (1995–2001) at National Industrial Research Institute of Nagoya, Japan. She has been Group Leader of the Tailored Liquid Source Research Group (2001–2005) and then of the Tailored Liquid Integration Research Group since 2005, and was promoted to a Prime Senior Research Scientist in 2013 maintaining the position as a Group Leader at AIST. During the period, she held several academic positions such as associate professors of Tokyo Institute of Technology, Nagoya Institute of Technology, and Hokkaido University. Dr. Kato works as a Prime Senior Research Scientist at AIST, and concurrently holds the positions of Group Leader and Deputy Director of Inorganic Functional Materials Research Institute at AIST.

Dr. Kato has served as a Key Research Scientist of AIST and Project Leader for many National R&D Projects (by Japanese Government):

- Ferroelectric Memories for Next Generation (1999–2003)
- Research and Development of Low-Emission Materials and Technology (2002–2007)
- Environmental-Friendly Sensors (2006–2010)
- Research and Development of Organic Optical Devices Using Self-Adaption Technology (2006–2007)
- Research and Development of Organic Optical Tape Modules Using Self-Adaption Technology (2008–2009)
- Research and Development of High-Performance Down-Sized Devices by Bottom-Up Fabrication Using Single Crystal Nanocubes (2011 to present)

Securing funding support for these projects signifies successful R&D efforts in Japan.

3 Recent Publications

Title: Nano-sized cube-shaped single crystalline oxides and their potentials; composition, assembly and functions
Author(s): Kato, K; Dang, F; Mimura, K; et al.
Source: Advanced Powder Technology; volume: 25; pages: 1401–1414; published: February 25, 2014

Title: Enhanced dielectric properties of $BaTiO_3$ nanocube assembled film in metal–insulator–metal capacitor structure

Author(s): Mimura, K; Kato, K
Source: Applied Physics Express; volume: 7; page: 160501-1-3; published: May 16, 2014

Title: Diversity in size of barium titanate nanocubes synthesized by a hydrothermal method using an aqueous Ti compound
Author(s): Ma, Q; Mimura, K; Kato, K
Source: CrystEngComm; volume: 16; pages: 8398–8405; published: August 6, 2014

Sources Used to Create This Profile

- Web of Science, http://thomsonreuters.com/thomson-reuters-web-of-science/, accessed December 9, 2014.
- AIST, e.g., https://unit.aist.go.jp/amri/group/talint/en/member.html, https://unit.aist. go.jp/amri/group/talint/en/result.html, accessed December 9, 2014.
- International Workshop for Women Ceramists, http://www.kcers.or.kr/home/kor/data/ %BF%A9%BC%BA%C0%A7%BF%F8%C8%B8/2007%B1%B9%C1%A6%BF% F6%C5%A9%BC%A5.pdf, accessed December 9, 2014.
- Japanese Journal of Applied Physics, http://www.ipap.jp/jjap/ed_board.html, accessed December 9, 2014.
- Penn State University, http://www.mri.psu.edu/news/events/ice-2015/bios.asp, accessed December 9, 2014.

M. Dresselhaus

J.M. Phillips

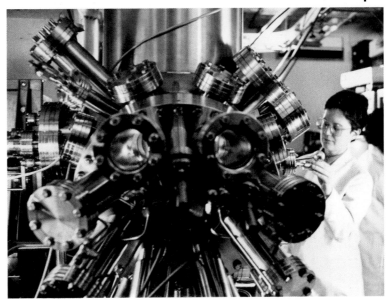

D. Rolison

M. Cole

N. Dubrovinskaia

D. Bonnell

L.G. Salamanca-Riba

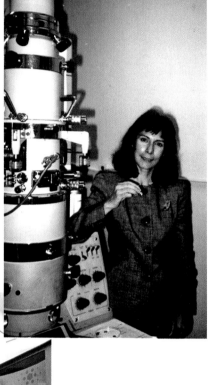

N.M. Rodriguez

D. Gerthsen

B. Noheda

M. Mayo

G. Kalonji

T.A. Prikhna

L.C. Chen

I. Talmy

Z. Chen

B. Dunbar

E. Williams

J.M. Phillips

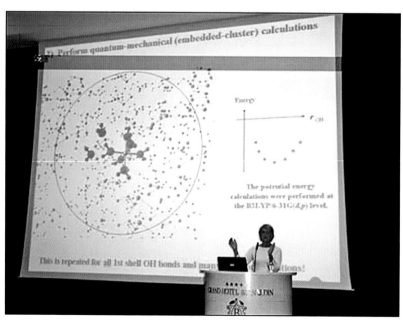

K. Hermannson

E.A. Carter

J.Y. Ying

D.D.L. Chung

D.D. Edwards

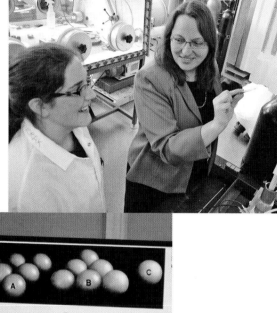

J.S. Chen

M. Ferraris

S.M. Johnson

L.C. Chen

C.A. Ross

M. Kosec

R.A. Gerhardt

E. Williams

G. Kalonji

G. Long

T.A. Prikhna

L. Nazar

S. Best

L. Folks

K. Rabe

J.L. MacManus-Driscoll

N. Spaldin

M. Kosec

Profile 51

Ruth H.G.A. Kiminami

Professor
Department of Materials Engineering
Federal University of São Carlos
São Carlos, SP
Brazil

e-mail: ruth@ufscar.br
telephone: +55 16 3351-8502

Publication/Invention Record

>150 publications: h-index 20
4 patents, 2 book chapters

Ruth Kiminami at about
30 years old, after receiving
her doctorate degree.

Proudest Career Moment (to date)

In 1986 I defended my doctoral thesis at the Rheinisch-Westfälische Technische Hochschule Aachen,
Germany; it is a moment in my career that I'm very
proud of. I had to overcome numerous difficulties and
challenges in my personal and professional life in
order to earn the title of Dr. Ing.

Tags

❖ Academe
❖ Domicile: Brazil
❖ Nationality: Brazilian
❖ Latina
❖ Children: 3

Academic Credentials

Ph.D. (1986) Ceramics, Rheinisch-Westfälische Technische Hochschule Aachen,
Germany.
M.Sc. (1981) Chemical Engineering, Federal University of Paraiba, Campina Grande,
PB, Brazil.
B.A.Sc. (1979) Materials Engineering, Federal University of São Carlos, São Carlos, SP,
Brazil.

Research Expertise: synthesis of ceramic powders and nanoceramic materials, and
microwave processing of ceramics, innovative materials with novel properties

Successful Women Ceramic and Glass Scientists and Engineers: 100 Inspirational Profiles. Lynnette D. Madsen.
© 2016 The American Ceramic Society and John Wiley & Sons, Inc. Published 2016 by John Wiley & Sons, Inc.

Other Interests: learning more about Women Engineering Leadership, fundraising activities, teaching, mentoring, international collaboration, and industrial applications

Key Accomplishments, Honors, Recognitions, and Awards

- Plenary or invited speaker in four countries (Argentina, Brazil, Germany, and the United States)
- Member of the Board of the Brazilian Ceramics Association, 2010–2016
- Award for the best postgraduate work presented at the 57th Brazilian Congress on Ceramics, 2013
- Award for the best work presented at the 56th Brazilian Congress on Ceramics, 2012
- Vice Coordinator of the Postgraduate Program in Engineering Science and Materials at the Federal University of São Carlos, 2008–2012
- Award for the best work presented at the 52nd Brazilian Congress on Ceramics, 2008
- Deputy Head of the Department of Materials Engineering, 2004–2006
- Award for the best work presented at the 48th Brazilian Congress on Ceramics, 2004
- Award for the best work presented at the 47th Brazilian Congress on Ceramics, 2003
- Member of the organizing committee of the Second Workshop on Metastable and Nanostructured Materials II, 2003 (NANOMAT 2003)
- University of Florida stipend for a postdoctoral researcher, 1998–1999
- Coordinator of the Montevideo Group, CCT/UFSCar, 1995–1997
- Elected Fellow of the American Ceramic Society, 1996
- Vice Coordinator of the area of Ceramic Materials of the Department of Materials Engineering at the Federal University of São Carlos, 1995–1996
- Honorable mention at the 31st Brazilian Congress of Ceramics, 1987

Biography

Early Life and Education

Ruth Herta Goldschmidt Aliaga Kiminami was raised in Bolivia by her Bolivian and German parents. She received her early education in the German private school system in Bolivia until the age of 10, after which she continued her early and higher education in the public school system of São Paulo, Brazil.

Ruth's undergraduate studies in Materials Engineering at the University of São Carlos were part of the university's education program. In those days, the role of women in engineering careers in Brazil was a challenging one and was rarely accepted. She received her B.A.Sc. degree in Materials Engineering in 1978. In those days, it was difficult for a female engineer to find an internship or job in the ceramics industry, so she chose an academic career (and she continued working in this vein with ceramic materials to this day).

Upon graduation from her first degree, Ruth continued her work on ceramic materials at the Institute of Atomic Energy (IEA) in São Paulo. She then continued her studies in Chemical Engineering with the help of a scholarship awarded by CNPq— Brazil's National Council for Scientific and Technological Development at the Federal

University of Paraiba, Brazil, which has since been renamed Federal University of Campina Grande. She completed her Master's degree in 1981 and then was immediately hired by the same university as an Assistant Professor.

In 1982, Ruth started her doctoral studies at the Rheinisch-Westfälische Technische Hochschule Aachen, in Germany, funded by a scholarship from the Deutscher Akademischer Austauschdienst (DAAD). She earned her Ph.D. in 1986.

Career History

After obtaining her doctorate degree, she continued teaching as Adjunct Professor at the Federal University of Paraiba for another 4 years. Then in July 1990, she was hired as an Adjunct Professor by the Federal University of São Carlos, Brazil. Finally, she carried out her postdoctoral studies at Florida University in 1998–1999—really, at this point, it was more like a sabbatical.

Her teaching at the Federal University of Paraiba and the Federal University of São Carlos included graduate

A recent image of Ruth Kiminami.

courses in Chemical Engineering, Production Engineering, and Materials Engineering. Her research was (and still is) conducted in the Laboratory of Ceramics Materials and the Laboratory of Microwave-Assisted Materials Development and Processing of the Federal University of São Carlos.

Ruth's primary research interests involve the synthesis of ceramic powders, nano-ceramic materials, and microwave ceramics processing. She has led more than 20 research projects, and has mentored 6 postdoctoral, 16 doctoral, 21 masters, and more than 90 undergraduate students. She acted as Deputy Head of the Department of Materials Engineering at the Federal University of São Carlos from 2004 to 2006, and was Vice Coordinator of the Graduate Program in Science and Materials Engineering at this institution from 2008 to 2012. She has participated in cooperative research programs with Argentina, Germany, and the United States. She has been a Fellow of the American Ceramic Society since 1996, and is an active member of the American Ceramic Society and of the Brazilian Ceramics Association.

She and her husband, Prof. Dr. Claudio Kiminami, a specialist in metals, have three children and work in the same department at the Federal University of São Carlos.

3 Most Cited Publications

Title: Synthesis, microstructure and magnetic properties of Ni-Zn ferrites
Author(s): Costa, ACFM; Tortella, E; Morelli, MR; Kiminami, RHGA
Source: Journal of Magnetism and Magnetic Materials, volume 256; issue: 1–3; pages: 174–182; published: January 2003, doi: 10.1016/S0304-8853(02)00449-3
Times Cited: 186 (from Web of Science)

Title: Combustion synthesized ZnO powders for varistor ceramics
Author(s): Sousa, VC; Segadães, AM; Morelli, MR; Kiminami, RHGA
Source: International Journal of Inorganic Materials; volume 1; issue: 3–4; pages: 235–241; published: September–October 1999; doi: 10.1016/S1466-6049(99)00036-7
Times Cited: 80 (from Web of Science)

Title: Combustion synthesis of aluminum titanate
Author(s): Segadães, AM; Morelli, MR; Kiminami, RHGA
Source: Journal of the European Ceramic Society; volume 18; issue: 7; pages: 771–781; published: 1998; doi: 10.1016/S0955-2219(98)00004-1
Times Cited: 64 (from Web of Science)

Researcher ID (E-8843-2012)

Challenges

I have faced numerous challenges in my professional career. For many years, the idea of a woman holding a degree in engineering was not readily accepted in Brazil. Among the many difficulties I have faced in my life, undoubtedly the one that required my greatest effort was to balance the goals of my professional career and family life. Although this involves a great deal of effort, I feel I have achieved this balance and I am very grateful for the unwavering support of my friends and loved ones.

On being a woman in this field . . .

At the beginning of my career, society, in general, and even industry, did not readily accept the idea of a woman with a degree in engineering. In my undergraduate courses, I was one of only 5 women in a class of 50 students. Even my own family was uncomfortable with the idea of having their youngest daughter studying Materials Engineering because they found it difficult to envision my professional role in this field. Moreover, Materials Engineering was a new field in Brazil, which made the acceptance of a female engineer even more difficult.

I was very lucky to have advisors in my master's and doctoral courses who treated me exactly as they would any male student. I found myself in the midst of people with a highly professional attitude, for whom the fact that I am a woman was not an issue. I had three young children to look after and it was not easy to prevent my personal life from interfering in my professional one, and vice versa. Luckily, however, I have a very understanding husband who has always been there for me, which has made my life much easier.

I believe that being a woman in this field enables me to multitask, which can be very helpful in a research environment, allowing me to achieve greater efficiency and results with different outcomes.

Words of Wisdom

Always fight for your dreams with a great dose of optimism, perseverance, and kindness.

Profile 52

Mareike Klee

Dr. Mareike Klee at Philips
Research Aachen 1994.

Fellow - Senior Director
Philips Research (Philips Group Innovation)
High Tech Campus 34-5.038
5656 AE Eindhoven
The Netherlands

e-mail: mareike.klee@philips.com
telephone: +31 610098879

Tags

❖ Administration and
   Leadership

Birthplace

Darmstadt, Germany

❖ Industry

❖ Domicile: Germany

Publication/Invention Record

❖ Nationality: German

>75 publications: h-index 11 (Web of Science)
42 U.S. granted patents and >175 published patents,
6 book chapters

❖ Caucasian

❖ Children: 0

Important Career Moment (to date)

In the late 1980s, I started investigations at Philips Research on complex oxide thin films such as $PbZr_xTi_{1-x}O_3$, $BaTiO_3$, and $SrTiO_3$, which exhibit ferroelectric and piezoelectric properties. Together with a small team, I developed a dedicated sol–gel process for these oxide thin films. At that time, a very limited number of groups worked in this innovative area, which transforms ceramic materials into thin films with thicknesses of several 100 nm to a few micrometers. Being the key person in this area, I established with my colleagues this new scientific area at Philips Research, which opened the way to integrate ceramic materials onto silicon substrates. We realized high-density capacitors with densities from 20 to $520\,nF/mm^2$. These high-density thin-film capacitors integrated with thin-film resistors and electrostatic discharge (ESD) protection diodes, formed a new class of integrated modules. The feature of these new modules was strong miniaturization compared to circuits realized with traditional ceramic capacitors, resistor, and ESD protection diodes. These high-K integrated discrete modules, based on my process, were mass manufactured by Philips Semiconductors. This is the highest reward a researcher in the industry can get: a new innovation, not available on the market, invented,

Successful Women Ceramic and Glass Scientists and Engineers: 100 Inspirational Profiles. Lynnette D. Madsen.
© 2016 The American Ceramic Society and John Wiley & Sons, Inc. Published 2016 by John Wiley & Sons, Inc.

Dr. Mareike Klee, Fellow—Senior Director at Philips Research Eindhoven invited in 2014 to the international symposium on sleep and respiratory care.

successfully transferred into development, and finally brought to the market by the business. This was a pivotal moment in my career. It was always my goal to create innovations in chemistry and successfully translate these with partners in the business into products.

Another important moment in my career was the move from microelectronic device research to research for patients with chronic diseases. Sleep and respiratory diseases are growing diseases and are affecting millions of patients worldwide. Innovations are needed to improve the quality of patient's lives and support the prevention of hospitalization. With a clear goal, I was able with my colleagues to create material and technology-related as well as service-related innovations for patients with chronic diseases such as obstructive sleep apnea (OSA) and chronic obstructive pulmonary disease (COPD). This outcome showed me that material expertise together with broad knowledge and enthusiasm for research forms the basis for successful innovations.

Academic Credentials

Ph.D. (1984) Chemistry, Technical University Darmstadt, Darmstadt, Germany
Diploma (1981) Chemistry, Technical University Darmstadt, Darmstadt, Germany
Vordiplom (1978) Chemistry, Technical University Darmstadt, Darmstadt, Germany.

Research Expertise: sol–gel and hydrothermal processing; single oxide films such as ZrO_2, Al_2O3, and Nb_2O_5 (both undoped and doped); complex thin films such as high-temperature superconductors in the system Y–Ba–Cu–O and Bi(Pb)–Sr–Ca–Cu–O, and ferroelectric as well as piezoelectric thin films in the systems $PbZr_xTi_{1-x}O_3$, $BaTiO_3$, and $SrTiO_3$ and layered perovskites such $SrBi_2Ta_2O_9$; microelectronic devices, integrated capacitors, and piezoelectric MEMS devices

Other Interests: create innovations and translate them into a business success

Key Accomplishments, Honors, Recognitions, and Awards

- Responsible for research portfolio Respiratory Care and Sleep Apnea Therapy Systems at Philips Research, since 2012
- Appointed Senior Director at Philips Research, 2007
- Coorganizer of Symposium B on "Passive Integration" at the MRS Fall Meeting, 2003
- First female researcher appointed Fellow at Philips Research, 2002

- First winner of the Eduard Pannenborg Award of Philips Research Aachen for outstanding research in the area of materials, devices, and systems, 2002
- Appointed Principal Scientist at Philips Research, 2001
- Appointed Senior Scientist at Philips Research, 1998
- Lecturer at the RWTH Aachen University, 1995–1996
- Several transfers of breakthrough innovations to business

Biography

Early Life and Education

Mareike Jakowski was raised in Germany. She has two sisters and one brother. Mareike received her high school education in Darmstadt. In 1976, she started studying chemistry at the Technical University Darmstadt, and she received a Vordiplom in 1978 and a Diploma in 1981. From 1981 to 1984, she carried out her doctoral studies at the Institute of Inorganic Chemistry and Crystal Chemistry in the area of single-crystal growth and X-ray diffraction analysis at the Technical University Darmstadt. She completed her doctorate in April of 1984. Since 1985, Mareike is married with Prof. Dr. Wilfried Klee.

Career History

After Mareike Klee finished her Ph.D. in April, she joined Philips Research Aachen directly in May 1984 and then in 2004, she moved to Philips Research Eindhoven. She is a trailblazer—she was the first female research scientist with a doctorate degree who started her career at Philips Research Aachen. At that time, her research focused on the inorganic chemistry of monosized TiO_2 particles and nanocrystalline $BaTiO_3$ powders with crystallite sizes of 10–100 nm processed by sol–gel and hydrothermal processes. She carried out this research for ceramic multilayer devices such as ceramic multilayer capacitors; it was a collaborative effort with Philips Passive Components.

Dr. Mareike Klee at Philips Research Aachen carrying out thin-film capacitor research.

Using her ceramic processing knowledge, Mareike extended her expertise into ceramic thin-film processes, which was a new research area in the late 1980s. She was a key person in strengthening Philips Research in this scientific area on complex inorganic thin-film materials for innovative microelectronic devices. She developed sol–gel thin-film processes for numerous inorganic materials comprising undoped and doped single-oxide films such as ZrO_2, Al_2O_3, and Nb_2O_5. In addition, she worked on complex thin films such as high-temperature superconducting films (Y–Ba–Cu–O and Bi (Pb)–Sr–Ca–Cu–O) and ferroelectric and piezoelectric thin films ($PbZr_xTi_{1-x}O_3$, $BaTiO_3$, and $SrTiO_3$, including layered perovskites films such as $SrBi_2Ta_2O_9$). The processes Mareike Klee developed enabled the deposition of oxide thin films on a number of substrates ranging from silicon to glass and ceramic. She and her team studied the materials and processes for a variety of innovative devices and systems. Ferroelectric $PbZr_xTi_{1-x}O_3$ and layered perovskite thin films were investigated for ferroelectric nonvolatile memories (FERAMs). Doped $PbZr_xTi_{1-x}O_3$ and $BaTiO_3$, $SrTiO_3$ thin films were examined for integrated capacitor/resistor networks on glass and ceramic substrates for miniaturized and integrated passive component modules.

The research on doped $PbZr_xTi_{1-x}O_3$ thin films on silicon substrates opened the way for several innovative applications. Mareike Klee and her team could realize high-density capacitors with capacitance densities of 20–520 nF/mm^2, which formed the basis for a new class of modules, comprising passive components (integrated capacitors, resistors) and active components (electrostatic discharge (ESD) protection diodes). These high-K integrated discrete modules, based on the Philips Research process, were mass manufactured by Philips Semiconductors.

In the area of piezoelectric thin films, Mareike Klee and her team realized piezoelectric MEMS devices such as piezoelectric switches and ultrasound transducers. The research resulted in a piezoelectric thin-film ultrasound transducer platform, which enables piezoelectric thin-film ultrasound transducers operating over a large frequency range from 50 kHz to beyond 4 MHz. Devices operating at 4 MHz were tested for medical imaging. The same platform was successfully used to process piezoelectric thin-film ultrasound transducer arrays for a new class of presence detectors. These piezoelectric thin-film MEMS devices performed well with features such as high output, low operating voltage, and high reliability. With the piezoelectric MEMS devices, more than 4900 hours continuous operation was demonstrated without any damage or degradation.

Mareike Klee was appointed Senior Scientist in 1998 at Philips Research Aachen and in 2001 she was appointed Principal Scientist. In 2002 she was the first female scientist who was appointed Fellow at Philips Research. She receives many invitations to discuss her research with other scientists outside Philips due to her broad expertise and intensive know-how and outstanding research in the area of materials, processes, devices, and systems. A significant number of her published papers and presentations are invited. On occasion she has delivered plenary lectures at major international conferences. In 1995–1996, she was also a lecturer at the RWTH Aachen University.

Dr. Mareike Klee, Fellow at Philips Research Eindhoven in 2011.

Since 2007, Mareike Klee has fulfilled the function of Senior Director at Philips Research Eindhoven. Her present research focuses on innovations for patients with chronic diseases such as sleep disordered breathing and respiratory diseases. Innovative materials, technologies, and systems are investigated for patients with OSA and COPD. To support patients with respiratory diseases from the hospital to the home, her research team also investigates new monitoring and coaching systems. In her function as Fellow and Senior Director, Mareike Klee is presently supporting the Philips Research management in the management of the respiratory care and sleep apnea therapy systems portfolio of Philips Research.

The goal of Mareike Klee has always been to make groundbreaking innovations and then successfully bring them to market with business teams. Philips Research is one of the largest private research organizations with the mission to create meaningful innovations that improve the quality of people's lives; so it is an excellent place to implement her ideas. The collaboration in multidisciplinary research teams and collaboration with development, marketing, and production sites on the one hand as well as universities, research institutes, and clinical sites on the other makes a vibrant, stimulating, and highly motivating environment, where Mareike is realizing her goals.

3 Most Cited Publications

Title: Electrode influence on the charge-transport through $SrTiO_3$ thin films
Author(s): Dietz, GW; Antpohler, W; Klee, M; et al.
Source: Journal of Applied Physics; volume: 78; issue: 10; pages: 6113–6121; published: November 15, 1995
Times Cited: 163 (from Web of Science)

Title: Processing and electrical-properties of $Pb(Zr_xTi_{1-x})O_3$ $(X = 0.2–0.75)$ films—comparison of metalloorganic decomposition and sol–gel processes
Author(s): Klee, M; Eusemann, R; Waser, R; et al.
Source: Journal of Applied Physics; volume: 72; issue: 4; pages: 1566–1576; published: August 15 1992
Times Cited: 158 (from Web of Science)

Title: Advanced dielectrics: bulk ceramics and thin films
Author(s): Hennings, D; Klee, M; Waser, R
Source: Advanced Materials; volume: 3; issue: 7–8; pages: 334–340; published: July–August 1991
Times Cited: 115 (from Web of Science)

Challenges

Research in industry has several challenges. One challenge is ensuring that the goals defined can be achieved with new innovations (e.g., with new devices or systems). This goes along with the next challenge—Can the specified technical performance be achieved through a new material, device, or system? Another important challenge in industry is meeting the timelines that were set. Realizing the new innovations on time as driven by the market is demanding and requires the full attention of a team. However, it is very exciting and motivating, to contemplate new innovations, not yet created by others, and then see these new products be realized and sold.

Another challenge is to unite the needed competences for innovation into multidisciplinary teams. This team effort requires researchers, who are team players having a broad knowledge and comprehensive competence, respect for other disciplines beyond the own capabilities, and interest, as well as capability, to integrate expertise of different disciplines into an overall system. Research teams with these capabilities have the power to realize breakthrough innovations and essentially to make "one plus one equal to three."

From a personal perspective, for many years, my husband and I have always been challenged to unite two positions, which are located ~130 km apart. Our solution is not ideal: We both travel every day 60–70 km twice a day between our home and our work. This routine is very time consuming for both of us—to compensate, we enjoy the remaining time each day, as much as we can, together. Life taught us that deep love and respect for each other can create unexpected energies and give us the power to overcome all challenges.

On being a woman in this field . . .

As a woman in industrial research, I experience an exciting and highly motivating research environment that offers collaboration with colleagues from numerous disciplines ranging from chemistry to materials science to electronic engineering, physics, and mechanical engineering. This multidisciplinary research on a day-to-day basis forms a strong environment to bring innovations forward. I enjoy the strong collaboration with business partners in the industry that enables one to successfully bring innovations from research into the marketplace. Next to an excellent industrial environment at Philips, which enables me to move new ideas forward, I am deeply grateful for the support my husband provides across the many years I have been active in this field. He shares my enthusiasm and motivation and supports my objective of pursuing a career as a female chemist in industry. It was always my goal to demonstrate that women are able to perform equally well to men in industry.

With respect to women in industry, in general I have observed a strong transformation over the past several years. Integration of business responsibility and private life is getting more attention in society and industry has started to create solutions that support women as well as men to take care of their loved ones in a more flexible way than existed in the late 1970s and 1980s. I am glad to see this shift.

Words of Wisdom

Social challenges and the needs that people have are a strong driver for innovations. I believe that industry as well as universities and institutes should take their responsibility in acting on these challenges and needs. Advancements that actively address these needs form a strong innovation foundation and will support mankind in challenging areas such as health care, healthy aging, environmental changes, and resource limitations. Innovations that support people and improve their quality of life are rapidly adapted. Mankind was always able and will be able to adapt quickly to high-speed technological developments: *tempora mutantur, nos et mutamur in illis* (the time is changing and we are changing in it).

Profile 53

Lisa C. Klein

Professor
Materials Science and Engineering
Rutgers, The State University of New Jersey
607 Taylor Road
Piscataway, NJ 08854-8065
USA

e-mail: licklein@rci.rutgers.edu
telephone: (848) 445-2096

Lisa Klein as a graduate student in 1975 with a stipend of $4800 per year, and able to rent an apartment on her own

Birthplace
Wilmington, DE, USA

Born
December 7, 1951

Publication/Invention Record

>100 publications: h-index 24 (Web of Science)
5 patents

Tags

❖ Academe
❖ Domicile: USA
❖ Nationality: American
❖ Caucasian
❖ Children: 1

Proudest Career Moment (to date)

In 2012, I was elected to membership in the World Academy of Ceramics, "for personal merits in the field." This organization founded in 1987 inducts members every other year in two categories: Academic and Industrial. I was in the category Academic. I was truly honored to be a part of this collection of scientists from around the world. Many of them serve on the Editorial Board of the Journal of the American Ceramic Society, which I have been affiliated with as an Editor since 1998. Several of the members also had contributed to a book that I coedited: Sol–Gel Processing for Conventional and Alternative Energy (ed. M. Aparicio, A. Jitianu, and L.C. Klein, Springer,

Lisa Klein being presented with election to World Academy of Ceramics, by Masahiro Yoshimura and Mrityunjay Singh.

Successful Women Ceramic and Glass Scientists and Engineers: 100 Inspirational Profiles. Lynnette D. Madsen.
© 2016 The American Ceramic Society and John Wiley & Sons, Inc. Published 2016 by John Wiley & Sons, Inc.

New York, 2012). This is the third book I have edited on the subject of sol–gel processing since my first book in 1988, which established the commercial viability of the sol–gel process. I have been involved in sol–gel processing research ever since.

Academic Credentials

Ph.D. (1976) Ceramics, Massachusetts Institute of Technology, Cambridge, USA.
S.B. (1973) Metallurgy, Massachusetts Institute of Technology, Cambridge, USA.

Research Expertise: glass chemistry, sol–gel processing, glass processing

Other Interests: encouraging women and other underrepresented groups to get involved in research through the American Chemical Society's SEED Program and the New York Academy of Sciences Science Research Training Program

Key Accomplishments, Honors, Recognitions, and Awards

- American Ceramic Society, Member of Coble Award Committee, 2011–present
- Journal of the American Ceramic Society, Editor with D.W. Johnson, D. Green, and J. Halloran, 1998–present
- Elected Member of the World Academy of Ceramics "for personal merits in the field," 2012
- Journal of Sol-Gel Science and Technology, Coeditor, 1993–2010
- Journal of Materials Research, Principal Editor, 1994–2006
- National Materials Advisory Board (NMAB) Panel Member, Hybrid Materials for "Defense-After-Next" Committee, 2000–2002
- Fellow of the New York Academy of Sciences "for contributions and service toward the advancement of science," 2001
- NMAB, a unit of the National Research Council (NRC), Member, 1999–2001
- National Academy of Sciences Materials Forum Organizer, "Materials in the New Millennium: Responding to Society's Needs," 2000
- Society of Women Engineers Achievement Award "for breakthrough contributions to sol–gel science and engineering, particularly sol–gel applications in electrolytes, electrochromics, membranes and nanocomposites," 1998
- The Graduate School, New Brunswick, Rutgers University Teaching Award, one of two faculty members honored for teaching excellence, 1997
- American Ceramic Society, Glass and Optical Materials Division: Program Committee Chairman, 1992–1993; Annual Meeting Program Chairman, 1992; Fall Meeting Program Chairman, 1990
- New Jersey Federation of Women's Clubs Women of Achievement Award "for innovative work in the field of ceramic engineering and pioneering efforts on behalf of women and engineering," 1992
- American Ceramic Society, Ceramic Educational Council (CEC): Trustee, 1989–1992; Chairman, 1986–1987; Chairman-Elect, 1985–1986; Secretary, 1984–1985; Treasurer, 1983–1984

- Fellow of the American Ceramic Society "in recognition of notable contributions to arts and sciences of ceramics," 1988
- National Institute of Ceramic Engineering (NICE) of the American Ceramic Society: Schwartzwalder-Professional Achievement in Ceramic Engineering (PACE) Award "for significant contributions to the profession and to the general welfare of the American People," 1987
- Central Jersey Engineering Council, representing 10 professional Societies, 1200 members: Engineer of the Year, "for outstanding achievement in engineering," 1986
- Society of Women Engineers, Distinguished New Engineer Award, 1984
- Society of Women Engineers, New Jersey Section: President 1980–1982; President-Elect 1979–1980
- Society of Women Engineers, National Student Conference Coordinator, Princeton University, 1980

Biography

Early Life and Education

Lisa Klein grew up in Wilmington, Delaware. She is the second of four children, with two sisters and one brother. She attended public schools in the Mount Pleasant District. She was influenced to study math and science in the backdrop of the NASA Space Program in the 1950s and the Race to the Moon in the 1960s. In particular, a summer program sponsored by the National Science Foundation in 1968 was the impetus to pursue engineering. This residential program at Brown University was an opportunity to be immersed in science and engineering for a period of 6 weeks at a nominal cost. Also, it was a chance to work in university-level laboratories with a cohort of students from across the United States.

She attended MIT beginning in Fall 1969. At first, she considered being a physics major. However, after taking a Freshman Seminar on "Air Pollution," which at the time was a new concern, from a chemical engineering professor, she realized that engineering was more compatible with her practical skills. She invested time in researching the available engineering disciplines and chose metallurgical engineering. This choice was based in part on the small size of the department and the opportunities to get involved in research projects. By junior year, she was participating in semiconductor crystal growth, when single crystals of doped germanium were being grown for studies in the Spacelab. In senior year, she had the chance to work on a NASA-sponsored research project that had access to recently returned moon rocks. Once rocks were brought back from several locations on the Moon, it was possible to do studies on analog materials to gain a better understanding of the thermal history of the surface of the Moon. Being able to contribute in some small way to the decades of space exploration was a fitting culmination to an undergraduate degree.

Fortunately, it was possible to extend the work on the moon rocks into a Ph.D. thesis on "Crystallization Behavior in Mineral Systems." She entered graduate school immediately after her B.S. By then, the department name had changed to Materials Science and Engineering. She specialized in ceramics, which was one of five options in the department. Her thesis research covered several important mineral systems that are relevant to both moon and earth geology. Her first publication in 1974 reported the crystallization rates in anorthite, a calcium aluminosilicate. In 1976, she gave her first oral presentation at a national meeting at the 7th Lunar Science Conference, at the Johnson Space Center in Houston, Texas.

After defending her thesis, she stayed at MIT for a 6-month postdoctoral period. The U.S. economy at the time was weak after the Vietnam War and the first Arab Oil Embargo. She interviewed with industrial labs and government labs, but an academic position was her first choice. Only two universities had openings, and she was offered the position in the Ceramics Department at Rutgers University. With a job lined up for June 1977, she took off for Europe for a 2-month trip, which included visits to six university and corporate labs, to establish some collaborations.

Career History

She has been at Rutgers University since 1977, during which time the Ceramics Department has changed to Materials Science and Engineering Department. She has supervised 23 Ph.D. theses and 19 M.S. theses. Her research has been funded by the National Science Foundation, Department of Energy, Office of Naval Research, The Department of Commerce Advanced Technology Program (ATP), and the New Jersey Science and Technology Commission. She has worked with many companies in collaborative projects, including Alcoa, DuPont, PPG, and Oil-Dri.

Lisa Klein supervising freshman introduction to dye-sensitized solar cells in the Sol-Gel Lab at Rutgers.

In addition, she has been a Visiting Faculty Member at Sandia Laboratories, Albuquerque, NM; Laboratoire D'Energetique Electrochimique, ENSEEG, Grenoble, France; and the Chemistry Department, The Hebrew University of Jerusalem, Israel. She has served on thesis committees for students in Australia, Egypt, and India.

During her 30+ years at Rutgers, she has taken temporary assignments in administration, serving as Graduate Director, Special Assistant in the Provost's Office, and as President of the Rutgers Chapters of AAUP-AFT, the union that represents 5000 full-time employees, including tenured, tenure-track, and nontenure-track faculty and graduate employees, on all three campuses. This elected position involves working with a

Prof. Klein addressing a rally in Trenton, New Jersey, in Summer 2008, on behalf of the members of the Rutgers AAUP-AFT.

professional staff of eight, negotiating a contract every 4 years, meeting with state representatives on matters affecting higher education, and giving testimony to the state assembly and senate on the higher education budget. Serving the union in this capacity has been particularly rewarding in keeping the focus of the university on education and the students.

Prof. Klein is one of the book's featured women that have a Wiki entry. Repeated from that site (http://en.wikipedia.org/wiki/Lisa_C._Klein) are some comments about her influence at Rutgers: "Klein is also noteworthy for her status at Rutgers University: she was the first female faculty member in the Rutgers School of Engineering (1977), the first female faculty member in the department of Materials Science and Engineering, the first female faculty member elevated to the distinguished PII rank, and, as of January, 2008, the only female professor in the department. As President of the Rutgers American Association of University Professors (AAUP) chapter, Klein also concluded ground-breaking negotiations that lead to novel methods for expanding the number of tenure track faculty at Rutgers that have been cited as potential models for faculty efforts to halt the attrition in numbers of tenure track faculty."

3 Most Cited Publications

Title: CoO_2, the end member of the Li_xCoO_2 solid solution
Author(s): Amatucci, GG; Tarascon, JM; Klein, LC
Source: Journal of the Electrochemical Society; volume: 143; issue: 3; pages: 1114–1123; published: March 1996
Times Cited: 457 (from Web of Science)

Title: Cobalt dissolution in $LiCoO_2$-based non-aqueous rechargeable batteries
Author(s): Amatucci, GG; Tarascon, JM; Klein, LC
Source: Solid State Ionics; volume: 83; issue: 1–2; pages: 167–173; published: January 1996
Times Cited: 250 (from Web of Science)

Title: Sol–gel processing of silicates
Author(s): Klein, LC
Source: Annual Review of Materials Science; volume: 15; pages: 227–248; published: 1985
Times Cited: 204 (from Web of Science)

Challenges

The challenges of an academic career are well known. At a research university, it is always a challenge to find funding for research, and to find motivated students who become your legacy. Since I have been flexible in forming collaborations and alliances, and willing to modify my research scope in response to funding, I have been able to maintain an interesting portfolio of projects. At times, I have scrambled to find funding, but certainly it was worth the effort.

320 LISA C. KLEIN

On being a woman in this field . . .

When I joined Rutgers, I was the only woman, not only in my department, but also in the entire School of Engineering. Slowly, each of the seven departments hired their first woman. Very gradually, they hired the second woman. In my department, it took 32 years before we hired the second woman. Currently, the percentage of women in Rutgers Engineering is about 17%, slightly below the national average for schools of engineering.

To find role models meant finding women in chemistry and physics. This was useful and led to some very productive collaborations. As the number of women in engineering increased, it became easier to build a network within the school. This network is important because you can rely on the network for support, for information, and for the recognition that we all deserve from our professional community. Often, women are overlooked for nominations and prizes, but the network of women can be used strategically to identify good candidates.

To be honest, faculty meetings in engineering are still locker room sessions with all the crudeness you can imagine. I have heard things and seen things that are outrageous. Since I tend to think that the climate at the university is reflective of the politics at large, we go through our ups and downs. In my opinion, what we need to make a permanent change is our first woman president.

Words of Wisdom

What has helped me over the years is the idea that we are all in it together. And if we stick together, everyone will move forward. I have never claimed that I did something all by myself. I could not have succeeded without the support of my parents, my siblings, my husband and daughter, and 23 nieces and nephews. My only advice to others is that they can count on me, when we are all going in the right direction.

Marija "Marička" Kosec
(1947–2012)

Former Head of Electronic Ceramics Department
Jožef Stefan Institute
Slovenia

Marija Kosec when she was a
young researcher. Credit:
IJS News.

Publication/Invention Record

345 publications: h-index 32 (Web of Science)
5 monograph articles, >100 technical reports
8 patents, editor of 14 books

Academic Credentials

Ph.D. (1982) Chemistry, University of Ljubljana, Slovenia.
M.Sc. (1975) Chemical Technology, University of Ljubljana, Slovenia.
B.S. (1970) Chemical Technology, University of Ljubljana, Slovenia.

Research Expertise: synthesis and characterization of electronic ceramics, in particular piezoelectrics and ferroelectrics

Other Interests: international collaborations, advising students

Tags

❖ Administration and Leadership
❖ Government
❖ Nationality: Slovenian
❖ Caucasian

Key Accomplishments, Honors, Recognitions, and Awards

• 3 plenary talks at international conferences
• 43 additional invited talks at international conferences
• 33 invited talks at institutions
• Past President of Society for Microelectronics, Devices and Materials of Slovenia (MIDEM)
• Past President of Ceramic Session at the Slovenian Chemical Society

Successful Women Ceramic and Glass Scientists and Engineers: 100 Inspirational Profiles. Lynnette D. Madsen.
© 2016 The American Ceramic Society and John Wiley & Sons, Inc. Published 2016 by John Wiley & Sons, Inc.

- Visiting Professor at the École polytechnique fédérale de Lausanne (EPFL), Switzerland
- Visiting Professor at the University of Shizuoka, Hamamatsu, Japan
- Adjunct Professor at Xi'an Jiaotong University, China
- Member of European Liaison Committee, International Microelectronics and Packaging Society (IMAPS) Europe, 1999–2012
- Head of the Center of Excellence: Materials for next-generation electronics and other emerging technologies, 2009–2012
- Ferroelectrics Recognition Award from the IEEE Ultrasonics, Ferroelectrics, and Frequency Control Society for her significant contributions to the processing science and technology of ferroelectric powders, bulk ceramics, thin and thick films, 2010
- Puh recognition for the implementation of research results in industry, 2009
- President of the Scientific Council of Jožef Stefan Institute, 2007–2009
- Elected Academician of World Academy of Ceramics, 2006
- Zois Prize for extraordinary achievements in science, 2006
- President of Slovenian Academy of Engineering Sciences, 2005–2006
- Elected member of the Academy of Engineering Sciences of Slovenia
- Ambassador of Science of the Republic of Slovenia, 2003/2004
- Senior member of IEEE, 2001

Biography

Early Life and Education

She received B.Sc. and M.Sc. degrees in chemical technology and a Ph.D. degree in chemistry from University of Ljubljana, Slovenia over the period of 1970 to 1982.

Career History

When Professor Marija Kosec passed away in December 2012, she was the Head of Electronic Ceramics Department of Jožef Stefan Institute. Her research interest was

in synthesis and characterization of electronic ceramics, particularly piezo-electrics and ferroelectrics. Her contribution to science is in understanding phenomena of the processing of bulk ceramics and thin films of practically important multicomponent systems. Recent results concern understanding of synthesis of ferroelectric thin films where she explained complicated reactions during solution synthesis and how to control these reactions. Practical results are significantly lower processing temperature. She had also

Marija Kosec's gravestone in Slovenia.
Credit: N. Setter.

been working on lead-free piezoelectrics, bulk ceramics, and thick and thin films. She tried to relate research in materials to their application. She authored and co-authored hundreds of scientific journal papers and conference proceedings and published five of these articles in monographs. She gave 46 invited talks on international conferences along with 3 plenary and 33 invited talks on institutions. She was an editor or co-editor of 14 conference proceedings. She was awarded 8 patents and had published more than 100 technical reports.

A recent photograph of Marija Kosec. Reproduction is permitted by Novice IJS 164, May 2013, courtesy of Mateja Jordovič Potočnik.

She was a member of various Ceramic Societies, a permanent member of IEEE Ferroelectric Committee, a Senior Member of IEEE, and a member of European Liaison Committee (ELC), IMAPS Europe. She was Past President of MIDEM (Society for Microelectronics, Devices and Materials of Slovenia), the Ceramic Session at the Slovenian Chemical Society, and Slovenian Academy of Engineering Sciences. In 2003, she was named Ambassador of Science of the Republic of Slovenia; in 2004, she became a member of World Academy of Ceramics; in 2006, she received the highest national award Zois Price for extraordinary achievements in science.

3 Most Cited Publications

Title: Effect of pH and impurities on the surface charge of zinc oxide in aqueous solution
Author(s): Degen, A; Kosec, M
Source: Journal of the European Ceramic Society; volume: 20; issue: 6; pages: 667–673; published: May 2000
Times Cited: 175 (from Web of Science)

Title: Dielectric dispersion of the relaxer PLZT ceramics in the frequency range 20 Hz–100 THz
Author(s): Kamba, S; Bovtun, V; Petzelt, J; et al.
Source: Journal of Physics: Condensed Matter; volume: 12; issue: 4; pages: 497–519; published: January 31, 2000
Times Cited: 129 (from Web of Science)

Title: Alkaline-earth doping in $(K, Na)NbO_3$ based piezoceramics
Author(s): Malic, B; Bernard, J; Holc, J; et al.
Source: Journal of the European Ceramic Society; volume: 25; issue: 12; pages: 2707–2711; published: 2005
Times Cited: 120 (from Web of Science)

Comments from her friends and colleagues

"Marička Kosec had an enquiring mind, a love for her work, and a passion for life. She liked nothing better than to engage in enthusiastic conversations about ceramics and ceramic processing, nature, and life, especially in exotic locations around the globe. She never stopped exploring, learning, and debating. She was determined, set ambitious goals, and almost always achieved them. She had to work hard to achieve her position as one of the most senior and highly respected scientists within a rather patriarchal

science community in Slovenia, beginning with a student exchange in Germany, establishing her own research interests in the processing of electronic ceramics, taking on and then expanding a large research group in at the Jožef Stefan Institute, building an international reputation, and then winning the highest honor for science in Slovenia. She was a remarkable woman who is missed by many colleagues and friends."

Marija Kosec (second from the left) having dinner in Slovenia, 2006. Credit: R. Waser.

Angus Kingon, Brown University, Providence, Rhode Island, USA

"It is always sad to lose a great scientist and wonderful person, but Marija represented ceramics research in Slovenia and was also one of the few senior women and role models in ceramics. Her membership in the World Academy of Ceramics was well deserved and she will be truly missed by many people from around the world!"

Lynnette D. Madsen, National Science Foundation and Svedberg Science, Inc., Virginia, USA

Safari in Tanzania in 2007: (from left to right) A. Kingon, E. Svedberg, S. Streiffer, M. Kosec, L. Madsen, and W. Wolny. Credit: R. Waser.

Marija Kosec (third from the left) in Tanzania in 2007.

"Marička was genuinely interested in people, science, and nature. She generously shared her knowledge at any moment, with everyone, without any hesitation. She actively listened to what you had to say and interacted with you constantly. This key element of her character made her liked by everyone. In the early 1990s, Marička spent a year in the EPFL, which I recall fondly. After a day in the lab, we would come back to my home and open a bottle of Cognac to start the evening. We would discuss intensively, for example, sintering mechanisms; she had so much knowledge all in her head. With joy she was able to formulate and predict how material will behave under certain conditions; the next day we would test her ideas in the lab. After some hours of discussing science and long after the shops had closed, we would begin to think of dinner. I would sneak with my fishing rod to the lake; in Switzerland, it is illegal to prey on sleeping fish. In parallel, Marička would light candles and explore my yard for mushrooms—she was a well-known mushrooms expert and expert in herbs. During the year Marička stayed with us, I did not sleep much, but I learned a lot about ceramics, mushrooms, and friendship. I will miss her capacity to undertake any subject and turn it into an interesting discussion.

Nava Setter, EPFL, Switzerland

"Marija was a joyous person with deep interests in ceramics processing. She will be remembered as a warm, gracious, energetic scientist who was both a mentor and a friend to many ceramists around the world. She was well known internationally for her work on processing of electroceramic materials. She held such a positive outlook on life, and was eager to explore and learn new things. I have many happy memories of times spent with her, including a lovely evening in her back garden with her husband, eating dinner, and listening to stories about life in Slovenia. We will miss her."

Susan Trolier-McKinstry, Penn State University, USA

Susan Trolier-McKinstry, Marija Kosec, and Hong Wang.

"In 2011, close to her birthday, we had some extended time together after attending a conference in Agadir. Marija and I alongside Nava Setter, Angus Kingon, and few other colleagues drove on a guided tour through the dessert in Morocco from Foum Zguid to M'hamid. Marija provided me with some of the background about multiferroics. Scientists (for the most part, physicists) had encountered extreme difficulties fabricating single-phase $BiFeO_3$ bulk ceramics; their issues ranged from low sinterability and resistivity, the introduction of secondary phases, to second-rate remanent ferroelectric polarization values. In detail, she explained how she solved these problems with her co-workers and colleagues. Despite the stifling temperature of 45 °C (the air-conditioning was off to prevent engine overheating), I enjoyed for hours listening to Marija and the ensuing stimulating discussions; all the time fascinating and varying exotic landscapes of the dessert, including few Arabian camels from time to time, passed in the background. She was excellent in telling stories, injecting her deep knowledge of the science of

Marija Kosec (in the black T-shirt) on a Moroccan tour
in 2011 with friends/colleagues.

ceramics into them. Marija often said: 'Life is great'. It is very unfortunate that her life didn't last longer; we are a little poorer without her presence. We have lost a leading scientist and a wonderful person."

Rainer Waser, RWTH Aachen, Germany

"Marička enjoyed life . . . Besides her professional activities (those she liked the most), she enjoyed many other activities. Throughout her life she enjoyed skiing, whether on groomed slopes or virgin natural mountains. She tortured herself with half a day uphill hiking (with her skis on her back), to have half an hour downhill satisfaction. She loved it. She liked trekking the most: she was very proud of visiting the Andes Highland Region in Peru, Kala Patthar near Mount Everest in Nepal, Popocatépetl in Mexico, and Mount Kenya and Rwenzori in Africa. She walked part of the Silk Route crossing a mountain chain between Kazakhstan and Kyrgyzstan. She climbed the active Avachinsky volcano on the Kamchatka-Far East Peninsula and explored Altai in Siberia. She was not afraid of carrying a 15 kg backpack, sleeping in a tent in quite cold or hot conditions, with billions of mosquitoes outside, with bears crossing the footpaths . . . she never put her knee down or gave up. Another real passion was collecting mushrooms and herbs: she knew many secret places where,

Marija Kosec in Morocco, 2011.
Credit: N. Setter.

in the deep forest, she always found them. Every summer she dried mushrooms (all around the house) to make a delicious soup or to give away. A special role in her life was her house, garden, flowers, and plants—with free moments she put on her boots and started to pick weeds, cut and trim plants, and take care of roses and flowers. She really enjoyed her life, but most of all she loved her profession: even when she was seriously ill, she liked to share the professional ideas with her colleagues, but she was courageous and too proud to bother anyone with her troubles. We will miss her smile and her good will forever."

Marija enjoying a hike. (See insert for color version of figure.)

Aljoša Lipovšek (husband), Slovenia

Marija Kosec with her husband, Aljoša, in Australia, 2012.
Credit: N. Setter. (See insert for color version of figure.)

Sources used to create this profile

- Discussions with Dr. Kosec and her wide circle of friends.
- Special Issue: Novice IJS 164, May 2013 (http://www.ijs.si/ijsw/Arhiv%20Novic, accessed February 17, 2015): although the initial pages are not in English, the latter part of the issue is and the commentaries are wonderful to read. Profs. Waser and Setter's commentaries are adapted from this source.
- ACerS Memoriam, http://ceramics.org/in-memoriam/marija-kosec, accessed February 17, 2015.
- IEEE Memoriam, http://www.ieee.org/ns/periodicals/uffc/may2013/InMemoriam. html, accessed February 17, 2015.

Disclaimer

This profile was prepared from various sources and is believed to be correct. However, the author assumes no responsibility for any errors or omissions; they were unintentional.

Waltraud (Trudy) M. Kriven

Professor of Materials Science and Engineering,
Affiliate Professor of Mechanical Science and Engineering
University of Illinois Urbana–Champaign
1304 W. Green St.
Urbana, IL 61801
USA

e-mail: kriven@illinois.edu
telephone: (217) 333-5258

Trudy Kriven at age
25 while she was a
Ph.D. student at
the University of
Adelaide, South
Australia.

Birthplace
Eisenstadt, Austria
Born
April 25, 1949

Publication/Invention Record

>275 publications: h-index 30
4 patents; editor of 24 books

Tags

❖ Academe
❖ Domicile: USA
❖ Nationality: Born in
  Austria; dual Australian
  and American citizen
❖ Caucasian
❖ Children: 4

Proudest Career Moment (to date)

My proudest moment was in 1985 when I was appointed as a Research Associate
Professor in a tenure track position at the University of Illinois at Urbana–Champaign,
and I received my first research contract from the Air Force Office of Scientific Research.
It was enough to support two Ph.D. students and I was honored and grateful that I had
gained such trust and respect as a research scientist and a university professor.

Academic Credentials

Ph.D. (1976) Physical and Inorganic Chemistry, University of Adelaide, Adelaide, South
Australia, Australia.
B.Sc. (1971) Biochemistry, University of Adelaide, Adelaide, South Australia, Australia.

Successful Women Ceramic and Glass Scientists and Engineers: 100 Inspirational Profiles. Lynnette D. Madsen.
© 2016 The American Ceramic Society and John Wiley & Sons, Inc. Published 2016 by John Wiley & Sons, Inc.

B.Sc. (1970) Physical and Inorganic Chemistry, University of Adelaide, Adelaide, South Australia, Australia.

Research Expertise: *in situ*, high temperature ($< 2000°C$) synchrotron studies of phase transformation and thermal expansion in ceramics; martensitic transformations and mechanisms of ferroelasticity and ferroelastic transformations in ceramics; microstructure characterization by scanning and transmission electron microscopy (SEM, TEM, EDS, HVEM, XPS); powder synthesis and processing of ceramics; structural ceramic composites and bioceramics (design, processing, characterization, and mechanical evaluation); geopolymers, their composites and geopolymer-derived ceramics; bioceramics and bioresorbable nanoceramics for gene/drug delivery

Other Interests: scientific innovation, working with students of all ages and their teachers, increasing diversity

Biography[1]

Early Life and Education

Waltraud ("Trudy") M. Kriven received Baccalaureate and Honors Baccalaureate degrees in Physical and Inorganic Chemistry and Biochemistry from the University of Adelaide, South Australia. Her doctorate was earned in Physical and Inorganic Chemistry also from the University of Adelaide.

Career History

Dr. Kriven spent 1 year as a Post-Doctoral Teaching and Research Fellow in the Chemistry Department at the University of Western Ontario in Canada. She then spent 3 years (1977–1980) jointly at the University of California at Berkeley, and at the Lawrence Berkeley Laboratory. There, Dr. Kriven conducted post-doctoral research in transmission electron microscopy of ceramics and was a Lecturer, teaching Phase Equilibria in the senior undergraduate Ceramics Program of the Department of Materials Science and Mineral Engineering. For almost 4 years (1980–1983) Dr. Kriven was a Visiting Scientist at the Max Planck Institute in Stuttgart, Germany. There she conducted research into the mechanism of transformation toughening of composite zirconia-based ceramics by 1 MeV HVEM.

Since February 1984, Professor Kriven has been at the University of Illinois at Urbana–Champaign. She is a Full Professor of Materials Science and Engineering, and an Affiliate Professor of Mechanical Science and Engineering.

Professor Kriven has internationally recognized expertise in the areas of phase transformations in inorganic compounds, their applications in structural ceramic composites, ceramic powder synthesis and processing, and geopolymers. In addition, she has made extensive contributions to oxide composites design, microstructure characterization by electron microscopy techniques, and phase equilibria. The Kriven group has

---

[1] Credit: http://www.matse.illinois.edu/faculty/Kriven.html

developed a new technique for *in situ*, hot stage (up to 2000°C) synchrotron studies of ceramics in air, including an image plate detector capable of taking a high resolution, X-ray diffraction spectrum within 20 s. She has written or co-authored more than 300 research publications, as well as given or co-authored more than 400 conference presentations. Prof. Kriven has edited or co-edited 24 books to date. She has given 195 invited lectures (including 28 keynote lectures at international meetings) in the United States and internationally. She has supervised 21 doctoral and 31 master's students and over 46 undergraduate students in her laboratory.

Prof. Trudy Kriven (right) and some of her research group of two post-doctoral research associates, four Ph.D., and one M.S. student. The research team was conducting around-the-clock synchrotron experiments *in situ*, at high temperature (to 2000°C) in air, in order to measure thermal expansions in 3D and discover new phase transformations in oxide ceramics. Experiments were carried out at the Advanced Photon (synchrotron) Source (APS) at Argonne National Laboratory (ANL) in August 2012.

In addition to being a great scientist and teacher, she is a wife and mother—she and her Australian-born husband, with significant help from her widowed mother, raised a family of three daughters and one son.

3 Most Cited Publications

Title: Understanding the relationship between geopolymer composition, microstructure and mechanical properties
Author(s): Duxson, P; Provis, JL; Lukey, GC; et al.
Source: Colloids and Surfaces A: Physicochemical and Engineering Aspects; volume: 269; issue: 1–3; pages: 47–58; doi: 10.1016/j.colsurfa.2005.06.060; published: November 1, 2005
Times Cited: 236 (from Web of Science)

Title: The effect of alkali and Si/Al ratio on the development of mechanical properties of metakaolin-based geopolymers
Author(s): Duxson, P; Mallicoat, SW; Lukey, GC; et al.
Source: Colloids and Surfaces A: Physicochemical and Engineering Aspects; volume: 292; issue: 1; pages: 8–20; published: January 5, 2007
Times Cited: 125 (from Web of Science)

Title: Polymerized organic-inorganic synthesis of mixed oxides
Author(s): Gülgün, MA; Nguyen, MH; Kriven, WM
Source: Journal of the American Ceramic Society; volume: 82; issue: 3; pages: 556–560;
published: March 1999
Times Cited: 118 (from Web of Science)

Challenges

I perceive that the following are challenges to a woman in a male-dominated professional field. First, there is the struggle to balance family and home, versus professional responsibilities. Second, I am expected, and expect myself, to be superior to my male counterparts, in order to be treated equally with them. This has been the case in the past with the older generation, but I perceive that the double standards against women are not applied so readily by the younger generation of men, except by those who are second-rate professionals. I strive to be a successful scientist, teacher, and researcher, but it is difficult for me to be aggressive, egotistical, and politically manipulative, as I perceive many of my male colleagues who get ahead financially, to be. Such character-istics are often evaluated negatively for women, but enable men to reap rewards. I keep striving to be a lady and a scholar, and when I receive unfair and unethical treatment, it is hard not to lose heart and interest in the specific workplace where such treatment occurred, and to search elsewhere for success and satisfaction.

On being a woman in this field . . .

In my youth I was totally dedicated to the study of science and I was extremely shy and somewhat of a "loner." As an undergraduate student at university where the handful of women were very much a minority, I had to be tough and strong, and able to work independently in a male chauvinistic Australian macho environment, which was very much the norm in the 1970s and, of course, why movements for women's equality in the workplace developed at that time. I left Australia and worked as a post-doctoral researcher for 8 years before receiving a faculty position. I was single until the age of 36 being rather shy and , suffering from a case of "continental drift," i.e., I drifted from continent to continent.

Shortly after we were married, my husband and I had four babies in 5 years between my ages 37 to 42. I was basically able to work full time because my dear widowed mother came from Australia and lived with us, intermittently due to visa restrictions, for about 20 years. I was extremely fortunate to have a strong backup team, my husband and my mother. It was also helpful that we lived in a small town with short commuting distances (~10 min by car to anywhere in town). In those days, the University of Illinois did not have maternity leave so I had to go back to work as soon as possible, being the only bread winner in the family, as my husband was a Ph.D. student at that time. I was fortunate that our first baby was born on November 30th, which gave me 3–4 weeks to recover over Christmas before the next semester. For the third baby, I was able to get one semester off teaching, but continued to mentor my research group of 12 students.

When I came up for promotion to Full Professor, which was 11 years after I joined the Department and after a 7-year Associate Professorship, I had 100 research publications, whereas some other male professors only had 35–40. Nevertheless, I was still asked to explain why my last 20 publications were in the previous year. Of course, the reason was simple—my four little children finally started sleeping through the night. Even with the much higher productivity compared to other senior colleagues in the department and a stronger reputation (nationally and internationally), I earned a salary that was about the same as that offered to a new assistant professor. My colleagues very grudgingly promoted me to Full Professor. Even though I noted these disparities to my colleagues, back in the 1980s such practices were not met with great scrutiny by the institution.

Upon completion of his Ph.D. in Theoretical and Applied Mechanics at our University, my husband decided to sacrifice his scientific career as a researcher and become a Lecturer in the same University, so as to allow me to continue in my career and keep our family together. As a faculty spouse, he was entitled to one-third of his salary to come from my home department, one-third from the University, and one-third from his home Department where he taught courses. However, my husband ended up each semester, teaching not only two or three courses in his department, but also teaching another course in our department. However, he did not receive any compensation from our department whatsoever, for teaching one extra course every semester for 18 years. Instead, his home department and the College of Engineering paid more than two-thirds of his salary while my department paid less, and he still had a significant overload of teaching, teaching four or five courses a semester instead of only three. At the time of this writing, 18 years after he started working for the University, he is responsible for teaching about 1000 undergraduate engineering students per semester.

The atmosphere for women in my department continued to be chilly. For example, there was one incident at a faculty meeting when I stated an unpopular but factually correct opinion. One of my colleagues responded to me in a loud and unpleasant tone of voice, telling everyone to listen to my "voice." At the end of the meeting, I quietly stated that I did not appreciate being angrily confronted and yelled at, just because I dared to have an opinion that was different. The Department Head apologized to me in front of everyone, saying that all of the faculty should be treated with respect. The aggressive professor later apologized to another woman faculty member in our department, who was not even present at the meeting, but not to me.

In summary, I came up into academia at a time when the situation for women was characterized by still blatant sexism. During that time, I felt I had to do much more than my colleagues to be accepted. Ironically, though, this pursuit of excellence in science has helped me get through the difficult times. In addition, my husband, who made career sacrifices, and my mother, who helped us raise our four children, enriched my life beyond the laboratory and provided emotional support to weather the negative incidents. My international recognition as a research scientist and university professor are accomplishments that nay-saying colleagues can never take away from me.

Profile 56

Michiko Kusunoki

Professor and Division Director
EcoTopia Science Institute (Division of Green
Materials)
Nagoya University
Furo-cho, Chikusa-ku
Nagoya 464-8603
Japan
e-mail: kusunoki@esi.nagoya-u.ac.jp
telephone: +81-52-789-3920

Michiko Kusunoki when she was
around 30 years old, working at
the Japan Fine Ceramics Center.

Birthplace

Yui-cho, Shizuoka, Japan

Born

December 8, 1952

Publication/Invention Record

>75 publications: h-index 15 (Scopus)

Tags

❖ Administration and Leadership
❖ Academe
❖ Domicile: Japan
❖ Nationality: Japanese
❖ Asian
❖ Children: 2

Proudest Career Moment (to date)

I deeply enjoyed my research life for 16 years at the Japan Fine Ceramics Center from
1990 to 2006. After moving to Nagoya University as a Professor, I am very happy to be
leading graphene researchers as a representative in the Synthesis Group with support
from "Science of Atomic Layer (SALT)" Grant-in-Aid for Scientific Research on
Innovative Areas, The Ministry of Education, Culture, Sports, Science, and Technology
(MEXT) in Japan, No. 2560 from 2013 to 2019.

Successful Women Ceramic and Glass Scientists and Engineers: 100 Inspirational Profiles. Lynnette D. Madsen.
© 2016 The American Ceramic Society and John Wiley & Sons, Inc. Published 2016 by John Wiley & Sons, Inc.

Academic Credentials

Ph.D. (1980) Science and Engineering, Tokyo Institute of Technology, Meguro, Japan.
M.Sc. (1977) Science and Engineering, Tokyo Institute of Technology, Meguro, Japan.
B.Sc. (1975) Engineering, Tokyo Institute of Technology, Meguro, Japan.

Research Expertise: transmission electron microscopy, electron diffraction, crystallographic analysis, ultrafine particles, phase transformation, carbon nanotubes, novel nanocarbon science, graphene from SiC and other carbides

Other Interests: international collaborations with graphene researchers

Key Accomplishments, Honors, Recognitions, and Awards

- Director of the Japanese Ceramic Society, 2009–2010 and 2015 to present
- Director of the Japanese Society of Microscopy, 2007–2011 and 2013–2014
- Director of the Japanese Society of Nano Science and Technology, 2013–2015
- The Chubu branch of the Surface Science Society of Japan, Academic Award 2014
- The Chubu branch of the Surface Science Society of Japan, Academic Award 2013
- The Chubu branch of the Surface Science Society of Japan, Academic Award 2012
- The Chubu branch of the Surface Science Society of Japan, Academic Award 2011
- The Chubu branch of the Surface Science Society of Japan, Academic Award 2010
- The Chubu branch of the Surface Science Society of Japan, Academic Award 2008
- The Japanese Society of Microscopy, Academic Award of the Society (The Setoh's Award), 2007
- The Ceramic Society of Japan, Academic Award of the Society, 2005

Michiko Kusunoki at approximately age 40.

- ASM International (previously known as the American Society for Metals), Metallographic Contest Third Place, "TEM *In Situ* Observation of ZrO_2 Particles at 1200°C," 1994

Biography

Early Life and Education

Michiko Kusunoki obtained all three of her academic degrees from the Tokyo Institute of Technology starting with the bachelor's degree in 1975, a master's degree in 1977, and then a doctorate in 1980.

Career History

Dr. Kusunoki worked at Tokyo Institute of Technology as an Assistant Professor from 1980 to 1982, and then had her first child (a boy) in 1982. From 1983 to 1986,

Dr. Kusunoki worked at the Japan Science and Technology Corp. in Nagoya as a researcher on the ultrafine particle project. In 1986, Dr. Kusunoki had her second son. From 1988 to 1990, Dr. Kusunoki moved north to Tsukuba and took a position at the Japan Science and Technology Corp. as a researcher on a solid surface project. In 1990, she was promoted to Chief Researcher and Group Manager at the Japan Fine Ceramics Center. After ~16 years, she moved to the EcoTopia Science Institute in Nagoya University as a Professor.

Michiko Kusunoki at about 60 years old.

3 Most Cited Publications

Title: A formation mechanism of carbon nanotube films on SiC(0001)
Author(s): Kusunoki, M; Suzuki, T; Hirayama, T; Shibata, N
Source: Applied Physics Letters; volume: 77; issue: 4; pages: 531–533; article number: PII S0003-6951(00)04330-8; published: July 24, 2000
Times Cited: 142 (from Web of Science)

Title: Epitaxial carbon nanotube film self-organized by sublimation decomposition of silicon carbide
Author(s): Kusunoki, M; Rokkaku, M; Suzuki, T
Source: Applied Physics Letters; volume: 71; issue: 18; pages: 2620–2622; published: November 3, 1997
Times Cited: 122 (from Web of Science)

Title: Formation of self-aligned carbon nanotube films by surface decomposition of silicon carbide
Author(s): Kusunoki, M; Suzuki, T; Kaneko, K; Ito, M
Source: Philosophical Magazine Letters; volume: 79; issue: 4; pages: 153–161; published: April 1999
Times Cited: 93 (from Web of Science)

Challenges

The discovery of novel carbon materials and their functionalities is a key challenge for me.

Words of Wisdom

Work should be exciting in a free and open-minded manner.

Jueinai Raynien Kwo

Distinguished Chair Professor of
Physics
National Tsing Hua University
(NTHU)
Hsinc hu
Taiwan

e-mail: raynien@phys.nthu.edu.tw
telephone: 886-3-574-2800

Dr. Raynien Kwo taken in front of Behlin lab
of Physics, University of Nebraska, 1988.

Birthplace
Taipei, Taiwan

Born
October 1, 1953

Publication/Invention Record

>350 publications: h-index 55 (Web of Science)
21 patents, 17 book chapters

Tags

❖ Academe
❖ Domicile: Taiwan
❖ Nationality: American and
  Taiwanese
❖ Asian
❖ Children: 0

Pivotal Career Moment (to date)

After having studied and worked in the United States for 28 years during the first half
of our lives, my husband, Dr. Minghwei Hong, and I reached the peak of our career in
advanced physics/materials research in the prime institute of Bell laboratories. We
decided jointly to return to our homeland, Taiwan, in June 2003. Bell Labs has graciously
donated our entire laboratory equipment to bless Taiwanese research and education.
We are committed to devote our second half of our careers to advanced education and
the promotion of excellent research in Taiwan at the forefront of technology.

Academic Credentials

Ph.D. (1981) Applied Physics, Stanford University, Stanford, CA, USA.
M.S. (1977) Applied Physics, Stanford University, Stanford, CA, USA.
B.S. (1975) Physics, National Taiwan University (NTU), Taipei, Taiwan.

Successful Women Ceramic and Glass Scientists and Engineers: 100 Inspirational Profiles. Lynnette D. Madsen.
© 2016 The American Ceramic Society and John Wiley & Sons, Inc. Published 2016 by John Wiley & Sons, Inc.

Research Expertise: three-dimensional topological insulators, spintronics, diluted magnetic oxides, novel high κ gate dielectrics, novel oxide materials, and heterostructures for electronic and photonic applications, transparent conductors, high-temperature superconductors, nanoelectronics, III–V metal–oxide–semiconductor field-effect transistors (MOSFETs), magnetic superlattices, low T_c superconductivity, superconducting tunneling

Other Interests: teaching, support of women in science, industrial applications

Key Accomplishments, Honors, Recognitions, and Awards

- National Tsing Hua Distinguished Chair Professor, 2008 to present
- Taiwan Outstanding Woman Scientist Award, 2015
- Distinguished alumni of the Department of Physics, National Taiwan University, 2014
- National Taiwan University Chair Professor, 2010–2012
- National Taiwan University Distinguished Professor, 2010–2012
- Elected Fellow of American Physical Society, 2009
- Endowed "Outstanding Scholar Chair Award," 2004–2009
- National Tsing Hua Natural Science Chair Professorship, 2004–2008
- Elected Fellow of the Republic of China Physics Society, 2005
- Taiwan Semiconductor Manufacturing Company (TSMC) endowed chair professorship in National Tsing Hua University, 2003–2004
- Distinguished Technical Staff Award, Agere Systems, 2001
- Institute for Scientific Information (ISI) Most Cited Physicists, ranked 746/1120, period 1981–1986/1997
- Outstanding Paper Award of the Center for Electronics and Electrical Engineering from National Institute of Standards and Technology, 1988

Prof. Kwo received "the Taiwan Outstanding Woman Scientist Award" in year 2015.

Biography

Early Life and Education

Raynien grew up in Taiwan, received her bachelor's degree from the Physics Department of NTU in 1975, and was first in her class. She then came to the United States to pursue her graduate studies in Applied Physics at Stanford University; at that time the advanced institutes of physics in Taiwan had not yet been established. To seek advanced graduate study in physics and to fulfill her passion and dream of having physics research as her future career, going abroad to the United States was the only choice for her and most Taiwanese physics B.S. graduates at that time. Her main research focus was the study of solid-state physics and low-temperature physics toward her Ph.D. at Stanford University. Therefore, she joined Prof. Ted Geballe's research group in low-temperature physics on superconductivity.

Career History

After obtaining her Ph.D. in Applied Physics in 1981, she joined Bell Laboratories in the Physics Research Division in Murray Hill, New Jersey, in the Department of Solid-State Materials and Physics as a member of technical staff. In 2000, she was promoted to distinguished member of technical staff of Agere Systems, formerly the Microelectronics Division of Bell Labs. Primarily, her research was exploring the frontier of physics in novel materials and devices by inventing advanced thin film growth techniques.

From 1981 to 1987, her best work was realized by inventing metal molecular beam epitaxy (MBE) to create metallic superlattices. Her work on synthetic rare earth magnetic superlattices was original and this meant that she was the first person to observe the modulation effect of magnetic properties by artificial layering and to demonstrate the concept of "magnetronics." Her pioneering work in the years 1985 and 1986 led to the subsequent discovery of giant magnetoresistance (GMR) effect in year 1988. Later, in 2007, Albert Fert and Peter Grünberg were awarded with the Nobel Prize for this discovery of GMR. This breakthrough has generated important applications, for example, it has had significant impact across the entire magnetic read/storage industry.

From 1987 to 1992, she devoted her research efforts to the newly discovered high-temperature superconducting perovskite oxide films. She made critical contributions to this field through pioneering efforts in oxide molecular beam epitaxy. The ability to produce single-crystal superconducting films using her approach has paved new ways of probing anisotropic superconducting properties and developing Josephson-coupled devices for superconducting electronics.

From 1993 until 2012, her interest was primarily focused on high dielectric constant κ oxide materials, specifically for electronic/photonic applications. By

Prof. Raynien Kwo with Prof. Minghwei Hong in their lab of molecular beam epitaxy in 2014.

Prof. Kwo gave a speech in a conference of "Woman Physicists in Taiwan" in 2004.

taking advantage of her expertise on oxide films, she invented two new transparent conducting oxides for display and electro-optical devices. She also spent a major effort on several high κ dielectric films to be used as microcapacitors in wireless communicators and in next-generation dynamic random-access memories (DRAMs). The discovery of a mixed oxide for compound semiconductors passivation (in collaboration with Dr. Minghwei Hong) allowed them to demonstrate the first GaAs-based MOSFET circuits in 1997. Since 1999, her goal has been to identify new high κ dielectrics that could be used as the gate oxide instead of silicon dioxide in Si-based transistors. Since their return back to Taiwan in the summer of 2003, she also continued the work on high κ dielectrics and nano Si complementary metal oxide semiconductor (CMOS) to develop high-mobility channel MOSFET based on III–V semiconductors. Her recent research interests are focused toward emergent quantum matters such as 3D topological insulators and spintronics, including spin pumping, spin torque ferromagnetic resonance (FMR), spin caloritronics, and spin transport.

In 2003 when they returned to Taiwan, her first appointment was a TSMC endowed Chair and Outstanding Scholar Chair Professor in NTHU. Subsequently, she served as the Physics Department Chair (2005–2008), and became the Director for Fundamental Science and Research Center in NTHU (2008–2009). She also served actively in the executive council of the Physical Society of Taiwan (2004–2008), and was elected as President of the Physical Society for the period of 2008–2009, as Taiwan's first female president. In 2008, she was elected a member of the International Union of Pure and Applied Physics (IUPAP) Commission C10 on Structure of Condensed Matter, and became the C10 Secretary in 2011. Now, she is the C10 Commission Chair and the vice president of the IUPAP in the executive committee. She was also elected as the delegate to the first Asia and Europe Physics Society Summit (ASEPS) in Tsukuba, Japan, in 2010. Prof. J. Raynien Kwo was the Director of the Center for Condensed Matter Sciences of National Taiwan University, Taiwan, during 2010–2012. At the beginning of 2013, she returned to Physics/NTHU to resume full-time research and teaching.

3 Most Cited Publications

Title: Systematic evolution of temperature-dependent resistivity in $La_{2-x}Sr_xCuO_4$
Author(s): Takagi, H; Batlogg, B; Kao, HL; Kwo, J; Cava, RJ; Krajewski, JJ; Peck, WF, Jr
Source: Physical Review Letters; volume: 69; issue: 20; pages: 2975–2978; published: November 16, 1992
Times Cited: 400 (from Web of Science)

Title: Epitaxial cubic gadolinium oxide as a dielectric for gallium arsenide passivation
Author(s): Hong, M; Kwo, J; Kortan, AR; Mannaerts, JP; Sergent, AM
Source: Science; volume: 283; issue: 5409; pages: 1897–1900; published: March 19, 1999
Times Cited: 292 (from Web of Sciences)

Title: Surface passivation of III–V compound semiconductors using atomic-layer-deposition grown Al_2O_3
Authors: Huang, ML; Chang, YC; Chang, CH; Lee, YJ; Chang, P; Kwo, J; Wu, TB; Hong, M
Source: Applied Physics Letters; volume: 87; issue: 25; pages: 252104; published: December 19, 2005
Times Cited: 226 (from Web of Science)

Challenges

A challenge for *being a woman scientist in this field* is because most of the peers are western men. It was not easy in the beginning to get used to, especially coming from an oriental cultural background. Also, just like most other female scientists, balancing the professional career with family has also been challenging. Luckily, I have a loving husband working in a similar field, who has given me very strong support and encouragements throughout my career.

Words of Wisdom

These words have inspired me greatly.
• Madam Curie: "Nothing in life is to be feared. It is to be understood."
• Albert Einstein: "The most incomprehensible about the universe is that the universe is comprehensible."
• Dr. Martin Luther King: "One has not lived a real life only until when he starts to live for the welfares of all others."

I am thankful to God to provide us a natural world as a school for us to learn continuously throughout our lifetime. I am devoted to carry out my life mission of "passion for knowledge, pursuit for excellence, and sharing of the knowledge," thus to influence young generations, especially many students that I have interacted with daily. Ultimately we will get to know Him better. For in Bible, the book of Proverb 9:10 says: "The fear of the LORD is the beginning of wisdom: and the knowledge of the holy is understanding."

Profile 58

Anne L. Leriche

Director
Laboratoire des Matériaux Céramiques et Procédés
Associés (LMCPA)
Laboratory of Ceramic Materials and Related Processing
University of Valenciennes and Hainaut-Cambresis
Valenciennes, Nord-Pas-de-Calais
France

e-mail: anne.leriche@univ-valenciennes.fr
telephone: +33(0)327531666

Anne Leriche when she was
about 30 years old.

Birthplace
Soignies, Belgium
Born
January 13, 1959

Publication/Invention Record

>150 publications: h-index 19 (Harzing's publish or perish)
3 patents, editor of 1 book, 3 book chapters

Tags

❖ Administration and
  Leadership
❖ Academe
❖ Domicile: Belgium
❖ Nationality: Belgian
❖ Caucasian
❖ Children: 0

Proudest Career Moments (to date)

I recall many great moments . . . I received the medal of the secondary school in 1981.
After I was awarded a M.Sc. Chemistry diploma with honors, I received a National Award
for my Master's research work. Then, I succeeded in obtaining a Belgian Ph.D. bursary
from IRSIA, a national foundation for the Ph.D. position (1981–1983). I achieved my
Ph.D. in a Research Centre (BCRC) and spent 1.5 years in industry before coming back to
university. I have been President of the French Ceramic Society since 2006. In 2011, I was
elected Member of the World Academy of Ceramics. However, I think that the proudest
moment of my career was the ceremony when I became President of the European Ceramic
Society in June 2013; it was in the presence of my Ph.D. supervisors.

Academic Credentials

"Habilitation à Diriger des Recherches" Accreditation to supervise research (1992),
University of Valenciennes, Valenciennes, France.
Ph.D. (1986) Chemistry, State University of Mons, Mons, Belgium.

Successful Women Ceramic and Glass Scientists and Engineers: 100 Inspirational Profiles. Lynnette D. Madsen.
© 2016 The American Ceramic Society and John Wiley & Sons, Inc. Published 2016 by John Wiley & Sons, Inc.

M.Sc. (1981) Chemistry, State University of Mons, Mons, Belgium.
B.Sc. (1979) Chemistry, State University of Mons, Mons, Belgium.

Research Expertise: synthesis and characterization of ceramic matrix composites (zirconia-toughened alumina, mullite–zirconia by reactive sintering, silicon carbide particles toughening of various matrices, and nanocomposites), assisted sintering methods by pressure, microwave, etc. of thermomechanical ceramics and bioceramics, porous ceramics, functional ceramics

Anne Leriche as a new academician of the World Academy of Ceramics with the two other Belgian academicians: Dr. Cambier and Professor Van der Biest.

Other Interests: development of a young ceramist network to favor the exchange between universities and industries and the organization of ceramic schools for students

Key Accomplishments, Honors, Recognitions, and Awards

- President, French Ceramic Society, since 2006 and Board Member since 1995
- Member of the Board of the Belgian Ceramic Society, 1985 to present
- President, Journal of European Ceramic Society Trust, 2009–2010 and 2015–2017
- Chair of the Advanced Ceramics Symposium of EUROMAT Conference in Cracow, Poland, 2015
- President, European Ceramic Society, 2014–2015
- Co-Organizer of the 13th ECerS Conference in Limoges, France, 2013
- Lectures for three international ceramic schools: Bardonecchia, 2012; Limoges, 2013; and Madrid, 2015
- Elected Academician, World Academy of Ceramics, 2011
- Organizer or Co-Organizer of the four conferences on ceramic composites in Mons, Belgium (1987, 1989, 1994, 1997) and of the Electroceramics Conference in Maubeuge, France, 1992
- Chevalier "Palmes académiques" French Education Award, 2005
- Author and co-author of several books edited by the French Ceramic Society
- Invited talks in many countries (Europe, China, and the United States)
- Participation in nine European research projects and administrative coordinator for the university of the three last ones on France–Belgium transborder projects INTERREG III and INTERREG IV
- Referee for ceramic journals (Journal of European Ceramic Society, Journal of Materials Science, Journal of Surface and Coatings, and Biomaterials)

Anne Leriche (front and right) meeting with American Ceramic Society and European Ceramic Society board members in Limoges, June 2013.

- Expert for various European national or regional research agencies to evaluate project submissions (French agency: ANR in 2007 and 2008, Science Foundation Ireland Basic Research Grant Programme, Wallonia Region, research bursary for Rhônes Alpes Region and Limoges Region: Oseo, French Ph.D. bursary: CIFRE)
- Evaluator of more than 40 Ph.D. theses in France, Italy, Spain, and Belgium

Biography

Early Life and Education

Anne Leriche was the second daughter of Belgian parents and spent her child time in Soignies, a Belgian town at south of Brussels. She followed the Latin–Greek program in secondary school and earned a diploma to attend university. She was always fascinated by the sciences (her father was doctor in medicine), the history of European civilization, and the philosophers of ancient Greece. That is why she wished to first study the classical Greek and Latin texts. At the end of her secondary education, she turned directly to the sciences and particularly to chemistry. She is convinced that the logical and synthesis thinking learnt by her classic studies helped in her scientific career. She was probably the only literary person in her university program. The first year was tough but she caught up quickly on the scientific knowledge.

In 1979, she obtained a basic degree in chemistry with honors. She continued her studies and received a Master's of Science in chemistry with honors in 1981. She received a Post-Graduate Award from National Research Council (IRSIA) to undertake doctoral studies. She completed her doctorate in materials science and engineering with honors in 1986 at the State University of Mons, Belgium. Her thesis was focused on the reactive sintering of mullite–zirconia composites under the supervision of Dr. Francis Cambier. During her Ph.D. studies, she spent a few months at Leeds University's Ceramic Department managed by Professor Richard Brook.

Career History

Upon graduation, she worked for a few years as a researcher at Belgian Ceramic Research Centre in Mons (Belgium) during which she participated in several European research projects on mullite, zirconia-based composites, silicon nitride composites reinforced by particles, platelets, fibers, and whiskers. During this time, she took responsibility for R&D at NEOCERAM, a start-up company involved in the processing of oxide advanced ceramics by pressing and slip casting. Following a meeting with Professor Garvie, well known for his research on zirconia, she realized that the industry was smaller than she originally thought and so decided instead to pursue a research career at a university.

In 1990, Anne Leriche got the opportunity to join the "Laboratoire des Matériaux Céramiques et Procédés Associés" in the University of Valenciennes as a research engineer. It is located at Maubeuge, just 35 km away from Valenciennes. With this move, she initiated a new research topic studying the relationship between microstructural characteristics and properties of ceramics and also extended the laboratory's expertise to oxide powder synthesis through liquid precursor routes. She was promoted during her first year to Lecturer in 1991. From Lecturer, she was promoted to Reader in 1994 and then to Full Professor in 2005. In 1999, she also became the Director of the laboratory leading 30 researchers. Today, the research carried out by the laboratory is mainly focused on bioceramics for bone substitutes, piezoelectric ceramics for actuator applications, and ceramic coatings by the sol-gel method for thermomechanical applications. Most of Ph.D. research topics are linked to industrial problems and financially supported by industry or regional/national bursaries.

Anne Leriche as a new academician of the World Academy of Ceramics with French academician colleagues: Professor Naslain, Professor Baumard, and Dr. Chartier.

During her career, she has supervised 15 students on their thesis research, participated in several European projects, and published about 150 papers or proceedings. She has served as President of the French Ceramic Society, President of the European Ceramic Society, and a Board Member of the Belgian Ceramic Society. Anne enjoys teaching; she really appreciates the human contact with the students and she tries to add humor in her lectures to create a relaxed atmosphere. She considers herself fortunate in that she teaches just a few students at a time (between 10 and 15) and that fosters communication and scientific exchange of information.

3 Most Cited Publications

Title: Influence of porosity on Young's modulus and Poisson's ratio in alumina ceramics
Author(s): Asmani, M; Kermel, C; Leriche, A; Ourak, M
Source: Journal of the European Ceramic Society; volume: 21; issue: 8; pages: 1081–1086; doi: 10.1016/S0955-2219(00)00314-9; published: August 2001
Times Cited: 78 (from Web of Science)

Title: Evidence of a dissolution–precipitation mechanism in hydrothermal synthesis of barium titanate powders
Author(s): Pinceloup, P; Courtois, C; Vicens, J; Leriche, A; Thierry, B
Source: Journal of the European Ceramic Society; volume: 19; issues: 6–7; pages: 973–977; doi: 10.1016/S0955-2219(98)00356-2; published: June 1999
Times Cited: 64 (from Web of Science)

Title: High temperature mechanical properties of reaction-sintered mullite/zirconia and mullite/alumina/zirconia
Author(s): Orange, G; Fantozzi, G; Cambier, F; Leblud, C; Anseau, MR; Leriche, A
Source: Journal of Materials Science; volume: 20, pages: 2533–2540; ISSN: 0022-2461 (print), 1573-4803 (online); published: 1985
Times Cited: 51 (from Web of Science)

Challenges

I have faced many challenges in my career:
- The first was to switch to a scientific academic background, after I started with a literary curriculum at the college. The first quarter was tough, but I persevered and finished the first degree satisfactorily. The studies that followed also went well.
- The second challenge was to leave the research community to participate in the creation of a start-up company producing technical ceramic parts. I was able to account for the problems of scale in the production of these products as well as to understand what human conflicts can occur between the production and R&D departments.
- The third challenge was to leave industry and my country to move into a researcher–teacher career in northern France. Certainly, the language was the same, but the approaches taken at work often differed. I had to adapt to the less organized lifestyle of the French public service. However, I think I managed to win the esteem of my

French colleagues because in 2006, I was elected President of the French Ceramic Society.

On being a woman in this field . . .

It is true that when I started my career in the field of the study of ceramic materials, there were fewer women than men among my colleagues and also few women in conferences abroad. I remember being one of the only Europeans to participate in the Mullite Conference in Tokyo in 1986. Due to this, I was personally invited to the home of the conference publisher. Fortunately, today women are more numerous as colleagues, especially in the field of bioceramics.

There is an advantage to being a woman in terms of establishing new contacts with foreign colleagues: as a woman you are more quickly spotted in a crowd. On the contrary, I am convinced that both in the industrial sector and in academe, a woman must have a stronger character than a man—both to be accepted and to defend the interests of her team or her laboratory.

Words of Wisdom

- I had no career plan at the end of my studies aside the fact that I wanted to be useful to society. If you do not have specific plans for the future, do not worry, opportunities will arise and you will discover that time what you like.
- Do not hesitate to change careers or research topic even if the new environment is unknown to you. The future belongs to the brave and those who want to invest heavily.
- It is difficult not to be overwhelmed by our research/work; ensure you keep outside activities and friends (non-professional relationships). To balance this aspect, our work often offers the opportunity to meet wonderful colleagues who can become friends for life.
- A research career in the field of ceramic materials is exciting because it offers endless possibilities for new applications and new investigations in order to develop new sustainable materials. New and continuously more sophisticated characterization techniques allow for a deeper understanding of the phenomena responsible for the ceramic properties. Advanced ceramics should respond to the new challenges of our civilization: clean energy and reduction in pollution and global warming.
- If you are passionate, you will never stop.

That's all the good that I wish you!

Claude Lévy-Clément

Emeritus Researcher
Centre National de la Recherche Scientifique (CNRS)
France

e-mail: claude.levy8@sfr.fr
telephone: 33 6 61330735

Claude Lévy-Clément when
she was 25 years old.

Birthplace
Versailles, France
Born
August 23, 1943

Publication/Invention Record

>200 publications: h-index 42 (Web of Science)
9 patents, editor of 2 books, 14 book chapters

Tags

❖ Government
❖ Domicile: France
❖ Nationality: French
❖ Caucasian
❖ Children: 1

Memorable Career Moments (to date)

As soon as I finished my Master's thesis, I was offered a permanent position at CNRS as a Junior Researcher. It was not necessary to complete a postdoc. I could read and write in English, but I could not deliver talks at international conferences. By 1981, I had become a Senior Scientist, and I took the opportunity that was offered to me to spend 7 months at Ames Laboratory, Ames, USA (in their Physics Department) as a Visiting Researcher. I improved my English further through short visits of 1 month each in both Canada and India, and by reading novels in English. To master English so that I could deliver a talk at an international meeting was the first pivotal point in my career.

The second pivotal moment came indirectly from my experience in the United States. From the time I spent in America, I knew it would also be necessary to secure funds to develop my own projects at CNRS. This insight allowed me to slowly build my own team (starting in 1986). My group leader was not altogether fair; he took a lot of the

Successful Women Ceramic and Glass Scientists and Engineers: 100 Inspirational Profiles. Lynnette D. Madsen.
© 2016 The American Ceramic Society and John Wiley & Sons, Inc. Published 2016 by John Wiley & Sons, Inc.

credit for our work. However, in the end, this situation was fortuitous since the Director of the Lab of Physics then encouraged me (in 1988) to start a second team of materials chemists. The initial years were severely stressful. In the end, I was successful in garnering support for numerous European, French, Israeli, and Japanese projects. My team grew to 10 members and I was also allotted three permanent researcher positions at CNRS for my best doctoral students and postdoctoral scholars. The research accomplishments I developed with my team put me at the forefront of materials research for energy conversion and storage.

My proudest moment was when I received the 2011 Energy Technology Division Research Award of the Electrochemical Society.

Academic Credentials

Ph.D. (1974) Solid State Inorganic Chemistry, University of Paris XI, Orsay, France.
M.Sc. (1967) Solid State Chemistry, University of Paris XI, Orsay, France.
B.Sc. (1965) Inorganic Chemistry, University of Paris XI, Orsay, France.

Research Expertise: electrodeposition of inorganic semiconductor thin films and nanostructures, (photo)electrochemistry of semiconductors, nanoporous silicon made by electrochemical and photoelectrochemical etching, nanowire arrays of silicon combining colloidal patterning and chemical etching, nanostructures of ZnO (arrays of nanowires, nanotubes, and urchins by electrochemical deposition) and lamellar transition metal chalcogenide compounds (inorganic fullerenes and nanotubes), photovoltaics (silicon- and semiconductor-sensitized inorganic nanostructured solar cells), remediation of pollutants in industrial polluted waters using diamond electrodes

Other Interests: international collaborations, working on innovation with industry, and mentoring

Key Accomplishments, Honors, Recognitions, and Awards

- Chairperson of the Heinz Gerischer Award of the Electrochemical Society, 1999–2014
- Vice Chair of the European local section of the Electrochemical Society, 1996–2014
- Laureate of the Energy Technology Division Research Award of the Electrochemical Society, 2011
- Vice Chair of the section of the Corrosion of the International Society of Electrochemistry, 2003–2005

Biography

Early Life and Education

From 1946 to 1949, I attended kindergarten school (starting at the age of 3). (In France, schooling is available for very young kids from 2 to 6 years old.) During this period,

I learned to read and count. From 1949 to 1954, I attended an elementary school dedicated to girls. I skipped the first grade and although I was a very good pupil, I was not the best.

From 1954 to 1958, I attended junior school, and it was during the ninth grade that I enjoyed so much the chemistry lessons that I decided to be a chemist. I come from a lower class family, e.g., my father was a blue-collar worker, and so I had the ambition to be a technician. I was prepared at the end of junior school to pass an exam to attend a Chemistry Technical School. My father was extremely smart (even though he stopped attending school at the age of 10), and he encouraged me to attend high school instead of entering a Chemistry Technical School. This choice was a turning point in my life, because at that time the majority of teenagers finished school at the age of 14 or 15.

From 1958 to 1961, I attended high school and at the end I received a French baccalaureate in mathematical sciences. From 1961 to 1975, I attended the University of Paris XI at the Faculty of Sciences in Orsay. I studied inorganic chemistry, physics, thermodynamics, and crystallography. In 1965, I was awarded a Bachelor's degree. For the next 2 years, I conducted

Claude Lévy-Clément is the recipient of the Energy Technology Division Research Award of the Electrochemical Society in Montreal, 2011.

research at Laboratory of Inorganic Chemistry, University Paris XI in France. Then, in 1967, I presented my Master's thesis on the "Préparation de la solution solide d'hydroxystannates de cadmium et de magnésium. Etude cristallographique et de la décomposition thermique" (Synthesis of the solid solution of cadmium and magnesium hydroxystannates and the crystallographic study of the thermal decomposition). Upon graduation, I became a Junior Researcher at CNRS in Laboratory of Inorganic Chemistry, University of Paris XI in Orsay. During this period (1967–1976), I presented my doctoral thesis (Thèse de Doctorat d'État ès Sciences Physiques) with the title "Sur de nouvelles phases oxygénées de plomb tétravalent et de cadmium ou de calcium" (New oxygenated mixed phases of tetravalent lead and cadmium or calcium).

Career History

I conducted research at CNRS for 42 years, at several laboratories near Paris, including the University of Paris XI (Faculty of Sciences at Orsay), the University of Paris V (Faculty of Pharmacology, Paris), CNRS Campus at Meudon-Bellevue, and CNRS Campus at Thiais (University of Paris XIII). Key points in my career include:

- 1965–1967: Master Grant from the government, Université de Paris XI—Faculté des Sciences d'Orsay, Laboratoire de Chimie Minérale.

- 1967–1976: Junior Researcher (Attachée de Recherche) at CNRS, University of Paris XI—Faculty of Sciences at Orsay, Laboratory of Inorganic Chemistry.
- 1976–1979: Senior Researcher (Chargée de Recherche) at University of Paris V, Faculty of Pharmacology, Laboratory of Inorganic Chemistry (associated with CNRS-LA 200).
- 1979–1999: Senior Researcher at Laboratory of Physics of Solids (UPR 1232), CNRS campus at Meudon-Bellevue.
- Since 1989: Research Director at CNRS second class and Head of Research Team on the photoelectrochemistry of semiconductors.
- Since 1999: Research Director at CNRS second class, Laboratory of Chemistry and Metallurgy of Rare Earth Materials (UPR 209), at CNRS Campus, Thiais. Head of the Research Team on the electrochemistry and nanomaterials with new morphologies.
- 2006–2008: Research Director at CNRS first class, CNRS East Paris Institute of Chemistry and Materials and Paris XIII University, Thiais.
- Since 2008: Emeritus Research Director at CNRS, Thiais.

Over the years, I have organized or co-organized 14 national and international conferences. I was an editorial board member of several international journals and was invited as a scientific expert in the fields of energy, solar hydrogen, and photovoltaics in French, European, and American panels. Notable among those were the International Conference on Photochemistry and Storage of Energy (IPS-15) (2004) and the Electrochemical Society (ECS) Heinz Gerischer Symposium (Physical Chemistry Symposium celebrating the first conferment of the Heinz Gerischer Prize to Honda and Fujishima) (2003) in Paris where I had a major role. The establishment of the most prestigious Heinz Gerischer Prize where I took a leadership role for many years is an illustration of my devotion to the research community. I also organized a number of European Union workshops dedicated to photovoltaic energy and to thin film deposition.

Claude Lévy-Clément chairing the IPS-15 Conference in Paris, 2004.

I took pride in educating students, particularly from the less developed Sahel countries (Senegal, Morocco, etc.). I used significant resources for this important mission and some of my students went back to their home countries and started their own research laboratories. I was the thesis director of 16 Ph.D. students and more than 30 undergraduate research students worked in my group.

I had the honor of hosting a steady stream of very highly regarded scientists (e.g., Prof. Tomkiewicz of Brooklyn College, USA; Prof. Reshef Tenne of the Weizmann Institute in Israel and Member of the Israel Academy of Sciences and Academia Europaea; Prof. Gary Hodes also from the Weizmann Institute; Dr. Amy Ryan of the

Jet Propulsion Lab, USA; and Prof. Santhanam of the Tata Institute of Fundamental Research (TIFR), India) who spent "sabbatical" periods in my lab.

My research in photoelectrochemistry included silicon as an electrode material. In these studies, I considered formation of porous silicon, and studied the conditions under which different morphologies of porous silicon could be formed. This work was done in close connection with a company Photowatt International. The company produced the material for solar cells, i.e., multicrystalline silicon by casting. The company also makes solar cells and modules. There is a demand for a more efficient method to decrease the reflectivity of the multicrystalline silicon solar cells. From the strong connection with the company, my group developed the acidic texturization method of silicon wafers to enhance its light-trapping ability, which in turn resulted in a patented method for improving the performance of polycrystalline Si solar cells. This patented method is widely used in the manufacture of polycrystalline Si solar cells.

I taught during Master's to undergraduate students laboratory experiments, and during my career at CNRS I also taught X-ray diffraction to Master's students at the University of Versailles and photovoltaic cells (silicon and thin film solar cells) at the Faculty of Sciences at Orsay. In the

Claude Lévy-Clément at the Porous Semiconductors Conference in Spain, 2008.

framework of "University Conference of Occidental Switzerland" (teaching Ph.D. students), I taught "Semiconductor Electrochemistry: Application to Photovoltaics" at Ecole Polytechnique Fédérale de Lausanne (EPFL) in Lausanne, Switzerland.

3 Most Cited Publications

Title: CdSe-sensitized p-CuScN/nanowire n-ZnO heterojunctions
Author(s): Lévy-Clément, C; Tena-Zaera, R; Ryan, MA; et al.
Source: Advanced Materials; volume: 17; issue: 12; pages: 1512–1515; published: June 17, 2005
Times Cited: 285 (from Web of Science)

Title: Metal-assisted chemical etching of silicon in $HF–H_2O_2$
Author(s): Chartier, C; Bastide, S; Lévy-Clément, C
Source: Electrochimica Acta; volume: 53; issue: 17; pages: 5509–5516; published: July 1, 2008
Times Cited: 211 (from Web of Science)

Title: Thin film semiconductor deposition on free-standing ZnO columns
Author(s): Konenkamp, R; Boedecker, K; Lux-Steiner, MC; et al.
Source: Applied Physics Letters; volume: 77; issue: 16; pages: 2575–2577; article number: PII S0003-6951(00)05142-1; published: October 16, 2000
Times Cited: 161 (from Web of Science)

Challenges

I had to face several challenges during my career. The first one was deciding whether to go to the middle of the United States (Iowa) alone with my 9-year-old daughter for an extended (7-month) sabbatical during 1981. She did not know one word of English and I only spoke a little. Upon arrival, I had to very quickly find a house/apartment to rent so that my daughter could go to school and I could continue my research. The university assisted in this process and I reached an agreement with a Professor at Ames who would be leaving shortly for a sabbatical to France. He rented part of his house to me, while I taught him French. This might sound easy, but after I finished working with him, I had to flip and teach some English to my daughter. It was not an easy period, but I have very good memories of that time.

The second challenge was to reorient my research from solid-state chemistry to electrochemistry of semiconductors and photoelectrochemistry and electrodeposition of semiconductors as thin films and nanomaterials. I attended several extra courses at the university on electrochemistry, quantum chemistry, and physics of semiconductors in order to broaden my knowledge. The third challenge was to build a team from the ground and to be successful in getting funding. I decided to focus my research on the much needed low-cost fabrication methods such as precipitation in aqueous medium, electrodeposition, electrochemical etching, and metal particle-assisted chemical etching. Although some characterization techniques could be expensive, they were generally available to me free of charge or for a small fee. Such a strategy allowed me to spend more money on the students and postdocs. The fourth challenge was to change laboratory/institute during my career at CNRS. It was not always my decision to do so, but it ended very well every time by changing little by little my research focus and giving me greater knowledge and confidence.

On being a woman in this field . . .

I found, as a woman in research, that it was a challenge to be visible in the scientific community. First, I had to fight the Cinderella complex, thinking that I may be not as good as my male colleagues. I have a few stories that illustrate how some male leaders behaved in the last century. When I received my junior position at CNRS in the 1960s, the Director of the lab congratulated me and at the same time advised me that as a woman I should know to stay in my place. Ten years later, my group leader told me that I was lucky to have a senior position at CNRS because his wife was only a primary school teacher. Until the end of 1990s, we had several secretaries to help us to type our documents and papers. However, as a woman, my requests were always completed after those for my male colleagues were finished. This led me to doing everything earlier on my computer, even if I had to work on weekends. In the 1990s, I was participating in a big conference in Japan. During the conference banquet buffet, a Japanese professor looked at my name tag and asked me if I was the spouse of the famous Professor Lévy-Clément. He was truly shocked when I told him that I was the professor. Please keep in mind that most of these experiences are unlikely to be repeated today. In contrast, I should also mention that I received very useful help from several male colleagues, especially to start writing software in crystallography in the 1970s and in introducing me to the European funding programs in the 1990s.

In addition, it took me considerable time as a woman to get promoted as Research Director at CNRS. I was told that to get promoted, more positions needed to be open at CNRS—at that time, it was considered that the salary of a woman was secondary, at least in France. I put all those "little" things out of my head; I had other matters to focus on. I know that to be recognized in research at the international level I would need to do the following: (i) work hard to develop a strategy in my research, (ii) gradually change my research as new discoveries were made, and (iii) ensure that my selected research directions were relevant (and yes, you may read "fundable" here). Overall in my career, I changed from producing finely divided mixed oxides to growing single crystals and characterizing them to the semiconductor electrochemistry interdisciplinary area. Initially, there is the impression of being a baby in your new research endeavors; it took me several years to become a master in these new research areas.

The other difficulty for me as a woman in research was also being a mother; my husband was also busy building his own future. After our daughter was born in 1971, I had difficulty spending enough time on my dissertation. Later on, after we divorced, I was fortunate to find a new partner who shared my passion for research and had two sons of his own. Once our children were older, I had more freedom to travel to meetings, conferences, and project meetings, and my research really started to bloom. Also changing laboratories was very important as it broadened my field of expertise. My view of research was different after my extended visit to Ames Laboratory in 1981; it was then that I understood that it is necessary to look for funding to develop your research. At the beginning of the 1990s when I returned to France, I was very successful in securing European projects.

To be a woman in research today is easier in some respects, but more difficult in others. It is easier because one's sex matters less and frequently now I meet women researchers at conferences and elsewhere. On the other hand, I found that it is more difficult to be visible because the competition to reach a high level in research is becoming much more intense. I also notice, these days, that students spend less time doing experiments (and more time with their computers)—in the long run, this may not be the most beneficial.

Words of Wisdom

My advice is to be sure to undertake new and exciting research, take calculated risks, take time to enlarge your areas of expertise, do not count your time in research, and go to international conferences to meet foreign colleagues who will help you to build numerous collaborations and perhaps also to obtain funding.

Profile 60

Jennifer A. Lewis

Professor of Engineering and Applied Sciences
Harvard University
Pierce Hall Rm. 221
29 Oxford St.
Cambridge, MA 02138
USA

e-mail: jalewis@seas.harvard.edu
telephone: (617) 496-0233

Jennifer Lewis as Hans Thurnauer Professor of Materials Science and Engineering. Credit: University of Illinois at Urbana-Champaign.

Birthplace

Daytona Beach, Florida, USA

Born

September 9, 1964

Publication/Invention Record

>150 publications: h-index 44 (Web of Science)
10 patents

Tags

❖ Academe
❖ Domicile: USA
❖ Nationality: American
❖ Caucasian
❖ Children: 0

Proudest Career Moment (to date)

I was honored to be selected as the first woman to receive the MRS Medal (2012) from the Materials Research Society (MRS).

Academic Credentials

Sc.D. (1991) Ceramics Science, Massachusetts Institute of Technology, Cambridge, Massachusetts, USA.
B.S. (1986) Ceramic Engineering (High Honors), University of Illinois, Urbana, Illinois, USA.

Research Expertise: colloidal processing, soft materials, 3D printing, bioprinting

Other Interests: entrepreneurship; education and outreach including teaching modules, science videos, and engagement with the popular media

Successful Women Ceramic and Glass Scientists and Engineers: 100 Inspirational Profiles. Lynnette D. Madsen.
© 2016 The American Ceramic Society and John Wiley & Sons, Inc. Published 2016 by John Wiley & Sons, Inc.

Key Accomplishments, Honors, Recognitions, and Awards

- Editorial Board, Advanced Materials, 2013 to present
- Editorial Board, Soft Matter, 2013 to present
- Editorial Board, Advanced Functional Materials, 2012 to present
- Physical Sciences Advisory Committee, Argonne National Laboratory, 2009 to present
- Robert B. Sosman Award, Basic Science Division, The American Ceramic Society, 2016
- MIT Technology cited our work as one of the "top 10 breakthrough technologies," 2014
- The National Geographic piece featured their printed rechargeable microbatteries (http://ngm.nationalgeographic.com/2014/12/3d-printer/smith-text), while the Foreign Policy magazine selected J. Lewis as one of the top 100 Global Thinkers (in the Innovation category, http://globalthinkers.foreignpolicy.com/#innovators/detail/lewis). Their new research on 3D bioprinting was featured in The New Yorker (tech issue, http://www.newyorker.com/magazine/2014/11/24/print-thyself), 2014
- Elected Academician, World Academy of Ceramics, 2014
- Editorial Board, 3D Printing and Additive Manufacturing, 2013
- Keynote Lecture, Science, Engineering, and Technology in the City (for High School Girls), Cambridge, MA, 2013
- MRS Medal, 2012
- Elected Fellow, American Academy of Arts and Sciences, 2012
- International Editorial Advisory Board Member, Soft Matter, 2005–2012
- Elected Fellow, MRS, 2011
- Plenary Lecture—Colloids, 2011
- Langmuir Lecture Award, American Chemical Society, 2009
- Associate Editor, Journal of the American Ceramic Society, 1997–2008
- Elected Fellow, American Physical Society, 2007
- Plenary Lecture, Composites at Lake Louise, 2007
- Plenary Lecture, Society of Rheology, 2007
- Penn Engineering Grace Hopper Lecture, 2007
- Featured Public Lecture, Boulder School for Condensed Matter and Materials Physics, 2006
- Plenary Talk, International Conference on Ceramic Processing Science, 2006
- DuPont Young Investigator Award, 2004–2005
- Fellow, American Ceramic Society, 2005
- Daily Illini's Incomplete List of Teachers Ranked Excellent by Their Students, 1992–1994 and 1999–2004
- Brunauer Award, American Ceramic Society, 2003

Jennifer Lewis (left in the front row) with her group at the Fall Materials Research Society meeting.

- College of Engineering Council's Outstanding Advisors List: 1994, 1995, and 2002
- Willett Faculty Scholar Award, College of Engineering, University of Illinois, 2002
- University Scholar, University of Illinois, 2001
- Xerox Award for Faculty Research, College of Engineering, University of Illinois, 2001
- Selected by the National Academy of Engineering, Frontiers of Engineering Meeting, 2000
- Allied Signal Foundation Award: 1998 and 1999
- Xerox Award for Junior Faculty Research, College of Engineering, University of Illinois, 1996
- Schlumberger Foundation Award, 1995
- MRS Travel Award for Young Scientists, International Conference on Advanced Materials, ICAM, 1995
- Arnold O. Beckman Research Award—Research Board, University of Illinois, 1994
- NSF Presidential Faculty Fellow Award, 1994
- Burnett Teaching Award—UIUC Materials Science and Engineering Department, 1994
- Gordon Research Conference in Ceramics, Graduate Student Scholar, 1989
- International Association for the Exchange of Students for Technical Experience (IAESTE) Fellowship, University of Leeds, 1986
- General Motors Scholar, 1984–1986
- Knights of St. Patrick Award, College of Engineering, University of Illinois, 1986
- A.W. Allen Award, Ceramic Engineering Department, University of Illinois, 1984

Biography

Early Life and Education

She was fortunate to have a wonderful younger sister and loving parents who valued education. She moved several times while growing up, living primarily in North Carolina, Ohio, and Illinois. She went to high school in Palatine, IL at William Fremd H.S., where she graduated near the top of her class and played two varsity sports (basketball and softball). At Fremd, she established a lifelong friendship with her english teacher, Mrs. Kolder, and her math teacher, Mr. Kolder. As an undergraduate student at University of Illinois, Urbana, she spent two summers working at General Motors in Flint, Michigan as part of the GM Scholars program. She also served as the President of Keramos, and was an avid fan of Illini sports. After completing her B.S. degree, she spent her summer at the University of Leeds, Leeds, England in the Department of Ceramic Engineering as part of the IAESTE program. From 1986 to 1990, she carried out her doctoral thesis research at the Massachusetts Institute of Technology under the direction of Prof. Michael J. Cima.

Career History

She joined the faculty of the Materials Science and Engineering Department at the University of Illinois at Urbana-Champaign in 1990, where she was ultimately appointed as the Hans Thurnauer Professor of Materials Science and Engineering and served as the Director of the Frederick Seitz Materials Research Laboratory from 2006 to 2012. In 2013, she joined the faculty of the School of Engineering and Applied Sciences as well as the Wyss Institute for Biologically Inspired Engineering at Harvard University.

Jennifer Lewis as the Hansjorg Wyss Professor of Biologically Inspired Engineering. Credit: Harvard University.

Throughout her career, she has consulted for various organizations: British Petroleum (2011), Dow Corning Technical Advisory Board Member (2010 to present), Oxane, Houston, TX (2005, 2007–2009), Unilever, The Netherlands (2003), Bosch, Germany (2002), BASF, Germany (2001), W.R. Grace, Cambridge, MA (2000 to present), Schlumberger, Inc., Paris, France (1999), Caterpillar Inc., Peoria, IL (1998–2000), Sandia National Laboratories (1997–1998), St. Gobain/Norton, Northborough, MA (1996), Argonne National Laboratory, special term appointment (STA) (1993–1996), Coors Electronic Packaging, Co., Chattanooga, TN (1992–1994), Gardner-Denver Machinery, Inc., Quincy, IL (1994, 1996, 2000, 2002), and Technology Management, Inc., Cleveland, OH (1995).

3 Most Cited Publications

Title: Colloidal processing of ceramics
Author(s): Lewis, JA
Source: Journal of the American Ceramic Society; volume: 83; issue: 10; pages: 2341–2359; published: October 2000
Times Cited: 634 (from Web of Science)

Title: Self-healing materials with microvascular networks
Author(s): Toohey, KS; Sottos, NR; Lewis, JA; et al.
Source: Nature Materials; volume: 6; issue: 8; pages: 581–585; doi: 10.1038/nmat1934; published: August 2007
Times Cited: 403 (from Web of Science)

Title: Omnidirectional printing of flexible, stretchable, and spanning silver microelectrodes
Authors: Ahn, BY; Duoss, EB; Motala, MJ; et al.
Source: Science; volume: 323; issue: 5921; pages: 1590–1593; published: March 20, 2009
Times Cited: 283 (from Web of Science)

On being a woman in this field . . .

I feel very fortunate to have had wonderful mentors in the field of ceramics, including most notably Michael Cima, David Payne, Gary Messing, and Fred Lange. I have also benefited from interacting with several leading women in ceramics, including Waltraud Kriven and Kathy Faber. They were pioneers! Watching them succeed helped me believe that anything was possible.

Words of Wisdom

You only get one life—make the most of it.

Profile 61

Hua Kun Liu

Distinguished Professor
Fellow of Australian Academy of Technological
Sciences and Engineering
Institute for Superconducting and Electronic Materials
University of Wollongong (UOW)
Australia

e-mail: hua_liu@uow.edu.au
telephone: +61 2 4221 4547

Hua Kun Liu as an
undergraduate student at
Jilin University, 1961.

Birthplace
People's Republic of China (PRC)

Publication/Invention Record

>675 publications: h-index 68 (Web of Science)
11 patents/provisional patents, editor of 2 books,
>10 book chapters

Tags

❖ Academe
❖ Domicile: Australia
❖ Nationality: Australian
❖ Children: 2

Important Career Moments (to date)

For Hua Kun Liu, the highlight of being a Professor is seeing herself and her students be successful in their careers.

Academic Credentials

Diploma (1980) Advanced Quantum Chemistry, Jilin University, Changchun, PRC.
M.S. equivalent (1968) Chemistry, Jilin University, Changchun, PRC.
B.S. equivalent (1965) Chemistry, Jilin University, Changchun, PRC.

Research Expertise: clean energy materials for use in batteries, supercapacitors, fuel cells, hydrogen storage, and superconductors

Other Interests: commercialization and industry development via technology transfer and spin-off companies; promotion of women postgraduates in science and technology

Successful Women Ceramic and Glass Scientists and Engineers: 100 Inspirational Profiles. Lynnette D. Madsen.

Key Accomplishments, Honors, Recognitions, and Awards

- Associate Editor, Frontiers in Energy Research, since 2013
- Editorial Board for The Open Electrochemistry Journal, since 2011
- Associate Editor for Advanced Science Letters, since 2007
- Editorial Board for Journal of Nanoscience and Nanotechnology, since 2000
- Advisory Board for Journal of New Materials for Electrochemical Systems, since 1998
- Honorary Professor, Kunming University of Science and Technology, 2015
- Honorary Professor, Southwest University, 2014
- Honorary Professor, Research Institute of Baili Mechanical and Electrical Ltd., 2014
- Board committee steering member of International Academy of Electrochemical Energy Science, 2014
- Fellow of Australian Academy of Technological Sciences and Engineering, 2013
- Vice-Chancellor's Research Excellence Award for Senior Researchers, 2013
- Honorary Professor, Ningbo Institute of Materials Technology and Engineering, Chinese Academy of Sciences, China, 2013
- Australian Professorial Fellowship within Australian Research Council (ARC) Centre for Electromaterials Science (ACES), UOW, 2006–2010, 2011–2013
- Honorary Professor, Hubei University, China, 2008
- Honorary Professor, Shanghai Institute of Microsystems and Information Technology, Chinese Academy of Sciences, China, 2007
- Australian Professorial Fellowship within ARC Centre for Nanostructured Electromaterials (ACNE), UOW, 2003–2005
- Honorary Professor, Shanghai University, China, 2004
- Australian Research Council Professorial Fellowship, 1999–2003
- Australian Research Council Senior Research Fellowship, 1994–1998

Biography

Professor Liu's research has been focused on clean energy materials. Her research covers materials science and engineering, electrochemistry, and applications. She has developed numerous advanced materials and novel technologies for use in lithium ion batteries (LIBs), supercapacitors, fuel cells, and hydrogen storage. Her research efforts address the challenge of clean energy storage for use in hybrid electric vehicles, portables, home, and aerospace. She has developed a number of strategies to enhance

A recent photo of Hua Kun Liu.

the electrochemical performance of LIBs. These approaches include strain engineering to control the negative strain in electrode materials by carbon coating or encapsulation, compositing, changing dimension and nanoparticles, the spray pyrolysis technique, the self-catalysis growth technique, liquid-assisted solid-state reaction, sulfur coating on mesoporous carbon, and high-voltage cathode materials.

Prof. Liu's contribution to education is demonstrated by her excellent postgraduate training outcomes. She has supervised 55 Ph.D. and 15 Master's students to completion and a number of postdoctoral and visiting fellows. A number of Ph.D. graduates have been accepted by prestigious overseas institutions, including Cambridge, Argonne Lab, Los Alamos Lab, Max Planck Institute, etc. These Ph.D. graduates and fellows are very well placed within the fields of science, technology, and industry worldwide, and have made significant contributions to the development of energy science and technology. More than 20 of these graduates have already been promoted to senior levels, including eight Professors and Associate Professors, five Lecturers/Senior Lecturers, three Deputy Directors of institutes, and five CEOs/CTOs for companies. A few examples are highlighted here.

Her former Ph.D. graduate Prof. Jun Chen is the Director for Energy Materials Institute in Nankai University, China, which has more than 50 researchers and doctoral students. He was awarded the Natural Science Prize in 2011, the highest award in China, prestigious NADO Fellowship in 2000 in Japan, and the highly competitive Changjiang Prof. Fellowship in 2005. Her former Ph.D. graduate Guo Xiu Wang was promoted to Professor and Director of Clean Energy Centre in University of Technology Sydney, Australia in 2011, won Australian Research Council Queen Elizabeth II (ARC QEII) Fellowship and Future Fellow in 2008 and 2013, respectively, and has become prominent researcher in LIBs. Her former Ph.D. graduate Wei Guo Wang was promoted to a Professor in 2006 and Deputy Director of Ningbo

Hua Kun Liu with her husband at the Australian Academy of Technological Science and Engineering (ATSE) 2013 Oration and New Fellows Dinner on November 22, 2013, Adelaide, Australia.

Materials Research Institute, Chinese Academy of Sciences (CAS) in 2010, and he is responsible for a large national program on fuel cells in China.

Promotion of Women in Science and Technology

Prof. Liu has worked tirelessly to train and promote female graduate students in science and technology. She has paid a special attention to their training. She understands their special needs and treats them as her friends and provides them opportunities for motivating themselves. Her students talk to her about everything from their hearts without any reservation. She has shown enormous patience and dedication to their personal and family's difficulties. As a result, all but one of her 20 Ph.D. and 4 Master's female students have successfully completed their degrees. This competition ratio is much higher than male counterparts and higher than national average in Australia. Some of her female Ph.D. graduates have also been promoted to senior positions. For example, her former Ph.D. graduate Zaiping Guo was awarded ARC Postdoctoral Fellow in 2004, QEII in 2010, and promoted to Professor in 2012, one of the most rapid promotions among all Ph.D. graduates at Institute for Superconducting and Electronic Materials (ISEM). She has become a group leader on her own right with more than 10 researchers and Ph.D. students. Rita Chen and Ling Yuan have been promoted as industry leaders in big business firms.

Much of Prof. Liu's satisfaction comes from seeing her students succeed in academic and business. A few examples follow.

- Establishment of a battery company by her Ph.D. graduate: The best example is Dr. Rita (Yao) Chen who is one of the key founders and the CEO of DLG Battery Co. Ltd. Rita was Prof. Liu's Ph.D. student during 2001–2005 at ISEM of UOW. DLG Battery was also registered in Australia with an office in Melbourne. Rita was well trained and she developed all of the necessary expertise, skills, experiences, and technical know-how on LIBs in much detail and with an in-depth understanding during her Ph.D. study with Prof Liu. This Ph.D. training process enabled her to transfer all the new technologies to production line in her company. After returning to China, she was awarded as an industry leading expert and excellent woman in commercialization in Shanghai. She has won a number of awards on development of high power, high charge rate, and safe LIBs. She has been granted 7 patents. Within a short period of 8 years, DLG Battery has grown to a world-class LIB company, is ranked at number one in Southeast China, and at the top 6 of LIB companies in China. It has several manufacturing plants around China (Shanghai, Shenzhen, Zhangjiagang, and Shzhou) with $60,000 \, \text{m}^2$ plant and more than 2000 workers, having more than 30 distribution agents around the world. Total capital value is A\$80,000,000 and turnover was \$50,000,000 in 2012. DLG has sponsored five ARC Linkage Projects since 2003 and participated in Auto CRC as an industry partner in 2011.

- Her Ph.D. graduates became industry leaders: Dr. Zhenwei Zhao, who did both Master's and Ph.D. with Prof. Liu on LIB topic during 2003–2008. The ability and expertise he obtained during his postgraduate study ensured his rapid promotion to a Deputy General Manager of Sinopoly Battery Ltd (Tianjin, China), also the Head of Technology Intelligence Centre of Bayer Co., responsible for business development, the corporate strategy and new business opportunity, and the new energy investment corporation. Sinopoly is one of the top 10 LIB companies in China.

- Prof. Liu's Ph.D. graduate Dr. Scott Needham worked on LIBs from 2003 to 2007. Armed with a strong research background, he was appointed as a Manager of Intven Ltd, a venture capital company in Sydney, in charge of \$100,000,000 in venture capital investment, making a significant impact on innovation and commercialization of the research outcomes.

- Another Ph.D. graduate Dr. Desmond Ng worked with Prof. Liu on LIBs from 2004 to 2007 and was appointed as Product Manager of Timchem in Singapore, which is one of the largest carbon materials supply companies for energy storage and other applications.

3 Most Cited Publications

Title: Enhancement of the critical current density and flux pinning of MgB_2 superconductor by nanoparticle SIC doping
Author(s): Dou, SX; Soltanian, S; Horvat, J; et al.
Source: Applied Physics Letters; volume: 81; issue: 18; pages: 3419–3421; published: October 28, 2002
Times Cited: 565 (from Web of Science)

Title: Preparation and electrochemical properties of SnO_2 nanowires for application in lithium-ion batteries
Author(s): Park, M-S; Wang, G-X; Kang, Y-M; et al.
Source: Angewandte Chemie, International Edition; volume: 46; issue: 5; pages: 750–753; published: 2007
Times Cited: 437 (from Web of Science)

Title: Highly reversible lithium storage in spheroidal carbon-coated silicon nanocomposites as anodes for lithium-ion batteries
Author(s): Ng, S-H; Wang, J; Wexler, D; et al.
Source: Angewandte Chemie, International Edition; volume: 45; issue: 41; pages: 6896–6899; published: 2006
Times Cited: 343 (from Web of Science)

ResearcherID (G-1349-2012)

Challenges

My main challenge as a woman and mother has been to be successful in both my career and family life.

Words of Wisdom

Have a dream, set a goal, and then work hard, and have self-confidence, a supportive husband, happy family, and good health.

Profile 62

Kathryn V. "Kaycee" Logan

Adjunct Professor and former Virginia Tech
Samuel P. Langley Professor
Virginia Polytechnic Institute and State University
Blacksburg, Virginia
and
Principal Research Engineer Emerita
Georgia Institute of Technology
Atlanta, Georgia
and
KCBEADS

Roswell, Georgia
USA

e-mail: kvlogan@bellsouth.net
telephone: (404) 401-2972

Kaycee Logan at age 23.

Birthplace

Atlanta, Georgia, USA

Publication/Invention Record

21 publications: h-index 5 (Web of Science)
19 patents, edited 7 books, 1 book chapter

Tags

❖ Administration and Leadership
❖ Academe
❖ Industry (Small Business)
❖ Domicile: USA
❖ Nationality: American
❖ Caucasian
❖ Children: 2

Proudest Career Moment (to date)

While participating as a member of the Board on Army Science and Technology (BAST) in a meeting (including a number of Army Generals) at the National Academies, I mentioned to another BAST member that I would really like to drive a tank. Suddenly the room became quiet and everyone was looking at me and smiling. Our next BAST meeting was held at the U.S. Army Research Lab. An Army Officer asked me to follow him without telling me where we were going. He escorted me to the driving field where an M1A1 Abrams tank was parked. I was instructed to climb into the driver's compartment just under the cannon. I was the only person in the compartment that had a T-bar to accelerate and decelerate, and a large power brake to stop the huge 70-ton

Successful Women Ceramic and Glass Scientists and Engineers: 100 Inspirational Profiles. Lynnette D. Madsen.
© 2016 The American Ceramic Society and John Wiley & Sons, Inc. Published 2016 by John Wiley & Sons, Inc.

vehicle. The top speed of the tank was 45 MPH. Yee-haw!!! I am now certified in driving an M1A1 Abrams tank!

Later, because of my membership in the BAST, I had supper with General Abrams, the son of the General Abrams the Abrams tank is named after.

Kaycee in the tank!

Academic Credentials

Ph.D. (1992), Georgia Institute of Technology, Atlanta, Georgia, USA.
M.S. (1980) Ceramic Engineering, Georgia
Institute of Technology, Atlanta, GA, USA.
B.Ceram.Eng. (1970) Ceramic Engineering, Georgia Institute of Technology, Atlanta, Georgia, USA.

Research Expertise: advanced synthesis and processing, high-temperature solid-state diffusion, refractory material development, analytical materials characterization, and mechanical properties of materials

Other Interests: working with professional societies effectively, using scientific and engineering skills to create wearable art; transferring, scale-up, and commercialization of technologies

Key Accomplishments, Honors, Recognitions, and Awards

- Distinguished Life Member, The American Ceramics Society (ACerS), 2014 to present
- Past Presidents Council Sub-Committee, ACerS, 2013 to present
- Greaves-Walker Award, National Institute of Ceramic Engineers (NICE), 2007
- Langley Colloquium and Sigma Series Lecture, 2005
- President, ACerS, 2004
- Who's Who in the World, 2004
- Fellow, NICE, 2002
- International Who's Who of Professional & Business Women, 2000
- James I. Mueller Award and Lecture, 1999
- Elected Academician of the World Academy of Ceramics, 1995
- Fellow, The American Ceramics Society, 1992 to present
- Professional Engineer, State of Georgia, since 1981

- Elected Chair of the Engineering Ceramics Division of ACerS, 1994
- Vice President of Programs, Meetings, and Expositions, 1998
- Vice Chair of the American Association of Engineering Societies, 2001
- Director on the Board of The American Ceramics Society, 2000–2005
- A member of the Board on Army Science and Technology, National Academies, 1997–2003
- President of the National Institute of Ceramic Engineers, 1997

Biography

Early Life and Education

Kaycee is an original Atlanta native. She grew up in Buckhead (metro Atlanta) and enjoyed being a rockhound beginning in her early years. She learned how to find gemstones, cut them into cabochons, and mount them in jewelry. She has always had a love for crystals and wanted to follow a career in gemology.

Kaycee was technical and mathematically inclined. In high school, she participated in some of the first invited advanced courses in physics and chemistry. She was also invited to participate in a special computer course at Georgia State University in her senior year (1963).

She graduated from high school in 1964 and in those days girls were not encouraged to attend Georgia Tech by the local high schools. Since Georgia Tech welcomed her with open arms, Kaycee was able to live at home and attend classes. She was one of 100 women attending Georgia Tech among 5000 men! The courses in ceramic engineering covered synthesis and processing of very high temperature materials (e.g., rocks, minerals, and crystals), so that became her choice of major.

Kaycee was often an island of one woman in the middle of a classroom of men because the men students thought she was a "curve breaker." When she failed the first test with the rest of the students, the guys relaxed and became compatriots. She obtained her Bachelor of Ceramic Engineering (the last one before Bachelor of Science degrees as well as the last one to be printed on sheepskin) in 1970. Kaycee was also awarded Sigma Xi's Outstanding Undergraduate Studies for her undergraduate thesis.

When she graduated with her Bachelor's degree from Georgia Institute of Technology, her job search was not too successful. Besides being an engineer, she could count and knew the alphabet, but she could not type. So her job opportunities in Atlanta were not very plentiful. In fact, her first job was working for a Temp Service that assigned her to "dance" at a Pepsi-Cola convention. It was "fun." However, Kaycee persisted and finally landed a job at the Georgia Tech Engineering Experiment Station as a Ceramic Engineer: a dream job synthesizing and processing ceramic materials.

Kaycee took time out to have her two children, Stephanie (1972) and Bill (1976). She had a 6-week maternity leave for Stephanie and a 1-year leave for Bill. She tried to return to Georgia Tech for her Master's degree when Bill was still an infant, but staying up at night with an infant and trying to study as well were too exhausting. Kaycee did return in 1978 and accomplished her Master of Science in Ceramic Engineering and again was awarded Sigma Xi's Outstanding Graduate Studies for her Master's thesis.

She returned in 1990 to complete her doctorate in civil engineering from Georgia Tech in 1992 as well. Kaycee had completed all of the courses in ceramic engineering that she could take, so she chose civil engineering (mechanics of materials) as her major.

Kaycee Logan at ~35 years old.

Career History

Dr. Logan has conducted research in the area of high-performance ceramic materials at the Georgia Institute of Technology since 1969. She worked at the Georgia Tech Research Institute for 22 years (1969–1992), the Office of Interdisciplinary Programs/ Computational Mechanics Center for 2 years (1992–1994), and then moved to the School of Materials Science and Engineering (1994–2001). She took two Inter-governmental Personnel Act assignments while at Georgia Tech to the government as a Materials Engineer: the first to the Materials Science Division of the Army Research Office (1998–2000) and second to the Army Corps of Engineers (2000–2001).

Her work involved the synthesis of refractory materials by using self-propagating high-temperature reactions and solid-state diffusion reactions above 2000 °C. Also included was manufacturing scale-up of high-performance ceramic materials. Dr. Logan's research involved the pilot plant scale-up of powder synthesis and forming processes using self-propagating high-temperature synthesis, development of high-performance materials, very high strain rate behavior of ceramic materials, development of refractory materials for application as a boiler liner, and mechanical and analytical microstructural characterization of materials. Dr. Logan has authored/co-authored over 40 major reports and publications and is inventor of 19 patents issued worldwide.

She retired from Georgia Tech in February 2001 and was selected the Virginia Tech Samuel P. Langley Professor in July 2001. Her offices were located at Virginia Tech in Blacksburg, VA and the National Institute of Aerospace (NIA) in Hampton Roads, VA. Her research laboratories were located on campus in the Department of Materials Science and Engineering and at NASA Langley Research Center. She taught both on the Virginia Tech campus and by distance from NIA. She is presently Adjunct Professor at Virginia Polytechnic Institute and State

Some of Prof. Logan's students at the NASA Langley Research Center, 2007.

University, and Principal Research Engineer Emerita at Georgia Institute of Technology. After her retirement, she has started her own small business, Powder Technologies, Inc. (KCBEADS), which focuses on creating/synthesizing unique ceramic, metal, polymer, and glass art for jewelry applications.

3 Most Cited Publications

Title: Aluminothermic reaction path in the synthesis of a TiB_2–Al_2O_3 composite
Author(s): Sundaram, V; Logan, KV; Speyer, RF
Source: Journal of Materials Research; volume: 12; issue: 7; pages: 1681–1684; doi: 10.1557/JMR.1997.0230; published: July 1997
Times Cited: 25 (from Web of Science)

Title: Reaction path in the magnesium thermite reaction to synthesize titanium diboride
Author(s): Sundaram, V; Logan, KV; Speyer, RF
Source: Journal of Materials Research; volume: 12; issue: 10; pages: 2657–2664; doi: 10.1557/JMR.1997.0355; published: October 1997
Times Cited: 23 (from Web of Science)

Title: Dynamic consolidation of combustion-synthesized alumina–titanium diboride composite ceramics
Author(s): Kecskes, LJ; Niiler, A; Kottke, T; et al.
Source: Journal of the American Ceramic Society; volume: 79; issue: 10; pages: 2687–2695; published: October 1996
Times Cited: 14 (from Web of Science)

Challenges

Mentors

Seeking mentors: One can never have too many mentors. Anyone who helps you in any aspect of your career is a mentor: family, professors, doctors, clergy, friends, etc.

- How to identify a mentor: Are they visible locally, nationally, internationally? Are they active in your chosen field? Do they publish in conferences, refereed journals? Are they cited, invited speakers? Do they conduct research; have contract funding, grad students, postdocs? Do they teach? Are they active in their professional society?
- Mentor availability: If a mentor has a lot of grad students and postdocs, you will be mentored more by the students or postdocs (which isn't necessarily bad).
- Ways to approach a potential mentor: Ask for his/her help to become active in a professional society, in your chosen field, solving a research problem, drafting a paper, and asking for guidance.
- Male and female mentors are all good to have.

Traveling

Travel for women today is not at all like it was 30 years ago; however, it would not be impossible to experience a few surprising events today. A few examples that I experienced, which are relatively humorous now, but disturbing to me then, are provided below:

- Airplanes: Numerous times I would be in my business suit and a man sitting next to me on the plane would be dressed in jeans. He would strike up a general conversation with me that would lead to telling me about his wife and love affairs. I would not of course encourage him, but he would treat me like he was confessing his sins.

- Trains: Invariably, my flight would land late at night. Since I rode the commuter train to my part of town, that time of night there were few riders; some like me, but others were denizens of the night. They would always sit close to me. I made no eye contact as best as I could, but they always felt a driving need to communicate with me. One such "gentleman" was dressed in motorcycle leathers and had a number of scrapes, scars, and a few bandages. He sat right across from me and kept looking at me. I was "reading" my book trying to ignore him. Well he made contact and began asking me about getting hurt and how the pain was lasting and hard to control. Uh-oh I thought. This is it for me. It turned out that he was a salesman on his way to a convention to sell magnet patches for pain and was trying to ask me where the convention center was. He had recently wrecked on his bike.

- Hotels: My secretary made my hotel reservations in New York in the hotel that had the same name as my hotel, but it was the hotel on the "wrong" side of town. When I arrived for check-in, there was an armed guard at the locked front door. The clerk looked like a person who was used to seeing people of a questionable intent. I asked to cancel, but he said "no" and laughed. When I went to my room, the hallway was lined with rooms having bare red light bulbs above the doors. My room had five chain locks on the door. I put a chair under the door knob for insurance.

- City streets: I had always been advised to not look a stranger in the street in the eye because that would encourage them to choose you as the one to give mischief. I was at the Boston Materials Research Meeting and for that one brief moment of time our eyes locked on a busy street at noon in front of Sax Fifth Avenue. I immediately looked down, but that didn't help. The person with a heavy growth of beard threw his face upward, let out an unearthly loud howl (yes, like a wolf), and headed straight toward me. I ran and melded into a group of women I didn't know and kept with them until the "excitement" was over. The wolfman continued past me and left me alone. Whew!

On being a woman in this field . . .

The opportunities are broad and endless, just like other career-minded individuals. Sometimes you are limited by being a woman; sometimes you have an advantage. Thirty years ago, it was considered bad luck for a woman to go down into a mine. So I could not follow my perceived dream of being a geologist. Because of that apparent roadblock, I followed a path into Georgia Tech and a wonderful career that included gemology, minerals, and much more than I expected. Since there were few women, I stood out and gained a lot of attention (good and bad) with subsequent opportunities... all my choice. I ALWAYS had a choice.

I was a pioneer and led the way in many instances. Of course, pioneers get the arrows.

As an example, since I was inventor in 19 patents, I began pathways for professors to commercialize their own technology and continue to teach without a conflict of interest. I set up a company in the on-campus incubator and proceeded step by step to commercialize my technology. I interacted with the University legal office and administration to convince them that it was okay. I was called to a "come to terms" meeting consisting of the Vice President and legal and contracting officers (about 12 people and me). The meeting began with the VP saying "First of all, it is illegal and cannot be done." One by one, the other participants said "But if she does it this way, it's not illegal." By the end of the 30-minute meeting, all were agreeing that it could be done. The VP then said "Go ahead Kathryn, it sounds like a great idea."

Many women are concerned about having a family and having to choose a family or career. It is possible to do whatever you wish: no family, no career, have children, don't have children, have a family, children, and career. It's your choice.

I had it all and still do. I have a wonderful husband, two beautiful children, five beautiful grandchildren, and a black lab boundless energy dog. I have an international career, numerous mentors, and many professional colleagues. I am a professor, research engineer, consultant, mentor, and advisor. I began my own company and have helped others begin their companies. It goes on and on and continues to do so.

Words of Wisdom

- Being a professional in your field is foremost over being a woman in your field. Others may try to have you feel otherwise.
- Try to ignore discrimination and harassment. Work around and through it. Don't be reluctant to obtain help if becomes necessary. It's a real continuing challenge!
- Don't have a baby while you are completing degrees. Complete the degrees first.
- Keep ethics foremost.
- Never believe that you do not deserve an accolade, that you received it just because you are a woman. A lot of times a woman will have to work harder to achieve a goal because she is a woman.

Kaycee Logan at 54.

Profile 63

Gabrielle Gibbs Long

Senior Advisor
Advanced Photon Source (APS)
Argonne National Laboratory (ANL)
9700 South Cass Avenue
Argonne, IL
USA

e-mail: gglong@aps.anl.gov
telephone: (630) 252-6012

Gabrielle Cohen (Long) at age 24, just before starting graduate school. Portrait by Bradford Bachrach, 1966.

Birthplace
New Brunswick, NJ, USA

Born
February 11, 1942

Publication/Invention Record

>120 publications: h-index 24 (Web of Science)
Editor of 1 book, 23 book chapters

Tags
❖ Administration and Leadership
❖ Government
❖ Domicile: USA
❖ Nationality: American
❖ Caucasian
❖ Children: 2

Memorable Career Moment

Gabrielle Long's pivotal career moment came very early in her career. Just after completing her Master's dissertation and publishing, her first research article in *Physical Review*, she traveled to Japan on vacation. While there, she looked in on some researchers at the University of Tokyo who were working in an area related to her own. To her surprise, they knew her name and her work, having just held a departmental seminar on her newly published paper. This moment was deeply inspiring. The lesson that she took from it is this: Science is important not only to those doing the work but also to a large and active community of investigators in related fields and beyond.

Academic Credentials

Ph.D. (1972) Physics, Polytechnic Institute of Brooklyn, New York City, NY, USA.
M.S. (1970) Physics, Polytechnic Institute of Brooklyn, New York City, NY, USA.
A.B. (1964) Physics, Barnard College, New York City, NY, USA.

Successful Women Ceramic and Glass Scientists and Engineers: 100 Inspirational Profiles. Lynnette D. Madsen.
© 2016 The American Ceramic Society and John Wiley & Sons, Inc. Published 2016 by John Wiley & Sons, Inc.

Research Expertise: microstructure characterization of ceramics, nanoscale materials, hyperuniform structures, metallic glasses, plasma-sprayed coatings, and fuel cells; materials deformation, small-angle X-ray and neutron scattering, X-ray optics, inelastic scattering, anomalous X-ray scattering, and dynamical diffraction by imperfect (real) crystals

Other Interests: working internationally

Key Accomplishments, Honors, Recognitions, and Awards

- Emeritus Scientist, National Institute of Standards and Technology, 2003–present
- Inducted into the NIST Portrait Gallery of Distinguished Alumni, 2014
- The Minerals, Metals and Materials Society Centennial Session in Honour of her 70th Birthday (Neutron and X-Ray Studies of Advanced Materials), 2012
- Appointed Member, Panel on Public Affairs of the American Physical Society, 2010–2012
- Appointed Member of the Science Council, Max Planck Institute at Stuttgart, Germany, 2008–2012
- Appointed Member of the Extended Science Council and Science Council of DESY, Hamburg, Germany, 2006–2011

Gabrielle Long in 2012. Photo by Knox Long.

- Elected Fellow, American Association for the Advancement of Science "for outstanding leadership in the development and application of advanced X-ray and neutron measurement techniques for materials analysis," 2004
- Appointed member of the Basic Energy Sciences Advisory Committee (BESAC), 2002–2007
- Maria Goeppert-Mayer Distinguished Scholar, Argonne National Laboratory, 2001–2002
- Department of Commerce Bronze Medal Group Award, 1998
- Elected Fellow, American Physical Society "for her sustained and significant contributions to the development and use of X-ray and neutron diffraction and spectroscopic techniques to studies of solids," 1991

Biography

Early Life and Education

Gabrielle Long was born and raised on a farm in New Jersey about 40 miles from New York City. She is the child of immigrant parents, who deeply valued learning and education. She was the first of her generation to want to go to college. After her freshman year at a public high school, her school lost its accreditation from the Mid-Atlantic States. Concerned that she would not be able to get into college, she asked her parents to send her instead to a nearby preparatory school. They agreed, and she continued her high school education at Rutgers Preparatory School, where she learned about the existence of the Seven Sister Colleges. She was accepted at Barnard (one of the Seven Sisters), where she decided to major in Physics, with a minor in French Literature. Campus life in physics at Columbia University in the 1960s was very exciting with lectures on the latest developments in particle physics. Through college and graduate school, her parents were consistently supportive. As she published research papers, and gave reprints to her parents, she found that her dad read every word, even though the topics and the discussion were totally foreign to him. She was deeply touched by his insistence that if she wrote research papers, he would read every one of them.

Career History

After obtaining her Ph.D. in Condensed Matter Physics, she took a postdoctoral appointment in X-ray astronomy at Columbia University from 1973 to 1976. She quickly returned to condensed matter topics, after she became an Assistant Professor at the Physics Department at Vassar College and she continued in this vein when she moved to the State University of New York at Stony Brook (now Stony Brook University). In 1980, she accepted a position in materials physics at the National Bureau of Standards, now the National Institute of Standards and Technology (NIST), where she spent most of her career. In 2003,

Gabrielle Long and long-time collaborator Lyle Levine, in front of the ultrasmall-angle X-ray scattering facility at the Advanced Photon Source. 2008. Photo by their collaborator Fan Zhang.

she retired from NIST to join Argonne National Laboratory and take on a position managing science at the Advanced Photon Source (APS).

While conducting an active research program at NIST, she also led the Materials Microstructural Characterization Group and later served as the (acting) Chief of the Ceramics Division. Her research projects covered a range of topics in materials science, unified by the goal of understanding the relationships between microstructure and materials' properties. Specific research interests included microstructure characterization of composite materials, ceramic and metallic alloys, glasses, nanoscale materials, plasma-sprayed coatings, and fuel cells. She also has long-standing interests in materials deformation, small-angle X-ray and neutron scattering, X-ray optics, X-ray inelastic scattering, anomalous X-ray scattering, and dynamical diffraction by imperfect (real) crystals.

Gabrielle Long is currently a senior advisor at Argonne's hard-X-ray synchrotron source, APS, where she started out as associate director of the Experimental Facilities

Division, and later became director of the new X-Ray Science Division in the APS. In these positions, she was responsible for all of the beam lines and experimental facilities that are operated by the U.S. Department of Energy.

During her career, she coauthored over 120 papers and has served as the lead editor of a volume on synchrotron radiation instrumentation. Her accomplishments include conceiving, building, commissioning, and operating unique national facilities at the National Synchrotron Light Source, Brookhaven National Laboratory, and the APS at Argonne National Laboratory. These facilities enable the atomic, molecular, microstructural, interfacial, and surface characterization of materials at levels previously unattainable. Notably, she has been a leader in the use of these facilities to further the understanding of materials.

When she is not working, Gabrielle has three main interests: fish-

Gabrielle Long with a mahi-mahi that she caught off the coast of Hawaii, 2012. Photo by husband Knox Long. (See insert for color version of figure.)

ing with her husband and family, traveling to distant and exotic parts of the world, and enjoying baroque and renaissance music.

3 Most Cited Publications

Title: Cavitation contributes substantially to tensile creep in silicon-nitride
Author(s): Luecke, WE; Wiederhorn, SM; Hockey, BJ; et al.
Source: Journal of the American Ceramic Society; volume: 78; issue: 8; pages: 2085–2096; published: August 1995
Times Cited: 102 (from Web of Science)

Title: Influence of spray angle on the pore and crack microstructure of plasma-sprayed deposits
Author(s): Ilavsky, J; Allen, AJ; Long, GG; et al.
Source: Journal of the American Ceramic Society; volume: 80; issue: 3; pages: 733–742; published: March 1997
Times Cited: 82 (from Web of Science)

Title: Microstructural characterization of yttria-stabilized zirconia plasma-sprayed deposits using multiple small-angle neutron scattering
Author(s): Allen, AJ; Ilavsky, J; Long, GG; et al.
Source: Acta Materialia; volume: 49; issue: 9; pages: 1661–1675; published: May 25, 2001
Times Cited: 79 (from Web of Science)

Sources Used for This Profile

- ANL: http://www.anl.gov/contributors/gabrielle-long, accessed December 9, 2014.
- Web of Science for publication data and information, accessed December 9, 2014.
- APS news: http://www.aps.anl.gov/News/APS_News/2004/20041014.htm, accessed December 9, 2014
- American Physical Society: http://www.aps.org/programs/honors/fellowships/archive-all.cfm?initial=L&year=2014&unit_id=&institution=, accessed December 9, 2014.
- Rochester Institute of Technology: http://www.rit.edu/cos/Long.php, accessed December 9, 2014.
- TMS, http://www.programmaster.org/PM/PM.nsf/ViewSessionSheets?Open Agent&ParentUNID=6F1EDAB10BA36A48852578E3006AD448, accessed December 9, 2014.

Challenges

The most significant challenge regarding careers was taken not by me, but by my husband. The problems we faced are common to two-career couples. We met when we both were postdoctoral associates at Columbia, and I was newly divorced with a young daughter. When I moved to the Washington, DC, area to take a position at the National Bureau of Standards, he left Columbia where he had become an Assistant Professor to take a job at Johns Hopkins University, so that we could marry and raise two children: my daughter and then later also our son. This meant that he had to change areas of specialization and leave a field where he had gained significant prominence. He eventually took a position with the Space Telescope Science Institute, which runs the Hubble Telescope, and rebuilt his career, but the sacrifice he made was significant.

Over the course of a long career, I have been fortunate to work with many superb collaborators, and some who were quite the opposite. In the latter cases, I found myself guided mostly by idealism, which could easily have gone very wrong. Instead, the directions I took turned out to be beneficial. One example involved a boss with whom I worked for many years, who was by turns very good and then the opposite. During the many good years, I included his name on the author list of every paper I wrote, whether he contributed a lot or almost nothing at all. Later on, he became jealous of my success and tried to discredit my work; unfortunately for him, all my papers included his name and so there was nothing he could do. After working for many years with people with large egos, I took the position that I do not need to work with difficult personalities as I have satisfied my lifetime quota requirement in that area. There is someone out there who is just as smart and is a pleasure to work with.

On being a woman in this field . . .

In a research discipline dominated by men, I have found that sometimes women are by turns very visible and overlooked. Sitting on committees, a lone woman's opinion may simply be ignored, if she is allowed to express it at all, but some of the banter is curbed by her presence. It's as if the other members of the committee know she is there, but do not recognize her as a scientist, only as a female.

Words of Wisdom

In the course of a science career, one's research may occasionally receive recognition in the form of awards. Such awards are, of course, very nice, and are certainly enjoyed by families and friends. But in the life of a scientist, the real excitement happens not with external recognition, but rather every day when we go to the laboratory and discover things that were never known before.

Profile 64

Judith L. MacManus-Driscoll

Professor of Materials Science
University of Cambridge
27 Charles Babbage Rd.
CB3 OFS Cambridge
Cambridgeshire
UK

e-mail: jld35@cam.ac.uk
telephone: +44 (0)1223 334468

Judith MacManus (as she was called before marrying) photographed in Oxford, UK in 1989.

Tags

- ❖ Academe
- ❖ Domicile: United Kingdom
- ❖ Nationality: British
- ❖ Caucasian
- ❖ Children: 2

Birthplace
London, UK
Born
May 25, 1966

Publication/Invention Record

>325 publications: h-index 44 (Web of Science)
10 patents

Proudest Career Moment (to date)

After having managed more than 50 researchers over the years, being contacted recently by my first ever researcher (a female research assistant), from over 20 years ago when I was a postdoctoral associate at Stanford, to work with her on a project closely related to what we began together all those years ago. She is a leading woman in a high tech company that is very satisfying to me. From the viewpoint of rekindling the research, it shows that you can never tell where research is going, that the work you do when you are at a junior level can be very important, and even if a research topic or paper remains in abeyance for a while, it may one day become a key, enabling technology.

Successful Women Ceramic and Glass Scientists and Engineers: 100 Inspirational Profiles. Lynnette D. Madsen.
© 2016 The American Ceramic Society and John Wiley & Sons, Inc. Published 2016 by John Wiley & Sons, Inc.

Academic Credentials

Ph.D. (1991) Materials Science, University of Cambridge, London, UK.
B.Sc. (1987) Engineering, Imperial College, London, UK.

Research Expertise: functional oxide materials for energy and electronics; oxide interfaces for electronics and photovoltaics; oxide nanocomposites for multiferroics, ferroelectrics, and superconductors

Other Interests: doing science shows and talks for girls (age 5–16)

Key Accomplishments, Honors, Recognitions, and Awards

- Editor, American Institute of Physics: Applied Physics Letters (APL) Materials, 2013 to present
- Long-term visiting staff member at Los Alamos National Laboratory, 2003 to present
- Fellow of Materials Research Society, 2015
- Joule Prize, Institute of Physics, 2015
- Fellow of the American Physical Society, 2011
- Fellow, Institute of Physics (UK), 2002

Judith Driscoll talking about the journal APL Materials, 2014.

Biography

Early Life and Education

She was educated at a single-sex Convent school in London from age 11 to 18. She thinks that single-sex education really favors girls in terms of building up their confidence, promoting them in maths and science, and allowing them to learn by a 'female' approach, which she believes to be very different from the male one. In addition, she believes the Christian Ethos of the school (delivered by women) was key to giving her purpose in her work and life, and the ability to be who and what she is

Judith Driscoll in 2004.

without worrying about what others think. With this background, she continued with her education.

She studied and completed her engineering degree at Imperial College in 1987 and then continued to graduate studies in materials science at the University of Cambridge. She was awarded a doctorate degree in 1991.

Career History

From mid-1991 to January 1995, she served as an IBM Postdoctoral Fellow at Stanford University and IBM Almaden Research Center. In February 1995, she accepted a position back in the United Kingdom as a Governor's Lecturer in the Department of Materials at Imperial College in London. She took two leaves of absence for maternity: September 1997 to March 1998 and July 1999 through January 2000. During her second leave period, in October 1999, she was promoted to Reader in Materials Science. In January 2003, she left Imperial to become a Lecturer in Materials Science at the University of Cambridge. Shortly thereafter, she took a leave of absence to go to work at the Los Alamos National Laboratory (LANL) as a Long Term Visiting Staff Member. She returned to Cambridge in October 2003, but remains a Visiting Staff Member at LANL to

Judith Driscoll hiking in Montana, summer 2005. (See insert for color version of figure.)

date. In October 2006, she was promoted to Reader in Materials Science at the University of Cambridge, and then, 4 years later, to Professor of Materials Science in October 2008. Along the way, in April 2005, she was appointed as a Fellow of Trinity College, where, for 7 years, she undertook much undergraduate student mentoring and intense small group teaching.

Her interests beyond work include outdoor activities such as hiking, skiing, running, and yoga, as well as spirituality.

3 Most Cited Publications

Title: Strongly enhanced current densities in superconducting coated conductors of $YBa_2Cu_3O_{7-x}+BaZrO_3$
Author(s): MacManus-Driscoll, JL; Foltyn, SR; Jia, QX; et al.
Source: Nature Materials; volume: 3; issue: 7; pages: 439–443; doi: 10.1038/nmat1156; published: July 2004
Times Cited: 594 (from Web of Science)

Title: ZnO: nanostructures, defects, and devices
Author(s): Schmidt-Mende, L; MacManus-Driscoll, JL

Source: Materials Today; volume: 10; issue: 5; pages: 40–48; doi: 10.1016/S1369-7021 (07)70078-0; published: May 2007
Times Cited: 593 (from Web of Science)

Title: Greatly reduced leakage current and conduction mechanism in aliovalent-ion-doped BiFeO$_3$
Author(s): Qi, XD; Dho, J; Tomov, R; et al.
Source: Applied Physics Letters; volume: 86; issue: 6; article number: 062903; doi: 10.1063/1.1862336; published: February 7, 2005
Times Cited: 535 (from Web of Science)

Challenges

As a young researcher, I found the culture very male, but I largely ignored it as I was so focused on my work. As soon as I had a tenured post, things were tougher as some men (but certainly not all) showed signs of suspicion of me, some made some negative comments, and some generally didn't seem to feel comfortable with an assertive/confident/capable woman around. I have found that most men born after around 1960 are much better and are not concerned with gender so much. So I think and hope things are improving for women a lot.

On being a woman in this field . . .

On the whole, we have to fit into a culture that was created by men. This means you won't always feel comfortable with it and often you will have to stand out as different, especially when you have small children and cannot be going to many conferences and do all the schmoozing that men like to do. But you have to simply stick with what you are doing and realize that being a good parent and having an exciting career is a real privilege—the combination that, in fact, is very challenging and requires determination to keep going. Personal sacrifices also need to be made, but you need to have a few relaxing treats to keep going.

Also, you have to love what you are doing. When you are young, you will have to put your head down and work very hard, follow your own ideas, don't become a side-kick to a more senior male, and be confident in what you are doing (even if it seems hard to be at the time). When you are older, you have to keep focused, remain confident as you take more leadership roles on, and realize that you need to change the culture to help the next generation of women to succeed.

Words of Wisdom

Be yourself, work on things that really interest you, follow your own ideas with passion, and as you move up the ladder, don't forget the challenges you faced along the way. A really great challenge is the childbearing years, but don't worry too much—do what you can and stick with it.

Profile 65

Lynnette D. Madsen

Vice-President
Svedberg Science, Inc.
and
Program Director
Division of Materials Research
National Science Foundation (NSF)
Suite 1065, 4201 Wilson Blvd.
Arlington, VA 22230
USA

e-mails: lynnette@svedbergscience.com;
lmadsen@nsf.gov
telephone: (703) 292-4936

Lynnette after receiving her
doctorate degree.

Tags

❖ Administration and
Leadership

❖ Government

❖ Domicile: USA

❖ Nationality: Canadian
and American

❖ Caucasian

❖ Children: 0

Publication/Invention Record

>75 publications: h-index 14 (Web of Science)
2 patents, 2 books, 2 book chapters

Proudest Career Moments (to date)

In 2007, I was selected for a National Science Foundation (NSF) Director's Award for Program Management Excellence for "outstanding performance in creating a strong portfolio of awards in the Division of Materials Research that reflects NSF's strategic goals." Although the prize money wasn't significant, it reflected on my job performance for the past 6 years, plus I had plans on continuing with NSF, so that boded well.

While working on this book (in 2013), I received the Alumni Achievement Medal for Professional Achievement from the Faculty of Engineering at the University of Waterloo for "vision and outstanding leadership in the areas of materials and nano-technology, as well as promotion of groups underrepresented in science and engineer-ing." It does not seem to matter how much time passes, it is always an honor to be recognized by your alma mater.

Successful Women Ceramic and Glass Scientists and Engineers: 100 Inspirational Profiles. Lynnette D. Madsen.
© 2016 The American Ceramic Society and John Wiley & Sons, Inc. Published 2016 by John Wiley & Sons, Inc.

Academic Credentials

Ph.D. (1994) Materials Science, McMaster University, Hamilton, Ontario, Canada.
M.Eng. (1988) Electronics, Carleton University, Ottawa, Ontario, Canada.
B.A.Sc. (1986) Electrical Engineering, University of Waterloo, Waterloo, Ontario, Canada.
B.A. (2004) Psychology, University of Waterloo, Waterloo, Ontario, Canada.

Research Expertise: glass, nitrides, carbides, intercalation compounds, complex oxides

Other Interests: increasing diversity in science and engineering

Key Accomplishments, Honors, Recognitions, and Awards

- Advisory Board of the Rosalind Franklin Society, 2008 to present
- The Minerals, Metals & Materials Society (TMS) Ellen Swallow Richards Diversity Award, 2016
- Author of this book, 2016
- Plenary or invited talks in eight countries (Canada, China, Czech Republic, Ireland, Italy, Sweden, Tanzania, and the United States)
- Elected American Vacuum Society (AVS) Trustee, 2014–2016
- Fellow, The American Ceramic Society, elected 2015
- Accreditation Board for Engineering and Technology (ABET) Claire L. Felbinger Award for Diversity, 2015
- NSF Performance Awards, 2005–2012 and 2014–2015
- Chair of Focus Topic "Accelerating Materials Discovery for Global Competitiveness" at AVS 60th International Symposium, 2013; Co-Chair, 2014 and 2015
- AVS Recognition for Excellence in Leadership, 2014
- Professional Achievement, University of Waterloo Faculty of Engineering Alumni Achievement Medal, 2013
- The Society of Hispanic Professional Engineers (SHPE) Junipero Serra Award, 2013
- Elected Chair of the Art, Archeology, and Cultural Science Division of The American Ceramic Society, 2013
- Lead speaker at the CAREER Workshop: International Conference on Advanced & Nano Materials, 2013
- Co-Chair of Focus Topic "Nanomanufacturing" at AVS 58th and 59th International Symposia, 2011–2012
- Plenary speaker at the Electronic Materials and Applications Meeting, 2011
- Chair of Focus Topic "Energy Science and Technology" at AVS 55th International Symposium, 2008
- NSF Director's Award for Equal Opportunity Achievement, 2008
- NSF Director's Award for Program Management Excellence, 2007
- Plenary speaker at the 1st International Congress on Ceramics, 2006
- NSF Incentive Award for Timely Program Management, 2004

- National Academy panelist, 2004
- Committee member for doctoral student (T. Jang) at Carnegie Mellon University, 1999–2001
- Chair of Ph.D. examination committee for doctoral student (Thomas Kallstenius) at Uppsala University, 2000; defense conducted in Swedish
- Docent in Materials Science, Linköping University, 1999
- Co-guest editor of special issue: Vacuum, 1998
- Co-guest editor of special issue: Microscopy Research and Technique, 1997
- Natural Sciences and Engineering Research Council (NSERC) Postdoctoral Fellowship, 1996–1997
- External examiner for Master's thesis (C. Cohen-Jonathon from Université Joseph Fourier/Ecole de Physique de Grenoble Magistére), 1994
- Nortel Networks Postgraduate Scholarship, 1989–1994
- Materials Research Society Presentation Award (Ferroelectrics Symposium), 1993
- Ontario Centre for Materials Research Industrial Cooperative Ph.D. Scholarship, 1989–1991, 1993
- NSERC Postgraduate Award, 1990–1991
- Nortel Networks Research Agreement for Graduate Studies, 1987–1988
- Carleton University Engineering Department Scholarship, 1987

Biography

Early Life and Education

Lynnette Madsen and her sister were raised in Canada by Canadian parents. Lynnette received her early education in the Ontario public school system and also continued with her higher education in Ontario.

Her undergraduate studies in electrical engineering at the University of Waterloo were part of the cooperative education program. Consequently, she enjoyed six work terms (totaling 2 years) at Canada Post Corporation, Transport Canada, Bell-Northern Research, and Nortel Networks while completing her first degree.

She continued to work at Nortel Networks and then completed her Master's degree in engineering electronics at the nearby Carleton University in Ottawa. This effort was supported through a Nortel Networks Research Agreement for Graduate Studies and a Carleton University Engineering Department Scholarship.

After working for an additional period, she secured three lines of financial support: a Nortel Networks Postgraduate Scholarship, an Ontario Centre for Materials Research Industrial Cooperative Ph.D. Scholarship, and a Natural Sciences and Engineering Research Council (NSERC) Postgraduate Award to undertake doctoral studies. She completed her doctorate in yet another discipline, materials science and engineering, and was awarded a Ph.D. from McMaster University in 1994. Her coursework included graduate courses in physics, mathematics, and materials science and engineering from Carleton University, The University of Ottawa, and The University of Toronto as well as McMaster University.

Her research was conducted in the laboratories of Nortel Networks and Queen's University. In total, from her undergraduate cooperative education days through to her doctoral studies, she spent roughly a decade working in industry at Nortel Networks in Canada.

It was after completing her doctorate that she finished her coursework for a B.A. in Psychology from the University of Waterloo; she specialized in industrial and organizational psychology and education.

Career History

After receiving her doctorate, she lived in Sweden for a number of years. Her initial appointment in 1994 was as a Postdoctoral Associate followed by a period as a NSERC Postdoctoral Fellow from 1996 to 1997 in conjunction with an appointment at the University of Illinois Urbana-Champaign. In 1997, she accepted a position as Assistant Professor at Linköping University. By the time she left Linköping University in 2001, she had been promoted to Docent (Associate Professor). From 1999 to 2002, she held a Visiting and Adjunct Faculty position at Carnegie Mellon University.

In 1994, I missed my Ph.D. convocation ceremony at McMaster University because I was hosting a workshop at the same time in Linköping, Sweden while I was a Postdoctoral Associate.

Dr. Madsen joined the NSF as a Program Director late in 2000. In addition to recommending the annual distribution of Ceramics Program budget (~$11 million), she conducts independent research and serves as a program representative for nanomanufacturing, graduate research supplements to broaden participation of underrepresented groups, and interactions with the National Institute of Standards and Technology (NIST). Previously, she led new cooperative activities with European researchers and was involved with energy, sustainability, and education issues.

Dr. Lynnette Madsen is also Vice President of Svedberg Science, Inc.—a small business that provides consulting about scientific research directions and trends.

She is an active member of the American Ceramic Society, AVS, American Association for the Advancement of Science (AAAS), and Materials Research Society (MRS).

In 1999, as a faculty member I was invited to participate in the Commencement Ceremony at Linköping University. I was wearing the McMaster academic hood and gown for the first time.

3 Most Cited Publications

Title: Surface damage formation during ion-beam thinning of samples for transmission electron microscopy
Author(s): McCaffrey, JP; Phaneuf, MW; Madsen, LD
Source: Ultramicroscopy; volume: 87; issue: 3; pages: 97–104; doi: 10.1016/S0304-3991(00)00096-6; published: April 2001
Times Cited: 110 (from Web of Science)

Title: Microstructure and electronic properties of the refractory semiconductor ScN grown on MgO(001) by ultra-high-vacuum reactive magnetron sputter deposition
Author(s): Gall, D; Petrov, I; Madsen, LD; et al.
Source: Journal of Vacuum Science & Technology A: Vacuum Surfaces and Films; volume: 16; issue: 4; pages: 2411–2417; doi: 10.1116/1.581360; published: July–August 1998
Times Cited: 79 (from Web of Science)

Title: Determination of the size, shape, and composition of indium-flushed self-assembled quantum dots by transmission electron microscopy
Author(s): McCaffrey, JP; Robertson, MD; Fafard, S; et al.
Source: Journal of Applied Physics; volume: 88; issue: 5; pages: 2272–2277; article number: PII S0021-8979(00)03217-5; doi: 10.1063/1.1287226; published: September 1, 2000
Times Cited: 43 (from Web of Science)

ResearcherID (F-8323-2012)

Challenges

There have been plenty of challenges in my career, beyond the scientific ones. When I look back, I recall that moving countries and changing jobs were both sources of joy and challenges. It was difficult to have a distance marriage for many years—if I had a "do-over," I would change this aspect. Rarely did I feel that I had all the mentors that I would have liked—there seemed to be a lot of breaking ground on my own. It is due to this lack of mentorship that most of my early publications appear in conference proceedings instead of journals. Writing every paper was (and still is) a challenge—it is the moment of truth where all is revealed and it is etched in stone (or so it seems at the time). Completing an article always brought great joy and pride.

On being a woman in this field . . .

I would like to say that being a woman is no different than being a man, but I know there have been both big and small differences throughout my career. For example, as an undergraduate student, I was one of six women in a class of 100 and consequently the professors always knew who I was. Throughout my career, visibility is a double-edged sword. While I am sure I have received breaks due to being a woman, I have also been

Dean Pearl Sullivan presenting Engineering Professional Achievement Alumni Award to Lynnette Madsen in 2013. Photo by Neil Trotter of Studio 66; courtesy of the University of Waterloo.

judged overly harshly at other times for precisely the same reason—it is with optimism that I believe that these two opposing factors balance. However, overall things are still skewed for women and it is one of the reasons motivating the writing of this book.

Often, it seems my voice is not heard—and yes, sometimes a man repeats what I said earlier, without acknowledgement, like a fairy had just whispered the brilliant idea into his ear. However, I do have a fond memory of one meeting at Nortel Networks where my senior, esteemed Ph. D. advisor kept repeating what I recently said, while both reinforcing the idea and giving me full credit. It was nice to see how a strong advocate ensured that my voice was heard.

Early on I knew I didn't want to have kids—this factored into my decision to leave Sweden (a simply superb country for raising children with more than a year of family leave per child) to take a U.S. federal government position (a country without a national law mandating paid time off for new parents). In addition, I have commuted long distances and moved often since getting married more than 15 years ago... and yes, so has my husband. The decisions are more complicated when two individuals and two careers are involved, but I would not have it any other way. The creative thinking that goes into managing the issues that arise with internationally mobile dual-careers sometimes has had surprisingly positive impacts.

Words of Wisdom

I participated in an American Ceramic Society outreach session for new graduates—postdoctoral associates and early career faculty and industry scientists and engineers—with Prof. Delbert Day, Dr. Marina Pascucci, and Dr. Jay Lane, and as well I recently attended a Women of Color Workshop at the National Academies. I am left with the following impressions:

- Believe in yourself and act accordingly.
- Work should be fun. Make changes in your career path to make it exciting and enjoyable. If you don't want to get up and go to your job for many days in a row, then it is time for a change—either with the same employer or a new one.

- Make sure you are living your dream and not just fulfilling someone else's.
- Risk is the zest of life.
- There is no fixed formula for success. Make a game plan and move forward with it.
- In terms of your career, it is important to hone into what is important to you and what is deemed important to those who assess your progression (success). If you are an academic, then publish great results regularly and each time in the best journal for that work. Don't assume people will see it—send your paper to them and/or present your work at conferences and other venues.
- Everyone knows something that you need to know; respect all of your coworkers.
- You are often told to network—what does it really mean? For me it entails taking the time and energy to interact with others in my field, i.e., getting to know them and helping where I can. In the long run, you will find yourself with new close friends and an array of acquaintances that can be called upon in times of need for support, advice, or help. It also means getting engaged in professional society activities where you have an interest and can make a contribution. These fantastic scientific conferences provide a way to maintain our focus on science beyond the day-to-day matters at work, while nourishing our professional network.

Profile 66

Merrilea J. Mayo

Mayo Enterprises, LLC consultancy
12101 Sheets Farm Rd.
North Potomac, MD 20878
USA

e-mail: merrileamayo@gmail.com
telephone: (301) 977-2599

Merrilea Mayo during her
years as an assistant
professor at Penn State.
From the MRS Bulletin,
January 2003 issue, p. 3,
in an editorial she wrote
as the then-MRS
president.

Birthplace
Syracuse, NY, USA
Born
September 21, 1960

Publication/Invention Record

>75 publications: h-index 23 (Web of Science)

Proudest Career Moment (to date)

Tags

❖ Administration and Leadership
❖ Non-Profit Organizations
❖ Domicile: USA
❖ Nationality: American
❖ Caucasian
❖ Children: 1

Co-founding two non-profits, Alliance for Research in Science and Technology for America
(ASTRA) and University-Industry Demonstration Partnership (UIDP); also, sparking the
creation of the Strange Matter museum exhibit (http://www.strangematterexhibit.com/),
which has traveled the United States for nearly a decade now.

Academic Credentials

Ph.D. (1988) Materials Science and Engineering, Stanford University, Stanford,
California, USA.
M.S. (1984) Materials Science and Engineering, Stanford University, Stanford,
California, USA.
B.S. (1982) Materials Science, Brown University, Providence, Rhode Island, USA.

Research Expertise: nanotechnology, development of fully dense bulk ceramics
with sub-100-nm grain sizes by sinter forging, pressureless sintering, and alternate

Successful Women Ceramic and Glass Scientists and Engineers: 100 Inspirational Profiles. Lynnette D. Madsen.
© 2016 The American Ceramic Society and John Wiley & Sons, Inc. Published 2016 by John Wiley & Sons, Inc.

densification routes, using a combination of experimental work and analytical modeling of the sintering/deformation processes

Other Interests: innovation, workforce, technology and the future of learning

Key Accomplishments, Honors, Recognitions, and Awards

- ACT Foundation Fellow, 2013
- Advisory Board, Center for Hierarchical Manufacturing, Amherst, 2006–2007
- Board of Advisors, Department of Chemical Engineering and Materials Science, 2005–2007
- Science and Engineering Research Council, University Space Research Association, 2005–2007
- President, Materials Research Society, 2003
- Optical Society of America (OSA)/ Materials Research Society (MRS) Congressional Science and Engineering Fellow, 1998–1999
- Multiple officerships/committeeships in the Materials Research Society: Foundation Task Force (2012), Nominating Committee (2009), Entrepreneurship Challenge (2005), Headquarters Building Task Force (1996–1997), Treasurer (1999–2002), Executive Committee (1999–2002), Long-Range Planning Committee (1996), External Affairs/ Public Affairs Committee Vice-President (1997), Councilor (1994–1996), Meeting Chair (1993)
- National Science Foundation Presidential Young Investigator Award, 1991–1996
- Fellowship: Japan Society for the Promotion of Science, 1993
- Fellowship: Exxon Foundation, 1982–1984

Merrilea Mayo at the reception to honor the ACT Foundation Fellows, in 2013. This appointment honored her work in "working learners." Much of Dr. Mayo's later career centered on using engineering analyses and tools to arrive at "elegant solutions to intractable [social] problems." (See insert for color version of figure.)

Biography

Early Life and Education

Though born in upstate New York, Merrilea Mayo was raised in the suburbs of Dallas, Texas. She attended a private all-girls' school (Hockaday) for 12 years and launched her technical career with a series of summer jobs, starting the year after high school graduation and continuing through the year after college. She began as a student research assistant at the University of Texas, Dallas; was subsequently a two-summer hire at Texas Instruments (where she wrote her first publication, on the computer-aided scheduling package she developed); spent another summer as a research assistant for Joe Gurland at Brown University, and ultimately was a summer hire at Exxon, working as a metallurgist in Exxon's research laboratories in Linden, New Jersey, before heading off to Stanford for her M.S. and Ph.D.

Career History

Dr. Mayo's work includes 2 years at Sandia National Laboratories, and 11 years on the faculty of Pennsylvania State University. She had a founding role in both ASTRA (an advocacy organization for the physical sciences) and the University-Industry Demonstration Partnership (an organization devoted to improving the ability of companies and universities to conduct joint research).

In 1998, she was named the OSA/MRS Congressional Science and Engineering Fellow. Her hope during her sabbatical year was to raise the level of political awareness within professional societies and to take her familiarity with the national laboratory and university systems, as well as technological concerns of the military and start-up companies to Washington, D.C.

From 2001–2006, Dr. Mayo was Director of the Government-University-Industry Roundtable (GUIRR) of the National Academies. GUIRR focused on trisector issues such as the science and engineering workforce, intellectual property rights, the impacts of globalization on national competitiveness, STEM education, deemed exports regulations, and national-laboratory university collaborations.

Dr. Mayo then took the position of Director of Future of Learning Initiatives at the Ewing Marion Kauffman Foundation. Her portfolio there included video game-based learning, educational virtual worlds, and cell-phone-delivered learning.

Dr. Mayo founded Mayo Enterprises, LLC consultancy in 2009. In this capacity, she has been assisting Innovate+Educate (I + E) and ROI Ventures (ROI) on workforce and education-related strategies for the W.K. Kellogg Foundation. Other projects undertaken include writing an Intel-funded science policy report for the Information Technology and Innovation Foundation (ITIF), providing advice on game-based learning to the Susan Crown Exchange, and assisting the Teaching Institute for Excellence in STEM (TIES) with an overarching strategy for education at NASA, under request of the Office of Administrator.

Her most time-consuming hobby, at about 4–5 h a day, is playing massive multiplayer online video games, a hobby that resulted in her authoring a *Science* article on game-based learning.

Sources: http://www.mrs.org/mrs-osa-fellows/and http://www.zoominfo.com/p/Merrilea-Mayo/8663776.

3 Most Cited Publications

Title: Processing of nanocrystalline ceramics from ultrafine particles
Author(s): Mayo, MJ

Source: International Materials Reviews; volume: 41; issue: 3; Pages: 85–115; published: 1996
Times Cited: 283 (from Web of Science)

Title: Mechanical-properties of nanophase TiO_2 as determined by nanoindentation
Author(s): Mayo, MJ; Siegel, RW; Narayanasamy, A; et al.
Source: Journal of Materials Research; volume: 5; issue: 5; pages: 1073–1082; published: May 1990
Times Cited: 234 (from Web of Science)

Title: A micro-indentation study of superplasticity in PB, SN, and SN-38 WT-percent-PB
Author(s): Mayo, MJ; Nix, WD
Source: Acta Metallurgica; volume: 36; issue: 8; pages: 2183–2192; published: August 1988
Times Cited: 161 (from Web of Science)

Challenges

The most frustrating challenge I had was being shut out socially, which made it just harder to do everything that relied on insider knowledge. It often took literally weeks or months to get information I needed to do my job, because very few people at the University would talk to me about anything meaningful, and back then there was no Internet. The upside of this challenge is that years later, I am very, very good at finding information that supposedly does not exist or that no one is talking about. This has led to an entirely different career path.

As one example, I arrived in academia realizing I had to write grants to survive. I also knew the National Science Foundation gave out grant money. I also knew that many of my colleagues had gotten money from NSF. So, I asked a few of these well-funded colleagues if they could give me NSF's phone number so I could call NSF and figure out what grant programs they had to offer (this was before the Internet). Each colleague I asked looked around frantically as if he were trying to escape the conversation and told me he had no idea and didn't know how to get it. How was this remotely possible if they had existing grants, as was clear from the department's funding profile, and had to deliver NSF grant reports? I was clearly getting the runaround. So, I called 411 (directory assistance) and asked for the phone number for the National Science Foundation, figuring I could at least get the main operator, and maybe that person could tell me something useful, like what kinds of divisions NSF had and therefore who might be appropriate to talk to. Oh, no! Even 411 says NSF doesn't exist (much later I realized NSF had just moved from Washington, DC to Arlington, VA, which is why they weren't in the DC directory anymore). But that was clearly ridiculous. So, then I called up our contracts and grants office. I asked them, "Do you process faculty grants to NSF?" They said yes. So I asked, "Would it be possible for you to tell me NSF's phone number? It's got to be on some of the grant applications." They said no, grant information like that was privileged. But, in the end, they were at least able to tell me NSF had moved to Arlington, VA, which then helped me to get NSF's number through 411. So, moral of the story, it took a week to get a simple piece of information like a phone number that would have taken 5 s of someone's time. But I didn't have the social standing to get it. Ergo, I needed some way to get into the National Science Foundation.

On being a woman in this field . . .

It is a bizarre experience to be the token anything. Your successes are so much more visible, your mistakes so much more visible also. As the only woman in, say, a conference of 3000 men, every single man there will remember you for years, and when they eagerly reintroduce themselves to you, years later, you're thinking: "Oh my gosh, I have no idea who you are; how can you possibly remember me from all those people at all those conferences?" But they will remember every aspect of your presentation and your delivery. People also take you as the "data point of one" and can't help themselves but extrapolate your specific idiosyncrasies to the entire gender. One day, after becoming a new faculty member at Penn State, I overheard the administrative staff discussing how "flat heeled shoes were 'in' as professional dress," and realized I was the only one I knew wearing such shoes—and that was only because I'm just klutzy in heels. My inability to wear heels translated to an unstated dress code for the entire department's female staff, without my even realizing it. My only hope is that I saved a few feet.

Words of Wisdom

You will have many responsibilities to balance, and sometimes it seems physically impossible to do them all. However, the energy you have at the end of the day is proportional to the RATIO of fun vs. slog you experienced in the day. If you have one absolutely awful task, and you do nothing else that day, you will have had an awful day and feel drained and miserable at the end of it. If you have an absolutely awful task, and you also do five other things that were interesting and fun, you will have a very upbeat day and feel charged at the end of it. Never give up the activities that give you energy, under the false assumption that doing so will save you time. Those are the only things that keep you going, and make it possible to accomplish the rest.

Profile 67

Janina Molenda

Professor
Department of Hydrogen Energy
Faculty of Energy and Fuels
AGH University of Science and Technology
al. Mickiewicza 30, 30–059 Krakow
Poland

e-mail: molenda@agh.edu.pl
telephone: +48 12 617 2522

Prof. Janina Molenda, in
her 30s.

Birthplace
Tarnów, Poland

Born
February 5, 1952

Publication/Invention Record
>150 publications: h-index 22 (Web of Science)
2 patents, editor of 5 special issues in scientific journals,
3 book chapters

Tags
❖ Administration and
  Leadership
❖ Academe
❖ Domicile: Poland
❖ Nationality: Polish
❖ Caucasian
❖ Children: 2

Pivotal Moment (to date)

A central moment of my career is tied to the development of the electronic model of the intercalation of alkaline ions into host transition metal oxides. The developed method, involving examination of the Li^+/Li_xMO_2 potential by means of measuring the electromotive force of the $Li/Li^+/Li_xMO_2$ cell, is an excellent experimental tool in solid-state physics, which allows for a direct observation of changes of the Fermi level in these systems, in relation to the lithium content. Today, the solid-state physics research community agrees that the step-like character of the discharge curve in the Na_xCoO_{2-y} system is purely electronic in nature and reflects abrupt changes in the position of the Fermi level in the anomalous electronic structure of Na_xCoO_{2-y} caused by the oxygen non-stoichiometry. This discovery has a universal character, and is of great importance for designing and searching for new electrode materials for Li and Na ion batteries.

Academic Credentials

D.Sc. (1988) Materials Science, AGH University of Science and Technology, Krakow, Poland.

Successful Women Ceramic and Glass Scientists and Engineers: 100 Inspirational Profiles. Lynnette D. Madsen.
© 2016 The American Ceramic Society and John Wiley & Sons, Inc. Published 2016 by John Wiley & Sons, Inc.

Docteur d' Universite (1984), Materials Science, Université Bordeaux I, Bordeaux, France
Ph.D. (1979) Chemistry, AGH University of Science and Technology, Krakow, Poland.
M.Eng. (1975) Materials Science, AGH University of Science and Technology, Krakow, Poland.

Research Expertise: solid oxide fuel cells, lithium ion batteries, nonstoichiometric compounds, fundamental studies of transport properties of solids, metal–insulator transitions, high-temperature superconductors, electrochromic effect, fundamental studies of intercalated cathode materials

Other Interests: teaching

Key Accomplishments, Honors, Recognitions, and Awards

- Representative of Poland to the States Representative Group to the Fuel Cells and Hydrogen 2 Joint Undertaking, 2014 to present
- Vice-President of Polish Hydrogen and Fuel Cell Association, 2013 to present
- Head of the Department of Hydrogen Energy in the Faculty of Energy and Fuels at AGH University of Science and Technology in Krakow, 2009 to present
- Coordinator of High-Temperature Fuel Cells Group in Polish Hydrogen and Fuel Cell Technology Platform, 2004 to present
- President of Polish Hydrogen and Fuel Cell Association, 2004–2013

Biography

Early Life and Education

When I was an undergraduate student at AGH University of Science and Technology, I made my first selection in terms of research areas. At that time I decided to focus my interest on semiconducting oxides. This choice was more intuitive than conscious; however, it turned out to be very fruitful since it marks the start of interest in looking at materials with a focus on their electronic properties. Later on, during my Ph.D. studies, I worked on semiconducting properties of ferrous sulfide at high temperatures under the supervision of Prof. S. Mrowec. That work revealed a fundamental correlation between electronic structure, nonstoichiometry, and ionic defect structure in ceramic compounds.

Career History

Upon completion of my Ph.D. thesis in 1979, I was employed as a researcher at AGH University. Soon after, I was

Prof. Janina Molenda, in her office.

accepted for a postdoctoral position at the Université de Bordeaux in the laboratory led by Prof. Paul Hagenmuller. I was assigned to a Solid State Electrochemistry group and I was inspired into new areas of research, especially, the intriguing topic of transition metal oxides intercalated with alkaline metal ions. My previous experience in studying electronic properties of solids allowed me to take a fresh approach, different than what was commonly presented in the literature. In searching for the mechanism of intercalation process and related phenomena, I developed an electronic model of electrochemical intercalation that linked variations of the cell's voltage with changes in the chemical potential of electrons in the cathode material. This model triggered a lively discussion and, finally, it was fully accepted by the scientific community.

Prof. Janina Molenda showing that her research still gives her a lot of excitement

After this success, I began intensive studies on numerous transition metal oxides in the context of physics and chemistry of intercalation process. My research was directed toward materials suitable to serve as electrode materials in Li or Na ion batteries. For example, this set included the following groups of compounds: Na_xCoO_2, Li_xCoO_2, Li_xNiO_2, $Li_xCo_{1-y}Ni_yO_2$, Li_xVO_2, Li_xWO_3, $Li_xYBa_2Cu_3O_{7-\delta}$, $LiMnO_2$, $LiMn_2O_4$, and $LiFePO_4$.

Using my previous experience, in the 2000s, I turned my attention to materials for high-temperature solid oxide fuel cells. Also, in this field, a detailed understanding of the ion and electron transport mechanisms contributed to significant advancements in developing perovskite-structured cathode materials.

At present, my interest is focused on layered transition metal oxides, in order to develop high-performance cathode materials for lithium and sodium batteries, as well as the technological scale-up of production of carbon-coated nano-sized $LiFePO_4$ as a cathode material for Li ion batteries.

Prof. Janina Molenda at the age of 60

In my free time, I enjoy floriculture, music, and art.

3 Most Cited Publications

Title: Modification in the electronic structure of cobalt bronze Li_xCoO_2 and the resulting electrochemical properties
Author(s): Molenda, J; Stoklosa, A; Bak, T
Source: Solid State Ionics; volume: 36; issue: 1–2; pages: 53–58; published: October 1989
Times Cited: 130 (from Web of Science)

Title: Electrical conductivity and reaction with lithium of $LiFe_{1-y}Mn_yPO_4$ olivine-type cathode materials
Author(s): Molenda, J; Ojczyk, W; Marzec, J
Source: Journal of Power Sources; volume: 174; issue: 2; pages: 689–694; published: December 6, 2007
Times Cited: 69 (from Web of Science)

Title: Functional materials for the IT-SOFC
Author(s): Molenda, J; Swierczek, K; Zajac, W
Source: Journal of Power Sources; volume: 173; issue: 2; pages: 657–670; published: November 15, 2007
Times Cited: 64 (from Web of Science)

Challenges

One challenge I face is to develop an efficient tool that will enable the design of high-performance lithium batteries.

On being a woman in this field . . .

In my view, in the field of science, differentiating between men and women is pointless. Women can make great discoveries in research, on par with men, or better.

Words of Wisdom

It is not your sex, but your wisdom, intelligence, and creativity that make you successful in both work and life in general.

Eliana Navarro dos Santos Muccillo

Senior Researcher
Center of Materials Science and Technology
Energy and Nuclear Research Institute
P.O. Box 11049
São Paulo 05422-970, SP
Brazil

e-mail: enavarro@usp.br
telephone: +55 11 31339203

Eliana Muccillo operating a thermal analysis instrument in the late 1990s.

Birthplace

São Paulo, SP, Brazil

Publication/Invention Record

>100 publications: h-index 17 (Web of Science)
2 patents, 1 book chapter

Tags

❖ Government
❖ Domicile: Brazil
❖ Nationality: Brazilian
❖ Children: 1

Proudest Career Moments (to date)

During my graduation in physics, I was selected to undertake research work with a renowned scientist, who was a member of the group responsible for the operation of the first nuclear research reactor in Brazil. After my Ph.D., I was invited to be member of an Excellence Center on Materials Science Research for research on nanostructured materials, sponsored by the Brazilian National Research Council. In 2013, I was invited to join a research group within the State of São Paulo recognized by the State Foundation as an Excellence Center on Multifunctional Materials.

Academic Credentials

Ph.D. (1993) Science, University of São Paulo, São Paulo, Brazil.
M.Sc. (1978) Nuclear Technology, University of São Paulo, São Paulo, Brazil.
B.Sc. (1974) Physics, Pontifical Catholic University, São Paulo, Brazil.

Research Expertise: solid electrolytes for energy production and sensors

Other Interests: advancing electroceramics for environmentally friendly energy production

Successful Women Ceramic and Glass Scientists and Engineers: 100 Inspirational Profiles. Lynnette D. Madsen.
© 2016 The American Ceramic Society and John Wiley & Sons, Inc. Published 2016 by John Wiley & Sons, Inc.

Key Accomplishments, Honors, Recognitions, and Awards

- Member of the Editorial Board of the journal Ceramica, 2008 to present
- Brazilian National Research Council Scholarship (Productivity in Research), 2000 to present
- Docent of the post-graduation program at the University of São Paulo, 1998 to present
- Coordinator of more than 20 research proposals for research and scholarships of the State Foundation (FAPESP) and National Research Council (CNPq), 1996 to present
- Guest Co-Editor of a special issue of Advanced Materials Research, 2014
- Reviewer of research proposals of National Science Foundation (USA), 2014
- Member of the organizing committee of the Brazilian Ceramic Congress, 2010–2014
- Vice Director, Center of Materials Science and Technology, Energy and Nuclear Research Institute, São Paulo, Brazil, 2009–2013
- Best work award of the 58th Brazilian Ceramic Congress, 2012
- Member of the Organizing Committee of the Latin American Conference on Powder Technology, 2005–2011
- State Governor Award of the national invention, 2001
- Energy and Nuclear Research Institute Scholarship, 1975–1976
- Pontifical Catholic University Scholarship, São Paulo, 1974

Biography

Early Life and Education

Eliana Navarro S. Muccillo was born in Brazil, and is the daughter of Spanish and Portuguese descendent father and mother, respectively. She has two brothers and one sister (now deceased). She attended fundamental and high school at public schools of the state of São Paulo.

During her undergraduate studies in physics at the Pontifical Catholic University in São Paulo, she started her scientific career at the Energy and Nuclear Research Institute (then called the Atomic Energy Institute). After obtaining her B.Sc. degree in physics, she received a scholarship for the M.Sc. program at the same institute.

In 1980, after completing her Master's degree, she developed technological research work on radiation dosimetry and nuclear fuels at the Center of Materials Science and Technology. About 8 years later, she refocused her career on the electrochemistry of solids. She spent 6 months undertaking research at the Max Planck Institut für Festkörperforschung (Solid State Research) in Stuttgart, Germany. She also collected data at the Laboratoire d'Ionique et d'Électrochimie des Solides (LIESG, now Laboratoire d'Électrochimie et de Physicochimie des Matériaux et des Interfaces (LEPMI)) in Grenoble, France to obtain her doctorate degree in sciences at the University of São Paulo, Brazil.

Career History

The most relevant work after her doctorate is related to teaching at the graduate level and conducting research in materials science. For the latter, she has had financial grants alone or from joint proposals.

Nowadays she is in charge of disciplines in the graduate courses of the University of São Paulo. She is also an active member of the Brazilian Ceramic Society, American Ceramic Society, and Materials Research Society. She is a reviewer of several journals in the materials science area and in the evaluation of Brazilian and foreign research proposals. She has already completed the supervision of 13 M.Sc. dissertations and 6 doctoral theses.

Eliana Muccillo, ca. 2005, after joining the Excellence Group on Nanostructured Materials of the Brazilian National Research Council.

3 Most Cited Publications

Title: Physical and chemical properties of nanosized powders of gadolinia-doped ceria prepared by the cation complexation technique
Author(s): Rocha, RA; Muccillo, ENS
Source: Materials Research Bulletin; volume: 38; issue: 15; pages: 1979–1986; published: December 10, 2003
Times Cited: 45 (from Web of Science)

Title: Electrical properties of YSZ/NiO composites prepared by a liquid mixture technique
Author(s): Esposito, V; de Florio, DZ; Fonseca, FC; et al.
Source: Journal of the European Ceramic Society; volume: 25; issue: 12; pages: 2637–2641; published: 2005
Times Cited: 41 (from Web of Science)

Title: Synthesis, sintering and impedance spectroscopy of 8 mol% yttria-doped ceria solid electrolyte
Author(s): Tadokoro, SK; Porfirio, TC; Muccillo, R; et al.
Source: Journal of Power Sources; volume: 130; issue: 1–2; pages: 15–21; published: May 3, 2004
Times Cited: 40 (from Web of Science)
ResearcherID (C-2959-2012)

Challenges

The main challenges were the adaptation of different steps in the career moving from the traditional/basic physics education to radiation physics for my M.Sc. work to undertaking a doctorate in materials science. Some experiments of a technological nature were conducted in the Department of Nuclear Metallurgy. Special efforts had to be made to overcome the usual difficulties when changing research areas. Teaching and guiding students in research across different fields—chemistry, physics, and engineering—represented another challenge. Finally, other challenges surfaced when supervising students with different backgrounds.

On being a woman in this field . . .

My work in the materials science area is very similar to the work of my colleagues in other areas of knowledge. However, we are faced with greater and tougher challenges than our female colleagues in other areas (such as the humanities) because this scientific area has traditionally been embraced and dominated by men.

Words of Wisdom

I always say to my students when they go abroad: "carry out your work thinking it is the last time; at the end you will reach great success (at least personally)."

Alexandra Navrotsky

Distinguished Interdisciplinary Professor of
Ceramic, Earth, and Environmental Materials
Chemistry
Edward Roessler Chair in Mathematical and
Physical Sciences
Peter A. Rock Thermochemistry Laboratory
and NEAT ORU
Department of Chemical Engineering and
Materials Science
University of California at Davis
Davis, CA 95616-8779
USA

e-mail: anavrotsky@ucdavis.edu
telephone: (530) 752–3292

Alex Navrotsky and her adviser, Ole
Kleppa, when she received her
doctorate in 1967.

Birthplace
New York, NY, USA

Born
June 20, 1943

Publication/Invention Record
>700 publications: h-index 58 (Web of Science)
3 patents, editor of 7 books, 1 book

Tags
❖ Academe: Research
❖ Domicile: USA
❖ Nationality: American
❖ Caucasian
❖ Children: 0

Proudest Career Moment

This is hard to choose, certainly the big awards, such as the Benjamin Franklin Medal,
Roebling Medal, and election to National Academy of Sciences (NAS) being the main
ones. On the other hand, when a Ph.D. student presents an excellent talk at a meeting, or
when the student's first paper is accepted for publication with almost no changes (after
we had played track-changes-tennis for 10 drafts to get to the point of submission), such
moments are very precious.

Academic Credentials

Ph.D. (1967) Physical Chemistry, University of Chicago, IL, USA.
M.S. (1964) Physical Chemistry, University of Chicago, IL, USA.
B.S. (1963) Physical Chemistry, University of Chicago, IL, USA.

Research Expertise: solid-state chemistry, thermodynamics, ceramics, physics and chemistry of minerals, geochemistry

Other Interests: development and improvement of instrumentation and techniques; application of thermochemistry to problems of energy; promoting solid-state chemistry, thermodynamics, and calorimetry

Key Accomplishments, Honors, Recognitions, and Awards

- William Mong Distinguished Lecture, University of Hong Kong, 2013
- Alfred R. Cooper Distinguished Lecture, The American Ceramic Society (ACerS), Materials Science & Technology Conference, 2013
- Cecil and Ida Green Senior Fellowship at the Geophysical Laboratory of the Carnegie Institute of Washington, 2012
- Honorary Professor, Three Gorges University, Yichang, China, 2012
- Featured Manuscript in the Journal of ACerS, Thermochemistry of Lanthana- and yttria-doped thoria, 2011
- Elected to the American Philosophical Society (APS), 2011
- International Union of Pure and Applied Chemistry, Fellow 2009
- Honorary Professor, Sichuan University, China, 2009
- Best University Paper Award, Department of Energy (DoE), Geoscience Grantee Meeting, 2009
- Roebling Medal, Mineralogical Society of America, 2009
- Honorary Professor at School of Environmental Sciences and Urban Studies, Shenzhen Graduate School, Peking University, China, 2008
- Sloan Faculty Distinguished Service Award, University of California at Davis, 2007
- Outstanding Engineering Senior Career Research Award, University of California at Davis, 2007
- Rossini Award, International Association of Chemical Thermodynamics (IACT), 2006
- Harry H. Hess Medal, American Geophysical Union (AGU), 2006
- Spriggs Phase Equilibria Award, ACerS, 2005
- Urey Medal, European Association of Geochemistry (EAG), 2005
- Fellow, The Mineralogical Society, Great Britain, 2004
- Highly Cited Researchers Award, ISI Thomson Scientific, 2002
- Benjamin Franklin Medal in Earth Science, 2002

- American Ceramic Society, Best Paper Award of the Nuclear and Environmental Technology Division, 2001
- Fellow of The American Ceramic Society, 2001
- Ceramic Educational Council Outstanding Educator Award, 2000
- Hugh M. Huffman Memorial Award, The Calorimetry Conference, 2000
- Alexander M. Cruickshank Award, Gordon Conference, 2000
- Kreeger-Wolf Visiting Scholar, Northwestern University, 1999
- Fellow of the Geochemical Society, 1997
- Doctor Honoris Causa, Uppsala University, Sweden, 1995
- Ross Coffin Purdy Award of the American Ceramic Society, 1995
- Elected to National Academy of Sciences, 1993
- President of the Mineralogical Society of America, 1992–1993
- Vice President of the Mineralogical Society of America, 1991–1992
- Fellow of the American Geophysical Union, 1988
- Arizona State University, Graduate College Distinguished Research Award, 1982–1983
- Mineralogical Society of America Award, 1981
- Alfred P. Sloan Fellowship, 1973–1975

Biography

Early Life and Education

Alexandra Navrotsky was educated at the Bronx High School of Science and the University of Chicago (B.S., M.S., and Ph.D. in physical chemistry). After completing postdoctoral work in Germany and at Pennsylvania State University, she joined the faculty in Chemistry at Arizona State University.

Career History

She remained at Arizona State University until 1985 when she moved to the Department of Geological and Geophysical Sciences at Princeton University. She chaired that department from 1988 to 1991 and has been active in the Princeton Materials Institute. In 1997, she became an Interdisciplinary Professor of Ceramic, Earth, and Environmental Materials Chemistry at the University of California at Davis and was appointed Edward Roessler Chair in Mathematical and Physical Sciences in 2001.

Her research interests have centered about relating microscopic features of structure and bonding to macroscopic thermodynamic behavior in minerals, ceramics, and other complex materials. She has made contributions to mineral thermodynamics; mantle mineralogy and high-pressure phase transitions; silicate melt and glass thermodynamics; order–disorder in spinels; framework silicates and other oxides; ceramic processing; oxide superconductors; nanophase oxides, zeolites, nitrides, perovskites; and the general problem of structure–energy–property systematics. The main technical area of her laboratory is high-temperature reaction calorimetry. She is director of the UC Davis Organized Research Unit on Nanomaterials in the Environment, Agriculture and Technology (NEAT-ORU).

She spent 5 years (1986–1991) as Editor, Physics and Chemistry of Minerals, and serves on numerous advisory committees and panels in both government and academe.

In her spare time, she enjoys dogs and in particular working with dog rescue.

Credit: http://navrotsky.engr.ucdavis.edu/pages/bioshort.html

3 Most Cited Publications

Title: Surface energies and thermodynamic phase stability in nanocrystalline aluminas
Author(s): McHale, JM; Auroux, A; Perrotta, AJ; et al.
Source: Science; volume: 277; issue: 5327; pages: 788–791; doi: 10.1126/science .277.5327.788; published: August 8, 1997
Times Cited: 443 (from Web of Science)

Title: Olivine-modified spinel–spinel transitions in the system Mg_2SiO_4-Fe_2SiO_4: calorimetric measurements, thermochemical calculation, and geophysical application
Author(s): Akaogi, M; Ito, E; Navrotsky, A
Source: Journal of Geophysical Research: Solid Earth and Planets; volume: 94; issue: B11; pages: 15671–15685; doi: 10.1029/JB094iB11p15671; published: November 10, 1989
Times Cited: 396 (from Web of Science)

Title: Progress and new directions in high temperature calorimetry revisited
Author(s): Navrotsky, A
Source: Physics and Chemistry of Minerals; volume: 24; issue: 3; pages: 222–241; doi: 10.1007/s002690050035; published: April 1997
Times Cited: 281 (from Web of Science)

Challenges

To stay ahead of the crowd and find new directions. To manage time, finish projects, and maintain funding.

On being a woman in this field . . .

I do not have the control experiment—what my career and life would have been like had I been a man. I have met many wonderful generous colleagues, some thoughtless ones, and a few who deserve to be called evil. Overall people treat me with respect, as I do them. I try to be logical and rational and often get to the core of an issue quickly, be it scientific, logistical, or simply human. With an active career in both materials science and geoscience, I often introduce one set of colleagues to another, leading to new interactions and collaborations. Sometimes being a woman helps in finding common ground between arguing factions in collaborative efforts, especially if I cook them a good dinner. I enjoy the window on the world that science gives me, and the globalization of the last few years lets me compare different cultures, including the roles and norms of women. I find it liberating when, as often happens, what I assumed was universal behavior turns out to be merely a local habit.

Words of Wisdom

- In science as in life, every day is a new adventure.
- Greet each new day with the same joy and enthusiasm as your dog.
- The goal isn't merely to survive, it is to thrive.
- You can do anything you want, but you cannot do everything you want, so choose wisely.

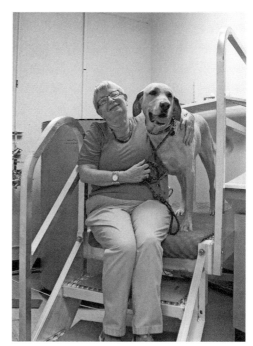

Alex Navrotsky and her lab, Tahoe, in her lab by the AlexSYS calorimeter.

Profile 70

Linda F. Nazar

Professor
Department of Chemistry
Cross-appointments with
Electrical Engineering and Physics
University of Waterloo
200 University Avenue West
Waterloo, Ontario N2L 3G1
Canada

e-mail: lfnazar@uwaterloo.ca
telephone: (519) 888-4637

Linda Nazar at about age 30 standing in front of her X-ray diffractometer.

Birthplace

Vancouver, B.C., Canada

Publication/Invention Record

>175 publications: h-index 60 (Web of Science)

Important Career Moments (to date)

Tags

❖ Academe

❖ Domicile: Canada

❖ Nationality: Canadian

❖ Caucasian

❖ Children: 0

Around 2009, I was quite happy to see our work on energy materials become better recognized on a larger front—it has helped in many ways, for example, in recruiting students. I believe that this sudden "notoriety" reflects a resurgence in the field—I was at the right place at the right time with the right stuff.

In addition, in 2013, I was honored by receiving the August Wilhelm von Hofmann Lecture at the Science Forum in Chemistry. Many impressive people have been recipients in the past, so I am keeping very good company. For more details, including a short video, see http://www.chemistryviews.org/details/video/1234291/Linda_Nazar_Developing_Materials_for_Energy_Storage_And_Conversion.html (accessed February 6, 2014).

Academic Credentials

Ph.D. (1984) Chemistry, The University of Toronto, Toronto, Ontario, Canada.
B.S. (1978) Chemistry (Honors), University of British Columbia, Vancouver, British Columbia, Canada.

Successful Women Ceramic and Glass Scientists and Engineers: 100 Inspirational Profiles. Lynnette D. Madsen.
© 2016 The American Ceramic Society and John Wiley & Sons, Inc. Published 2016 by John Wiley & Sons, Inc.

Research Expertise: solid-state chemistry; energy storage materials and electrochemistry (in particular lithium-, sodium-, and magnesium-ion batteries, and lithium–sulfur and metal–oxygen batteries); photovoltaic materials

Other Interests: communicating results to the public and clarifying facts in legal disputes

Key Accomplishments, Honors, Recognitions, and Awards

- Officer, Order of Canada, 2015 to present
- Member of the Scientific Advisory Board, State University of New York (SUNY) Energy Frontiers Research Center, USA, 2011 to present
- Member of the Editorial Advisory Board, Angewandte Chemie, 2011 to present
- Member of the Scientific Board for the International Lithium Battery Meeting including IMLB-2010, 2010 to present
- Member of the Editorial Advisory Board, Journal of Materials Chemistry A, 2010 to present
- Member of the Natural Sciences and Engineering Research Council of Canada (NSERC) Advisory Council for Industry–University Grants, 2001 to present
- Member of the Editorial Boards, Chemistry of Materials; and Solid State Sciences, 2000 to present
- Selected as a Thomson Reuters Highly Cited Researcher, 2014
- Member of the Scientific Advisory Board, Cornell Energy Frontiers Research Center, USA, 2012–2014
- August Wilhelm von Hofmann Lectureship (German Chemical Society), 2013
- Scientific Leader, Chemical Transformation Thrust: Energy Storage Hub, USA, 2013
- Member of the BASF International Battery and Electrochemistry Network, Germany, 2013
- Consultant, Blackberry Battery Division, 2013
- Senior Canada Research Chair in Solid State Energy Materials, Tier 1, 2004–2012, renewed 2012
- International Union of Pure and Applied Chemistry (IUPAC) Distinguished Women in Chemistry/Chemical Engineering, 2012
- Department of Energy (DOE) Panel Review Member for the Batteries for Advanced Transportation Technologies (BATT) Program, 2010–2012
- Served as reviewer for two international academic promotion cases (Israel, USA), 2012
- Chair, The Electrochemical Society (ECS) Battery Division Research Award Committee, 2011–2012
- Board Member and Advisor, Northeastern Energy Frontiers Research Center, USA, 2008–2012
- Fellow of the Royal Society of Canada, 2011
- Distinguished Women in Chemistry or Chemical Engineering Award, 2011

- Rio Tinto Alcan Award winner for her research in inorganic electrochemistry, Canadian Society for Chemistry, 2011
- International Battery Association Award, 2011
- Member of the Editorial Advisory Board, Angewandte Chemie, 2011
- Served as an expert for international tenure review, two cases, 2011
- Quorum Leader, Waterloo Global Science Initiative (WGSI) Global Equinox Summit: Energy 2030, 2011
- Consultant for BP International; Honeywell UOP; Planar Energy; and DuPont Central Research and Development (CR&D), 2011
- Member of the Advisory Board of the Journal of Materials Chemistry, 2008–2011
- Chair and formerly Member of the Chemistry/Biology/Physics Science Funding Panel of NATO, Brussels; 2009–2010
- Global Climate Energy Project Panel Review Member (Stanford University), 2010
- Member, National Science Foundation (NSF) Funding Panel for Energy Storage (USA), 2010
- ECS Battery Division Research Award, 2009
- Moore Distinguished Scholar at the California Institute of Technology (USA), 2009–2010
- Guest Editor for a special edition of the Materials Research Society (MRS) Bulletin on Electrical Energy Storage to Power the 21st Century, 2010–2011
- Guest Editor for a themed edition of the Journal of Materials Chemistry on Advanced Materials for Lithium Batteries, 2010–2011
- Co-organizer for the 2011 European MRS Symposium on Large-Scale Energy Storage, 2010–2011
- Expert Witness/Technical Expert on two U.S. lawsuits involving cathode materials in lithium-ion batteries, 2006–2010
- Member of the NSERC Discovery Grants Funding Panel (Canada), 2006–2009
- Expert Witness/Technical Expert on a U.S./worldwide lawsuit involving oxide materials, 2007
- Member of a workshop to chart new directions for electrochemical energy storage, DOE Workshop in Arlington, USA, 2007
- Member of the funding review panel for the Division of Materials Research, NSF, USA, 2007
- Member of the Scientific Board for the International Lithium Battery Meeting, IMLB-2006, Biarritz, France, 2005–2006
- Member of the Planning Committee for the Nanotechnology Program at the University of Waterloo, 2004–2006
- Member of the DOE panel to review the Energy Technologies Division, Lawrence Berkeley Labs, USA, 2006
- Member of the Scientific Board for the 15th Solid State Ionics Meeting, Stuttgart, Germany, 2005

- Member of the Scientific Board for the International Lithium Battery Meeting, Nara, Japan, 2004
- Member of the Editorial Board, Journal of Solid State Chemistry, 2000–2003
- Professeur Invité: Centre national de la recherche scientifique (CNRS), Laboratoire de Cristallographie, Grenoble, France, 2002
- Member of the Scientific Board for the 8th Asian Conference on Solid State Ionics, 2002
- Professeur Invité: Université de Nantes/Institut des Matériaux, France, 2000
- NSERC Site Visit Committee Chair, Collaborative Research and Development (CRD) Grant, University of Montreal, 2000
- Member of the Jury for the Diplôme d'Habilitation à Diriger des Recherches for Dr. Annie Le Gal La Salle, Université de Nantes and the Institut des Matériaux de Nantes, 2000
- Chair, NSERC Strategic Grants Review Panel, 1999
- Organizer, MRS Symposium, Advanced Materials for Batteries and Fuel Cells (MRS Spring Meeting, San Francisco, USA), 1999
- Member of 12-person DOE (USA) workshop team—Partially Disordered Materials: Frontiers of Science, 1998
- NSERC Strategic Grants Review Panel Committee, 1997–1998
- Visiting Research Professor, Department of Materials Science, University of California, Los Angeles (UCLA), Los Angeles, USA, 1995
- Exxon Postdoctoral Fellowship, Exxon Corporate Research Labs, USA, 1984; NSERC of Canada Postdoctoral Fellowship (declined)
- NSERC Graduate Scholarship, 1978–1982
- Royal Chemical Society Undergraduate Award, University of British Columbia, 1978

Biography

Early Life and Education

Linda was raised and educated in Canada. She had always planned on attending university and was excited by it. Initially she thought about majoring in physics, but a first-year chemistry class greatly impressed her and she switched her major to chemistry. She completed two degrees in chemistry—a Bachelor's degree at the University of British Columbia followed by a doctorate at the University of Toronto.

Career History

Upon graduation, she turned down a fully funded international postdoctoral scholarship (from NSERC) to undertake a postdoctoral fellowship at Exxon Corporate Research. After this period, she accepted a position at the University of Waterloo as an Assistant Professor, while first spending 6 months at Rutgers, The State University of New Jersey, in their Center for Ceramics Research as a Research Scientist in order to "learn to speak ceramic engineering." At the University of Waterloo in the Department of Chemistry, she progressed through the ranks of Associate Professor in 1992 and then to Full Professor in 1998. She held a fellowship as a Professeur Invité at the Université de Nantes (Institut des Matériaux), France (April to July 2000) and a CNRS fellowship as a Professeur Invité at the CNRS Cristallographie, Grenoble, France (September to December 2002). Prof. Nazar was also the recipient of one of two Distinguished Moore Fellowships at California Institute of Technology (Caltech) in 2009–2010.

Prof. Nazar sits on the Advisory Boards of the Northeastern Energy Frontiers Research Center, Angewandte Chemie, and the Journal of Materials Chemistry, and on the Executive Board of the International Meeting for Lithium Batteries since 2005.

Linda Nazar at the University of Waterloo, 2011. Photo by Mike Brown.

She is currently a member (worldwide) of two battery networks: the BASF International Battery and Electrochemistry Network and a Scientific Leader for the Chemical Transformation Thrust in the U.S. DOE funded Energy Storage Hub.

She was a member of a group of 15 international participants in a BP Workshop on Energy Storage in 2011 that presented the cutting-edge science and technology and addressed the potential for breakthroughs that could lead to disruptive change in energy storage. She has served on many funding panels (NSF, DOE, the Global Climate and Energy Project (GCEP), NATO, NSERC) as both member and chair, and co-organized several symposia and journal editions devoted to focus topics on energy storage.

Her excellence in science has resulted in Prof. Nazar being in much demand as a speaker. She presents approximately 10–15 invited lectures at universities and international meetings each year, including about three to four keynote or plenary lectures annually. Recently, these include a plenary talk at the BASF "Smart Energy" Science Symposium in Germany in 2014, the August Wilhelm von Hofmann Distinguished Lecture to the German Chemical Society in 2013, Distinguished Lecture at the Lawrence Berkeley National Labs/University of California Berkeley in 2011, the Moore Lecture at

Linda Nazar in her new office in the Mike and Ophelia Lazaridis Quantum Nano Centre (QNC) building at the University of Waterloo, 2013. Photo credit: Erik B. Svedberg.

Caltech in 2010, a plenary lecture at the Solid State Ionics-17 meeting, a keynote lecture at the IBM Almaden Research Labs Battery-500 conference, and the ECS Award Lecture in 2009. In the past, she has presented keynotes at the Lithium Battery Discussion meetings, the International Battery meetings, the Solid State Ionics conferences, the Florida Power Sources conferences, the Japanese Battery meetings, the Korean Battery Society meetings, a keynote lecture as part of a 12-part "Discover Chemistry Series" hosted by DuPont de Nemours in 1996, and lectures at many Gordon Conferences.

3 Most Cited Publications

Title: A highly ordered nanostructured carbon–sulphur cathode for lithium–sulphur batteries
Author(s): Ji, X; Lee, KT; Nazar, LF
Source: Nature Materials; volume: 8; issue: 6; pages: 500–506; doi: 10.1038/NMAT2460; published: June 2009
Times Cited: 1071 (from Web of Science)

Title: Approaching theoretical capacity of $LiFePO_4$ at room temperature at high rates
Author(s): Huang, H; Yin, SC; Nazar, LF
Source: Electrochemical and Solid State Letters; volume: 4; issue: 10; pages: A170–A172; doi: 10.1149/1.1396695; published: October 2001
Times Cited: 756 (from Web of Science)

Title: Nano-network electronic conduction in iron and nickel olivine phosphates
Author(s): Herle, PS; Ellis, B; Coombs, N; et al.
Source: Nature Materials; volume: 3; issue: 3; pages: 147–152; doi: 10.1038/nmat1063; published: March 2004
Times Cited: 685 (from Web of Science)

ResearcherID (H-2736-2014)

Challenges and on being a woman in this field . . .

The science, itself, has not been my main challenge. Rather, interactions with some colleagues have proved to be challenging—some in the "old boys" network have old-fashioned ways and are not so keen on accepting a strong headed woman into their world. In my experience, most women know this situation (all too well) and are reminded of it at many levels all the time. Fortunately, the atmosphere is more welcoming in my specific area of research and I hasten to point out that, like most women, I have survived and find tremendous satisfaction in the work that I do.

Words of Wisdom

First, let me say that I wish I knew then what I know now . . . just accept that you will continuously learn things along the way. One bit of advice I can provide is to network, early on in your career, with both men and women. And what I mean by net-working is that you should establish a good solid scientific network—make connections, and especially develop strong scientific collaborations from a position of your own strength. The second bit of advice is to not take scientific criticism personally, don't be defensive. It's all easy to say, but one needs to develop a thick skin, and learn positively from criticism. The third bit—always reach high!

Diving in the tropics. (See insert for color version of figure.)

Profile 71

Tina M. Nenoff

Senior Scientist
Sandia National Laboratories
Albuquerque, NM
USA

e-mail: tmnenof@sandia.gov
telephone: (505) 844-0340

Tina M. Nenoff, Sandia
National Laboratories.

Birthplace
Orange, NJ, USA

Born
December 7, 1965

Publication/Invention Record

>150 publications: h-index 31 (Web of Science)
12 patents, 4 book chapters

Tags

❖ Government
❖ Domicile: USA
❖ Nationality: American
❖ Caucasian
❖ Children: 0

Proudest Career Moment (to date)

My proudest career moment was learning that the crystalline silicotitanate materials we developed at Sandia were successfully being used at Fukushima Daiichi Nuclear Power Plant accident site. The CSTs, now commercially produced by Honeywell UOP LLC, were being used to clean up the radiological cesium ions from the coolant seawater pooled in the exploded reactor buildings. That seawater had been pumped onto the accident site to cool the damaged reactors. It makes me proud that basic/fundamental research was being used in an important real-world application.

Academic Credentials

M.A./Ph.D. (1993) Chemistry (Inorganic/Solid State), University of California, Santa Barbara, CA, USA.
B.A. (1987) Chemistry, University of Pennsylvania, Philadelphia, PA, USA.

Research Expertise: research spans the basic to applied research areas of materials directed toward energy and environment. Particular focus is given to understanding the structure–property relationship between confinement of ions or molecules in nanoporous crystalline materials. Areas of research interest include (i) the synthesis

Successful Women Ceramic and Glass Scientists and Engineers: 100 Inspirational Profiles. Lynnette D. Madsen.
© 2016 The American Ceramic Society and John Wiley & Sons, Inc. Published 2016 by John Wiley & Sons, Inc.

and characterization of novel inorganic molecular sieves, zeolites, metal–organic frameworks (MOFs), and ceramics for the sequestration and storage of radioactive cations and volatile gases (Hanford; Fukushima); (ii) the synthesis, characterization, and sintering of alloy nanoparticles by room temperature radiolysis; (iii) the synthesis and testing of inorganic crystalline porous bulk phases and membranes for catalysis and separations of light gases and organic molecules in biofuel processes; and (iv) the energy and cost-efficient catalytic reactive separations of industrial feedstock chemicals. Investigator of research programs and multiple national lab–industry–university, and cooperative research and development agreements (CRADAs)

Tina Nenoff in the lab in front of the powder X-ray diffractometer.

Other Interests: responsibilities include direction of research, partnering with industrial and governmental customers, staff and program management, grant proposal writing, and patents/publications/presentations. Interests also include broader-scale, program development and leadership environments

Key Accomplishments, Honors, Recognitions, and Awards

- Subcommittee Chair for Science and Technology, Committee on Science, 2012 to present
- Member American Chemical Society (ACS) Committee on Science, 2009 to present
- Elected Councillor ACS/Colloid (COLL) Division, 2006–2008, 2009 to present
- Elected Member, International Zeolite Association, Commission on Synthesis, 2004 to present
- Member Editorial Advisory Board Industrial Engineering & Chemical Research Journal, 2012–2014
- Appointed Member/Lead of ACS Awards Selection Committee, by ACS President N. Jackson, 2012–2014
- Team Member Federal Laboratory Consortium (FLC) Award for "Removal of Radioactive Cesium from Seawater Using Crystalline Silico-Titanates," 2013
- Appointed Member of Board of Directors of the Malta Conferences Foundation, 2011–2013
- Scientific Consultation to the U.S. Air Force, 2011
- Subcommittee Co-Chair for Science and Technology, Committee on Science, 2011
- Appointed Fellow, American Chemical Society, 2011
- Invited Panel Member, DoE Nuclear Separations Technology Workshop, Bethesda, Maryland, 2011

- Guest Editor for Industrial & Engineering Chemistry Research Special Issue on Nuclear Energy (ACS Journal), 2011
- Chair Nanoporous Materials Gordon Research Conference (GRC), 2011
- Subcommittee Chair for Awards, Committee on Science, 2010–2011
- Member Editorial Advisory Board Chemistry of Materials Journal, 2006–2011
- Subject of KNME Science, Central TV interview, 2010
- Invited Panel Member, The National Academies "Waste Form Technology and Performance," 2009–2010
- Review Panel Member, Lawrence Livermore National Laboratory "Transformational Materials Initiative," 2009
- Subject of a Miller-McCune Magazine feature article, 2009
- Elected Chair Nanoporous Materials Gordon Research Conference (GRC), 2008
- Vice-Chair Nanoporous Materials Gordon Research Conference (GRC), 2008
- Associate Member ACS Committee on Science, 2007–2008
- Review Panel Member, Los Alamos National Laboratory "Materials Science and Technology Capabilities," 2007
- Review Panel Member, DOE-BES Site Review, Oak Ridge National Laboratory Chemicals Program, 2007
- Interview subject for TechComm Magazine feature article, 2006
- Guest Editor, MRS Bulletin: Membranes for H2 Purification, 2006
- Elected Vice Chair Nanoporous Materials Gordon Research Conference (GRC), 2005
- Invited Presentation Zeolite and Layered Materials GRC, 2005
- Interviewed for nationally aired PBS program "Roadtrip Nation," 2005
- Elected Alt-Councillor for ACS/COLL Division, 2004–2005
- Elected Representative to Materials Secretariat for ACS/COLL Division, 2002–2004
- Recipient of the U.S. Rep. Heather Wilson Women's History Month Award, 2003
- Organizer Fall ACS Symposium "Modeling and Simulation for Colloid Materials," 2003
- Team Recognition Award, SOMS—Sandia National Laboratories, 2002
- Invited Presentation Inorganic GRC meeting, 2001
- Invited Presentation Solid State GRC meeting, 2001
- Semi-Annual Organizer of the MRS "Women in Science and Engineering" Symposium, Fall and Spring National Meetings. 2001
- Recipient of the NM-YWCA Women on the Move Award, 2001
- Individual Employee Recognition Award, Sandia National Laboratories, 1999
- Elected Secretary, ACS Colloid (COLL) Division. 1998–2000
- Nominated for the NSTC Presidential Early Career Award, 1998
- Organizer Fall ACS Symposium: "Catalysis with Designed Materials," 1998
- U.S. R&D100 Award, Crystalline Silicotitanate (CSTs) Ion Exchangers, 1996

- Team Recognition Award, CSTs—Sandia National Laboratories. 1996
- Recipient of the University of California President's Fellowship, 1988–1989

Biography

Early Life and Education

Tina is the oldest of three children to Dr. Vladimir Nenoff and Mrs. Lydia Nenoff, immigrants from Bulgaria. She grew up and attended K-12 schools in New Jersey. She graduated valedictorian from Immaculate Heart Academy High School.

For college, Tina attended the University of Pennsylvania and obtained a B.A. with honors in Chemistry and a minor in Archaeology. During college, she did undergraduate research in the lab of Nobel laureate Dr. Alan McDiarmid's lab. After graduation, she moved to Manhattan and worked as a research chemist at Ciba-Geigy Chemicals.

In 1989, Tina entered the University of California at Santa Barbara (UCSB) on a UC President's Fellowship. At UCSB, Tina obtained her M.S. and Ph.D. in the lab of Dr. Galen Stucky in the synthesis and crystallographic study of zeolites and molecular sieves. Part of her Ph.D. research was performed at DuPont, at the Experimental Station, under the guidance of Dr. David Corbin.

Career History

Tina was recruited from her UCSB Ph.D. directly to Sandia National Laboratories to become a research staff scientist. The focus of the hiring program was the synthesis, characterization, and testing of novel molecular sieves and porous oxide materials for the separations of radiological ions from complex aqueous systems at the Hanford nuclear waste site. The project evolved to include the one-step thermal transformation of those loaded porous materials into durable ceramic waste forms for the eventual long-term and safe storage of the radiological materials.

Tina M. Nenoff, Sandia National Laboratories, 20th Anniversary of working at Sandia.

3 Most Cited Publications

Title: Membranes for hydrogen separation
Author(s): Ockwig, NW; Nenoff, TM
Source: Chemical Reviews; volume: 107; issue: 10; pages: 4078–4110; published: OCT 2007
Times Cited: 237 (from Web of Science)

Title: A general synthetic procedure for heteropolyniobates
Author(s): Nyman, M; Bonhomme, F; Alam, TM; et al.
Source: Science; volume: 297; issue: 5583; pages: 996–998; published: August 9, 2002
Times Cited: 138 (from Web of Science)

Title: Room-temperature synthesis and characterization of new ZnPO and ZnAsO sodalite open frameworks
Author(s): Nenoff, TM; Harrison, WTA; Gier, TE; et al.
Source: Journal of the American Chemical Society; volume: 113; issue: 1; pages: 378–379; published: January 2, 1991
Times Cited: 118 (from Web of Science)

Challenges

Challenges on becoming a staff scientist directly from graduate school included developing skills not taught in graduate school, including building and designing labs, successful grant writing skills, building a scientific reputation independent of well-known graduate advisors, learning to ride the cycles of funding as tied to the political landscape (how events in Washington, DC, directly affect funding to me), how to be diversified in research projects, and funding grants so as to ride the cycles of funding.

On being a woman in this field . . .

Challenges on being a woman in this field include learning to be confident in expressing my views when being the only woman (and usually the youngest person) in the room; dressing appropriately for business and professional careers in the sciences, picking the right people to confide in when new to the work/location/corporation, and not always fighting to be heard but correctly presenting myself and being heard.

Words of Wisdom

My career on its path to today was a marathon not a sprint, so learning to pace myself through research, funding, team building, personal versus professional life, etc. was very important.

Profile 72

Beatriz Noheda

Professor
Solid State Materials for Electronics
Zernike Institute for Advanced Materials
University of Groningen
Nijenborgh 4, 9747AG Groningen
The Netherlands

e-mail: b.noheda@rug.nl
telephone: +31 (0)50 3634565

Beatriz Noheda setting a sample at the powder diffraction beamline at Brookhaven National Laboratory, 2001. Credit: Brookhaven National Laboratory, http://www.bnl.gov/bnlweb/pubaf/pr/2001/bnlpr031601.htm (accessed February 8, 2015).

Birthplace

Cuenca, Spain

Born

September 23, 1968

Publication/Invention Record

>75 publications: h-index 30 (Web of Science)
1 book chapter

Tags

❖ Academe
❖ Domicile: The Netherlands
❖ Nationality: Spanish
❖ Caucasian
❖ Children: 2

Proudest Career Moment (to date)

There have been a handful of proud moments, but I have the most pride about the setup of my current lab. Every time that a new instrumentation has been installed, I have felt a wave of pride. Looking at my lab, I see not only the state-of-the-art equipment that enables us to work at the front of the field, but I also recall all of the support obtained throughout the years to reach this state. In particular, the support within the Zernike Institute from my colleagues is greatly appreciated. They believed in the dreams of this young Spanish scientist who wanted to achieve atomic control of ferroelectrics and they have kept supporting me in times of crisis (when grants were lacking): Their support is always invaluable.

I also felt very proud when I received the letter of the American Physical Society (APS) awarding me a Fellowship. I had no clue that someone had nominated me. When I learned that the nomination was submitted by my former colleagues at Brookhaven National Lab, that I had left 10 years before, I felt fantastically honored and rewarded for all the tough moments in my career. I doubt that other jobs can supply such a feeling.

431

Academic Credentials

Ph.D. (1996) Physics, Universidad Autonoma de Madrid (UAM), Madrid, Spain.
M.Sc. (1992) Physics, Universidad Autonoma de Madrid, Madrid, Spain.

Research Expertise: functional/smart materials, in particular, perovskites, ferroelectrics, piezoelectrics, and multiferroics (materials that are simultaneously both ferroelectric and magnetic)

Other Interests: mentoring students and teaching; participating in forums on the future of the energy use, new energy markets, the energy transition, energy harvesting, and how new materials can contribute to a greener (energy) future.

Key Accomplishments, Honors, Recognitions, and Awards

- Editorial board of the journal Materials, 2015 to present
- Advisory committee for the series of International and European Meetings on Ferroelectricity (IMF and EMF), 2013 to present
- Editorial board of the journal Phase Transitions, 2007 to present
- Chair of the selection committee of seven new Rosalind Franklin Fellows (tenure track positions for women) at the Faculty of Sciences (University of Groningen) with 170 excellent applicants from all over the world and from all science disciplines, 2015
- Editorial board of *Applied Physics Letters* and *Journal of Applied Physics*, 2013–2015
- Fellow of the American Physical Society "for fundamental structural studies of new phases in perovskite-type ferroelectric materials and of domain nanostructures in epitaxial films of multiferroics," 2012
- Teacher of the Year award (in Chemistry) by the Faculty Sciences, University of Groningen, 2009
- Editorial board of European Journal of Physics: Applied Physics, 2008–2010
- VIDI Personal Grant of the Dutch funding organization (Netherlands Organisation for Scientific Research—NWO), 2009

Beatriz Noheda stands under the poster of herself advertising Groningen as the city of talent. The text in the poster in Dutch means "Charging your phone for free with the energy of people's steps? Later it will be possible!" Credit: http://www.cityoftalent.nl/. (See insert for color version of figure.)

- Rosalind Franklin Fellowship at the University of Groningen: 2004–2009
- Deutscher Akademischer Austauschdienst (DAAD) Fellowship for a research stay at Saarlandes University, Saarbrücken, Germany, 1994

Biography

Early Life and Education

I am the oldest of four sisters. My parents were both schoolteachers and we were raised with their passion for educating and their decided approach to make us as independent as possible. At the same time, we have very strong family roots and ties. We have been and still are (despite the distance and the sad loss of our father) a very close family. Among my sisters, there is an economist (Marta), a lawyer (Isabel), and a publicist/photographer (Cristina); they are very talented and make me feel very proud.

My first memories of wanting to become a scientist are from primary school, a catholic nun's school, where my sisters and I studied. In the sports hall, the nuns had organized a film session projecting a classical movie about Madam Curie (Mervyn Leroy, 1944). As a girl of 9 or 10 years old, I saw the intelligence, but also the superhuman physical efforts of the experimental chemist and her drive to achieve—these things impressed me deeply; I already wanted to become like her. At school, I liked literature, geography, and history, but I also really enjoyed mathematics, physics, and chemistry because I saw them as tools to understand nature and its mysteries and they did not require memorizing. How much of this preference developed because of Madam Curie, it is difficult to tell, but it is to me clear that we make decisions at an early age. Interestingly, at that time I never thought that being a scientist was not a female occupation. I guess this comes from the fact that I was raised among girls (I studied in a girls-only school until the age of 16 and I had no brothers). I was never confronted with the self-confidence of the boys of the class. I believe that this difference in self-confidence between male and female may be highly detrimental for the self-image developing in young girls.

Additionally, all my teachers were female. It seems logical to me that the explanations of a woman on a challenging subject, with subtle emphases and connections, will appeal more or be more enlightening to female students than to male students. This probably makes a very small difference that can be compensated by other means, but it may be enough to put some younger people off certain subjects. As much as I hated not to have boys in the class during my puberty, I think that it was a very good thing for my self-confidence and, thus, for my career.

For a year or two, I debated whether to study chemistry or physics at university. Finally I chose physics because I thought it would give me the most flexibility (I thought I could study chemistry afterward, but believed that the other way round would probably be tougher). As an experimental physicist, I focused on research in materials science. For my doctoral dissertation, I worked at the ferroelectrics laboratory of Prof. Julio Gonzalo in Madrid. He introduced me to the field of ferroelectrics and to phase transitions, helped me to see the world for the first time by sending me to (many) international conferences. He helped me to obtain a grant for my first (6 month) stay abroad. My world enlarged and my passion for research grew day by day. Many of the things that I have achieved later on are a subset of the dreams and wishes from those days. All these aspects were crucial, but if I had to choose the single most important thing that Julio Gonzalo did, it was to introduce me to Dave Cox and Gen Shirane who later became quite important in my career.

I had met them in Madrid, when they went to visit Julio. Dave gave me a private master class of several hours on powder diffraction. Gen taught me how to analyze some neutron data that he had taken for us at Brookhaven National Lab. He would speak to me in short and cryptic sentences requesting quick answers. Initially puzzled, I then realized that he was guiding me into the essence of the problem. To have the possibility to learn from them and their way of working and thinking is the best thing that has ever (scientifically) happened to me.

Career History

Motivated by these two inspiring scientists, I did my postdoctoral studies at the lab of Mike Glazer (Clarendon Laboratory) in Oxford to learn more about powder diffraction. Then, I obtained a grant to investigate the so-called morphotropic phase boundary in lead zirconate titanate (PZT), a technologically important piezoelectric material with Gen and Dave at Brookhaven National Lab (NY). This research on bulk piezoelectric materials showed that lowering of the crystal symmetry can give rise to huge electromechanical and dielectric responses via the polarization rotation effect. These low-symmetry phases were thought to

Beatriz Noheda
dressed for the
university's
graduation
ceremony, 2011.

be scarce in nature, but nowadays the improved analysis (mainly diffraction) techniques and the possibility to induce low symmetries using thin-film epitaxy are revealing polarization rotation in different oxides and the concept and association with giant piezoelectric responses is now clearly established.

I was truly happy at Brookhaven. We had exciting results and good publications, and I was the local contact at one synchrotron beamline and had the chance to meet many users who broadened my scientific horizon. I was appointed as an Assistant Physicist. In addition, I met this handsome Dutch theoretical physicist at the lab's laundry room, so I had all I needed in Long Island.

We thought about our future and feared not being able to find positions together in Europe quickly if we needed to be close to our families, so together we started looking for positions in Europe. Almost immediately he obtained a prestigious tenure-track position in Amsterdam, but the situation for me was not easy since the two groups in the Netherlands working on ferroelectrics discontinued their work due to lack of funding. In the meantime, Brookhaven had offered me a promotion to Associate Physicist (with tenure) if I would stay. For the first time in my life, at 33, I decided to prioritize my relationship against my career. It was not a logical step to take; it was just a strong gut feeling. I decided to take a risk with my career, but not with my relationship. I chose to use the situation as an excuse to prove to myself how good of a scientist I was. I thought that if I did not succeed in changing topics at this stage in my career, it probably meant that I was not a good scientist and should do something else.

I really wanted to learn how to grow thin films, and then I wanted to make the materials myself, and make them better and with more control, so why not use this chance to learn from scratch. I applied for a postdoctoral position in Amsterdam in the group of Ronald Griessen to work on thin films of metal hydrides (switchable mirrors). Amazingly, he gave me the position (although I had no experience in the field). He had a very nice, diverse, and strong group and it was another great learning experience for me. It was a team of competitive and driven young people working on (and solving) a problem from different angles. My research was going very well, but my future in metal hydrides after such short experience was uncertain. However, being there I learned about the Rosalind Franklin Fellowships (RFF) that were being offered for the first time in Groningen. All fields of research were welcome in order to increase the chances of enticing the best women to apply. That was my only possibility to stay in the Netherlands and I was very honored to get one of the six positions among 120 applicants from all over the world. Now, a decade later, I serve, maybe not coincidentally, as the chair of the selection committee of the fifth round of RFF.

In general, I find materials showing ferroelectric, magnetic, or other types of ordering fascinating. Low-symmetry (monoclinic) phases we first observed (and linked to polarization rotation and to giant piezoelectric response) are bound to appear as very small (nano) domains that are regions of a few nanometers in size with slightly different crystal orientations. Due to the very small length scales, smaller than those defined by the coherent length of the X-rays used to characterize them, there is still controversy in the field (after 15 years) as to whether these are true monoclinic distortions at the unit cell level or if the twining of higher symmetry nanodomains produces low-symmetry diffraction patterns. What makes it relevant for my research is the realization that nanodomains and strain effects are essential to obtain new highly piezoelectric materials. The presence of phase transitions makes the physical properties of these materials highly sensitive to external stimuli, such as magnetic and electric fields, temperature, and stress, and they can, therefore, become functional. The manipulation of the physical responses requires deep understanding of the materials at the atomic level and a very fine control of their nanoscale. At the same time, the measurement of their response is challenging and often involves novel approaches and deep physical insight.

I plan to continue in my wonderful position as leader of the Functional Nano-materials team to play an active role in attracting students and promoting this exciting area of nanoscience.

3 Most Cited Publications

Title: A monoclinic ferroelectric phase in the $Pb(Zr_{1-x}Ti_x)O_{-3}$ solid solution
Author(s): Noheda, B; Cox, DE; Shirane, G; et al.
Source: Applied Physics Letters; volume: 74; issue: 14; pages: 2059–2061; published: April 5, 1999
Times Cited: 635 (from Web of Science)

Title: Origin of the high piezoelectric response in $PbZr_{1-x}Ti_xO_3$
Author(s): Guo, R; Cross, LE; Park, SE; et al.

Source: Physical Review Letters; volume: 84; issue: 23; pages: 5423–5426; published: June 5, 2000
Times Cited: 513 (from Web of Science)

Title: Tetragonal-to-monoclinic phase transition in a ferroelectric perovskite: the structure of $PbZr_{0.52}Ti_{0.48}O_3$
Author(s): Noheda, B; Gonzalo, JA; Cross, LE; et al.
Source: Physical Review B; volume: 61; issue: 13; pages: 8687–8695; published: April 1, 2000
Times Cited: 507 (from Web of Science)

On being a woman in this field . . .

As a mother of two young boys (a 7- and 3-year olds), I cannot think of a bigger challenge than managing and being satisfied in my role as mother, in addition to my role as scientist/teacher. I now travel to conferences much less often and for shorter periods because of our kids. I have fewer chances to have dinner or relaxing conversation with my colleagues overseas. I think that women in this situation miss chances for collaboration, to be visible and noted, and also possibilities to generate new ideas that often come through meetings and during discussions with colleagues. It is true that many of my male colleagues also have children and also reduce their trips frequency and duration, they also leave the office earlier in the evening than when they had no children, and they also come into the office in the morning with signs of having had little sleep. However, no matter how equitable I try to be, most of the time it's me who has more trouble ignoring the kids weeping before leaving on a trip and thus, I am the one that makes the bigger career sacrifices. Despite these remarks, I have already given about 70 invited seminars and talks in international conferences. Would my career have been the same if I had had my kids at a younger age?

Words of Wisdom

Listen carefully to everybody, but only follow your own advice. We often try to follow the path that other successful role models opened. However, women walk often very unique paths that lead to original and creative solutions, we should not lose this advantage.

Profile 73

Soon Ja Park

Emeritus Professor
Department of Materials Science and Engineering
Seoul National University
Seoul 151-744
Korea

e-mails: sunja5237@hanmail.net;
psj077650@gmail.com

Prof. Soon Ja Park.

Birthplace
Sinuiju, Korea
Born
January 7, 1933

Publication/Invention Record

ca. 75 publications: h-index 15 (Scopus)

Tags
❖ Academe
❖ Domicile: Korea
❖ Nationality: Korean
❖ Asian

Academic Credentials

Ph.D. (1971) Inorganic Materials Engineering, Seoul National University, Seoul, South Korea.
M.S. (1958) Analytical Chemistry, Seoul National University, Seoul, South Korea.
B.S. (1954) Chemical Engineering, Seoul National University, Seoul, South Korea.

Research Expertise: electronic ceramics, sensors

Other Interests: educating students of ceramics, fostering international relations

Key Accomplishments, Honors, Recognitions, and Awards

• Korean Academy of Science and Technology Award, 2008
• President, Materials Research Society of Korea, 1997–1998
• Outstanding Engineering Professor Award, Seoul National University, 1996
• Elected Academician of the World Academy of Ceramics, 1995
• Vice President, Materials Research Society of Korea, 1991–1995
• Elected Member of Korean Academy of Science and Technology, 1994
• Vice President, Korean Ceramic Society, 1992–1993

Successful Women Ceramic and Glass Scientists and Engineers: 100 Inspirational Profiles. Lynnette D. Madsen.
© 2016 The American Ceramic Society and John Wiley & Sons, Inc. Published 2016 by John Wiley & Sons, Inc.

- Vice President, Korean Sensors Society, 1992–1993
- Vice President, Korean Magnetics Society, 1991–1993
- Editorial Board Member, Journal of Materials Science: Materials in Electronics
- Advisory Board Member, Korea–Japan Joint Symposium on Chemical Sensor, 1991
- Program Committee Member, International Conference of Chemical Sensor, 1991
- Program Committee Member, Korea–Japan New Ceramic Seminar, 1990
- Korean Ceramic Society Award, Korean Ceramic Society, 1986

Biography

Early Life and Education

Soon Ja Park was raised and educated in Korea. She completed three degrees at Seoul National University—first, a bachelor's degree in chemical engineering; second, a master's degree in analytical chemistry; and finally, a doctorate in inorganic materials engineering.

Career History

She served at the National Industrial Research Institute as a ceramic scientist for nearly 20 years (1955–1973).

In the 1970s through to the mid-2000s, she returned to the Department of Inorganic Materials Engineering of Seoul National University for a faculty position. She is widely recognized as the first and one of the greatest woman ceramic educators in Korea. With a great deal of research and development experience in ceramic engineering, she has devoted herself to educating students in ceramics. More than 100 students (half at the master's and doctoral levels) have graduated under her professorship and guidance.

Significantly, given the history of Korean and Japanese engagement, she has played a key role in establishing Korea–Japan joint ventures in exchanging researchers, technical information, and business interests in various areas of advanced ceramics.

3 Most Cited Publications

Title: Influence of ZnO evaporation on the microwave dielectric properties of $La(Zn_{1/2}Ti_{1/2})O_3$
Author(s): Cho, S-Y; Seo, M-K; Hong, KS; Park, SJ; Kim, I-T
Source: Materials Research Bulletin; volume: 32; issue: 6; pages: 725–735; published: June 1997
Times Cited: 57 (from Scopus)

Title: Preparation of $BaTiO_3$ thin films by metalorganic chemical vapor deposition using ultrasonic spraying
Author(s): Kim, I-T; Lee, C-H; Park, SJ
Source: Japanese Journal of Applied Physics, Part 1: Regular Papers, Short Notes & Review Papers; volume: 33; issue: 9B; pages: 5125–5128; published: September 1994
Times Cited: 38 (from Scopus)

Title: Preparation of spherical SnO$_2$ powders by ultrasonic spray pyrolysis
Author(s): Lee, J-H; Park, S-J
Source: Journal of the American Ceramic Society; volume: 76; issue: 3; pages: 777–780;
published: March 1993
Times Cited: 34 (from Scopus)

Profile 74

Marina R. Pascucci

President
CeraNova Corporation
Marlborough, MA
USA

e-mail: mpascucci@ceranova.com
telephone: (508) 460-0300

Marina Pascucci preparing
for her first job after
graduate school. She
purchased her first business
suit for interviews; this
portrait was included as a
gift with the purchase!

Birthplace

Deer Park, Long Island, NY, USA

Publication/Invention Record

13 publications: h-index 5 (Web of Science)
2 patents

Tags

❖ Administration and Leadership
❖ Industry
❖ Domicile: USA
❖ Nationality: American
❖ Caucasian
❖ Children: 2

Proudest Career Moment (to date)

The proudest period in my career was serving as President of the American Ceramic Society (ACerS). I have been an active member of ACerS for close to 40 years, from when I was an undergraduate student. As graduate students at Case Western Reserve University (CWRU), we supported the local section in Cleveland. Later, I was an officer in both the Central Ohio and New England sections. I first became involved at the national level when I was employed at General Telephone and Electric (GTE) Laboratories and was mentored by William Rhodes, a colleague at GTE who was ACerS President at the time. I have served on numerous committees at the national level and on the Board of Directors. I have participated in significant changes within the Society—in terms of the Society's structure, physical moves of headquarters (twice), selections of

Successful Women Ceramic and Glass Scientists and Engineers: 100 Inspirational Profiles. Lynnette D. Madsen.
© 2016 The American Ceramic Society and John Wiley & Sons, Inc. Published 2016 by John Wiley & Sons, Inc.

At the PACRIM9 conference dinner in Cairns, Australia, July 2011 during my term as ACerS President. From left: Professor Akio Makishima (Past President of the Ceramic Society of Japan) with his wife, Marina Pascucci, and Charlie Spahr (Executive Director of ACerS).

three new Executive Directors, and changes in the Society's technical publications. I was honored and thrilled when I was nominated to serve as Society President. Although the actual term was 1 year (October 2010 to October 2011), it was a 3-year cycle since the President-Elect position and the Immediate Past-President position also involve significant responsibility. It was a very exciting time for me and I especially enjoyed traveling: internationally—meeting colleagues from all over the world and experiencing different cultures; and within the United States—meeting industry professionals, faculty members, and students who taught me so much about the breadth and depth of the activity within the ceramics community. Importantly, I learned that ACerS, while highly diverse, is also a close-knit, welcoming, supportive, and technically vibrant community.

Academic Credentials

Ph.D. (1984) Materials Science, Case Western Reserve University, Cleveland, Ohio, USA.
M.Sc. (1980) Ceramic Science, Case Western Reserve University, Cleveland, Ohio, USA.
B.Sc. (1977) Ceramic Science, Alfred University, Alfred, New York, USA.
B.A. (1977) Chemistry, Alfred University, Alfred, New York, USA.

Research Expertise: processing and characterization of transparent ceramics and composites for a wide variety of applications, mechanical properties characterization, transmission electron microscopy, examination and mitigation of radiation damage

With my friend, Katherine Faber (also a Past President of ACerS), and two of our Korean hosts during our trip to South Korea as keynote speakers for the International Workshop for Women Ceramists, November 2007, Seoul, South Korea. From left: Jong-Sook Lee (Professor at Chonnam National University, Gwangju, Korea), Marina Pascucci, Katherine Faber (Professor at CalTech), and Younghee Kim (Principal Research Scientist, Korea Institute of Ceramic Engineering and Technology).

Other Interests: project management and new business development, promoting diversity in the workplace, encouraging/mentoring students and young professionals

Key Accomplishments, Honors, Recognitions, and Awards

- President of CeraNova Corporation since 1997
- Keynote Speaker, "Facilitating Diversity—The Importance of Mentoring and Net-working," ASME 2014 8th International Energy Sustainability/Energy Sustainability (ES)–Fuel Cell 2014 Conference, Boston, MA, 2014
- Invited Speaker, "Don't Burn Those Bridges! The Impact of Networking and Professional Development," Emerging Professionals Symposium, MS&T, Montreal, Canada, 2013
- National Science Foundation, Committee of Visitors for the Division of Materials Research, 2011
- Keynote Speaker for the Opening Ceremony of the Inamori Kyocera Fine Ceramics Museum—Technical Ceramics: Past, Present and Future, Alfred University, 2011
- Invited Speaker, "Women in the Engineering World—Opportunities and Challenges," Alfred University Women of Influence Seminar Series, Judson Leadership Center, 2011
- American Ceramic Society (ACerS) President, 2010–2011

- Invited Keynote Speaker "Women in the Engineering World—Opportunities and Challenges," International Workshop for Women Ceramists, organized by the Committee for Women in Ceramics of the Korean Ceramic Society, Seoul, South Korea (in conjunction with the 50th Anniversary of the Korean Ceramic Society), 2007
- ACerS Fellow, 1999
- Alfred University Career Achievement Award, presented by the Alfred University Alumni Association, 2000
- F.H. Norton Distinguished Ceramist Award, presented by the New England Section of the American Ceramic Society, 1992

Biography

Early Life and Education

Marina Pascucci was born and raised in Deer Park, Long Island, NY, where she attended public schools and graduated from high school with a New York State (NYS) Regents Diploma. She is the granddaughter of Italian immigrants and the oldest child in her immediate family, both of which influenced her desire to excel.

Having been awarded a NYS Regents Scholarship covering full tuition, she elected to attend college in New York State. Although she had done well in both math and science, the option of majoring in engineering was never suggested to her by her high school guidance counselor. She applied to several colleges in New York as a potential math major. At her on-campus interview at Alfred University, she was encouraged to enroll in the NYS College of Ceramics dual-degree program. She graduated 4 years later with a B.S. degree in ceramic science and a B.A. degree in chemistry. While at Alfred, she also participated in the Study Abroad Program attending a Goethe Institute for intensive study in German language, working for 2 months at Siemens Kraftwerk Union (KWU) in Erlangen, and attending a semester of ceramic engineering and science classes at the University of Erlangen-Nürnberg.

After graduating from Alfred, Pascucci attended graduate school at CWRU in Cleveland, Ohio in the Department of Materials Science and Engineering. Her research focus was radiation damage in ceramics, in particular the amorphization of crystalline quartz upon exposure to ionizing radiation. Her major research tool was transmission electron microscopy (TEM). At CWRU, she had the opportunity to perform some of her graduate research at Imperial College (London), Oxford University, and Bristol University, using specialized TEM facilities and learning cutting-edge techniques, and interacting with world-renowned researchers and graduate students at each of those institutions.

Career History

Dr. Pascucci's first position after completing graduate school was with Battelle Laboratories in Columbus, Ohio, where she worked primarily on evaluating ceramics for nuclear waste storage, densification of ultrapure silica lenses, and processing of non-oxide ceramic–ceramic composites. She was also employed as a technical writer/reviewer for the Metals and Ceramics Information Center (MCIC), a U.S. Department of Defense sponsored center at Battelle, where she prepared feature articles and technical reviews for the MCIC Current Awareness Bulletin.

Marina Pascucci and Mark Parish (founder of CeraNova) in front
of the CeraNova booth at the American Ceramic Society New
England Section Expo Night, Marlborough, MA, October 2013.
Courtesy of NEACerS.

After several years at Battelle, she moved to the Boston area to work for GTE Laboratories in Waltham, MA, where she was a Senior Member of the Technical Staff in the Materials Science Laboratory. Her focus included processing and characterization of various oxide and non-oxide ceramic materials, both monolithic and composite. While at GTE, she began working on transparent ceramics for infrared domes, and was responsible for Taguchi experiment design, mechanical properties testing, and failure analysis of transparent yttrium oxide. She also interacted regularly with several GTE/Sylvania manufacturing groups specifically supporting testing and characterization efforts in polycrystalline alumina for advanced lighting products.

In 1992, Pascucci accepted a position as an Assistant Professor of Materials Science and Engineering in the Mechanical Engineering Department at Worcester Polytechnic Institute (WPI), located in Worcester, MA. In addition to teaching undergraduate and graduate classes, she advised both undergraduate projects and graduate theses in both the Mechanical Engineering Department and the Chemical Engineering Department. She also began collaborating with local industry, obtaining support for undergraduate and graduate projects. These collaborations resulted in joint presentations, publications, and proposals.

After collaborating with CeraNova Corporation for several years as a faculty member, Pascucci was asked to join the company as President in 1997. At CeraNova, she has been responsible for project management and contract administration for numerous Department of Defense contracts in the area of piezoelectric ceramics for active vibration control and, more recently, transparent polycrystalline ceramics for mid-wave infrared applications. Other responsibilities include proposal preparation, financial management, and new business development. Pascucci plays an active role in developing company strategy and direction. She has been instrumental in securing several large private sector contracts with the objective of applying CeraNova technology to meet the needs of both defense prime contractors and non-defense commercial customers.

It is not all work and no play. Marina Pascucci began playing the flute when she was 12 years old. She continues to play regularly in her church and with various wind ensembles in the Worcester, MA area. She and her husband, an avid sailor, have been sailing with their family around Narragansett Bay for the past 20 years: first on a Cal 25 and more recently on a Hunter 30. They have also sailed with her husband's family through the Florida Keys, the Abaco Islands, and the Bahamas. She participates regularly in an Italian conversation group, although admits that her comprehension is much better than her conversation! She also enjoys cooking (really a lot like chemistry!), traveling, and attending live theater—especially community theater performances, where her husband is often in the cast. She enjoys watching live professional dance, and has been a proud subscriber to and supporter of the Boston Ballet for nearly 30 years.

3 Most Cited Publications

Title: Dense perovskite, $La_{1-x}A'_xFe_{1-y}Co_yO_{3-\delta}$ ($A' =$ Ba, Sr, Ca), membrane synthesis, applications, and characterization
Author(s): Tsai, CY; Dixon, AG; Ma, YH; et al.
Source: Journal of the American Ceramic Society; volume: 81; issue: 6; pages: 1437–1444; published: June 1998
Times Cited: 160 (from Web of Science)

Title: Active PZT fibers, a commercial production process
Author(s): Strock, HB; Pascucci, MR; Parish, MV; et al.
Editor(s): Wuttig, M
Source: Proceedings of the Society of Photo-Optical Instrumentation Engineers (SPIE); volume: 3675; pages: 22–31; published: 1999
Times Cited: 28 (from Web of Science)

Title: Lateral cracks and microstructural effects in the indentation fracture of yttria
Author(s): Cook, RF; Pascucci, MR; Rhodes, WH
Source: Journal of the American Ceramic Society; volume: 73; issue: 7; pages: 1873–1878; published: July 1990
Times Cited: 26 (from Web of Science)

Challenges

The biggest challenge for me has been balancing career and family. When my children were toddlers, my husband and I were both on the faculty at WPI. Our schedules were hectic, but flexible, and we could arrange our time so that one of us could be home with the kids for dinner and bedtime. Since leaving WPI, we have had to deal with the challenges of longer commutes, business trips, snow days when school was cancelled, orthodontist appointments, and children at home from school when they were ill. Without immediate family members nearby, we were very fortunate to have found a wonderful daycare near our home and, later, several warm, reliable individuals who helped take care of our children. We could not have continued in our careers successfully without them. When you have two working parents with school age children, there is always some "juggling" and compromise necessary. Life can get pretty hectic, even chaotic at times, and we've had our share of very stressful situations. However, our kids are young adults now, and doing well, so we have all managed to survive—and thrive!

I could have chosen a different career path that involved more time "in the office," more travel, and higher compensation. But working in a small business has been challenging and rewarding and has afforded me the opportunity to control my own schedule and allowed me to spend time with my family.

On being a woman in this field . . .

I have been fortunate to have had many wonderful mentors and supporters throughout my academic and professional life, beginning with my parents and continuing to this day with many of the wonderful colleagues I have met through my membership and participation in the ACerS. Although female students were scarce when I was an undergraduate and graduate student, the faculty members at both Alfred University and CWRU were very encouraging and pushed all of us (women and men) to achieve. I occasionally had difficulties with male graduate students who came from cultures where women were not encouraged to attend engineering school or to join the workforce—they could not understand why I was not married and having children.

For the most part, my employers and fellow employees have all been supportive and fair. There were a few who were chauvinistic and some who were only interested in building their own kingdoms. In those cases, my philosophy has been to ignore/avoid people I don't enjoy working with, and associate with people who respect each other, praise individual efforts, and embrace teamwork. There were situations where I could have used a mentor, and I should have been more proactive about finding some.

I have always believed that balancing work and family is not a "women's issue"—it is a family issue. Family leave policies should apply to both women and men. I am in favor of policies that promote workplace flexibility—flextime, part-time, or work from home. Employers need to recognize that talented, productive employees have much to contribute—even if they are not physically in the office with regular hours. I believe that flexible workplaces contribute to economic competitiveness.

Words of Wisdom

You don't have to be the best, but you need to be the best you can be. Believe you can do anything you set your mind to. The only thing worse than failing is not even trying. Often the most successful people are those who learn best from their mistakes. Try not to listen to naysayers: "success is the best revenge."

Don't be afraid to move outside your comfort zone. It will force you to learn new things and continually challenge yourself. Don't be afraid to be a leader. Standing out in a crowd can be a good thing, if done correctly. Learn some leadership skills. Always remember that self-confidence is not the same as arrogance.

Find people you respect and admire—get to know them, observe them, and learn from them. Network, network, network. Learn how to shake hands and engage people in conversations. Learn and practice communication skills—writing, speaking, and most importantly *listening*—all are very important.

Work with people who enjoy their work and respect each other. The ability to get things done in any organization is all about relationships. Learn how to work with people from diverse backgrounds. The ability to understand and appreciate the perspective of others is useful in cooperative/team environments and negotiations.

Join a professional society and get involved. This is a great way to meet new people, network, acquire leadership skills, practice communication skills, and move outside your comfort zone. And, it also provides opportunities for giving back, e.g., through mentoring.

Julia M. Phillips

Previously Vice President and Chief
Technology Officer
Sandia National Laboratories
Albuquerque, NM 87185-0351
USA

e-mail: jmphill76@gmail.com

Julia Phillips in her molecular beam
epitaxy laboratory at AT&T Bell Laboratories
in about 1985.

Birthplace
Freeport, Illinois, USA

Born
1954

Publication/Invention Record
>150 publications: h-index 35 (Web of Science)
5 patents, editor of 2 books, 3 book chapters

Proudest Career Moment (to date)

I am particularly proud to have led the development of a focused research strategy for
Sandia National Laboratories.

Tags
❖ Administration and Leadership
❖ Government
❖ Domicile: USA
❖ Nationality: American
❖ Caucasian
❖ Children: 2

Academic Credentials

Ph.D. (1981) Applied Physics, Yale University, New Haven, Connecticut, USA.
B.S. (1976) Physics, College of William and Mary, Williamsburg, Virginia, USA.

Research Expertise: epitaxial insulating and metallic films on semiconductors; high-
temperature superconducting, ferroelectric, and magnetic oxide thin films; and novel
transparent conducting materials

Successful Women Ceramic and Glass Scientists and Engineers: 100 Inspirational Profiles. Lynnette D. Madsen.
© 2016 The American Ceramic Society and John Wiley & Sons, Inc. Published 2016 by John Wiley & Sons, Inc.

Other Interests: management of research, bringing together different research efforts into a coherent whole, science education, science and technology policy, professional society engagement, connecting science and society

Key Accomplishments, Honors, Recognitions, and Awards

- Chair Succession, Panel on Public Affairs, American Physical Society (APS) (Chair in 2016), 2015–2017
- Chair, Publications Quality Subcommittee, Materials Research Society (MRS), 2012–2015
- Member of the National Academy of Engineering (NAE) Council, 2008–2014
- Chair Succession, APS Topical Group on Energy Research and Applications, 2010–2013
- Board of Directors of the American Association for the Advancement of Science (AAAS), 2009–2012
- Editorial Board, MRS Communications, 2011–2012
- Elected Fellow of the MRS, 2010
- Chair of the APS Division of Condensed Matter Physics, 2005–2009
- George E. Pake Prize from the APS for leadership and pioneering research in materials physics for industrial and national security applications, 2008
- Elected Fellow of the American Academy of Arts and Sciences, 2005
- Principal Editor, Journal of Materials Research, 1990–2005
- Elected member of the NAE, 2004
- Wilbur Cross Medal, Yale University (awarded to distinguished graduate alumni), 2004
- Elected Fellow of AAAS, 2002
- U.S. Department of Labor Women's Bureau Horizon Award, 2002
- Lifetime National Associate of the U.S. National Academies, 2002
- Editorial Board of the Applied Physics Reviews, 1998–2001
- Public Broadcasting Service (PBS) Advisory Committee, 1 hour documentary on the invention of the transistor, 1998
- MRS Woody Award (in recognition of outstanding service and dedication on behalf of MRS), 1996

Julia Phillips in her office around February 2004, shortly after her election to the National Academy of Engineering was announced. Credit: Randy Montoya, Sandia National Laboratories.

- President of the MRS, 1995
- Editorial Board of the Applied Physics Letters and Journal of Applied Physics, 1992–1994
- Elected Fellow of APS, 1993

Biography

Early Life and Education

There are several factors that likely influenced Julia into a career in science. First, her father and both grandfathers were doctors and overall this gave a technical bent to her family. She believes that her mother might well have become a scientist if she had grown up in a different time and place. Her mother was the source of a constant supply of science kits for both her and her brother and her mother encouraged them to participate in science fairs. In addition, Julia came from a small town where the stargazing was excellent; she was further influenced by an outstanding teacher in fifth grade—together these things pointed her to a future in astronomy. She was also attracted to science by its quantitative and objective, rather than qualitative and subjective, attributes.

Julia had decided in high school that she wanted to attend the College of William and Mary, attracted by the extremely high admission standards for out-of-state women and its status as the second-oldest college in the United States. Since W&M did not have astronomy as a major, she selected physics. The most important thing was that it was science, something quantitative with answers, and not so subjective. She recognized that physics was hard, but she liked challenges, so this was fine. She was not disturbed or put off by the fact that there were not so many women in physics, rather she kind of liked it and described herself as a maverick (at that time). Also, at that point, the "boy" personalities seemed much more straightforward with fewer (or less complex) hidden agendas.

Career History

Upon receipt of her Ph.D., Julia moved to AT&T Bell Laboratories where she spent 14 years conducting research. From 1981 to 1988, she was a member of technical staff conducting research on epitaxial thin films. She called her research "lunatic fringe molecular beam epitaxy," since she was growing single crystalline films of metals, insulators, and semiconductors on each other, in contrast to most of the then current work that concentrated only on semiconductor thin film

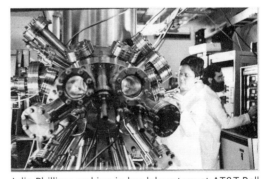

Julia Phillips working in her laboratory at AT&T Bell Laboratories in about 1987, growing crystalline insulating CaF_2 on semiconducting Si by molecular beam epitaxy. Credit: AT&T Bell Laboratories. (See insert for color version of figure.)

growth. From 1988 to 1995, she was the technical manager of the thin film research group. During this period, she worked on superconductors, ferroelectrics, magnetic oxides, optical materials, and transparent conductors. She examined film growth by pulsed laser deposition, off-axis sputtering, and evaporation. She correlated the structural and electrical/magnetic properties and evaluated the device possibilities of these new materials.

In 1995, she came to Sandia National Laboratories where she held various positions with successively more responsibility:

- 1995–2000: Manager, Materials & Process Computation & Modeling Department (1998–2000), Surface & Sensor-Controlled Processes Department (1995–1998)
- 2000–2001: Deputy Director, Materials & Process Sciences Center
- 2001–2010: Director, Physical, Chemical, & Nano Sciences Center
- 2005–2007: Director, Center for Integrated Nanotechnologies (CINT) (Sandia and Los Alamos National Laboratories)
- 2010–2011: Director, Nuclear Weapons Science & Technology Programs
- 2011–2013: Deputy Chief Technology Officer and Director, Laboratory Research Strategy & Partnerships
- 2013–2014: Vice President and Chief Technology Officer

In her most recent position, her responsibilities included leadership of Sandia's $165 million Laboratory Directed Research and Development (LDRD) program, providing research strategy development and implementation, and ensuring intellectual property protection and deployment.

Julia Phillips discussing Sandia's new research strategy, ~2014. Credit: Randy Montoya, Sandia National Laboratories. (See insert for color version of figure.)

3 Most Cited Publications

Title: Very large magnetoresistance in perovskite-like La-Ca-Mn-O thin films
Author(s): McCormack, M; Jin, S; Tiefel, TH; et al.
Source: Applied Physics Letters; volume: 64; issue: 22; pages: 3045–3047; published: May 30, 1994
Times Cited: 435 (from Web of Science)

Title: Single-crystal epitaxial thin films of the isotropic metallic oxides $Sr_{1-x}Ca_xRuO_3$
Author(s): Eom, CB; Cava, RJ; Fleming, RM; et al.
Source: Science; volume: 258; issue: 5089; pages: 1766–1769; published: December 11, 1992
Times Cited: 406 (from Web of Science)

Title: Fabrication and properties of epitaxial ferroelectric heterostructures with (SrRuO$_3$) isotropic metallic oxide electrodes
Author(s): Eom, CB; Van Dover, RB; Phillips, JM; et al.
Source: Applied Physics Letters; volume: 63; issue: 18; pages: 2570–2572; published: November 1, 1993
Times Cited: 351 (from Web of Science)

Challenges

I had an unfortunate experience when I was in graduate school. About 9 months before I was ready to defend my thesis, my advisor had a nervous breakdown and ceased to function effectively as my advisor. These were certainly difficult and trying circumstances in which to finish up one's doctorate. However, I have a takeaway from this experience. At the time, I didn't know where to turn, but people (seeing and recognizing my predicament) reached out to me. Now, I appreciate that it would have been better to have reached out to them and talked to people and simply asked for help.

On being a woman in this field . . .

Now with a career's worth of experience, the biggest thing is to really understand that there are strengths and tendencies that women bring to the field that differ from the dominant male culture that are beneficial—new ways of looking at problems. The challenge is to recognize those strengths and be constructive in moving forward (complementing rather than railing against the dominant culture). I add that most often the burden is still on the women to ensure that this is all done in the best way possible.

The nature of women's early career contributions may be different from those of men, but these are also significant. It is important to make sure they are recognized, valued, and rewarded, even in a male-dominated organization.

Women need a champion, more so than men, but don't always get one. You need opportunities to show what you can do and someone farther along in their career who believes in you and will help with self-confidence, visibility, and mentoring and support.

Julia Phillips with her daughters at "Take Our Daughters to Work Day," April 22, 2004. Bridget Connor (on the left) is now a graduate student pursuing her Ph.D. in chemistry at Stanford University. Julia Connor (right) is pursuing a Master's degree in violin performance. Incidentally, the round things on the door in the background are from my MBE days—Si wafers that were failed experiments (so bad, it wasn't even worth cutting them up to analyze!). Credit: Sandia National Laboratories.

Words of Wisdom

I would say that there are two key things: (1) Stay positive about your work (while being realistic). Focus on what you *can* do. Pick your battles (choose the important and winnable battles), take action, and don't be afraid to take some chances. *You* are the only one who is really responsible for your career and your success. That said, recognize that you don't have to do everything alone. (2) Build your network so that there are seemingly invisible hands that will help you. Keep in mind, you must use your network—this works in both directions—pay into it, so benefits can be reaped at another time.

Tatiana Alexeevna Prikhna

Head of the Department of Technologies of Superhigh
Pressures, Functional Structured Ceramic Composites and
Dispersed Nanomaterials
V. Bakul Institute for Superhard Materials (ISM)
National Academy of Sciences of Ukraine
Professor of Chemistry
Kiev National University of Building and Architecture
Kiev
Ukraine

e-mail: prikhna@mail.ru
telephone: +38 (044) 430-11-26

Tatiana Prikhna when she
was around 25 years old.

Birthplace
Kiev, Ukraine (former USSR)
Born
September 22, 1957

Publication/Invention Record
>175 publications: h-index 11 (Web of Science)
24 patents, 5 books, and book chapters

Tags

- ❖ Administration and Leadership
- ❖ Government
- ❖ Domicile: Ukraine
- ❖ Nationality: Ukrainian
- ❖ Caucasian
- ❖ Children: 1

Proudest Career Moments (to date)

I am very proud of each accomplishment in my career: (i) In the early days of winning a gold medal when I graduated from secondary school in Kiev to more recently such as when (ii) I became the head of one of the largest research departments in the world-renowned V. Bakul Institute for Superhard Materials of the National Academy of Ukraine, (iii) when I was awarded a Diploma of conferred title of Professor from the Chair of Chemistry at the Kiev National University of Building and Architecture, (iv) when I received the Diploma as Academician of the World Academy of Ceramics (in Italy), and (v) when I was elected as Corresponding Member of the National Academy of Sciences of Ukraine.

Successful Women Ceramic and Glass Scientists and Engineers: 100 Inspirational Profiles. Lynnette D. Madsen.
© 2016 The American Ceramic Society and John Wiley & Sons, Inc. Published 2016 by John Wiley & Sons, Inc.

Academic Credentials

Equivalent to Habilitation (1998) Materials Science, National Academy of Sciences of Ukraine, Kiev, Ukraine.
Diploma (1990) Physics, Kiev State University, Kiev, Ukraine.
Equivalent to Ph.D. (1986) Materials Science, National Academy of Sciences of Ukraine, Kiev, Ukraine.
Diploma (1980) Mechanical Engineering, Kiev Polytechnical Institute, Kyiv, Ukraine.

Research Expertise: superconducting materials (bulk, thin films, wires), MAX phases, ceramic composites, superhard materials, dispersed nanomaterials, single crystals of aluminium dodecaborides, techniques of high pressure, spark plasma sintering, hot pressing, vacuumed and pressureless sintering, Auger spectroscopy, scanning electron microscopy (SEM), transmission electron microscopy (TEM), X-ray diffraction, measurement of superconducting and mechanical properties

Other Interests: international collaborations (e.g., with France, Germany, Austria, Sweden, Hungary, Greece, Israel, the United States, and Russia)

Key Accomplishments, Honors, Recognitions, and Awards

- Member of the Editorial Board of the journal Superhard Materials (Springer), 2010 to present
- Member of the Editorial Board of Ceramics International, 2007 to present
- Elected Corresponding Member of the National Academy of Sciences of Ukraine (NASU), 2006 to present
- Elected Academician of the World Academy of Ceramics, 2006 to present
- Member of the International Board of European Conference of Applied Superconductivity (EUCAS), 2009 and 2013
- Order of Princess Olga of III Grade, 2012
- Honorary Mark "Ushinskiy K.D." of Academy of Pedagogical Sciences of Ukraine, 2011
- Member of the Editorial Board of the European Society for Applied Superconductivity (ESAS), 2005–2010
- Honorary Diploma of Counsel of Ministers of Ukraine, 2008
- Honorary Medal for the 90th Anniversary of the National Academy of Sciences of Ukraine, 2008
- Conferred the title of Professor from the Chair of Chemistry at the Kiev National University of Building and Architecture, 2007
- Conferred the title of Assistant Professor from the Chair of Chemistry at the Kiev National University of Building and Architecture, 2004
- Bronze Medal of the Exhibition of Achievements of the National Economy of the Soviet Union, 1987
- Awarded a gold medal upon graduation from secondary school in Kiev, 1974

Biography

Early Life and Education

In 1974, Tatiana finished secondary school in Kiev and was awarded a gold medal. She continued her studies at Kiev Polytechnical Institute and received a Diploma (honors degree) in Mechanical Engineering in 1980; her focus was the *Machinery and Process Technology of Polymeric Materials.*

She undertook postgraduate courses at the ISM (materials in machine building) from 1982 to 1985. In 1986, she became a candidate of engineering sciences (equivalent to a Ph.D.) in materials science of machine-building from ISM. From 1989 to 1990, she studied at the Special Faculty of the T.G. Shevchenko Kiev State University (in parallel with her work in the ISM NASU) and in 1990 she was awarded a second Diploma of the University for high education in high-temperature superconductivity (physics). In 1990, she was conferred the title of

Tatiana Prikhna in the laboratory. (See insert for color version of figure.)

Senior Scientist. She then continued her studies with a period of postdoctoral courses at the Institute for Superhard Materials (in Materials Science in Machine Building) from 1992 to 1995. In 1998, she became Doctor of Engineering Sciences (in Materials Science) at ISM.

Career History

In 2000, in parallel with research work in ISM NASU, she started lecturing at the Chair of Chemistry at the Kiev National University of Building and Architecture in the position of Assistant Professor and since 2001 she occupies a Professor position at this University. She was conferred the title of Assistant Professor by the Chair of Chemistry at the Kiev National University of Building and Architecture in 2004, and then the title of Professor in 2007. One-quarter of her time is dedicated to this position.

Tatiana Prikhna in 2009.

Dr. Tatiana Prikhna (a former Champion of Ukraine in windsurfing) near the coast of Hawaii during 1997 International Workshop on Superconductivity, Big Island, Hawaii. (See insert for color version of figure.)

Concurrently, during 2005–2012, she served as the Head of the Department of Advanced High-Pressure Technologies, Dispersed Materials and Sinter Processes of Ceramics at ISM. She is currently Head of the Department № 7: Technologies of superhigh pressures, functional structured ceramic composites and dispersed nanomaterials at ISM where she is responsible for 51 personnel.

Through these years, she received several recognitions, including the title Corresponding Member of the National Academy of Sciences of Ukraine (in 2006) and election to the World Academy of Ceramics as an Academician (also in 2006).

In her spare time, Tatiana windsurfs and she is quite excellent at it. In 1982, she won both the Gold Medal Cup of Ukraine and the Gold Medal Championship of the Ukraine in Windsurfing.

3 Most Cited Publications

Title: Batch-processed melt-textured YBCO with improved quality for motor and bearing applications
Author(s): Gawalek, W; Habisreuther, T; Zeisberger, M; et al.
Source: Superconductor Science & Technology; volume: 17; issue: 10; pages: 1185–1188; article number: PII S0953-2048(04)73137-1; published: October 2004
Times Cited: 31 (from Web of Science)

Title: High-pressure synthesis of MgB_2 with addition of Ti
Author(s): Prikhna, TA; Gawalek, W; Savchuk, YM; et al.
Source: Physica C-Superconductivity and Its Applications; volume: 402; issue: 3; pages: 223–233; published: February 15, 2004
Times Cited: 29 (from Web of Science)

Title: High-pressure synthesis of a bulk superconductive MgB_2-based material
Author(s): Prikhna, TA; Gawalek, W; Savchuk, YM; et al.
Source: Physica C-Superconductivity and its Applications; volume: 386; pages: 565–568; published: April 15, 2003
Times Cited: 18 (from Web of Science)

Profile 77

Karin M. Rabe

Board of Governors Professor of
Physics and Astronomy
Rutgers, The State University of
New Jersey
136 Frelinghuysen Road
Piscataway, NJ 08854-8019
USA
e-mail: rabe@physics.rutgers.edu
telephone: (848) 445-9030

Karin Rabe at age 35, hiking in the Swiss Alps in the summer, in a week between two conferences in Europe. (See insert for color version of figure.)

Birthplace

New York, NY, USA

Born

April 1, 1961

Publication/Invention Record

>150 publications: h-index 46 (Web of Science)
1 book (edited), 2 book chapters (coauthored)

Tags

❖ Academe
❖ Domicile: USA
❖ Nationality: American
❖ Caucasian
❖ Children: 1

Proudest Career Moment (to date)

One moment that stands out for me was at the American Physical Society (APS) March Meeting at which I received the David Adler Award in Materials Physics. One of my former postdocs, Nicola Spaldin (now at ETH Zurich), organized a celebration dinner with about 20 of my past and present group members and close collaborators. I regard the mentoring of my group members as one of my most significant contributions to the field, and it was a very happy moment for me to see all the success and energy that was gathered together for that dinner.

A 2013 good-bye lunch for a postdoc working with Karin Rabe in the "electronic structure" group at Rutgers.

Successful Women Ceramic and Glass Scientists and Engineers: 100 Inspirational Profiles. Lynnette D. Madsen.
© 2016 The American Ceramic Society and John Wiley & Sons, Inc. Published 2016 by John Wiley & Sons, Inc.

Academic Credentials

Ph.D. (1987) Physics, Massachusetts Institute of Technology, Cambridge, MA, USA.
A.B. (1982) Physics, Princeton University, Princeton, NJ, USA.

Research Expertise: theoretical investigation of ferroelectrics and related materials and magnetic and nonmagnetic martensites; use of first-principles density functional methods both directly and in the construction of first-principles effective Hamiltonians for theoretical prediction and analysis of properties of materials, both real and as-yet hypothetical, in bulk and thin film forms

Other Interests: beyond doing good science myself, my main interest is in facilitating good science by others. I strongly believe the best teaching is one-on-one, where the teacher guides and supports the student and targets individual weaknesses and strengths. In mentoring Ph.D. students and postdocs, while still providing needed guidance, I aim to promote independence by providing resources and opening the doors of opportunity. Through my involvement in the Aspen Center for Physics, the primary function of which is to create an environment that nurtures great science, I can work to do this on an even larger scale.

Key Accomplishments, Honors, Recognitions, and Awards

- President of the Aspen Center for Physics, 2013 to present
- National Academy of Sciences, elected member, 2013
- American Academy of Arts and Sciences, elected member, 2013
- Fellow of the AAAS, 2011
- David Adler Lectureship Award in Materials Physics, 2008
- Promotion from Professor I to Professor II, Rutgers University, 2004
- Fellow of the American Physical Society, 2003
- Arthur Greer Memorial Prize, Yale College, 1994
- Alfred P. Sloan Research Fellowship, 1993
- Junior Faculty Fellowship in the Natural Sciences, Yale University, 1991
- NSF Presidential Young Investigator, 1990
- Clare Boothe Luce Professorship (5-year term), 1989
- Phi Beta Kappa, 1982
- George B. Wood Legacy Prize (first in junior class), 1982
- NSF Graduate Fellowship, 1982
- Ida M. Green Fellowship, 1982
- AT&T Graduate Research Program for Women, fellowship, 1982

Biography

Early Life and Education

I grew up in New York City, with one sister 2 years younger. My dad was an engineer and enjoyed teaching me maths; I remember him teaching me algebra in fourth grade. I tried many different types of schools—a Catholic school, an all-girls private school, Hunter College High School—ending up happily at the Bronx High School of Science, which even with the New York City near-bankruptcy of the 1970s was a great place to develop an interest in science. At the Science Honors Program, a Saturday program for high school students at Columbia University, I took a course in modern physics and even though I didn't understand much, it intrigued me enough that I decided to start my undergraduate studies at Princeton as a physics major, and see how I liked it.

At Princeton, I took the honors physics course first semester, and really struggled with mechanics (I still find the word "torque" scary even though I now can teach the course myself). I was able to pull though, thanks to support from the section instructor, Robert Schrock (now at SUNY Stony Brook), and help from the smartest guy in the class, Greg Moore, who turned out to be a great teacher. As it happens, I am still learning from him, as we have now been married for over 26 years. That spring, I also fell in love with electricity and magnetism, and everything started to fall into place.

My graduate work at MIT is the cornerstone of my career. I was lucky to join the group of John Joannopoulos, who has a well-deserved reputation as one of the greatest thesis advisors of all time, and to get in almost on the ground floor with first-principles calculations of the structure and properties of crystals, in my case specifically of structural phase transitions. Over the last 30 years, these techniques have developed to the point that we can use them to make predictions of interesting functional properties of materials that are welcomed as guidance by experimentalists in the discovery and development of new materials.

Career History

The start of my postdoc at Bell Labs almost exactly coincided with the discovery of high-Tc cuprate superconductors. Everyone dropped everything they were doing to apply their technique to the cuprates. It was a unique opportunity to learn about a wide variety of experimental characterization techniques, although the capability of first-principles calculations for studying cuprates was limited. This forced me to learn other approaches, including "structural diagram" data mining techniques and phenomenological modeling of complex crystal structures.

As a junior faculty member in Applied Physics at Yale, I started to build a group, returning to first-principles studies of structural phase transitions, specifically ferro-electrics. I started with two outstanding students, Umesh Waghmare, now a professor at the Jawaharlal Nehru Centre for Advanced Scientific Research in Bangalore, and Serdar Ogut, now a professor at University of Illinois at Chicago Circle, and then got up the courage to hire postdocs: first Eric Cockayne (now at National Institute of Standards and Technology (NIST)) and then Nicola Spaldin (now at ETH Zurich) and Philippe Ghosez (now at the University of Liege in Belgium). Just before I left for Rutgers University, I started working on understanding ferroelectricity in thin films and superlattices, which turned out to be an important problem that opened up a whole experimental/theoretical field for productive interactions.

My move to Rutgers was the result of their offer to my string-theorist husband, but scientifically it was one of the best moves I could have made. I went from relative isolation at Yale to join a dynamic "electronic structure" group with David Vanderbilt and David Langreth, with lots of students, postdocs, and visitors getting together at lunch and weekly group meetings. As David Langreth unfortunately passed away a few years ago, we are now down to a group of two, but my ongoing collaborations with David Vanderbilt on a variety of projects related to first-principles calculations, modeling, and predictions of functional materials are a great source of scientific pleasure.

3 Most Cited Publications

Title: Epitaxial $BiFeO_3$ multiferroic thin film heterostructures
Author(s): Wang, J; Neaton, JB; Zheng, H; et al.
Source: Science; volume: 299; issue: 5613; pages: 1719–1722; published: March 14, 2003
Times Cited: 2609 (from Web of Science)

Title: Structural anomalies, oxygen ordering and superconductivity in oxygen deficient $Ba_2YCu_3O_x$
Author(s): Cava, RJ; Hewat, AW; Hewat, EA; et al.
Source: Physica C; volume: 165; issue: 5–6; pages: 419–433; published: February 15, 1990
Times Cited: 993 (from Web of Science)

Title: Physics of thin-film ferroelectric oxides
Author(s): Dawber, M; Rabe, KM; Scott, JF
Source: Reviews of Modern Physics; volume: 77; issue: 4; pages: 1083–1130; published: October 2005
Times Cited: 881 (from Web of Science)

Challenges

My biggest challenge is to juggle teaching, administration, outreach, and home responsibilities to carve out extended periods to concentrate on physics research.

On being a woman in this field . . .

The best advice I got on this was from Jim Phillips, a senior colleague at Bell Labs: To not worry about any extra advantage one might have by being a woman, since everyone has some kind of advantage of which one would be foolish not to make the most—in any case, there are plenty of difficulties to compensate for any advantages one might receive.

Words of Wisdom

The older I get, the more I find I enjoy giving advice. For new women faculty, the single most important piece of advice is that sometimes you have to be tough to be nice; in particular, it's painful to cut an unsuccessful research student loose, but every time it turns out that this made her or him find the thing they were good at. The other thing is to remember that the reason we do science is for that feeling you get when after puzzling for a long time over a problem, you figure something out and it all starts to become natural and simple.

Profile 78

Nelly M. Rodriguez

President
Catalytic Materials LLC
325 Heartland Drive
Pittsboro, NC 27312
USA

e-mail: rodriguez@catalyticmaterials.com
telephone: (919) 704-5736

N.M. Rodriguez attending a
NATO meeting in the
Algarve, Portugal.

Birthplace
Topaga, Colombia
Born
July 5, 1953

Publication/Invention Record

>100 publications: h-index 27 (Web of Science)
28 patents, 1 Ed. book

Tags
❖ Administration and
Leadership
❖ Industry
❖ Domicile: USA
❖ Nationality: American
❖ Hispanic
❖ Children: 0

Proudest Career Moment (to date)

When I walked away from an academic position to head my own research company and making this venture a success.

Academic Credentials

Ph.D. (1986) Chemistry, University of Newcastle upon Tyne, Tyne and Wear, UK.
B.S. (1978) Chemistry, Universidad Nacional de Colombia, Bogota, Colombia.

Research Expertise: *in situ* electron diffraction, graphene nanofibers, carbon, catalysis, carbon nanotubes, materials for hydrogen storage, materials for energy storage

Other Interests: technology transfer, mentoring of women and girls to follow science and engineering careers

Key Accomplishments, Honors, Recognitions, and Awards

• National Science Foundation Committee of Visitors for the Division of Chemistry, 2012–2013

Successful Women Ceramic and Glass Scientists and Engineers: 100 Inspirational Profiles. Lynnette D. Madsen.
© 2016 The American Ceramic Society and John Wiley & Sons, Inc. Published 2016 by John Wiley & Sons, Inc.

- Winner of the outstanding Commercial Achievement Award for participating in the DoE assistance program, 2006
- General Electric Award for the retention of engineering students, 2000
- NATO Fellowships for participation in Workshops in Europe, 1987 and 1989
- Overseas Students Fellowship—Principals of British Universities, 1984–1986
- Colombian Institute of Educational Credit and Technical Studies Abroad (ICETEX) Scholarship for advanced studies abroad, 1983–1985
- Universidad Nacional de Colombia Chemist's Thesis—Meritorious, 1979

Biography

Early Life and Education

N.M. Rodriguez was born in a small coal mining village in the mountains of Colombia. After enrolling in the local primary school for a year, her family moved to a larger town where opportunities for education were vastly improved. After graduating from high school at the age of 16, she was admitted to the School of Chemistry at the largest state University. Her thesis was entitled "Studies of the $OH-\pi$ Interaction by Infrared Spectroscopy and Solution Calorimetry." Following several years of teaching at the Universidad Industrial de Santander, she was then accepted at the graduate school at the University of Newcastle upon Tyne in England in 1983. As part of her studies, she was invited researcher and worked at the University of Hokkaido, Japan, Airco Carbon, Niagara Falls, NY, and at the Corporate Research Laboratories at Exxon R&Eng Company in Clinton, NJ. She obtained her Ph.D. in Chemistry of Materials in 1986.

N.M. Rodriguez with colleague and former graduate students at an American Chemical Society meeting; from left, Dr. Chris Marotta, Prof. Terry Baker, Dr. Nelly Rodriguez, Dr. Olinda Carneiro, Dr. Paul Anderson, Dr. Kate Laubernds, and Dr. Elizabeth Engel.

Career History

After obtaining a Ph.D. at the University of Newcastle upon Tyne, Rodriguez joined the research group of Dr. Terry Baker who is considered by many as the father of carbon nanotechnology as a postdoctoral fellow in 1986. She then moved to the Pennsylvania State University as an Assistant and later Associate Professor of Materials. In 1995, Rodriguez cofounded Catalytic Materials LLC, with the mission of becoming an organization devoted to the commercial development of the most advanced carbon nanomaterials. She then accepted a position of Associate Professor of Chemistry at Northeastern University in Boston. Following a successful academic career, she subsequently took the full-time position as President of Catalytic Materials. Using very advanced techniques and methods, Catalytic Materials under her leadership has acquired a formidable patent portfolio with technologies not only for the manufacture but also for applications of advanced carbon nanomaterials in areas such as advanced Li^+ batteries, fuel cells, and graphene production. She developed a method for the manufac-

N.M. Rodriguez working at the controlled atmosphere, electron microscope in 1990. The photograph was taken by visitors of Tonen Corporation of Japan. (See insert for color version of figure.)

ture of very thin, long multiwalled carbon nanotubes that are free of impurities and crystalline defects. The company has reached commercialization with these technologies being implemented in Asia and very soon in the United States.

3 Most Cited Publications

Title: A review of catalytically grown carbon nanofibers
Author(s): Rodriguez, NM
Source: Journal of Materials Research; volume: 8; issue: 12; pages: 3233–3250; published: December 1993
Times Cited: 705 (from Web of Science)

Title: Hydrogen storage in graphite nanofibers
Author(s): Chambers, A; Park, C; Baker, RTK; et al.
Source: Journal of Physical Chemistry B; volume: 102; issue: 22; pages: 4253–4256; published: May 28, 1998
Times Cited: 704 (from Web of Science)

Title: Graphite nanofibers as an electrode for fuel cell applications
Author(s): Bessel, CA; Laubernds, K; Rodriguez, NM; et al.
Source: Journal of Physical Chemistry B; volume: 105; issue: 6; pages: 1115–1118;
published: February 15, 2001
Times Cited: 442 (from Web of Science)

Challenges

Having been born in a small village, the greatest difficulty was in not only gaining acceptance but also finding the funds to attend a suitable school. I had to travel a great distance at a high cost to my family in order to reach the dream of obtaining a degree in science. The quest for a Ph.D. then became the next challenge as I had to master and defend my thesis in a second language and once again find personal funding for this opportunity.

On being a woman in this field . . .

It has become evident through my many professional years that being a woman is perhaps the greatest handicap that anyone working in science and engineering can experience. Engineering in the United States is still considered a man's world. Women often have to work harder in order to attain the same degree of recognition to that of men under equivalent circumstances. During my academic years, I found that agencies very often select panels consisting of men who do not believe women are capable of being competitive with men on an equal basis.

Words of Wisdom

It is possible to overcome most difficulties in life and in one's career through education and hard work. There still is a long way to convince the world that women are quite capable of taking on challenges of science and engineering, but we are making progress.

Profile 79

Debra R. Rolison

Section Head, Advanced Electrochemical Materials
U.S. Naval Research Laboratory (NRL)
Washington, DC 20375
USA

e-mail: rolison@nrl.navy.mil
telephone: (202) 767-3617

An electrochemist views the
world from inside a fullerene.
Credit: Robert J. Nowak. (See
insert for color version of
figure.)

Birthplace
Sioux City, IA, USA
Born
May 26, 1954

Publication/Invention Record
>125 publications: h-index 45 (Web of Science)
34 patents, 1 book (coauthored), 6 book chapters

Tags
❖ Administration and Leadership
❖ Government
❖ Domicile: USA
❖ Nationality: American
❖ Caucasian
❖ Children: 0

Proudest Career Moment (to date)

My pivotal career moment predated my life as a professional scientist—emphasis on "professional" because observant children are also scientists. I trace my willingness to fly in the face of expectations (in picking scientific problems, in challenging the too-white, too-male paradigm in science) to when my mother divorced my father in the early 1960s—a time when that schism was barely accepted by society and even less accepted when the family on both sides was Catholic. To recognize at age 7 that societal and religious authority figures—all male—did not have a grasp of the reality on the ground freed me forever from investing blind acceptance in their dictates. A scientist embarking on a life in science without carting about such baggage is lucky indeed.

Academic Credentials

Ph.D. (1980) Chemistry, University of North Carolina (UNC), Chapel Hill, NC, USA.
B.S. (1975) Chemistry, Florida Atlantic University (FAU), Boca Raton, FL, USA.

Research Expertise: design, synthesis, characterization, and application of nano-structured materials with emphasis on ultraporous nanoarchitectures for rate-critical

Successful Women Ceramic and Glass Scientists and Engineers: 100 Inspirational Profiles. Lynnette D. Madsen.
© 2016 The American Ceramic Society and John Wiley & Sons, Inc. Published 2016 by John Wiley & Sons, Inc.

applications such as catalysis, energy storage and conversion, porous compliant magnets, and sensors

Other Interests: equity in science and Title IX, writing for the public, mentoring

Key Accomplishments, Honors, Recognitions, and Awards

• Editorial Advisory Board, Advanced Energy Materials, 2012 to present

• Editorial Advisory Board, Journal of Electroanalytical Chemistry, 1995–2004 (1st woman appointed to the Board) and 2007 to present

• Appointed Adjunct Professor of Chemistry, University of Utah, Salt Lake City, UT, 2000 to present

• Award in Electrochemistry from the Division of Analytical Chemistry of the American Chemical Society (ACS), 2014

• Alan Berman Publication Award for "Dual-function air cathode nanoarchitectures for metal–air batteries with air-independent pulse power capability," Advanced Energy Materials (2013), NRL to J.W. Long, C.N. Chervin, N.W. Kucko, E.S. Nelson, and D.R. Rolison, 2014

Dr. Debra Rolison received the C.N. Reilley Award in 2012. Credit: Jonathan Howell.

• NRL Technology Transfer Royalty Award, 2012

• C.N. Reilley Award, Society for Electroanalytical Chemistry, 2012

• Hillebrand Prize, Chemical Society of Washington, 2012

• Elected Fellow, ACS, 2011

• Award in the Chemistry of Materials, ACS, 2011

• NRL Edison Patent Award, NRL, 2011

• Editorial Advisory Board, Annual Review of Analytical Chemistry, Inaugural Board, 2006–2010

• NRL Edison Chapter of Sigma Xi Award in Pure Science (first woman to have been presented the award, given annually at the NRL since 1955), 2009

Being the first woman to receive the American Chemical Society Award in the Chemistry of Materials in 2011 was a thrilling honor made even more significant by receiving the award from (right) Uma Chowdhry (representing the award sponsor E.I. du Pont de Nemours and Company) and (left) Nancy Jackson, then ACS President. Credit: John N. Russell, Jr.

- A.K. Doolittle Award for the best paper presented before the Polymer Science & Engineering (PMSE) Division at the 236th American Chemical Society Meeting, Philadelphia, PA, 2009
- Elected Materials Research Society (MRS) Fellow (Inaugural Class), 2008
- R.A. Glenn Award for the best paper presented before the Division of Fuel Chemistry at the 231st ACS Meeting, Atlanta, GA, 2007
- Elected Association for Women in Science (AWIS) Fellow, 2006
- Distinguished Alumna of the Charles E. Schmidt College of Science, Florida Atlantic University, 2005
- Editorial Advisory Board, Nano Letters, Inaugural Board, 2000–2004
- Editorial Advisory Board, Encyclopedia of Nanoscience and Nanotechnology, Inaugural Board, 2001–2004
- Distinguished Graduate Alumna of the University of North Carolina (UNC) in honor of the 100th Anniversary of the UNC Graduate School, 2003
- Editorial Advisory Board, Langmuir, 2000–2002
- Elected American Association for the Advancement of Science (AAAS) Fellow, 2001
- Alan Berman Research Publication Award for "Electronic connection to the interior of a mesoporous insulator with nanowires of crystalline RuO_2," Nature (2000), NRL to A.D. Berry, J.W. Long, R.M. Stroud, V.M. Browning, C.I. Merzbacher, and D.R. Rolison, 2001
- Editorial Advisory Board, Analytical Chemistry, 1990–1992
- American Institute of Chemists Distinguished Senior Award, 1975
- Best Analytical Chemistry Senior Student at Florida Atlantic University, 1975
- Best Student Paper at the Florida ACS Meeting-in-Miniature, Orlando, FL, 1975

She has also served on numerous advisory boards for conferences, university departments, and on advisory committees for professional organizations and federal agencies.

Biography

Early Life and Education

Being born and schooled from kindergarten through 8th grade in Iowa prepared me well for intellectual pursuits. Moving to south Florida in 1968 and entering high school revealed that I had been tracked into advanced classes in Iowa without such labels being applied. I still thank the Iowa public school system for such a rigorous educational start— and the 8th-grade teachers who lent me their college texts to read. In Florida in the late 1960s/early 1970s and as a young woman coming from a family of limited means, high-school guidance counseling was minimal and I was not steered onto a college track of coursework. Being an avid reader and good writer, I thought I might ultimately teach English or join the Peace Corps. But once I recognized, thanks to my junior-year chemistry class, that I had an affinity for chemistry—and an instinctive understanding of redox (reduction/oxidation) chemistry—my sense of the future changed. My senior-year coursework was hastily rescheduled to include some of the math and physics I needed to pursue a degree in science. (I still had to teach myself trigonometry the summer after I graduated so that I could take calculus my first year at college.)

I also had to find a way to fund my higher education as my single mother could not afford to send me to university. A last-minute notice that my grade point average (GPA) allowed me to take the Educational Testing Service (ETS) exam administered one Saturday at my high school became that way forward. Those of us who tested in the 90th percentile were granted admission as Faculty Scholars to Florida Atlantic University in Boca Raton and could earn a bachelor's degree in 3 years, not 4, having just tested out of our freshman course requirements. So, thanks to some scholarships, work–study to cover living expenses, and a National Defense Student loan, in the Fall of 1972 I could immerse myself in the scholarly joys of a still-new, technologically modern college that had no freshman or sophomores, only graduate programs and junior/senior courses for students holding associate degrees—and better yet: no varsity sports, fraternities, or sororities. A campus that emphasized scholarship first was ideal for me. I began independent research in electrochemistry my second year at FAU—with my own small office/lab—and spent my second summer at the University of Florida participating in their undergraduate summer research program and doing organic synthesis and analysis.

Even with such autonomy and ultimately deriving three peer-reviewed papers from my various undergraduate research experiences, I was concerned about my "worthiness" for graduate school. I so revered the creativity at the heart of research that I questioned whether I could bring the necessary quality of ideas to a life in research. Once I saw my male cohort—all of whom I was outperforming—assume as a given that (of course!) they would go to graduate school, off I went to the University of North Carolina at Chapel Hill. As I finished my doctoral work at UNC with Prof. Royce Murray and experienced the heady camaraderie as our group helped create the field known as chemically modified electrodes—an area that blended electrochemistry, materials chemistry, surface science, and analysis—I recognized that the way I wanted to approach research would be valued by an academic department 10 years into my independent career, but would not be deemed tenure-worthy at year 6. That recognition profoundly directed my next steps toward a life in research.

Career History

With academics out of the running, I did not turn to a postdoctoral position, but looked to industry for a potential career home. With a Ph.D. about-to-be conferred from a top five department in my degree specialty (analytical chemistry), I was interviewed in Winter/ Spring 1980 by many of the top chemical firms in the country . . . only to hear a uniform refrain: I would be on the bench for only a year and then be moved into management. I instinctively recognized that such a plan was actually a box in which I would not be happy. When one of my hosts described the comparability between the technical and managerial ladders of advancement, but then slipped by noting that all of their best people went into management, my instinctive sense was validated—I wanted to stay technical, but if I did so, I would then be identified as not one of the company's "best."

Now wary of an industrial research home, I was rescued by the opportune timing of the network: A former Murray postdoctoral associate fortuitously reached out to seek candidates for a new hire as a staff scientist at the U.S. Naval Research Laboratory in Washington, DC. One visit confirmed that the middle ground was immediately compelling—hands-on research and collaboration with NRL staff and with students and postdocs, rather than classroom lecturing or immediately overseeing a research team.

My first 10 years at NRL were just that: hands-on, at the bench, just as I had wished, because I enjoy the craft of designing and performing experiments. But when my first postdoctoral associate joined me in 1990, expanding my ability to put more of my ideas into play, followed by even more postdocs over time, I learned the even greater, albeit selfish, joy of being handed great results and intriguing, yet-to-be-resolved unknowns with which I could let my mind run free at play, thanks to the hard, careful work of my excellent junior colleagues. By the end of the decade, the success of our multiple research endeavors, including launching a program in multifunctional ultraporous materials, and the papers of import and patented inventions that ensued led lab management to insist I form a section and hire people. Thus I started the Advanced Electrochemical Materials section at NRL in 1999, ultimately hiring over the next 15 years eight of my former postdoctoral associates as staff scientists and colleagues at the NRL.

Baker Street Café is one of our team's secrets of success; here awaiting afternoon tea and cookies at our ad hoc group café, which we set up—as needed—in the hallway between our labs. *Credit: Jeffrey W. Long.*

Although the bulk of our team at the NRL comprises Ph.D. research staff and postdoctoral associates, one of the keys to our success is the undergraduate researcher—someone who interrupts her or his on-campus course-taking to spend 6–12 months with us doing full-time research. As *valuable* as a summer research experience may be for an undergraduate, we find an extended full-time research experience is *invaluable*—for the undergraduate, for the Ph.D.s in the section, and for the science. The energy and sense of adventure our undergraduate colleagues bring to the lab help to revitalize our post-doctoral colleagues (still a touch jaded from the joys of the graduate-student experience, even at the nondysfunctional science and engineering departments) and spur "what happens when" explorations that frequently sprout new research offshoots. The undergraduates tell us they earn a deeper appreciation of the why-and-what their faculty cover in lectures, but more importantly come to recognize that they are not students, but scientists and the future avenues they can take in science and engineering fields are many, not few. To date, 46 of my 150 NRL-derived peer-reviewed papers have undergraduate coauthors. And although I could have retired 5 years ago (with a full pension because I was hired as a government scientist when only 26), too many interesting questions remain to be asked and answered by our team.

3 Most Cited Publications

Title: Catalytic nanoarchitectures: the importance of nothing and the unimportance of periodicity
Author(s): Rolison, DR
Source: Science; volume: 299; issue: 5613; pages: 1698–1701; published: March 14, 2003
Times Cited: 658 (from Web of Science)

Title: Charge transfer on the nanoscale: current status
Author(s): Adams, DM; Brus, L; Chidsey, CED; et al.
Source: Journal of Physical Chemistry B; volume: 107; issue: 28; pages: 6668–6697; published: July 17, 2003
Times Cited: 650 (from Web of Science)

Title: Three-dimensional battery architectures
Author(s): Long, JW; Dunn, B; Rolison, DR; White, HS
Source: Chemical Reviews; volume: 104; issue: 10; pages: 4463–4492; published: October 2004
Times Cited: 597 (from Web of Science)

Challenges

I posit that there is no academic freedom for science: Academic freedom in research equals what one can convince some entity to fund. That fact is true whether one works in industry, government labs, or academe. For me, scientific research that opens new doors is driven by an intuitive sense that "there really is a there, there." Try getting leaps into the unknown funded—anywhere—even with a record of succeeding at nonbandwagon science. Research funds are too precious and currently too scarce to be left to supporting merely sound science when we need to invest in truly innovative science.

On being a woman in this field . . .

I love being a woman in science—I just want more company!

Words of Wisdom

Science is too important to be left just to men.

Profile 80

Caroline Anne Ross

Associate Head of the Department of
Materials Science and Engineering
Toyota Professor of Materials Science
and Engineering
Massachusetts Institute of Technology
77 Massachusetts Ave.
Cambridge, MA 02139
USA

e-mail: caross@mit.edu
telephone: (617) 258-0223

Caroline Ross as an assistant
professor. Credit: MIT DMSE.

Birthplace
London, England

Publication/Invention Record

>325 publications: h-index 46 (Web of Science)
19 patents, edited 2 books

Tags

❖ Academe
❖ Domicile: USA
❖ Nationality: British
❖ Caucasian
❖ Children: 2

Proudest Career Moments (to date)

Giving a seminar or lecture and seeing a room full of people who are interested in what I
am saying always makes me feel proud. Having papers accepted in good journals is also a
source of pride, when I feel as though my guidance of the project and mentoring of the
students and researchers made the difference.

Academic Credentials

Ph.D. (1988) Materials Science, Cambridge University, Cambridge, UK.
B.A. (1985) Materials Science, Cambridge University, Cambridge, UK.

Research Expertise: magnetic materials, photonics, nanotechnology, self-assembly

Other Interests: teaching and mentoring students

Successful Women Ceramic and Glass Scientists and Engineers: 100 Inspirational Profiles. Lynnette D. Madsen.
© 2016 The American Ceramic Society and John Wiley & Sons, Inc. Published 2016 by John Wiley & Sons, Inc.

Key Accomplishments, Honors, Recognitions, and Awards

- Editorial board member of the Journal of Magnetism and Magnetic Materials, 2009 to present
- Editorial board member of Applied Physics Letters/Journal of Applied Physics, 2009–2014
- Fellow of the IEEE, 2013
- Listed in the Top 100 Materials Scientists of the past decade (2000–2010), Times Higher Education Supplement, 2012
- Chair of the Magnetism and Magnetic Materials Conference, 2011
- Editorial board member of Journal of Physics D: Applied Physics, 2004–2010
- Institute of Physics Wohlfarth lecture, 2007
- Fellow of the Materials Research Society for contributions in templated self-assembly methods for the fabrication of nanostructures, and the magnetic properties of thin films and nanopatterned materials, 2009
- Fellow of the American Physical Society for innovative research into the magnetic properties of thin film and nanoscale structures, and for the development of novel lithographic and self-assembly methods for nanostructure fabrication, 2004
- Fellow of the Institute of Physics (UK), 2004
- Chair of the Materials Research Society Spring Meeting, 1998

Biography

Early Life and Education

My parents were scientists, so we always had a lot of scientific activities at home. I became interested in geometrical crystal structures when I was in school and started to collect mineral specimens and learn about their classification. My sister and I constructed dozens of paper models of polyhedra, from the easy ones (the five Platonic solids) through to complicated Archimedean polyhedra and stellated structures with hundreds of tiny facets. I won a poetry competition when I was 8, and spent the prize money on an optical microscope.

I went to a girls' high school in London that had an excellent science and mathematics program, and then attended Cambridge University to obtain an undergraduate degree and Ph.D. in materials science. My Ph.D. was on the topic of electromigration in thin metal films, a diffusion process driven by high current densities and at the time an important failure mechanism in integrated circuits. This started a long-term interest in the properties of thin films. I was interested in working abroad, so went to Harvard University as a postdoctoral fellow, working on deposition and interdiffusion in multilayer films made by electrodeposition.

Career History

From 1991 to 1997, I worked as an engineer at Komag, Inc., a manufacturer of hard disks, in Silicon Valley. This was a research and development position, involving everything from developing next-generation heads and disks to troubleshooting production problems.

In 1997 I joined the Department of Materials Science and Engineering in MIT as an Assistant Professor. I was promoted to Associate Professor in 3 years and to full Professor in 2004. I am now serving as Associate Department Head. I teach classes in electrical, optical, and magnetic materials and devices, and in magnetism and magnetic materials. I run a thin film deposition lab with multiple users, and work on magnetic and photonic oxides, magnetic nanostructures, and self-assembly processes with applications in nanofabrication. I have a group of about 12 students and postdocs.

3 Most Cited Publications

Title: Patterned magnetic recording media
Author(s): Ross, C
Source: Annual Review of Materials Research; volume: 31; pages: 203–235; doi: 10.1146/annurev.matsci.31.1.203; published: 2001
Times Cited: 501 (from Web of Science)

Title: Formation of a cobalt magnetic dot array via block copolymer lithography
Author(s): Cheng, JY; Ross, CA; Chan, VZH; et al.
Source: Advanced Materials; volume: 13; issue: 15; pages: 1174+; doi: 10.1002/1521-4095(200108)13:15<1174::AID-ADMA1174>3.0.CO;2-Q; published: August 3, 2001
Times Cited: 472 (from Web of Science)

Title: Templated self-assembly of block copolymers: top-down helps bottom-up
Author(s): Cheng, JY; Ross, CA; Smith, HI; et al.
Source: Advanced Materials; volume: 18; issue: 19; pages: 2505–2521; published: October 4, 2006
Times Cited: 402 (from Web of Science)

Challenges

My career included many challenges, including moving from one country and culture to another, and moving between academia and industry. Every move caused disruptions in my social as well as professional life, but I now have a network of friends around the world. Moving from an industrial position, with limited opportunities to publish, back to academia was difficult. I found it was helpful to form collaborations and networks and to stay involved in professional activities such as

Caroline Ross attended a scientific conference at Easter Island, Chile, in 2013. (See insert for color version of figure.)

societies and conferences. Professionally, the hardest change came from having a child, when I suddenly became responsible for another person and my time was no longer my own. This required a dramatic change in work habits in order to stay productive.

On being a woman in this field . . .

Materials science and engineering tends to attract more women than other engineering disciplines. In MIT, over half our undergrad MSE students are women. However, the proportion diminishes as we move through graduate school, postdoctoral work, and into faculty positions. A key challenge is retaining women in science and engineering in order to ensure a diverse faculty in universities and a high proportion of women engineers in industry. I was fortunate to have excellent, mostly women, science teachers at school, and supportive parents, so I was not discouraged from following my interests. I was the only woman Ph.D. in Komag for part of the time I worked there, but I had many great people to work with.

Words of Wisdom

You will find many generous people who can provide mentorship and guidance, but ultimately you need to take charge of your own destiny. Try to work out where you want to go, in terms of career and family and geographical location, and follow a plan to get there, whether it involves publishing research papers, learning new skills, volunteering to help in a professional society, going back to school, or changing between different types of employment. The plan may just be for a couple of years or it might represent a lifetime's goal. It may not always work, but following a plan (or failing that, a backup plan) and being proactive is more likely to lead to success than drifting and hoping for the best.

Della Martin Roy

Research Professor
Materials Science & Engineering
Arizona State University
Tempe, AZ 85287
USA
Professor Emeritus
Materials Science and Engineering
The Pennsylvania State University
University Park, PA 16802
USA

e-mail: Della.Roy@asu.edu
telephone: (818) 865-1196; (814) 237-7261

Della Roy, ACerS Distinguished Life Member, 2008. Credit: The American Ceramic Society.

Birthplace
Merrill, OR, USA

Born
November 3, 1926

Publication/Invention Record
>425 publications: h-index of 37
4 patents, editor of 9 books

Tags
❖ Academe
❖ Domicile: USA
❖ Nationality: American
❖ Caucasian
❖ Children: 3

Academic Credentials
Ph.D. (1952) Minerals, Pennsylvania State University, State College, PA, USA.
M.S. (1949) Pennsylvania State University, State College, PA, USA.
B.S. (1947) University of Oregon, OR, USA.

Research Expertise: crystal chemistry and phase transitions, crystal growth, cement chemistry, hydration and microstructure, concrete durability, biomaterials, special glasses, radioactive waste management, geologic isolation, and chemically bonded ceramics, materials synthesis, processing characterization in inorganic, ceramic, cement and mineral systems; phase equilibria and thermodynamic properties; hydrothermal synthesis; chemically bonded ceramics; biomaterials synthesis; and nano- and microstructural design

Other Interests: addressing environmental issues

Successful Women Ceramic and Glass Scientists and Engineers: 100 Inspirational Profiles. Lynnette D. Madsen.
© 2016 The American Ceramic Society and John Wiley & Sons, Inc. Published 2016 by John Wiley & Sons, Inc.

Key Accomplishments, Honors, Recognitions, and Awards

- Corecipient of a Golden Goose Award, for a 1968 discovery (award honors obscure studies that have led to major breakthroughs with a significant, and often serendipitous, impact on humans and the economy), 2012
- The American Ceramic Society (ACerS) Bleininger Award, 2004
- The Della Roy Lecturer Award was started in 2000. The lecturer is selected by consensus of the division executive committee of the Cements Division of ACerS in consultation with Elsevier Science, who sponsors the lecture. The person selected does not need to be a member of ACerS or the Cements Division. An honorarium of $1000 and a certificate are awarded to the lecturer.
- Hosler Alumni Scholar Medal, 1999
- Trustee, the American Ceramic Society, 1992–1995
- Elected to the World Academy of Ceramics, 1991–the first woman
- American Concrete Institute Slag Award, 1989
- Honorary Fellow, Institute of Concrete Technology, 1987
- Elected as a member of the National Academy of Engineering, 1987–the first woman Materials Scientist. Citation: "For internationally recognized contributions to the applied science and engineering of cement and concrete."
- The American Ceramic Society L.E. Copeland Award, 1987
- Honorary Fellow, Institute of Concrete Technology, UK, 1987
- Sir Frederick Lea Memorial Lecture, 1987
- The American Ceramic Society John Jeppson Award, 1982
- Founding Editor, Cement and Concrete, 1971–2005
- Founded the journal, Cement and Concrete, 1971
- Fellow, the American Ceramic Society, 1971
- The mineral, dellaite, is named for her because she synthesized the mineral, 1965.

Biography

Early Life and Education

In June 1948, while Della was a student at Pennsylvania State University, she married Rustum Roy shortly after he received his Ph.D.

Career History

Upon receiving her doctorate, Della Roy was hired at Pennsylvania State University (PSU) as a Senior Research Associate. In 1959, she began as an Associate Professor of Materials Science at PSU and was promoted to full professor in 1969. Throughout her research career, Della Roy published more than 400 publications, was granted 4 patents, edited 9 books, founded a research journal, mentored 36 graduate students, coauthored 55 major reports to government agencies, and chaired numerous conferences and committees. Della became an Emeritus Professor at PSU in 1975 and officially retired in 2012.

In addition to her own research, Della and her husband, Rustum, collaborated on many research projects and together raised three sons (Neill, Ronnen, and Jeremy). They both led exciting careers and were well known for their work in the scientific community and with professional societies. They were also controversial, for example, they coauthored a book entitled, Honest Sex, in 1969. At the age of 86, in 2010, Rustum passed away.

The Roys, both PSU graduates, endowed the annual Rustum and Della Roy Innovation in Materials Research Award. Recipients are full-time graduate students, postdoctoral researchers, and/or junior faculty members at Penn State who are conducting interdisciplinary materials research. The award recognizes and honors interdisciplinary materials research that is considered to be valuable, likely to yield unexpected results, and demonstrates genuine innovation.

3 Most Cited Publications

Title: Hydroxyapatite formed from coral skeletal carbonate by hydrothermal exchange
Author(s): Roy, DM; Linnehan, SK
Source: Nature; volume: 247; issue: 5438; pages: 220–222; published: 1974
Times Cited: 389 (from Web of Science)

Title: Ettringite and C-S-H Portland cement phases for waste ion immobilization: a review
Author(s): Gougar, MLD; Scheetz, BE; Roy, DM
Source: Waste Management; volume: 16; issue: 4; pages: 295–303; doi: 10.1016/S0956-053X(96)00072-4; published: 1996
Times Cited: 183 (from Web of Science)

Title: The system $Mgo-Al_2O_3-H_2O$ and influence of carbonate and nitrate ions on the phase equilibria
Author(s): Roy, DM; Roy, R; Osborn, EF
Source: American Journal of Science; volume: 251; issue: 5; pages: 337–361; published: 1953
Times Cited: 140 (from Web of Science)

Sources used to create this profile

Penn State University (http://www.gradschool.psu.edu/graduate-funding/fellowships/programs/roymatres/, accessed February 20, 2015)
Arizona State University (https://webapp4.asu.edu/directory/person/888980, accessed February 20, 2015)
University of Oregon (http://pages.uoregon.edu/wits/wits/news/, accessed February 20, 2015)
ACerS Lifetime Member (http://ceramics.org/acers-awards-and-honors/distinguished-life-membership/2008-distinguished-life-members, accessed February 20, 2015)

ACerS Della Roy Lecture (http://ceramics.org/?awards=della-roy-lecture, accessed February 20, 2015)

National Academy of Engineering (http://www.nae.edu/28044.aspx, accessed February 20, 2015)

American Physical Society (http://physicsfrontline.aps.org/tag/della-roy/, accessed February 20, 2015)

Rustum and Della Roy, Honest Sex (1969), Signet Press, Author's Choice Press, 2003 reprint: ISBN 0-595-27213-4

Tiffany K. Wayne, American Women of Science Since 1900: Essays A–H, volume 1, ABC-CLIO, 2011.

Dellaite (http://webmineral.com/data/Dellaite.shtml#.VOd-6C6UJ-8, accessed February 20, 2015).

Disclaimer

The woman featured in this profile could not be reached. This profile was prepared from various sources and is believed to be correct. However, the author assumes no responsibility for any errors or omissions; they were unintentional.

Profile 82

Tanusri Saha-Dasgupta

Professor of Condensed Matter Physics and Material Science
and Associate Dean
S.N. Bose National Centre for Basic Sciences
Kolkata 700 098
West Bengal
India

e-mail: t.sahadasgupta@gmail.com and tanusri@bose.res.in
telephone: +91-33-2335 5706/7/8

Prof. Saha-
Dasgupta at the
age of 35 years,
immediately after
joining S.N. Bose
Centre as a young
faculty, (Credit:
S.N. Bose Centre).

Birthplace
Kolkata, India

Born
November 12, 1966

Publication/Invention Record

147 publications: h-index 25 (Web of Science)
5 book chapters

Tags

❖ Administration and Leadership
❖ Academe
❖ Domicile: India
❖ Nationality: Indian
❖ Asian
❖ Children: 0

Proudest Career Moments (to date)

My proudest career moment was in January 2005, when I was given the certificate in a ceremony in IIT Delhi that says that I have been selected as Head of one of the five Max Planck-India partner groups that have been established. I felt deeply honoured, especially since the certificate was given to me by none other than Prof. Ole Andersen, who at that time was the Director of Max Planck Institute for Solid State Physics in Stuttgart, and who also happened to have been my post-doctoral supervisor.

In the same year, I was also awarded the prestigious Swarnajayanti award, which is given to Indian scientists below the age of 40 years for their outstanding scientific contribution. I was the first woman scientist to receive this award in physics since its inception.

Academic Credentials

Ph.D. (1995) Physics, University of Calcutta, West Bengal, India.
M.Sc. (1989) Physics, University of Calcutta, West Bengal, India.
B.Sc. (1987) field, Presidency College, Kolkata (also known as Calcutta), West Bengal, India.

Successful Women Ceramic and Glass Scientists and Engineers: 100 Inspirational Profiles. Lynnette D. Madsen.
© 2016 The American Ceramic Society and John Wiley & Sons, Inc. Published 2016 by John Wiley & Sons, Inc.

Research Expertise: development of electronic structure method, electronic structure of strongly correlated electron systems, low-dimensional quantum spin systems, double perovskites, spinels (interplay between charge, orbital and spin), nanomaterials

Other Interests: teaching

Key Accomplishments, Honors, Recognitions, and Awards

- Head of Thematic Unit of Excellence on Computational Materials Science, funded by Nano Mission, 2012 to present
- Member of Editorial Board, Pramana Journal of Physics, 2012 to present
- International Advisory Board member, International Conference on Materials and Technologies, Italy, 2014
- Member, American Physical Society (APS)–Indo-US Science and Technology Forum (IUSSTF) Program for Professorship and Student Visitation Program, 2009–2013
- Head of Advanced Materials Research Unit, funded by Department of State and Technology, India, 2008–2012
- Director of the Research Program on "Novel Materials," International Institute of Physics, Natal, Brazil, 2012
- Recipient of Dr. P. Sheel Memorial Lecture Award, National Academy of Sciences, 2012
- Organizer of the focused session in APS March meeting, 2011
- Elected Fellow of Indian Academy of Sciences, Bangalore, 2010
- Elected Fellow of National Academy of Science, India, 2010
- Appointed Head of Max-Planck-India Partner Group, 2005
- Swarnajayanti Fellowship, 2005

Biography

Early Life and Education

Tanusri Saha was born in Kolkata (formerly known as Calcutta), a city in the eastern part of India. She was an only child. Her father was a physicist and later became an administrator in the Examination Department in Calcutta University.

She began her schooling locally in the Bagbazar Multipurpose Government School, which was rather close where she lived. After completion of this schooling, she was admitted to Bethune College also located in northern part of the city Kolkata; subsequently she moved to the prestigious Presidency College for a bachelor's degree with honours in Physics. She was awarded a Master's degree from Science College in Physics, with a specialization in condensed matter physics. After a brief stay in Saha Institute in Kolkata, she joined the group of Prof. Abhijit Mookerjee in S.N. Bose Centre to obtain her doctorate. Her Ph.D. research involved computational study of alloy phase stability. After being awarded her doctorate, Tanusri married and began publishing under her new name, Saha-Dasgupta.

Career History

Soon thereafter, Dr. Saha-Dasgupta went to Paris to Le Centre Français de Recherche Aérospatiale (ONERA) for post-doctoral study in the group of Profs. F. Ducastelle and A. Finel. She examined the statistical mechanics of alloy phase transition. Then, following a brief stay in University of Cergy-Pontoise, she moved to Max-Planck Institute (MPI) for Solid State Physics in Stuttgart in Germany for her second post-doctoral position. At MPI, Stuttgart, in the group of Prof. O. K. Andersen, she was involved in the development of an electronic structure method with the aim of it being accurate, fast, and intelligible. She finally returned to India in 1999 and joined the Indian Institute of Science Bangalore as a research associate.

Dr. Saha-Dasgupta with Prof. O. K. Andersen during the inaugural session of Advanced Materials Research Unit, (Credit: S.N. Bose Centre).

Then, she returned to her hometown Kolkata and joined S.N. Bose Centre as a permanent faculty member in 2000. Prof. Saha-Dasgupta continued her research career in S.N. Bose Centre and became full Professor in 2011. She is currently also the coordinator of Thematic Unit of Excellence on Computational Materials Science within the S.N. Bose National Centre. Prof. Saha-Dasgupta so far has supervised and guided seven students to their doctorates. Her present research group consists of six doctoral students and two post-doctoral fellows. Prof. Saha-Dasgupta works in the area of computational condensed matter physics. A major portion of Prof. Saha-Dasgupta's research is devoted to the application of first principles electronic structure calculations to understand the physics and chemistry of novel and complex materials.

Prof. Saha-Dasgupta giving a lecture in Indian Institute of Science, Bangalore, India, (Credit: Organizers of the Indo-Japan conference).

Prof. Saha-Dasgupta is also engaged in various International research programs including Indo-German, Indo-European, Indo-US, and Indo-Swedish programs. She has visited and delivered lectures over much of the world.

Prof. Saha-Dasgupta in her office, (Credit: S.N. Bose Centre).

3 Most Cited Publications

Title: Band-structure trend in hole-doped cuprates and correlation with T-c (max)
Author(s): Pavarini, E; Dasgupta, I; Saha-Dasgupta, T; et al.
Source: Physical Review Letters; volume: 87; issue: 4; article number: 047003; published: July 23, 2001
Times Cited: 324 (from Web of Science)

Title: Electronic structure of Sr_2FeMoO_6
Author(s): Sarma, DD; Mahadevan, P; Saha-Dasgupta, T; et al.
Source: Physical Review Letters; volume: 85; issue: 12; pages: 2549–2552; published: September 18, 2000
Times Cited: 314 (from Web of Science)

Title: Muffin-tin orbitals of arbitrary order
Author(s): Andersen, OK; Saha-Dasgupta, T
Source: Physical Review B; volume: 62; issue: 24; pages: 16219–16222; published: December 15, 2000
Times Cited: 266 (from Web of Science)

Challenges

My family encouraged my school and college education. However, they were rather conservative and not very comfortable with the idea of a girl going to a far-off place for study. Their reluctance to have me move was one of the primary reasons I stayed in Calcutta after finishing my master's degree and joined the Ph.D. program at a research Institute in Calcutta. Afterward, I realized the importance of going out and seeing the world to broaden my vision. I fought a lot to convince my parents about the importance of going abroad for post-doctoral studies. After returning to India, there was constant pressure from my family to take a job in the same city where my husband is, but I felt it was important to have my own independent research career. My husband and I lived separately for 7 years; we put our careers first. We sacrificed the family life for the sake of science and our careers.

Prof. Saha-Dasgupta with two of her collaborators (Prof. K. Held and his student) in Vienna, (Credit: T. Saha-Dasgupta).

On being a woman in this field . . .

I do feel that a woman in this field can be as important and as well recognized as a man. However, it is also indeed true that women have to work harder to achieve the same status and recognition, and often they also have a need to balance home and work. This field of condensed matter physics is creative and demands dedicated attention that often makes the balancing difficult. A woman also has to be mentally strong to take the challenges from the profession as well as family, handling them together.

There is a general trend of not taking a woman seriously. One has to take extra effort to make their voice heard. Unfortunately, I felt this extends to the issues such as recognitions and awards. Nevertheless, this should not get you down; rather, it should make you stronger and able to face the harsh reality more efficiently.

Words of Wisdom

- Be mentally strong to take on new challenges.
- Know your balance between home and career.
- Clearly understand the priorities in your life.
- Remain focused.

Lourdes G. Salamanca-Riba

Materials Science & Engineering
A. James Clark School of Engineering
University of Maryland
College Park, MD 20742-2115
USA

e-mail: riba@umd.edu
telephone: (301) 405-5220

Lourdes Salamanca-Riba at age 30 in her home.

Birthplace

Mexico City, Mexico

Born

December 12, 1955

Publication/Invention Record

>150 publications: h-index 35 (Web of Science)

Tags

❖ Academe
❖ Domicile: USA
❖ Nationality: dual American and Mexican
❖ Latina
❖ Children: 2

Proudest Career Moment (to date)

It is hard to pick one moment as the one I am most proud of. I always feel proud when any of my students wins an award or gets a very good job. This is, of course, because of their hard work and ability, but I feel that I contributed to their success. There have been several moments that have made me feel proud. The first one was when I got the MRS Graduate Student Award for my Ph.D. thesis project. I investigated a crystalline to glass

Successful Women Ceramic and Glass Scientists and Engineers: 100 Inspirational Profiles. Lynnette D. Madsen.
© 2016 The American Ceramic Society and John Wiley & Sons, Inc. Published 2016 by John Wiley & Sons, Inc.

Prof. Lourdes Salamanca-Riba at the University of Maryland operating the transmission electron microscope. (See insert for color version of figure.)

phase transition in $SbCl_5$ intercalated graphite in which the glass phase was the low temperature phase. By very careful transmission electron microscopy (TEM) investigation we found that the phase transition was induced by the electron beam because the sample was more sensitive to the beam at low temperatures than at room temperature. Much later in my career, another proud moment was when I was an Assistant Professor and my first Ph.D. student got his degree. Other proud moments were when we obtained a nanocomposite of $BaTiO_3$ and $CoFe_2O_4$ that self-assembles into columns of $CoFe_2O_4$ in a $BaTiO_3$ matrix when deposited in thin film form. A more recent proud moment was when I started working with a small company that discovered a new method of incorporating carbon in metals for which the metal-carbon phase diagram indicates that C is not soluble in the metal. I discovered that the carbon is not only incorporated in the metal lattice but it forms nanostructures such as graphene ribbons. Last summer I felt very proud when I visited Seoul, South Korea and met with 8 alumni from the University of Maryland (five of whom I was their advisor, and including my first Ph.D. student). I was proud to see how successful they are now and seeing how much they appreciate what I taught them both in courses as well as while working on their thesis projects.

Academic Credentials

Ph.D. (1985) Massachusetts Institute of Technology (MIT), Cambridge, Massachusetts, USA.
B.S. (1978) Universidad Autónoma Metropolitana, Mexico City, Mexico.

Research Expertise: Hybrid solar cell (photovoltaics), light emitting diodes (LED) and high temperature devices, structural studies of thin film semiconductor heterostructures and superlattices, superconductors, nanocomposites of ferroelectric-/-magnetic oxides, metals containing high C in the form of nanostructrues

Other Interests: mentoring of students, nanodevices, traveling

Key Accomplishments, Honors, Recognitions, and Awards

• Elected member of the Air Force Studies Board of the National Academies, 2002–2008
• University of Maryland Most Outstanding Advisor, 2007
• Materials Research Society (MRS), Graduate Student Award, 1984

Biography

Early Life and Education

She was born and raised in Mexico City. She has four sisters and one brother. She obtained a B.S. degree in Physics from the Universidad Autónoma Metropolitana in Mexico City. She completed her doctorate degree in Solid State Physics under the supervision of Mildred Dresselhaus, who is also featured in this book. Her Ph.D. thesis project was on Graphite Intercalation Compounds (GIC) where she used transmission electron microscopy to investigate structural phase transitions in GICs and the recrystallization process of ion implanted graphite and carbon fibers.

Career History

From September 1985 to January 1987, she was a Senior Research Scientist at GM Research Laboratories where she performed research in semiconductor quantum wells for light emitting diodes. In February 1987, she joined the University of Maryland as a faculty member and she has been there ever since. Her research projects have varied from semiconductor nanostructures to ferroelectric and ferromagnetic oxide films and carbon nanostructures in metals where carbon is not soluble. She serves a point of contact for the MRS University Chapter for the University of Maryland. She is fully fluent in Spanish as well as English.

3 Most Cited Publications

Title: Multiferroic $BaTiO_3–CoFe_2O_4$ nanostructures
Author(s): Zheng, H; Wang, J; Lofland, SE; et al.
Source: Science; volume: 303; issue: 5658; pages: 661-663; doi: 10.1126/science.1094207; Published: January 30, 2004
Times Cited: 1163 (from Web of Science)

Title: On the origin of high-temperature ferromagnetism in the low-temperature-processed Mn-Zn-O system
Author(s): Kundaliya, DC; Ogale, SB; Lofland, SE; et al.
Source: Nature Materials; volume: 3; issue: 10; pages: 709–714; doi: 10.1038/nmat1221; published: October 2004
Times Cited: 350 (from Web of Science)

Title: Heteroepitaxy of ZnO on GaN and its implications for fabrication of hybrid optoelectronic devices
Author(s): Vispute, RD; Talyansky, V; Choopun, S; et al.
Source: Applied Physics Letters; volume: 73; issue: 3; pages: 348–350; doi: 10.1063/1.121830; published: July 20, 1998
Times Cited: 334 (from Web of Science)
ResearcherID (B-3785-2009)

On being a woman in this field . . .

It is frequently difficult to be taken seriously when you are a woman in Physics. Being a Hispanic woman makes it even more difficult. For example, when I tell people that I work at the University of Maryland, they frequently assume that I teach Spanish. It is also hard to make people listen to you at meetings. So, many times I have suggested something at a meeting and then someone else says the same thing without acknowledging that I said it first.

Words of Wisdom

I like to advise young female students to be persistent and work hard on what they believe is right. Don't let others tell you that you cannot do what you want, or that it is not a job for a woman. Be true to yourself and work hard.

Maxine Lazarus Savitz

Former Vice President
National Academy of Engineering
Keck Center of the National Academies
500 Fifth Street, NW
Washington, DC 20001
USA

e-mail: maxinesavitz@aol.com
telephone: (310) 271-0874

Maxine Savitz on her MIT
graduation day in June 1961.

Birthplace
Baltimore, MD, USA
Born
February 13, 1937

Publication/Invention Record
~20 publications: h-index 4

Proudest Career Moment (to date)[1]

Tags
❖ Administration and Leadership
❖ Government
❖ Domicile: USA
❖ Nationality: American
❖ Caucasian
❖ Children: 2

My proudest moment is getting the field of energy efficiency started, in terms of both the technologies and the policies for the efficient use of energy. As a result of these efforts, in the 1970s and 1980s, energy use in the United States did not increase, but the gross domestic product did. In addition, I am proud of getting these materials in the ceramics area from the laboratory into production; to actually have your materials as part of the airplanes flying is harder than one thinks.

[1]Credit: http://sandt.blogs.brynmawr.edu/2009/07/13/qa-with-maxine-savitz-%E2%80%9958/.

Successful Women Ceramic and Glass Scientists and Engineers: 100 Inspirational Profiles. Lynnette D. Madsen.
© 2016 The American Ceramic Society and John Wiley & Sons, Inc. Published 2016 by John Wiley & Sons, Inc.

Academic Credentials

Ph.D. (1961) Organic Chemistry, Massachusetts Institute of Technology (MIT), Cambridge, MA, USA.
B.A. (1958) Chemistry, Bryn Mawr College, Bryn Mawr, PA, USA.

Research Expertise: development and manufacturing of innovative materials for the aerospace, transportation, and industrial sectors; energy efficiency (R&D, policies, and programs); distributed energy resources (gas turbines, microturbines, and fuel cells); high-temperature materials and application

Other Interests: making and influencing policy decisions on a national level

Key Accomplishments, Honors, Recognitions, and Awards

- Board of Trustees, Keck Graduate Institute, 2012 to present
- Senior Fellow of the California Council on Science and Technology, 2005 to present
- Board of Directors, Institute of Industrial Productivity, 2011 to present
- Director's Advisory Committee, Jet Propulsion Laboratory, 2010 to present
- President's Council of Advisors for Science and Technology (PCAST) (co-vice chair), 2009 to present
- Visiting Committee, Sponsored Research, MIT, 2009 to present
- Board of Directors, Energetics, Inc., 2009 to present
- Sandia Science Advisory Board, 2005 to present
- Advisory Board, Pacific Northwest National Laboratory, 2002 to present
- Board of Directors, American Council for an Energy-Efficient Economy, 1985 to present
- Vice President, National Academy of Engineering, 2006–2014
- Clean Energy Education & Empowerment (C3E) Lifetime Achievement Award Winner, 2013
- Elected fellow to the American Academy of Arts and Sciences, 2013
- Keynote speech "Materials: An Enabler" at Fourth International Congress on Ceramics, 2012
- Member of the Department of Energy's Energy Efficiency and Renewable Energy Advisory Board, 2010–2012
- Member of the committee for the American Physical Society's study: Energy Future: Think Efficiency, 2008
- Member of the California Council on Science and Technology, 1997–2000
- American Ceramic Society, Orton Memorial Lecturer Award, 1998
- Elected Member, National Academy of Engineering, for technical developments contributing to national initiatives in energy conservation and energy efficiency, 1992
- Department of Energy Outstanding Service Medal, 1981

- President's Meritorious Rank Award, 1980
- Recognition by the Engineering News Record for Contribution to Construction Industry, 1979 and 1975
- US Mobility Equipment Research & Design Command (MERDC) Commander Award for Scientific Excellence, 1967
- National Science Foundation Fellow at the University of California, Berkeley, 1961–1962

Biography

Early Life and Education

She completed her first degree in 1958 in Chemistry at Bryn Mawr College, a woman's college that first opened in 1885. Although Bryn Mawr offers Ph.D.'s, Maxine decided to undertake her graduate studies at MIT. By 1961, she had completed her doctorate in organic chemistry and was wed to Alan.

Career History

After completing a postdoctoral fellowship at the University of California, Berkeley, Dr. Savitz moved to the Washington, DC, area with her husband (a physician with the U.S. Army), and launched her technical career working on fuel cells at the U.S. Army Laboratory at Fort Belvoir, Virginia.

A recent image of Dr. Maxine Savitz.
Credit: National Academy of Engineering.

Next, she worked at the National Science Foundation (NSF) on Research Applied to National Needs. She then directed her expertise to the Department of Energy in the area of buildings energy efficiency. In total she worked at the Department of Energy and its predecessor agencies under three presidents, Nixon, Ford, and Carter, served as Deputy Assistant Secretary for Conservation, and won the department's outstanding service medal.

In 1985, Dr. Savitz joined Garrett Corporation and in 1987 became director of the Garrett Ceramic Components Division of the Allied-Signal Aerospace Company (that later became Honeywell Inc.). From 1987 until mid-2000, she was the general manager, which at the time was the only U.S.-owned silicon nitride structural ceramic manufacturer for gas turbine application. In this capacity, she oversaw the development and manufacturing of innovative materials for the aerospace, transportation, and industrial sectors. She subsequently retired from being the general manager of Technology Partnerships at Honeywell, Inc. Her recent position at the National Academy of Engineering was as Vice President.

She has more than 40 years of experience managing research, development, and implementation programs for the public and private sectors, including the aerospace, transportation, and industrial sectors. This expertise has led to her service on numerous boards and committees, the most prominent being the Vice Chair of PCAST; she was appointed to this Council by President Obama in 2009.

Maxine and her late husband had two children—Adam and Alison. Now, as well, there are five grandchildren—four boys and one girl—none have expressed an interest in science or engineering, but all are very good at quantitative thinking and games.

3 Cited Publications

Title: Decomposition of tertiary alkyl hypochlorites
Author(s): Greene, FD; Lau, HM; Smith, WN; et al.
Source: Journal of Organic Chemistry; volume: 28; issue: 1; page: 55; published: 1963
Times Cited: 165 (from Web of Science)

Title: Photochemistry of stilbenes. I.
Author(s): Mallory, FB; Gordon, JT; Wood, CS; et al.
Source: Journal of the American Chemical Society; volume: 84; issue: 22; page: 4361; published: 1962
Times Cited: 100 (from Web of Science)

Title: Rethinking energy innovation and social science
Author(s): Fri, RW; Savitz, ML
Source: Energy Research and Social Science; volume: 1; issue: 1; pages: 183–187; published: 2014; doi:10.1016/j.erss.2014.03.010
Times Cited: 14 (from Elsevier)

Challenges

There have been many challenges for me throughout the years. Certainly, there were more challenges in the 1960s since women were not (often) hired into professional positions. MIT was always welcoming to women; so there were no problems there and just being at MIT gave me an instant stamp of competence. However, it was still difficult to quickly move positions with my spouse. In one case when I applied for a position, it was clear from the letter I received that they had never hired a woman. At that time, it was the matter of finding a good place—welcoming, with a great mentor, and so on—and landing that position was a bit of the "luck of the draw." To a certain extent, this challenge remains today for everyone. Looking back, I can see remarkable strides for women. For example, even in the civil service, in the army at one time, it was questioned if women could travel . . . a lot has changed since then! I would say that the 1970s was the period when things really changed significantly, and for the better, for women.

When I worked at NSF on Research Applied to National Needs, it was truly exciting and the team I worked with was very interdisciplinary. Then, the oil embargo came in the early 1970s; the United States was very dependent on oil at that time. Consequently, it was easy to implement new policy. I learned on the fly that it was important to talk with industry. Overall this experience helped me to learn about the political processes and navigate them.

The key challenges that remain today occur in two areas: (i) with work and family balance, and (ii) the promotion of women into leadership positions. I serve on several boards; these issues are raised together or in tandem. Certainly today, promoting women is "in." We are continuously pushing for nominations of woman to become members of both the National Academy of Engineering (NAE) and the National Academy of Sciences (NAS).

On being a woman in this field . . .

Women should be engaged in the science and engineering of ceramics—it is an exciting area since materials enable everything—this excitement is pervasive no matter which sector you work in (academe, industry, or a national laboratory). New materials and areas of interest are popping up all the time, for example, due to nanotechnology. Changes also occur due to advances in instrumentation. In addition, materials science combines areas such as physics, chemistry, engineering, and sometimes also biology, very nicely. Certainly, many problems exist that need solving, which can be addressed through research in ceramics.

Words of Wisdom

First and foremost, get a good education as it will provide you with the ability to think analytically and that will be your most important skill in moving forward. Whether you decide to delve deeply (into one area) or to address a broad swath of the field (which has to do with how you think about things and what opportunities exist), work in this area can be both meaningful and exciting. I caution you not to get wedded to one path, be willing to try different things and to take risks. Ensure you have mentors (both female and male mentors). And, if you decide to marry, ensure you have a supportive husband (and family), so you are not held back in your career.

Profile 85

Nava Setter

Professor
Director of the Ceramics Laboratory
École Polytechnique Fédérale de
Lausanne (EPFL)
Laboratoire de Céramique
Station 12, EPFL
1015 Lausanne
Switzerland

e-mail: nava.setter@epfl.ch
telephone: +41-21-693-2961

Nava Setter at 18 (at the right) with two friends during their military service. They were doing periodic shifts in the kitchen (male soldiers did as well; no discrimination in this sense). They were peeling vegetables for a soup for dinner for about 500 people.

Birthplace

Haifa, Israel

Born

September 13, 1949

Publication/Invention Record

>475 publications: h-index 63 (Web of Science)

Tags

❖ Leadership
❖ Academe
❖ Domicile: Switzerland
❖ Nationality: Swiss and Israeli
❖ Caucasian

Proudest Career Moment

In 1988, the head of the department where I was working told me, second year in a row, that I will not be promoted yet in spite of what I considered an excellent performance and a very strong recommendation by my direct superior. To ease my disappointment, I checked an alternative position and found that the EPFL, a prestigious university known at that time mostly in Europe, had an opening for professor of ceramics engineering. I postulated, got shortlisted, was invited to present myself, apparently impressed the selection committee, and was offered at once to join as a tenured professor. All this was unexpected, as my true motivation was only to overcome the disappointment due to the delay in my promotion at my place of work, where I was highly content with my work, feeling it was a mission and not just a job. The offer was irresistible and I decided to take it. So I made a phone call from Switzerland to Israel to my parents. My father answered. I told him I was nominated. Quiet. Then I heard my mother asking my father: "What is the matter?" My father answered (in a doubting voice): "She is becoming a professor." I heard my mother replying my father: "Congratulate her," and my father told me: "Mazel Tov!" (congratulations, in Hebrew). I was very proud at that moment.

Successful Women Ceramic and Glass Scientists and Engineers: 100 Inspirational Profiles. Lynnette D. Madsen.
© 2016 The American Ceramic Society and John Wiley & Sons, Inc. Published 2016 by John Wiley & Sons, Inc.

Academic Credentials

Ph.D. (1980) The Pennsylvania State University, University Park, PA, USA.
M.Sc. (1976) The Technion—Israel Institute of Technology, Technion City, Haifa, Israel.
B.Sc. (1973) The Technion—Israel Institute of Technology, Technion City, Haifa, Israel.

Research Expertise: functional ceramics focusing on piezoelectric and related materials: ferroelectrics, dielectrics, pyroelectrics, and also ferromagnetics

Other Interests: lecturing/teaching; enabling high-school science learning, in particular in rural areas in developing countries

Key Accomplishments, Honors, Recognitions, and Awards

- The American Ceramic Society (ACerS) Robert B. Sosman Award and Lecture, 2013
- Japanese Ferroelectric Materials Association Award, 2013
- AVS Excellence in Leadership, 2013
- Recipient of W.R. Buessem Award (which is presented annually at a meeting of the Center for Dielectric Studies to a member of the dielectrics community with a lifetime of achievement in the field), 2011
- IEEE-Ultrasonics, Ferroelectrics and Frequency Control (UFFC) Society Achievement Award, 2011
- European Union European Research Council (ERC) Advanced Investigator Grant, 2010
- Fellow of the Institute of Electrical and Electronic Engineers (IEEE), 2007
- Academician of the World Academy of Ceramics, 2006
- Distinguished lecturer of the UFFC-IEEE, 2004–2005
- Ferroelectrics-IEEE Recognition Award, 2004
- Founder and President of the TALENT Foundation (supporting education of talented youth in developing countries), since 2002
- Swiss-Korea Research Award, 2001
- International Symposium on Integrated Functionalities (ISIF) Outstanding Achievement Award, 2000
- Fellow of the Swiss Academy of Technical Sciences, since 1995

Biography

Early Life and Education

Nava Setter was raised up in Haifa and then in Raman-Gan in Israel by a loving and caring mother who was a librarian and father who was a truck driver and provided her with best conditions for studies.

After 2-year military service, she completed both Bachelors and Masters in Civil Engineering in the Technion followed by completion of a doctorate in solid state science in the Pennsylvania State University. During her studies, she worked part-time as a shopseller, waiter, high-school replacement-teacher, engineering-apprentice, engineering-drafter, and university teaching-assistant.

Nava Setter at 18 (in the middle) with two friends during their
military service "posing" in front of pots of soup.

Career History

She undertook postdoctoral studies at the Universities of Oxford (UK) and Geneva (Switzerland), and then she joined an R&D institute in Haifa (Israel), where she became the head of the Electronic Ceramics Lab (1988).

She began her affiliation with EPFL (the Swiss Federal Institute of Technology) in Lausanne in 1989 as the Director of the Ceramics Laboratory. In 1992, she was promoted to full Professor of Materials Science and Engineering. She had been Head of the Materials Department in the past and, more recently, she has served as the Director of the Doctoral School for Materials.

In 2004, she initiated and began supporting a high school for science and humanities that has since become one of the most highly performing schools in Tanzania. Two issues are known to impede Tanzania's efforts in science education: (i) the scarcity of trained science teachers, and (ii) the lack of science laboratories and other teaching resources. To forward the overall effort, she has focused on the training and education of high-school science teachers; the development of physics and chemistry laboratories; and the production of "integrated science modules" that include classroom experiments and use of videos relating the curriculum to industrial and daily life applications.

Nava Setter in her office in 2011.
Photograph by J. Vainunska.

In 2012, she established and began supporting a minicenter for in-service training of high-school science teachers in Tanzania.

3 Most Cited Publications

Title: Ferroelectric materials for microwave tunable applications
Author(s): Tagantsev, AK; Sherman, VO; Astafiev, KF; Venkatesh, J; Setter, N
Source: Journal of Electroceramics; volume: 11; issue: 1–2; pages: 5–66; doi: 10.1023/B: JECR.0000015661.81386.e6; published: September–November 2003
Times Cited: 778 (from Web of Science)

Title: The role of b-site cation disorder in diffuse phase-transition behavior of perovskite ferroelectrics
Author(s): Setter, N; Cross, LE
Source: Journal of Applied Physics; volume: 51; issue: 8; pages: 4356–4360; doi: 10.1063/1.328296; published: 1980
Times Cited: 673 (from Web of Science)

Title: Ferroelectric thin films: review of materials, properties, and applications
Author(s): Setter, N; Damjanovic, D; Eng, L; et al.
Source: Journal of Applied Physics; volume: 100; issue: 5; article number: 051606; published: September 1, 2006
Times Cited: 588 (from Web of Science)

On being a woman in this field . . .

When I started my appointment at the EPFL, I was the only female among the EPFL faculty. Success came easily because the appointment gave me a lot of exposure. Over the years, the number of female faculty grew substantially. Only then I discovered the hurdles women undergo in the academic world: For example, typically women go through the tenure track, while many men are often parachuted to a tenured position; I observe that certain characteristics, positive ones on the absolute scale, such as shyness, a soft voice, giving priority to pleasing and helping others, hinder women. Embedded biases in the academia against women are difficult to eliminate because it is still men's world. I find it helpful that we have now a network of women faculty at our university. It is also a pleasure because our monthly meetings are stimulating intellectually. During my travel, I meet women in other continents and see how much is yet to be done, in particular in Africa and Asia but also everywhere else, till women will get the conditions to reach their full potential.

Words of Wisdom

I think it is important to have dreams and ambitions. Generalizing, based on myself, I believe women tend to dream more than men. So, my two-pence of wisdom is to say that all one has to do is to *want* to fulfill one's own dreams and take first steps toward it; the rest will come by itself.

Susan B. Sinnott

Professor
Department Head
Materials Science and Engineering
The Pennsylvania State University
111 Research Unit A
University Park, PA 16802
USA

e-mail: sinnott@matse.psu.edu
telephone: (814) 863-3117

Susan Sinnott photographed
as part of a 1999 article titled
"Thinking big by thinking
small" by Debbie Gibsen as
part of the University of
Kentucky (UK) magazine
"Odyssey." The article
discussed research on carbon
nanotubes by Sinnott and
others taking place at the
United Kingdom as part of a
National Science Foundation
Materials Research Science
and Engineering Center.

Birthplace
Manhattan, KS, USA
Born
August 14, 1966

Publication/Invention Record

>200 publications: h-index 35 (Web of Science)
9 book chapters

Tags

❖ Academe
❖ Domicile: USA
❖ Nationality: American
❖ Caucasian
❖ Children: 2

Proudest Career Moment (to date)

When my first Ph.D. student successfully defended his dissertation, it was a time of celebration and reflection as I witnessed the outcome of the combination of his hard work and my guidance of his professional development.

Academic Credentials

Ph.D. (1993) Physical Chemistry, Iowa State University, Ames, IA, USA.
B.S. honors (1987) in Chemistry, University of Texas, Austin, TX, USA.

Research Expertise: investigate the properties and processing of materials using computational tools, including molecular dynamics simulations and Monte Carlo simulations with reactive, many-body empirical potentials, and first principles, electronic-structure calculations using density functional theory and quantum chemical methods. Current problems of interest include the development of inventive new computational methods; design of new metal alloys and ferroelectric materials by combining first-principles calculations with materials informatics; electronic structure and stability of defects in ceramics; physical, chemical, and electrical properties of nanostructures; atomic and molecular scale friction and wear; structure and properties of heterogeneous interfaces; catalysis at metal clusters supported on oxide surfaces; and the oxidation of metals and nitrides

Other Interests: volunteering with professional societies, faculty advisor for student chapters of professional societies

Key Accomplishments, Honors, Recognitions, and Awards

- Editor-in-Chief, Computational Materials Science, 2014 to present
- Chair, MRS Theory of Materials Award Subcommittee, 2014 to present
- Principal Editor, Journal of Materials Research, 2012 to present
- American Institute of Physics (AIP) Advisory Committee on Physics Today, 2011 to present
- Materials Research Society (MRS) Academic Affairs Committee, 2010–2014
- Divisional Associated Editor, Physical Review Letters, 2009 to present
- Editorial Board, Current Opinion in Solid State and Materials Science, 2009 to present
- Surface, Interface and Atomic-Scale Science Editorial Board of Journal of Physics: Condensed Matter, 2009 to present
- AVS President Elect, 2012; President, 2013; Past President, 2014
- American Ceramic Society (ACerS) Nominations Committee member, 2010–2013
- Fellow American Physical Society, 2013
- MRS Theory of Materials Award Subcommittee, 2011–2013
- University of Florida Research Foundation Professor, 2011–2013
- Fellow MRS, 2012
- Selected as one of the Top 25 Women Professors in the state of Florida, 2012
- Fellow ACerS, 2011
- Program Chair, AVS International Symposium and Exhibition, 2011
- University of Florida Blue Key Distinguished Professor, 2010–2011
- AVS Director, 2009–2011
- Fellow AAAS, 2010

Susan Sinnott moderating a session on professional development during the Cyberinfrastructure for Atomistic Simulation (CAMS) 2013 Summer School in Gainesville, FL, May, 2013.

- Program Vice-Chair, AVS International Symposium and Exhibition, 2010
- ACerS Member Services Committee member, 2004–2010
- AVS Chapters, Groups, and Divisions Chair, 2006–2008, and member, 2005–2010
- University of Florida College of Engineering Doctoral Dissertation Mentoring Award, 2009
- ACerS Member Services Committee Chair, 2006–2009
- AVS Trustee, 2006–2008; Trustee Chair, 2008
- University of Florida Faculty Excellence Award, 2002–2006 and 2008
- AVS Committee on Student Issues: Chair, 1998–2000 and member, 1997–2007
- Editorial Board Member, Nano Letters, 2004–2006
- North American Editor, Journal of Nanoscience and Nanotechnology, 2001–2006
- Fellow AVS, 2005
- University of Florida College of Engineering Teacher/Scholar of the Year Award 2005
- NSF Award for Special Creativity, 2005
- Chair of the Florida Section of the American Ceramic Society, 2005
- AVS Topical Conferences Chair, 2003–2005
- Vice-Chair of the Florida Section of the American Ceramic Society, 2003
- AVS Marketing Committee member, 2001–2003
- Secretary of the Florida Section of the American Ceramic Society, 2002
- Japan Society for the Promotion of Science Fellow, 2001
- Treasurer of the Florida Section of the American Ceramic Society, 2001
- Chair of the AVS Surface Science Division Nominations Committee, 2001
- AVS Information Technology Committee member, 1998–2000
- AVS Long Range Planning Committee member, 1998–2000
- University of Kentucky College of Engineering, Tau Beta Pi, and Department of Chemical and Materials Engineering Outstanding Teaching Award (Materials Engineering Program), 1998 and 2000
- Chair of the AVS Surface Science Division Morton M. Traum Award Committee, 2000
- ASM Bluegrass Chapter: Chair, 1999–2000; Vice-Chair, 1998–1999, and Member of the Executive Committee, 1997–1998
- Oak Ridge Associated Universities Ralph E. Powe Junior Faculty Enhancement Award in Engineering, 1996
- National Research Council Postdoctoral Fellowship Award, 1993–1995
- Iowa State University Proctor and Gamble Fellowship Award, 1992–1993
- Iowa State University Premium Academic Excellence (PACE) Award, 1988
- The University of Texas Robert A. Welch Undergraduate Fellowship Award, 1987

Biography

Early Life and Education

Susan is a chemist by training, completing her bachelor's degree at the University of Texas at Austin followed by a doctorate at Iowa State University.

Career History

She completed a National Research Council Postdoctoral Associate at the Naval Research Laboratory, Surface Chemistry Branch in Washington, DC, from 1993–1995. She then became an Assistant Professor in the Department of Chemical and Materials Engineering at the University of Kentucky in Lexington. In 2000, she was snatched away by the University of Florida in Gainesville and promoted to Associate Professor in the Department of Materials Science and Engineering. In 2005, she was promoted to full professor. In 2007, she became an Affiliate Professor in the Department of Mechanical and Aerospace Engineering. In 2011, she joined the Quantum Theory Project.

3 Most Cited Publications (excluding review articles)

Title: A second-generation reactive empirical bond order (REBO) potential energy expression for hydrocarbons
Author(s): Brenner, DW; Shenderova, OA; Harrison, JA; Stuart, SJ; Ni, B; Sinnott, SB
Source: Journal of Physics-Condensed Matter; volume: 14; issue: 4; pages: 783–802; article number: PII S0953-8984(02)31186-X; doi: 10.1088/0953-8984/14/4/312; published: February 4, 2002
Times Cited: 1626 (from Web of Science)

Title: Model of carbon nanotube growth through chemical vapor deposition
Author(s): Sinnott, SB; Andrews, R; Qian, D; Rao, AM; Mao, Z; Dickey, EC; Derbyshire, F
Source: Chemical Physics Letters; volume: 315; issue: 1–2; pages: 25–30; doi: 10.1016/S0009-2614(99)01216-6; published: December 17, 1999
Times Cited: 372 (from Web of Science)

Title: Carbon nanotubes: synthesis, properties, and applications
Author(s): Sinnott, SB; Andrews, R
Source: Critical Reviews in Solid State and Materials Sciences; volume: 26; issue: 3; pages: 145–249; published: 2001
Times Cited: 233 (Web of Science)
Researcher ID (P-8523-2014)

Challenges

Achieving work–life balance is always a challenge. Support from an understanding spouse and family at home, and productive group members and like-minded colleagues at work, are key to achieving this balance.

On being a woman in this field . . .

When I first started in the field, there were very few women, especially at the senior level. This remains true today, although the numbers are increasing. It is therefore common for me to be one of only a handful of women at workshops of about 25–50 people. Consequently, people tend to remember me, which is positive. The challenge lies in the fact that expectations from students and colleagues seem to be distinctly different for women; I have always felt that the bar was set higher for us, with perhaps a greater variety of pathways available for us to get over the bar. Interestingly, some of the harshest criticisms I have received over the course of my career have been from more senior women.

Susan Sinnott and colleague Prof. Ronald Akers taking part in the University of Florida 2010 Homecoming Parade, an honor awarded to faculty named University of Florida Blue Key Distinguished Professors. Photo Credit: Prof. W. Gregory Sawyer, University of Florida.

Words of Wisdom

Having access to one or more really good mentors is extraordinarily helpful to launching and building a successful career.

Nicola A. Spaldin

Professor and Chair of Materials Theory
ETH Zürich
Wolfgang-Pauli Strasse 27
CH-8093 Zürich
Switzerland

e-mail: nicola.spaldin@mat.ethz.ch
telephone: +41 (0) 44 633 37 55

Spaldin in 2000 during her first sabbatical that she spent at the Jawaharlal Nehru Centre for Advanced Scientific Research in India.

Birthplace

Sunderland, UK

Born

Mar. 1, 1969

Publication/Invention Record

>125 publications: h-index 57
(Web of Science)
2 books

Tags

❖ Administration and Leadership
❖ Academe
❖ Domicile: Switzerland
❖ Nationality: British
❖ Caucasian
❖ Children: 0

Proudest Career Moment (to date)

Without question, every time a graduate student completes her/his Ph.D.! It's the most wonderful experience for an educator to follow the development of a young scientist from beginning researcher to independent professional, and to help her/him a little to make that transition. It was also a tremendous privilege to win the American Physical Society's McGroddy Prize for New Materials together with two fantastic experimental scientists, Ramamoorthy Ramesh and Sang-Wook Cheong. The award was a fabulous endorsement for the many theoreticians who have contributed to the development of electronic structure methods over the decades in affirming that they are now really useful

Successful Women Ceramic and Glass Scientists and Engineers: 100 Inspirational Profiles. Lynnette D. Madsen.
© 2016 The American Ceramic Society and John Wiley & Sons, Inc. Published 2016 by John Wiley & Sons, Inc.

for designing and predicting new materials. Oh, and recently I had my first
invitation to a conference to provide the
dinner music rather than to speak about
my work (I'm a slightly obsessive
chamber musician) that was great fun!

Academic Credentials

Ph.D. (1996) Chemistry, University of
California, Berkeley, CA, USA.
B.A. (1991) Natural Sciences (First
Class Honours), Cambridge University,
Cambridge, UK.

Spaldin in 2010 at the Happy Boulders in the
Sierra Nevada, celebrating winning the
McGroddy Prize for New Materials. (See insert
for color version of figure.)

Research Expertise: development
and application of first-principles theoretical techniques to study the fundamental physics
of novel materials that have potential technological importance: systems of interest
include "multiferroic" materials (which are simultaneously ferromagnetic, ferroelectric,
and ferroelastic) and multiple functionalities at oxide interfaces

Other Interests: fostering international collaborations, teaching

Key Accomplishments, Honors, Recognitions, and Awards

- Körber European Science Prize, 2015
- American Physical Society (APS) Outstanding Referee, 2014
- ETH Zürich Golden Owl Award for Teaching Excellence, 2014
- Fellow, American Association for the Advancement of Science, 2013
- Max Rössler Prize of the ETH, 2012
- European Research Council Advanced Grant, 2012
- Fellow, Materials Research Society, for fundamental theoretical contributions leading
  to the emergence of the field of multiferroics, 2011
- American Physical Society McGroddy Prize for New Materials, for groundbreaking
  contributions in theory and experiment that have advanced the understanding and
  utility of multiferroic oxides, 2010
- NSF American Competitiveness and Innovation Fellow, for her creative integration of
  fundamental science and modern computational methods for the discovery of new
  materials and for elucidating phenomena at the nanoscale, and her role in promoting
  education and international activities, 2009
- Fellow, APS, for her development and implementation of new computational and
  theoretical tools for computing the properties of complex solids and their application to
  the rational design and understanding of new multifunctional materials, and for her
  profound and diverse contributions to physics education, 2008

- Miller Institute Research Professorship, 2007
- Alfred P. Sloan Foundation Research Fellowship, 2002
- University of California, Santa Barbara, Distinguished Teaching Award, 2001
- Office of Naval Research Young Investigator Award, 2000
- Technology Review Magazine Young Innovator Award, 1999
- National Science Foundation (NSF) Professional Opportunities or Women in Research and Education (POWRE) Award, 1999
- Regents Junior Faculty Fellowship, University of California, 1998
- Fulbright Scholarship from the US–UK Fulbright Commission, 1991–1996
- Electronic Materials Conference Graduate Student Award, 1995
- Cray Research Fellowship in Computational Chemistry, 1995
- Materials Research Society Graduate Student Award, 1994
- Ephraim Weiss Fellowship, Graduate Study at U.C. Berkeley, 1993–1994
- Outstanding Graduate Student Instructor Awards, U.C. Berkeley, 1991 and 1992
- Dora Garibaldi Fellowship, Graduate Study at U.C. Berkeley, 1991–1992
- Ronald Norrish Prize for Physical Chemistry, U. Cambridge, 1991
- Mineralogical Society of Great Britain Student Award, 1989
- Undergraduate Scholarship, General Electric Company (GEC) Hirst Research Centre, 1987–1991
- Royal Society of Chemistry A-Level Chemistry Award, 1987

Biography

Early Life and Education

Spaldin grew up and attended school in the north of England. A "gap year" at General Electric Company sparked her interest in applying fundamental science to tackling technological problems, and led to a Bachelors Degree in Natural Sciences with an emphasis on Chemistry at Cambridge University. Motivated in part by the need to be closer to mountains, she moved to California for her doctoral work, where she completed her Ph.D. in Theoretical Chemistry at the University of California Berkeley.

Career History

A postdoctoral appointment in Applied Physics at Yale University in the group of Karin Rabe gave Spaldin the opportunity to move into new research directions, and in particular to learn the tool—density functional theory—that she uses most often today in her research. She then moved to the Materials Department at the University of California, Santa Barbara, where

Nicola Spaldin at ETH Zurich, 2012.

she established her independent research career and progressed from Assistant through Associate to Full Professor. In 2011, after spending half of her life in California, she returned to her native Europe to take the Chair of Materials Theory at the ETH in Zürich. Again close to beautiful mountains of course.

3 Most Cited Publications

Title: Epitaxial $BiFeO_3$ multiferroic thin film heterostructures
Author(s): Wang, J; Neaton, JB; Zheng, H; et al.
Source: Science; volume: 299; issue: 5613; pages: 1719–1722; published: March 14, 2003
Times Cited: 2915 (from Web of Science)

Title: Why are there so few magnetic ferroelectrics?
Author(s): Hill, NA
Source: Journal of Physical Chemistry B; volume: 104; issue: 29; pages: 6694–6709; published: July 27, 2000; doi: 10.1021/jp000114x
Times Cited: 1685 (from Web of Science)

Title: Multiferroics: progress and prospects in thin films
Author(s): Ramesh, R; Spaldin, NA
Source: Nature Materials; volume: 6; issue: 1; pages: 21–29; published: 2007; doi: 10.1038/nmat1805
Times Cited: 1648 (from Web of Science)

ResearcherID: (A-1017-2010)

Challenges

Having enough time for everything that I want to do.
Figuring out what to *not* do when I realize there isn't enough time for everything.
Finding a nice way to tell people that I am not going to do everything that they would like for me to do.

On being a woman in this field . . .

I and my peers certainly owe a huge thank you to the previous generation of pioneering women scientists for the absence in large part of overt gender discrimination in the sciences, at least in the United States and Europe where I have most experience. I wonder though whether we have done such a good job in continuing to improve the situation for the next generation who I would like to have a more gender-balanced and gender-neutral work environment than we have managed to shape for them.

Words of Wisdom

Oh dear, this is a tough one. In spite of a recent birthday that prevents me from still pretending I'm in my early 40s, I don't feel very qualified to give out words of wisdom. Can I plagiarize the words of Martin Lovett, the cellist of the famed Amadeus String Quartet, who when asked for advice for young quartets suggested, "Always say thank you very much." I think you can't go wrong with that.

Profile 88

Susanne Stemmer

Professor
Materials Department
University of California
Santa Barbara, CA 93106-5050
United States

e-mail: stemmer@mrl.ucsb.edu
telephone: (805) 893-6128

Susanne at about 31 years of age.

Tags

❖ Academe
❖ Domicile: United States
❖ Nationality: dual German & American
❖ Caucasian
❖ Children: 0

Publication/Invention Record

>200 publications: h-index: 42
Co-editor 2 books, author 2 book chapters

Proudest Career Moment (to date)

The latest manuscript. One of the nicest aspects of a research career for me is that the current research problem always seems to be the most exciting and interesting in all of science. Tackling a scientific question (and, eventually, publishing it) is hugely rewarding (and a lot of fun).

Academic Credentials

Dr. rer. nat. (1995) University of Stuttgart, Stuttgart, Germany.
Diploma (1992) Materials Science and Engineering, Friedrich-Alexander Universität, Erlangen-Nürnberg, Germany.

Research Expertise: oxide molecular beam epitaxy, functional oxide thin films and heterostructures, novel dielectric thin films and dielectric/semiconductor interfaces, quantitative atomic resolution scanning transmission electron microscopy

Other Interests: teaching, educating the next generation of scientists

Key Accomplishments, Honors, Recognitions, and Awards

• Fellow of the Materials Research Society, 2014
• Matsen Lectureship, University of Texas, Austin, 2013

Successful Women Ceramic and Glass Scientists and Engineers: 100 Inspirational Profiles. Lynnette D. Madsen.
© 2016 The American Ceramic Society and John Wiley & Sons, Inc. Published 2016 by John Wiley & Sons, Inc.

- Fellow of the American Physical Society, 2012
- Fellow of the American Ceramic Society, 2011
- CAREER Award, National Science Foundation, 2000
- Training and Mobility of Researchers (TMR) grant, European Commission, 1996–1997

Biography

Early Life and Education

Susanne Stemmer was born and raised in Germany. She has three younger brothers. She attended high school in Roth (Bavaria) and the nearby Friedrich-Alexander Universität Erlangen-Nürnberg to study Materials Science. She received her university Diploma (which is approximately equivalent to a Masters degree in the United States) in 1992. She sent a job application to the Max Planck Institute for Metals Research in Stuttgart (Germany), as this sounded like an interesting place to do materials research, and was told that if she wanted to work as a researcher she should first get a doctoral degree. With this insight, she joined Professor Manfred Rühle's group at the Max Planck Institute for

Susanne Stemmer in her office; photo taken by her colleague Ram Seshadri.

Metals Research in 1992 as a doctoral student, focusing on a project in transmission electron microscopy of ferroelectric thin films. She enjoyed her time there, as support for research was exceptional, she was given lots of freedom to pursue interesting questions, and the group hosted many international visitors (including several from University of California at Santa-Barbara (UCSB), who later became her colleagues). She graduated in 1995, with the Dr. rer. nat. degree awarded by the University of Stuttgart.

Career History

From 1995 to 1996, she was a Research Associate at Case Western Reserve University in the Department of Materials Science and Engineering. Still uncertain about her future career, she moved to the Departement Metaalkunde en Toegepaste Materiaalkunde at Katholieke Universiteit in Leuven Belgium with support from the European Commission through a Training and Mobility of Researchers fellowship. Because the U.S. academic system allows for research independence of younger scientists, and exciting developments in transmission electron microscopy, she decided to join Nigel Browning's group at the University of Illinois at Chicago as a Visiting Assistant Professor in 1998. At that time, his group was in the process of setting up one of the first conventional transmission electron microscopes capable of atomic resolution scanning transmission electron microscopy. During this period, she also made up her mind to pursue an academic career, and joined the Department of Mechanical Engineering and Materials Science at Rice University in Houston as an Assistant Professor in 1999. In 2002, she moved to the University of California, Santa Barbara, as an Assistant Professor. Three years later she was promoted to Associate Professor and in another 3 years (2008), promoted again, this time to Full Professor. Since establishing an independent research group, she and her group have worked in a number of different research areas (though most of them related to oxide thin films) on problems spanning from applied to very basic science.

3 Highly Cited Publications

Title: Low-loss, tunable bismuth zinc niobate films deposited by RF magnetron sputtering
Author(s): Lu, JW; Stemmer, S
Source: Applied Physics Letters; volume: 83; issue: 12; pages: 2411–2413; doi: 10.1063/1.1613036; published: September 22, 2003
Times Cited: 137 (Web of Science)

Title: Quantitative atomic resolution scanning transmission electron microscopy
Author(s): LeBeau, JM; Findlay, SD; Allen, LJ; Stemmer, S
Source: Physical Review Letters; volume: 100; issue: 20; article number: 206101; doi: 10.1103/PhysRevLett.100.206101
Times Cited: 135 (Web of Science)

Title: Atomistic structure of 90° domain walls in ferroelectric $PbTiO_3$ thin films
Author(s): Stemmer, S; Streiffer, SK; Ernst, F; Rühle, M
Source: Philosophical Magazine A; volume: 71; issue: 3; pages: 713–724; published: March 1995
Times Cited: 122 (Web of Science)
ResearcherID (H-6555-2011)

Challenges

I think an academic career is no more and no less challenging and demanding than one in industry or government. Specific challenges are that funding for basic research is tight—finding sufficient funds for research requires a never-ending effort and is a constant source of worry. The many different demands on one's time (reports, administration, teaching, service, . . .) make it very difficult to find quality time to think deeply about science.

On being a woman in this field . . .

I think we are in the midst of a sea change. For example, most younger male colleagues have professional spouses, which didn't seem to be the case for the generation that is older than I, and are dealing equally with issues such as child care and balancing family and career. Still, it's hard to miss the fact that there are far fewer women in faculty positions in Materials Science and Engineering than men. Like other women I know,

I often find myself counting the number of female invited speakers at conferences. I have never given much thought as to what being female has meant to me personally in terms of my career, as statistics are bad, and differences are probably subtle. I have blissfully ignored many of the advices given especially to women (on speaking more assertively, on dress codes, . . .), and I think the encouraging lesson learned is that one can stay oneself and have a successful career.

Words of Wisdom

I found that early in my career, friends and peers were much more helpful in terms of career advice than advisors or supervisors, probably because they knew me much better. I think it is best to pursue what one loves doing without too much second-guessing—the rest will fall into place (eventually). Making some career mistakes is normal and not the end of the world as long as one corrects them. As a researcher, good science should be the first priority, which requires learning to ignore a lot of "noise."

Profile 89

Wei-Ying Sun

Shanghai Institute of Ceramics
Chinese Academy of Sciences
1295 Ding Xi Road
Shanghai
200050, P. R. China

e-mail: sun100@sh163.net
telephone: +21-62512990

Wei-Ying Sun at Newcastle University, 1984.

Birthplace
Shanghai, P. R. China
Born
March 1, 1938

Tags

❖ Academe
❖ Domicile: China
❖ Nationality: Chinese
❖ Asian
❖ Children: 1

Publication/Invention Record

>50 publications

Academic Credentials

B.Sc. (1962) Chemistry, Fudan University, Yangpu, Shanghai, China.

Research Expertise: studies of phase relationships in the M-Si-Al-O-N where M=Y, Ln, Ca, Mg, Li, Na, and Sr; new SiAlON composites

Key Accomplishments, Honors, Recognitions, and Awards

• First prize of Nature Science by the Chinese Academy of Science for the achievement in the studies of phase equilibrium and compositional design in complicated nitrogen ceramics, 1997
• Elected to the World Academy of Ceramics, 1995
• First prize of Science and Technology by the Chinese Academy of Science for the achievement in the studies of phase relationships in M-Si-AI -O-N systems, 1987

Biography

Early Life and Education

She earned one higher education degree at Fudan University in Chemistry in 1962.

Successful Women Ceramic and Glass Scientists and Engineers: 100 Inspirational Profiles. Lynnette D. Madsen.
© 2016 The American Ceramic Society and John Wiley & Sons, Inc. Published 2016 by John Wiley & Sons, Inc.

Career History

Upon graduation, and until 1983, Wei-Ying Sun became an Assistant Professor at Shanghai Institute of Ceramics (SIC). She then became a Visiting Scientist at the University of Newcastle for 2 years. Upon her return to SIC, she was promoted to Associate Professor and became the Project Director for studying the phase relationships in the M-Si-AI-O-N (M = Y, Ca, Mg, Li) systems. In 1989, she served as a Visiting Professor again, this time at the University of Michigan. Returning to SIC, she was again promoted, this time to Professor and Project Director for the research on materials design in the Y-Si-AI-O-N system. Her most recent Visiting Professor period was from 1992 to 1993 at the University of Newcastle. When she returned to SIC, she became Project Director for the research on phase relationships and materials design in the Ln-Si-AI-O-N systems. She retired in 1998.

3 Most Cited Publications

Title: Solubility limits of alpha'-SiAlON solid-solutions in the system Si,Al,Y/N,O
Author(s): Sun, WY; Tien, TY; Yen, TS
Source: Journal of the American Ceramic Society; volume: 74; issue: 10 pages: 2547–2550; published: October 1991
Times Cited: 101 (from Web of Science)

Title: Phase-relations of the Si_3N_4-AlN-CaO System
Author(s): Huang, ZK; Sun, WY; Yan, DS
Source: Journal of Materials Science Letters; volume: 4; issue: 3; pages: 255–259; published: 1985
Times Cited: 66 (from Web of Science)

Title: Subsolidus phase-relationships in part of the system Si,Al,Y/N,O: the system Si_3N_4-AlN-YN-Al_2O_3-Y_2O_3
Author(s): Sun, WY; Tien, TY; Yen, TS
Source: Journal of the American Ceramic Society; volume: 74; issue: 11; pages: 2753–2758; published: November 1991
Times Cited: 56 (from Web of Science)

Profile 90

Inna G. Talmy

Retired Distinguished Ceramic Scientist; now Consultant
Naval Surface Warfare Center (NSWC)
Carderock Division
Maryland, USA

Dr. Talmy, NSWC,
1980.

e-mail: igtalmy@gmail.com
telephone: (301) 610-7468

Birthplace
Moscow, Russia
Born
January 14, 1935

Publication/Invention Record
>100 publications: h-index 12 (Web of Science)
20 patents; editor of 2 books

Tags
❖ Administration and
   Leadership
❖ Government
❖ Domicile: USA
❖ Nationality: American
❖ Caucasian

Proudest Career Moments (to date)

In her 55 years career in ceramic research and development, Dr. Talmy is very proud of numerous projects. Below are the most important ones:

- At the Prague Institute of Chemical Technology, Dr. Talmy conducted research on large electrical porcelain insulators. Her goal was to eliminate the deformation of the insulators during firing. She used a nonconventional (in porcelain research) compression creep method to determine the effect of porcelain composition on the rate of creep. The modification of the composition by an increase in a clay component-to-feldspar ratio and the addition of $BaCO_3$ resulted in a significant decrease of the rate of creep. The effect was related to an increase in glass phase viscosity. The research resulted in a dramatic increase in the structural stability of the insulators during firing.
- At NSWC, one of the projects of Dr. Talmy's was to develop new ceramic materials for radomes for future classes of advanced tactical missiles as a substitution for slip-cast fused silica. She proposed high-purity monoclinic celsian ($BaO \cdot Al_2O_3 \cdot 2SiO_2$) as a candidate for this application. Celsian exists in two main crystalline modifications: monoclinic and hexagonal. Only monoclinic modification has the required properties

---

Successful Women Ceramic and Glass Scientists and Engineers: 100 Inspirational Profiles. Lynnette D. Madsen.
© 2016 The American Ceramic Society and John Wiley & Sons, Inc. Published 2016 by John Wiley & Sons, Inc.

for radomes. However, the hexagonal modification tends to be the first product of solid-phase reaction and has a strong tendency to persist metastably. The preparation of the monoclinic modification is a challenging task. Dr. Talmy developed two methods for the preparation of monoclinic celsian: first, the introduction of monoclinic seeds in the raw materials; and second, the formation of solid solutions of $BaO \cdot Al_2O_3 \cdot 2SiO_2$ with analogous $SrO \cdot Al_2O_3 \cdot 2SiO_2$. The latter exists only in the monoclinic form, thus promoting the formation of monoclinic celsian (U.S. Patents 5695725, 1997 and 5642868, 1997, respectively).

Dr. Talmy and Mr. Martin demonstrate a principle for beneficiation of superconducting powders to the patent attorney Mr. Ken Walden (1988). The method was patented (U.S. Patent 5047387, 1991).

- The preparation of loose mullite ($3Al_2O_3 \cdot 2SiO_2$) whiskers and rigid mullite whisker felt structures from AlF_3 and SiO_2 mixtures was another significant development. The whiskers can be used as a reinforcement in ceramic, metal, and other composite matrices. Felt consisted of single-crystal mullite whiskers that are uniformly distributed and randomly oriented in three dimensions and are mechanically interlocked to form a rigid felt structure capable of maintaining

Dr. Talmy, Mr. Martin, and Ms. Haught are quite pleased with the laboratory results received on the utilization of fly ash for the production of ceramic materials. The patent was awarded (U.S. Patent 5521132, 1996).

its shape without binders. The felt can be used as preforms for ceramic–matrix or metal–matrix composites or by itself as thermal insulation and filters for powders and liquids including liquid metals. The developed processes resulted in three patents, which were licensed by Dow Chemical Company (U.S. Patents 5521132, 1996; 4911902, 1990; and 4910172, 1990).

- The discovery of new compounds with the formulas $Ti_{0.5}Ta_{0.5}B$, $Zr_{0.5}Ta_{0.5}B$, and $Hf_{0.5}Ta_{0.5}B$ of orthorhombic structure and melting temperature about 2300 °C. Ceramics based on the new compounds were prepared by hot pressing. The microstructure of the ceramics consists of interlocking bar-like crystals. This microstructure resulted in a high flexural strength of about 500 MPa and high fracture toughness.

These properties make the materials useful for high-temperature structural applications (U.S. Patent 7854912, 2010).

Academic Credentials

Ph.D. (1965) Ceramic Science and Engineering, Mendeleev Institute of Chemical Technology, Moscow, Russia.
M.S. (1957) Ceramic Science and Engineering, Mendeleev Institute of Chemical Technology, Moscow, Russia.

Research Expertise: electrical porcelain, celsian and phosphate ceramics as candidates for next-generation tactical missile radomes, dielectric ceramics, superconductors, non-oxide structural high-temperature materials with improved oxidation and thermal stress resistance for hypersonic and strategic missile applications; ceramic–matrix composites, cermets, and creep

Other Interests: mentoring high school and college students

Key Accomplishments, Honors, Recognitions, and Awards

- Fellow of the American Ceramic Society, 2011
- Scientific and Engineering Advisory Board for 3G-HTS Corporation, 2011
- The position of Distinguished Ceramic Scientist, 2000
- NAVSEA Scientist of the Year Award, 1996
- Science and Technology Excellence Award, 1993
- Adolphus Dahlgren Award, 1991
- U.S. Navy Meritorious Civilian Award, 1990

Biography

Early Life and Education

Dr. Talmy was born in Moscow, USSR. She started her schooling in Kazakhstan where she lived with her mother during WWII. Back in Moscow in 1944, she continued her school education. During her school years, she was interested in many subjects, especially, biology, history, and chemistry. However, in the senior year her interest in chemistry was strongly encouraged by the chemistry teacher and became predominant. After graduation from High School in 1952, she entered the Mendeleev Institute of Chemical Technology in Moscow, one of the best schools of chemistry in USSR, which gives very deep knowledge in all fields of chemistry and chemical technology. Her major was "Chemistry and technology of ceramic materials" with emphasis on technical (engineering) ceramics. The 1950s were a time of worldwide research and development of pure oxide ceramics based on Al_2O_3, ZrO_2, MgO, spinel, and other oxides. After graduation in 1957, she was offered a job in the Ceramic Materials Department as an assistant of prominent professor R.Ya. Popilsky. In 1965, she defended her Ph.D. dissertation "Sintering and properties of ceramics in the $MgO–MgAl_2O_4$ system."

Career History

Dr. Talmy and her family (husband Vladimir, daughter Dasha, and two mothers) came to the United States in February 1980. In addition to science, Dr. Talmy is interested in art, literature, and music.

Among other places, she worked at the Institutes of Chemical Technology in Moscow and Prague (Czechoslovakia), and at Chemical Abstract Service in Columbus, OH.

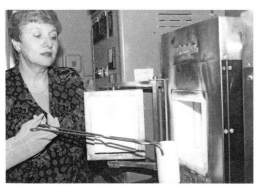

Dr. Talmy joined the Naval Surface Warfare Center in 1983. She led the Advanced Ceramics Group, which grew under her leadership to be one of the primary groups for research and development of ceramics for radomes and high-temperature materials with improved oxidation and thermal stress resistance for hypersonic and strategic missile applications.

Dr. Talmy is active in the Engineering Ceramics Division of the American Ceramic Society. In 2011, she became a Fellow Member

Dr. Talmy removing the crucible containing mullite whiskers from a furnace. (See insert for color version of figure.)

of the Society. She has presented numerous papers at Society meetings and reviewed manuscripts for the Journal of the American Ceramic Society and other journals. Her work has resulted in over 100 publications in peer-reviewed journals and conference proceedings and 20 patents. Dr. Talmy is an internationally recognized expert in ultra-high-temperature ceramics and radome materials.

Credits: http://ceramics.org/acers-community/award-winners-resources?bio=20837; http://www.af.mil/news/story.asp?id=123043316.

3 Most Cited Publications

Title: Refractory diborides of zirconium and hafnium
Author(s): Fahrenholtz, WG; Hilmas, GE; Talmy, IG; et al.
Source: Journal of the American Ceramic Society; volume: 90; issue: 5; pages: 1347–1364; doi: 10.1111/j.1551-2916.2007.01583.x; published: May 2007
Times Cited: 303 (from Web of Science)

Title: Mechanical, thermal, and oxidation properties of refractory hafnium and zirconium compounds
Author(s): Opeka, MM; Talmy, IG; Wuchina, EJ; et al.
Conference: Meeting on New Developments in High Temperature Materials; location: Istanbul, Turkey; date: August 12–15, 1998
Source: Journal of the European Ceramic Society; volume: 19; issue: 13–14; pages: 2405–2414; doi: 10.1016/S0955-2219(99)00129-6; published: 1999
Times Cited: 256 (from Web of Science)

Title: Oxidation-based materials selection for 2000 °C plus hypersonic aerosurfaces: theoretical considerations and historical experience
Author(s): Opeka, MM; Talmy, IG; Zaykoski, JA
Conference: Workshop on Ultra-High Temperature Ceramic Materials; location: Wintergreen, VA; date: November 5–7, 2003; sponsor(s): AFOSR
Source: Journal of Materials Science; volume: 39; issue: 19; pages: 5887–5904; doi: 10.1023/B:JMSC.0000041686.21788.77; published: October 1, 2004
Times Cited: 242 (from Web of Science)

Challenges

My challenge was to occupy a place in American ceramic science corresponding to my abilities, knowledge, and qualification.

On being a woman in this field . . .

I have never had any problems of being a woman in this field. There was no woman discrimination in the Soviet Union where I started my career. I came to America being an established scientist. I did not have any problems in finding job and making a career in this country.

Words of Wisdom

- You cannot make a significant career working only 8 hours a day.
- If you watch time working at the laboratory, do not expect any success.
- If you hate your job, you hate two thirds of your life. You better think about changing your profession.

Susan Trolier-McKinstry

Director, W.M. Keck Smart Materials Integration
Laboratory
Director, Nanofabrication Facility
Professor of Ceramic Science and Engineering
The Pennsylvania State University
N-227 Millennium Science Complex
University Park, PA 16802
USA

e-mail: STMcKinstry@psu.edu
telephone: (814) 863-8348

Susan Trolier-McKinstry as an
Assistant Professor at
Penn State.

Birthplace
Syracuse, NY, USA
Born
March 3, 1965

Publication/Invention Record
>350 publications: h-index 45 (Web of Science)
3 patents

Tags
❖ Academe
❖ Domicile: USA
❖ Nationality: American
❖ Caucasian
❖ Children: 2

Memorable Career Moment (to date)

The times I remember most clearly at work are those where I am teaching, and someone's eyes light up as understanding dawns. Those are the moments that make teaching the right career for me. I am extremely proud of all of the wonderful students that I have mentored over the years, and like to think that I have a small hand in shaping the way in which they think. Of these students, 20 have gone on to academic appointments at universities around the world. Others have gone on to careers in industry or at national laboratories.

Academic Credentials

Ph.D. (1992) Ceramic Science, The Pennsylvania State University, University Park, Pennsylvania, USA.

Successful Women Ceramic and Glass Scientists and Engineers: 100 Inspirational Profiles. Lynnette D. Madsen.
© 2016 The American Ceramic Society and John Wiley & Sons, Inc. Published 2016 by John Wiley & Sons, Inc.

M.S. (1987) Ceramic Science, The Pennsylvania State University, University Park, Pennsylvania, USA.
B.S. (1987) Ceramic Science and Engineering, The Pennsylvania State University, University Park, Pennsylvania, USA.

Research Expertise: electroceramic materials, ferroelectrics, thin films for dielectric and piezoelectric applications, piezoelectric microelectromechanical systems, structure–processing–property relations in thin films, spectroscopic ellipsometry

Other Interests: serving as a mentor and educator

Key Accomplishments, Honors, Recognitions, and Awards

- American Ceramic Society (mentor) member of Futures Committee: 2006 to present
- Associate Editor, Applied Physics Letters, 2008 to present
- Host of the U.S. Navy Workshop on Acoustic Transduction Materials and Devices, 2002 to present
- Distinguished Mentor Program of the American Ceramic Society, 1998 to present
- Co-Chair and Secretary for IEEE Standard for Ferroelectricity, 1996 to present
- Materials Research Society (MRS) Vice President/President Elect, 2016
- Associate Editor, IEEE Transactions on Ultrasonics, Ferroelectrics, and Frequency Control, 2002–2015
- Associate Editor, Journal of the American Ceramic Society, 1999–2015
- President of IEEE Ultrasonics, Ferroelectrics, and Frequency Control Society, 2008–2009; President-Elect, 2006–2007, Past President, 2010–2013

Trolier-McKinstry group, January 2013.

- Materials Research Society Board of Directors, 2011–2013
- Technical Program Committee for the 1996, 2004, and 2006 IEEE International Symposium on Applications of Ferroelectrics; Program Co-Chair in 2000 and 2013
- IEEE Ultrasonics, Ferroelectrics, and Frequency Control Distinguished Lecturer, 2015–2016
- Fellow, Materials Research Society, 2015
- Outstanding Service Award, IEEE Ultrasonics, Ferroelectrics, and Frequency Control Society, 2012
- Outstanding Educator Award, Ceramic Education Council, 2011
- U.S.–Japan Meeting on Dielectric and Piezoelectric Ceramics; Program Chair, 2001; General Chair, 2003 and 2011
- Technical Program Subcommittee of the Materials Research Society, 2004–2010
- Fellow, IEEE, 2009
- Outstanding Faculty Award, Department of Materials Science and Engineering, Penn State, 2009
- Faculty Mentoring Award, College of Earth and Mineral Sciences, Penn State, 2009
- Adjunct Professor of Xi'an Jiaotong University, June 2005 to May 2009
- Strategic Planning Committee, American Ceramic Society, 2006–2009
- Inaugural Class of National Security Science and Engineering Fellows, 2008
- Academician, World Academy of Ceramics, 2008
- IEEE Ultrasonics, Ferroelectrics, and Frequency Control Society Ferroelectrics Achievement Award, 2008
- Adjunct Professor of Shanghai University, June 2005 to May 2008
- Associate Editor, Journal of Electroceramics, 2004–2008
- Richard M. Fulrath Award of the American Ceramic Society, 2006
- Penn State University College of Earth and Mineral Sciences Wilson Award for Excellence in Research, 2006
- Co-Organizer for "Heterogeneous Integration of Materials for Passive Components and Smart Systems" Symposium, at the Fall Materials Research Society Meeting, 2006
- Editorial Board of Journal of Applied Physics and Applied Physics Letters, 2004–2006
- Vice President for Ferroelectrics, IEEE Ultrasonics, Ferroelectrics, and Frequency Control Society, 2000–2005
- Editorial Board Member, Materials Research Innovations, 2000–2005
- International Symposium on Integrated Ferroelectrics Advisory Committee member, 2000–2005
- Fellow of the American Ceramic Society, 2004
- Co-Chair of the Fall Materials Research Society Meeting, 2003
- Keramos (Past President, 2000–2002; President, 1998–2000; Vice President, 1996–1998; Treasurer, 1994–1996; Secretary, 1992–1994)

- Member of the IEEE Ultrasonics, Ferroelectrics, and Frequency Control Administrative Committee, 1999–2001
- Co-Organizer for Symposium on "Ferroelectric Thin Films," Fall Materials Research Society Meeting, 2001
- Penn State University College of Earth and Mineral Sciences Wilson Award for Outstanding Teaching, 2000
- Robert L. Coble Award for Young Scholars (given by the American Ceramic Society), 2000
- Materials Research Laboratory Faculty Achievement Award, 2000
- Ceramics Education Council (President, 2001; Vice President, 2000; Treasurer, 1999; Secretary, 1998)
- Organizing Committee, 5th International Symposium on Ferroic Domains and Mesoscopic Structures, 1998
- Co-Organizer for "Materials for Smart Systems" Symposium for the Fall Materials Research Society Meeting, 1996
- NSF CAREER Award, 1995
- Lead Organizer for Symposium II: "Materials for Smart Systems" for the Fall Materials Research Society Meeting, 1994

Biography

Early Life and Education

Susan Trolier-McKinstry is the daughter of James A. Trolier and Sarah Jane Trolier; she has two elder brothers and a younger sister. The family was a close one, in which perseverance and hard work were encouraged. Education was also a high priority—ultimately all four of the children received doctorates in their chosen fields.

Susan was educated in public school systems in Fayetteville, NY, New Holland, PA, and Wayne, PA in the United States. Following graduation from high school, she entered the Pennsylvania State University, one of few schools with strong programs in both ceramic science and engineering and aeronautical engineering. She rapidly switched her full attention to ceramics on (a) figuring out that space shuttle tiles had already been invented, and (b) receiving outstanding mentoring and guidance from faculty in the ceramics program at Penn State. She started her M.S. degree in her sophomore year, under the guidance of Prof. Robert E. Newnham. In 4 years, she received both B.S. and M.S. degrees, with a thesis aimed at chemical etching of piezoelectric ceramics for miniaturized sensors and actuators. Given the 18″ difference in their heights, it is perhaps not surprising that Prof. Newnham assigned Susan a thesis topic on miniaturization.

As part of her Ph.D. program (also at Penn State, under the supervision of Prof. Newnham), she participated in a 6-month internship at Hitachi Central Research Laboratory in Kokubunji, Tokyo. This was her first experience at living in another culture, and the first time since that age of 5 or so that she was functionally illiterate. This was a fantastic experience, and initiated a series of professional friendships that have lasted a lifetime.

Susan co-taught a class for Corning Incorporated with Prof. Newnham nearly a dozen times between 1990 and 1997. The class was quite diverse: participants ranged from Ph.D.'s in other fields to people working on factory floors while trying to earn their high school equivalency diplomas at night. Prof. Newnham was widely acknowledged as one of the finest teachers in the field, for his eloquent way of creating simple-to-remember analogies that were broadly applicable. Watching the way in which he tailored course content and pace to the participants was an incredible practical education in the art of teaching.

Career History

Her first appointment as a faculty member was as an Assistant Professor of Ceramic Science and Engineering at the Pennsylvania State University in August 1992—an appointment begun the same date as her Ph.D. graduation. She has been at Penn State since that time, with breaks as a summer Research Faculty Fellow at the Electronic and Power Sources Directorate of the U.S. Army Research Laboratories, in the summer of 1993, and a visiting scientist appointment at the Swiss Federal Institute of Technology during the summer of 1996. In June 1998, she was promoted to Associate Professor; in 2002, she was promoted to Professor of Ceramic Science and Engineering. For an additional year, she served as an Adjunct Professor of Xi'an Jiaotong University (i.e., from June 2005 to May 2009).

In a laboratory at the Millennium Science Complex at Penn State with Dan Marincel and Ryan Keech.

Her current research portfolio revolves around the use of thin films for dielectric, piezoelectric, and resistive applications. Her research group spans the range from fundamental studies of structure and the mechanisms that contribute to the available property coefficients, to the processing science of film growth and patterning, to piezoelectric microelectromechanical systems.

She is a Fellow of IEEE and the American Ceramic Society, and an academician in the World Academy of Ceramics. She is also an active participant in the Materials Research Society, with whom she recently completed a term on the board. She currently serves as Associate Editor for Applied Physics Letters.

3 Most Cited Publications

Title: Thin film piezoelectrics for MEMS
Author(s): Trolier-McKinstry, S; Muralt, P
Source: Journal of Electroceramics; volume: 12; issue: 1–2; pages: 7–17; published: January–March 2004
Times Cited: 303 (from Web of Science)

Title: The properties of ferroelectric films at small dimensions
Author(s): Shaw, TM; Trolier-McKinstry, S; McIntyre, PC
Source: Annual Review of Materials Science; volume: 30; pages: 263–298; published: 2000
Times Cited: 288 (from Web of Science)

Title: Domain wall motion and its contribution to the dielectric and piezoelectric properties of lead zirconate titanate films
Author(s): Xu, F; Trolier-McKinstry, S; Ren, W; et al.
Source: Journal of Applied Physics; volume: 89; issue: 2; pages: 1336–1348; published: January 15, 2001
Times Cited: 267 (from Web of Science)

Challenges

There have been numerous challenges over the course of my life and career. One of the ongoing challenges is balancing time between a professional career and spending time with my family. There is no magic wand here that makes all things possible, just a realization that if I am not there to watch my children grow up, that is a time that I will never get back later. So, I try and copy the women who seem to handle things more gracefully than me. There is, still, alas a professional cost to the choice to have children. Hopefully, mine will be the last generation that needs to pay it.

On being a woman in this field . . .

The field of ceramics has changed considerably during my career. I taught my first class to an all-male student body; now my classes are generally on the order of 25% women. Living in Japan as a young women working with older men (where respect scales with both age and (male) gender) was also educational. Fortunately, both in Japan and in the United States, I have had colleagues who have been supportive.

Curiously enough, I am de facto the senior woman in the field of ferroelectrics in the United States. I am not that old, there are just few women who have persisted in the field. The situation is markedly different in Europe, where there are numerous women at the highest academic and professional ranks. It is an inspiration.

I will probably never know what the balance is between the advantages and disadvantages I have experienced as a result of being a woman in the field. My responsibility is to make sure that I leave the field better than I found it as a place for both men and women.

Words of Wisdom

As a professional, I believe it is imperative to have deep intellectual ownership and understanding of at least one area. If the understanding is deep enough, and if there is adequate breadth, then a reasonably small core knowledge can be parleyed into the intellectual flexibility and depth required to adapt to new problems and new funding environments.

Have a life. Work isn't everything.

Papers should be about the science, not the scientist. Write in the third person and avoid excessive hype.

Profile 92

María Vallet-Regí

Professor
Inorganic and Bioinorganic Chemistry Department
Faculty of Pharmacy
Universidad Complutense de Madrid
Spain

María Vallet-Regí in the
1970s.

e-mail: vallet@ucm.es
telephone: +34 913941843

Birthplace
Canary Islands, Spain
Born
April 19, 1946

Publication/Invention Record
>625 publications: h-index 62 (Web of Science)
10 patents, 7 books (edited), 7 books (authored), 32 book chapters

Tags
❖ Academe
❖ Domicile: Spain
❖ Nationality: Spanish
❖ Children: 3

Proudest Career Moments (to date)

In 2001 I demonstrated that it is possible to load silica mesoporous materials with biologically active molecules such as drug peptides, proteins, and growth factors for releasing them, later on, in a controlled manner. This development opened a new line of research that is nowadays followed by many hundreds of researchers who investigate these materials for use in new biomedical applications, including the cancer treatment, of the design of new biomaterials resistant to the infection.

An endearing moment was receiving the Spanish Research Award in 2008, as well as the entrance into the Spanish Royal Academies of Engineering and Pharmacy, and to be invested as Doctor Honoris Causa for the Basque Country University and Jaume I Universities.

Academic Credentials

Ph.D. (1974) Chemistry, Universidad Complutense de Madrid (UCM), Madrid, Spain.
M.Sc. (1969) Inorganic Chemistry, UCM, Madrid, Spain.
B.Sc. (1964) Chemistry, UCM, Madrid, Spain.

Research Expertise: bioceramics with clinical applications: production and study of bioceramic systems, implant bioceramics for bone tissue replacement as well as

Successful Women Ceramic and Glass Scientists and Engineers: 100 Inspirational Profiles. Lynnette D. Madsen.
© 2016 The American Ceramic Society and John Wiley & Sons, Inc. Published 2016 by John Wiley & Sons, Inc.

coatings and nanoparticles for injectable solutions, incorporation of biologically active species, surface functionalization, preparation of porous pieces, manufacture of three-dimensional scaffolds for biocompatible nanoparticles and matrices, silica-based ordered mesoporous materials (i.e., systems such as SiO_2-ZrO_2, $SiO_2-P_2O_5$), organic–inorganic hybrid materials, and smart and reinforced tissue scaffolds for regenerative biomedicine

Other Interests: international cooperation

Key Accomplishments, Honors, Recognitions, and Awards

- Doctor Honoris Causa for the Jaume I University, 2015
- International Union of Pure and Applied Chemistry (IUPAC) Distinguished Woman of the Year in Chemistry and Chemical Engineering, 2013
- Research Prize in Sciences "Miguel Catalan" of the Autonomous Community of Madrid, 2013
- Doctor Honoris Causa for the Basque Country University, 2013
- Fellow of Biomaterials Science and Engineering of International College of Fellows of Biomaterials Science and Engineering (ICF-BSE), 2012
- Member of the Spanish Royal Academy of Pharmacy (Medal Number 42), 2011
- Research Award from Federación Empresarial de la Industria Química Española (FEIQUE), 2011
- Spanish Royal Society of Chemistry (RSEQ) Gold Medal Award, 2011
- National Research Award L. Torres Quevedo in Engineering, 2008
- Spanish Royal Society of Chemistry (RSEQ) Prize in Inorganic Chemistry, 2008
- Comisión Nacional Evaluadora de la Actividad Investigadora (CNEAI) National Committee, 2004–2008
- Member of the Spanish Royal Academy of Engineering (Medal Number LII), 2004
- Steering Committee of NATO "Science for Peace" Programme, 1999–2005
- Vice President of the Spanish Royal Society of Chemistry, 1999–2007
- Prix Franco-Espagnol awarded by Societé Française de Chimie, 2000
- Honorary member of the Materials Research Society in India, 1997
- Founding member of the Instituto de Magnetismo Aplicado U.C.M.-RENFE-CSIC, 1988

A recent image of María Vallet-Regí in the laboratory.

Biography

Early Life and Education

María Vallet-Regí obtained a degree in Chemistry from the Universidad Complutense de Madrid, Spain, in 1969. Then, she worked as an Assistant Professor in the same university. During this period, she also continued her education and in 1974 was awarded a Ph.D. in Chemical Sciences from UCM.

Career History

María Vallet-Regí occupied positions of Chercheur Associé, Maitre de Conferences, Chercheur Asocié de C.R. 1ª classe and Professeur Associé de la Université J. Fourier of the Institute National Polytechnique of Grenoble (France) and worked as an invited researcher at both the National Institute for Research in Inorganic Materials (NIRIM) of Tsukuba (in Japan) and the University of Stockholm (in Sweden).

Since 1990 she has served as the Chair of Inorganic Chemistry in the Faculty of Pharmacy of UCM.

María Vallet-Regí
in the 1990s.

3 Most Cited Publications

Title: Mesoporous materials for drug delivery
Author(s): Vallet-Regí, M; Balas, F; Arcos, D
Source: Angewandte Chemie International Edition; volume: 46; issue: 40; pages: 7548–7558; published: 2007; doi: 10.1002/anie.200604488
Times Cited: 1052 (from Web of Science)

Title: A new property of MCM-41: drug delivery system
Author(s): Vallet-Regí, M; Ramila, A; del Real, RP; et al.
Source: Chemistry of Materials; volume: 13; issue: 2; pages: 308–311; published: February 2001; doi: 10.1021/cm0011559
Times Cited: 917 (from Web of Science)

Title: Metal–organic frameworks as efficient materials for drug delivery
Author(s): Horcajada, P; Serre, C; Vallet-Regí, M; et al.
Source: Angewandte Chemie International Edition; volume: 45; issue: 36; pages: 5974–5978; published: 2006; doi: 10.1002/anie.200601878
Times Cited: 654 (from Web of Science)
ResearcherID (M-3378-2014)

Challenges

One challenge is to continue learning. Conducting research is a continuous challenge; however, it is a privilege to work in what you like. My greater challenge is to work in science that is the gateway to knowledge. Everything is based on a reason, but often it is not known. Science allows us to find those reasons and to improve the world in which we live by contributing with solutions and new technologies for the well-being of our society. What greater challenge can be had?

On being a woman in this field . . .

Being a woman in science has not been an obstacle. It was only a problem of choice and organization of time. Early in my career I dedicated more time to my family (I have three children) and in the later years I shifted my focus and prioritized my research.

Words of Wisdom

It is important to work systematically and every day; talent is not enough. Conduct your research with an optimistic and social vision; be opened to new research lines, particularly if society will benefit. There are always new problems to

María Vallet-Regí with her three sons: Ignacio, Natalia, and Alvaro (the last two are twins).

solve, so your skills can be put to good use. What impels me to continue investigating is the continuous learning and mainly, at this moment, that the results of the investigation of my group address real problems of our society. We are working on issues with bones and teeth (particularly aging effects) and in designing nanoparticles loaded with cytotoxic agents to fight against cancer.

María Vallet-Regí with her research group
(Credit: www.valletregigroup.esy.es).

Research is often collaborative; both you and the research will benefit from working with brilliant people. Although I appreciate the need for management, I enjoy it less than research. In the laboratory, I feel joy in being surrounded by very valuable and clever people. It is very stimulating to be continuously learning and discovering things.

Research is one of the best jobs, but it is not idyllic. It has "dark parts" like the bureaucracy and the need to coexist with certain levels of injustice (e.g., sometimes in the assignation of job positions and research projects). Fortunately, these frustrations progressively decrease (as society matures and as you establish yourself as a researcher and become better networked); keep going.

Paula Maria (Lousada Silveirinha) Vilarinho

Associate Professor
Department of Materials and Ceramic Engineering (DEMaC)
CICECO—Aveiro Institute of Materials
University of Aveiro
3810-193 Aveiro
Portugal

e-mail: paula.vilarinho@ua.pt
telephone: +351 234 370354/259

Paula Vilarinho in the summer of 1985. Photo taken by Pedro Vilarinho.

Birthplace
Coimbra, Portugal

Born
September 30, 1959

Publication/Invention Record

>250 publications: h-index 33 (Web of Science)
9 patents, editor of 4 books, 5 book chapters

Tags

❖ Academe
❖ Domicile: Portugal
❖ Nationality: Portuguese
❖ Caucasian
❖ Children: 0

Proudest Career Moments (to date)

It is difficult to identify just one single moment where I am proud. There are several that I proudly remember, those in which I felt that my contribution to the field of materials science or to the education of young engineers has been recognized. In addition, there are indeed several that I am able to identify as pivotal in my career: (i) Ph.D. degree; (ii) first sabbatical in North Carolina State University, USA (2001); (iii) organizing the Advanced Study Institute (ASI) NATO 2002; and (iv) organizing the Materials 2005 (International Meeting of the Portuguese Materials Society). However, there is one event in my life that I feel is unique. In February 2014, I was invited to talk to 300 recent graduate female engineers from a college of 3000 female engineering students in India, the VINS Women Christian College of Engineering in Vins Nagar, Chunkankadai Tamil Nadu. I was asked to address these young women as a role model, as a female professor. In a country, such

Successful Women Ceramic and Glass Scientists and Engineers: 100 Inspirational Profiles. Lynnette D. Madsen.
© 2016 The American Ceramic Society and John Wiley & Sons, Inc. Published 2016 by John Wiley & Sons, Inc.

Paula Vilarinho at the University of Aveiro receiving her Ph.D. degree in July 1995 during the official award ceremony of diplomas. Photo taken by official photographer of the ceremony.

as India, in which the discrimination of girls and women is still a reality, I felt a significant responsibility. This experience is dear to me.

Academic Credentials

Ph.D. (1994) Materials Science and Engineering, University of Aveiro, Portugal.
1st Degree (1983) Ceramics and Glass Engineering, University of Aveiro, Portugal.

Research Expertise: perovskite-type ferroelectrics, piezoelectrics, dielectric and multiferroic oxides, tellurium-based oxides, electroceramics, relaxor ferroelectrics, solution-based processing (sol–gel and electrophoretic deposition), single crystals, ceramics and films

Other Interests: promoting public awareness and understanding of materials science and engineering and science education, technology transfer, contribution to the development of Portuguese industries toward the growth of the Portuguese economy

Key Accomplishments, Honors, Recognitions, and Awards

- President of the Portuguese Society of Materials (SPM), 2013 to present
- Coordinator (Portugal) of the Emerging Technologies Program within the University of Texas at Austin, USA—Fundação para a Ciência e Tecnologia (FCT)

Portugal Program (Collaboratory for Emerging Technologies, CoLab), 2013 to present

- Editorial Board of Scientific Reports (from Nature Publishing Group), 2012 to present
- Editorial Board of International Scholarly Research Network (ISRN) Ceramics, 2011 to present
- Vice President of the Portuguese Society of Materials (SPM), 2009–2013
- 1st Place Concurso de Ideias da UATEC—Empreende+, Empreendedorismo de Base Tecnológica, da Universidade de Aveiro, Aveiro, Portugal, with the technology Thin Film Tec—Advanced Processing Solutions, Paula M. Vilarinho (team leader) and Luís Amaral, 2011
- 2nd Place of Idea to Product (I2P)—COTEC, Portugal, with the technology ThinFilm Tec—Advanced Processing Solutions, Luís Amaral, Monika Tomczyk, Paula M. Vilarinho (team leader), 2011
- 2nd Poster Prize at 2nd Joint Congress of the Portuguese and Spanish Microscopy Societies, Aveiro, with the work Electron Microscopy Study of Porous $BaTiO_3$ and Co-Functionalised $BaTiO_3$ Thin Films, P. Ferreira, A. Castro, Paula M. Vilarinho, M.-G. Willinger, J. Mosa, C. Laberty, C. Sanchez, 2011
- President of the Portuguese Society of Microscopy, 2010–2011
- Member of the Board of the Portuguese Materials Society, 2007–2009
- Best work of symposium Multiferroic and Magnetoelectric Materials, E-MRS Spring Meeting, Strasbourg, with the work Electric and Magnetic Properties of Incipient Ferroelectric Films Doped with Manganese, A. Tkach, O. Okhay, Paula M. Vilarinho, S. Bedanta, V. V. Shvartsman, P. Borisov, and W. Kleemann, 2008
- Excellent paper from Symposium F: Functional Materials, Materials Research Society International Meeting, Chongqing, China, with the work PFN/PZT Multi-Phase Composite Thick Films Prepared by Hybrid Sol–Gel Process, Aiying Wu, Paula M. Vilarinho, Andrei Kholkin, Janez Holc, and Marija Kosec, 2008
- "Estímulo à Excelência" (Stimulus to Excellence) from the Portuguese Foundation for Science and Technology (FCT), 2007
- Best Poster, Electroceramica, Universidade de Aveiro, Portugal, with the work Growth and Properties of Sol–Gel Derived $BiScO_3$-$PbTiO_3$ Thin Films with Oxide Electrode Layers, Jingzhong Xiao, Aiying Wu, and Paula M. Vilarinho, 2007
- Best Poster, XXXIX Reunião da Sociedade Portuguesa de Microscopia Electrónica e Biologia Celular, Universidade de Aveiro, Aveiro, Portugal, with the work Micro-structural Studies of Mg Doped $SrTiO_3$ Films, O. Okhay, A. Wu, Paula M. Vilarinho, I. M. Reaney, A.R.L. Ramos, and E. Alves, 2004
- 3rd Best Poster, International Meeting of the Portuguese Society of Materials, Materiais, Caparica, Portugal, with the work Perovskite Phase Stabilization of Pb $(Zn_{1/3}Ta_{2/3})O_3$ Ceramics Induced by $PbTiO_3$ Seeds, Li Zhenrong, Aiying Wu, and Paula M. Vilarinho, 2003
- Best paper, 34th International Symposium on Microelectronics, Baltimore, MD, USA, with the work Thin Film Capacitors Embedded into High Density Printed Circuit

Boards, A. Kingon, Taeyun Kim, Paula M. Vilarinho, Jon-Paul Maria, and Robert T. Croswell, 2001

Biography

Early Life and Education

I was born in Coimbra in a hot September, as the middle child and the first daughter of Lindonor and Rodrigo Silveirinha. We lived in Coimbra until my father got a position in the most famous porcelain company in Portugal, Vista Alegre, as an engineer. I was just 5 years old when we moved to Aveiro. In Aveiro I received all my education from the basic school to the highest degree (Ph.D.).

Even before attending primary school, due to the background and degree of my mother, I started learning foreign languages. My first foreign language was French; it was taught in the Catholic kindergarten that I was attending before the primary school. My second foreign language was English; however, that was soon replaced by German. I returned to working on my English during my doctoral studies.

I attended the primary school of Glória in Aveiro, the Female High School of Aveiro, and then after the revolution (on April 25, 1974), I attended José Estevão High School, also in Aveiro. It was during this transition that it became clear to me that I would like to study applied sciences and become an engineer. It was also at this time that I met and engaged with the love of my life (Pedro, my husband). Many other things were of interest to me during that time, for example, acting, dancing, singing, and practicing gymnastics (I still do some of these activities today). After the revolution (changing from a fascist republic to a democratic one), Portugal underwent enormous societal and

Paula Vilarinho (center) at the University of Aveiro in 1981. Group photo with her undergraduation course mates.

economic changes, including with its education system. Several new universities were created (e.g., the University of Aveiro sprung up during 1973–1974), education began in new areas for the country, and a focus was placed on the crucial support of the country's economic development.

Consequently, courses were created at the new University of Aveiro in Electronics and Telecommunication Engineering, as well as Ceramics and Glass Engineering. So, it was easy to stay in Aveiro and enroll at a new university in a new field. I entered the University of Aveiro to get a degree in Ceramics and Glass Engineering in the fall of 1979. Five years later I was awarded a degree. During that period, I had several internships in ceramic companies and after my graduation I started looking for a position in a ceramic company. Fortunately, at that time, the Department of Ceramics and Glass Engineering of the University of Aveiro also had new open positions—I applied and was appointed Assistant Lecturer in 1984. As an Assistant Lecturer, I was in charge of teaching, mainly practical classes. But, at the same time, I discovered my research interests in electroceramics. In 1988 I was promoted to Lecturer. While teaching, I started my doctoral research under the guidance of Professor João Lopes Baptista in the field of relaxor ferroelectric materials. Based on the strategic interest of Portugal in tungsten, I studied the relaxor behavior of lead-based tungstate ceramics for multilayer capacitors and related applications. These years were very enriching and inspiring. Under his guidance I was taught many different things, but above all the importance of trying to be excellent in everything we do.

Paula Vilarinho (center) at the University of Aveiro after the viva voce of her recent Ph.D. student, now Dr. Luis Amaral, in July 2012. Photo taken by an Electroceramics group member.

Though I was awarded a Ph.D. in 1994, I had already been supervising for a year my first Ph.D. student, Liqin Zhou, also the first Chinese Ph.D. student in the Department.

Career History

In 1994, after being awarded a doctorate degree, my R&D interests were broadened considerably, moving first from bulk ceramics to thin and thick films and more recently to single crystals and additive manufacturing, and from lead-based perovskite to lead-free materials, dielectrics to semiconductors, and microscale phenomena to nanosciences and nanotechnology. Exploring electromechanical platforms for tissue engineering is one of my newest R&D challenges.

In 1995 in our Electroceramics group, we began preparing polycrystalline thin films by sol–gel, aiming for lower temperatures and sustainable processing. Later, this work evolved to a combination of knowledge between our group and the group of Dr. Lourdes Calzada, CSIC, Madrid, Spain. In 2009 a novel solution method was developed (and patented) that enabled the processing of functional oxides under low-temperature conditions so that direct large-area integration of active layers with flexible electronics may be turned into reality.

Supported by the British Council, in March 1998, we organized the United Kingdom–Portugal Workshop on Advanced Ceramics and Glass Materials, where I met Dr. Ian Reaney and discovered our common interest and complementarities. Quickly I realized that excellent research in materials always requires some excellent electron microscopy. Since the first project on microstructure and properties of electroceramics (in 1999–2000), we keep in regular contact that has evolved from the research level to include friendship as well.

Soon I understood that excellence in R&D depends also on the ecosystem you live in. Although I live in Portugal and work in Aveiro, I am a citizen of the world. My first sabbatical was spent in North Carolina State University, USA, in the group of Dr. Angus Kingon in 2001. I was involved in a project with Motorola in the preparation of ferroelectric thin films in metallic foils. During this period, I was exposed to a very different R&D ambience, for example, to alternative processing and characterization techniques. I met Dr. Alexei Gruverman who introduced me to piezoresponse force microscopy (PFM) and with whom I organized the first ASI NATO on Scanning Probe Microscopy in Portugal in 2002. All of this triggered more travel and contact with additional groups. I was also exposed to the entrepreneurial and technology transfer ambience of Kingon´s group. This sabbatical was certainly a turning point in my career. When I returned to Portugal, I assumed the leadership of the Electroceramics Group of the Associated Laboratory CICECO at the University of Aveiro.

Later, during my second sabbatical, I spent 1 year (from 2007 to 2008) in Sheffield, UK, in the group of Ian Reaney. Our common interest at that time was the electrophoretic deposition (EPD) of thick films for microelectronics applications and the exploitation of EPD for patterning structures. During that period, we conducted the first experiments for the industrial development of high-frequency antennas by electrophoretic deposition, and we demonstrated the role of constrained sintering in tuning the properties of thick dielectric films for high-frequency applications.

Establishing collaborations has been a constant in my career. Always aiming to conduct excellent R&D and to contribute to the field of electroceramics, I have been collaborating with groups all over the world from China to the United States, via Europe.

As I grew as a material scientist, it became very clear that materials science and engineering is a very interdisciplinary field and encompasses key technologies for the industrial development of Europe and Portugal that lack recognition within society. With this in mind, in 2013, I accepted the invitation (from Drs. Manuela Oliveira and João Moura Bordado) to be President of the Portuguese Materials Society (SPM) and more recently (2013–2014) the invitation from Dr. Brian Korgel to be the Coordinator (Portugal) of the Emerging Technologies Program within the University of Texas at Austin, USA—FCT Portugal Program (CoLab). Also, in 2013, as a part of a long-term strategy aiming to promote teaching, learning, and dissemination of Materials and Materials Engineering, I was in charge of launching an institutional partnership between the High School José Estevão (of Aveiro) and the University of Aveiro, the LinkinMat project.

3 Most Cited Publications

Title: Effect of different dopant elements on the properties of ZnO thin films
Author(s): Nunes, P; Fortunato, E; Tonello, P; Fernandes, FB; Vilarinho, P; Martins, R
Source: Vacuum; volume: 64; issue: 3–4; pages: 281–285; published: January 2002
Times Cited: 214 (from Web of Science)

Title: Dependence of the structural and dielectric properties of $Ba_{1-x}Sr_xTiO_3$ ceramic solid solutions on raw material processing
Author(s): Zhou, LQ; Vilarinho, PM; Baptista, JL
Source: Journal of the European Ceramic Society; volume: 19; issue: 11; pages: 2015–2020; published: 1999
Times Cited: 91 (from Web of Science)

Title: (Sr, Mn)TiO_3: a magnetoelectric multiglass
Author(s): Shvartsman, VV; Bedanta, S; Borisov, P; Kleemann, W; Tkach, A; Vilarinho, PM
Source: Physical Review Letters; volume: 101; issue: 16; article number: 165704; published: October 2008
Times Cited: 81 (from Web of Science)

Challenges

Challenges in my personal and professional life have always been present. I think that in terms of my career, I have been facing two different types of challenges: those more related with my personality and those more related with external aspects. My lack of confidence was holding me back at the beginning of my career. In addition, time has always been an issue. Most Portuguese professors have heavy teaching load (in comparison to many U.S. or UK professors) that does not leave sufficient time to be focused and to do excellent research. Additionally, the lack of funding, a well-defined countrywide research and development strategy, and infrastructure has hampered my research development and career. More recently, being part of a system that lacks incentives and rewards for excellent performers has been a key challenge.

On being a woman in this field . . .

I feel very good about being a woman in Materials Science and Engineering. Looking back, I think that my career and my personal life, more than being a woman, have been mainly affected by my personality and background. I was born Portuguese and now I am working in Portugal; these aspects have helped shaping who I am and my career.

However, as I grew older and matured, I have become more aware that even in the 21st century, being a working woman remains a challenge. Going back to the nineteenth century, Susan Brownell Anthony (1820–1906) wrote that ". . . The day will come when men will recognize woman as his peer, not only at the fireside, but in councils of the nation. Then, and not until then, will there be the perfect comradeship, the ideal union between the sexes that shall result in the highest development of the race." Now I realize that this day hasn't come yet. Overall, gender equality still doesn't exist, and so this inequality evades careers in scientific research. On a competitive basis, I feel that we (women) have to be always much better than men; at least twice as good as men, to be thought half as good. "Boys clubs" do exist. Otherwise, what would be the point of the so-called positive discrimination? Though I feel it as a discrimination itself, I hope that it will somehow prove that we (women) are "mentally capable of holding any position of authority",[1] against very old beliefs.

Besides, paraphrasing Lois P. Frankel,[2] as working women, we have been making unconscious mistakes that sabotage our career. We are often easily more critical of ourselves than necessary and reluctant to take credit where it is due. This self-sabotage prevents women from reaching their career goals and excelling in terms of their performance.

In addition, the lack of female role models, needed for identity construction, has been considered as the second biggest constraint (after sex-role stereotyping of leadership) to women's career success.[3]

I consider the role of women in science and technology to be the same as it is for men. But, in reality, it has not been the same. Women have been engaged in science since very ancient times, but are often not properly acknowledged. Creating knowledge, solving problems, being the driving force of the development, and finally contributing to a better life are the role/duty of both women and men researchers. Currently, it is being said that complementarities/interdisciplinarities are key factors for the 21st century science and development. Is there any better complementarity than combining the diversity that both women and men bring (and as Susan Brownell Anthony suggested in the nineteenth century)?

To conclude, I believe that women in science have an additional role or mission and that is demonstrating and proving women's competencies.

---

[1] Adapted from Leigh A. Whaley, *Women's History as Scientists*, ABC-CLIO, Inc., 2003.

[2] Lois P. Frankel, Nice girls don't get the corner office, *Business Plus*, 2004.

[3] Ruth H.V. Sealy and Val Singh, The importance of role models and demographic context for senior women's work identity development, *International Journal of Management Reviews*, 2009, *12*, 284–300.

Words of Wisdom

To Discover: Discover yourself; identify yourself, identify your principles, know who you are, know what you believe in, and where you want to go. Set your own standards.
To Be: Be true to yourself, listen to your inner self, follow your intuition, simply be yourself, and above all feel good with yourself.
To Live: Do not be afraid and never give up. Stick to your principles. Be generous—give and do not wait for the return—simply give. Be an optimist; say yes and build optimism. Be flexible and adapt to the changes; life is not a straight line. Be different; if you are different, you will stand out. Smile, the power of smile is unbelievable. Have fun, life is too short . . .
To Excel: Be excellent in everything you do, work hard, try even harder, be passionate, find a mentor, and be emotionally intelligent since it determines how outstanding you will be.

Eva Milar Vogel

Retired: Distinguished Member of Technical Staff
Lucent Technologies—Bell Labs
Murray Hill, NJ
USA

e-mail: eva.vogel@gmail.com

Picture of Eva Vogel
from 1988.

Birthplace
Kežmarok, Czechoslovakia
Born
August 5, 1946

Publication/Invention Record

>75 publications: h-index 26 (from google scholar)
7 patents

Tags

❖ Industry
❖ Domicile: USA
❖ Nationality: born in Czechoslovakia; now American
❖ Caucasian
❖ Children: 2

Proudest Career Moment (to date)

In my 35-year career, I was fortunate to have several proud moments. It started with a job offer at the prestigious Bell Telephone Laboratories just few months after my emigration to the United States. Many years later I received an invitation to write review articles and to give invited talks that established my reputation in glass community. Another memorable moment was when I received my Ph.D. 25 years after I finished required courses, but before starting the thesis research. It happened late in my career as the result of political turmoil in former Czechoslovakia. Also, I was honored to be elected to Fellow of The American Ceramic Society and Division Chair of the Electronics Division of The American Ceramic Society. Reflecting back, the most memorable moments are when I overheard my sons talking with pride about my career.

Academic Credentials

Ph.D. (1994) Chemistry within Technology of Silicates, Slovak Technical University, Bratislava, Slovakia.
M.S. (1969) Physical Chemistry within Chemical Engineering, Slovak Technical University, Bratislava, Czechoslovakia.

Successful Women Ceramic and Glass Scientists and Engineers: 100 Inspirational Profiles. Lynnette D. Madsen.
© 2016 The American Ceramic Society and John Wiley & Sons, Inc. Published 2016 by John Wiley & Sons, Inc.

B.S. (1969) Chemical Engineering, Slovak Technical University, Bratislava, Czechoslovakia.

Research Expertise: glass, nonlinear optical phenomena in glass, reliability and quality assurance of fiber optic components for telecommunications including high-power fiber devices, extensive experience in many facets of materials science, including discovery, synthesis, and reliability of new glass and ceramic compositions for electronic and optical devices

Other Interests: making science accessible to general public. Encouraging young people's interest in science and technology

Key Accomplishments, Honors, Recognitions, and Awards

- Division Chair of the Electronics Division of The American Ceramic Society, 1993–1994
- Elected Fellow of The American Ceramic Society, 1992
- Invited to deliver a series of lectures as a visiting professor at University of Bordeax, Laboratoire de Chimie du Solide du CNRS, France, 1992
- Invited to deliver series of lectures at Kyoto University, Mie University, Hoya Glass, Nippon Sheet Glass and International Meeting on New Glass Technology, NGF 90, Tokyo, Japan, 1990
- Member, speaker, organizer, and chair of many symposia and sessions in The American Ceramic Society, Materials Research Society, International Conference on Ferrites, Gordon Research Conference on Glass, Conference on Solid State Chemistry, National Fiber Optic Engineers Conference, and International Wire and Cable Symposia, 1978–2000
- Ph.D., M.S., and senior thesis advisor to Massachusetts Institute of Technology (MIT), New Jersey Institute of Technology (NJIT), and Rutgers University students, 1988–1998

Biography

Early Life and Education

I was born in the High Tatry region of Slovakia, in the former Czechoslovakia right after World War II. Both my parents were Holocaust survivors and a sense of life's precariousness was always present. Hence, I understood from the early age that you can lose everything, but never your education. Also, I knew I had to choose "a portable" career—fortunately, I liked and I was good in science that is independent of the political system. I was admitted to Slovak Technical University and graduated in 5 years with the equivalent of M.S. degree. Inorganic chemistry was my favorite from the beginning and later I majored in physical chemistry. The political situation changed with the arrival of Prague Spring in 1968. I graduated in May and by August 1969 I was in the United States, just 1 year after the invasion of Czechoslovakia by the Russian and Warsaw Pact tanks.

I started working in materials science research at AT&T Bell Labs, Murray Hill, NJ, in 1970. W.G. Pfann inventor of zone refining interviewed me and hired me into his group. The first 10 years of my career I had a support position in a research group. In 1972 I married an engineer who also worked at Bell Labs and we have two sons. The affirmative action environment and a supportive husband helped me to manage a balance of family–work life. After my first son was born in 1977, I started to work 4 days a week. The breakup of the AT&T in 1984 (Bell Labs was an R&D arm of AT&T) gave me an opportunity to start as an independent researcher at Bellcore (the R&D organization serving the telephone operating companies). The political system changed once again in the country of my birth, Czechoslovakia ceased to exist and Slovakia became country of my origin and education. These changes provided me with an opportunity to apply and to be admitted to the Ph.D. program in Silicate Technology (Ceramics) at Slovak Technical University. I was granted a doctorate after a 25-year hiatus following the qualifying exam and the defence of my Ph.D. thesis. I was lucky that some of my old professors were still active and they understood why I interrupted my education. Twenty-five years after completing my B.S./M.S. degree, I finally completed my Ph.D.

As a consequence of my life journey, I learned to speak an array of languages: English, Slovak, and Czech, and even some Polish and Russian, all with an accent.

Career History

I was one of very few technical women hired in 1970 in the materials research department of AT&T Bell Labs at Murray Hill, NJ. During the first 10 years, I moved from amorphous solids to ceramics research in a support research role. After the breakup of the AT&T, a new research company, Bellcore, was created in 1984. I joined this new company and I had the freedom and funding to choose a new direction of research. I focused on nonlinear optical properties of glasses. That was a very active and exciting part of my career with great collaborations especially with two women colleagues, a physicist, Dr. Denise Krol, and a device expert, Dr. Janet Jackel.

During that period, I presented graduate student seminars at Brown University, Vanderbilt University, University of Arizona, Rutgers University, University of Illinois at Urbana-Champaign, Georgia Tech, and University of Rochester. I was invited to participate on review panels for the National Science Foundation (NSF), Department of Energy (DoE), and National Research Council (NRC). I was also an active member of The American Ceramic Society as a speaker, organizer, and chair of many symposia and sessions.

Dr. Eva Vogel in 2006.

This time my career was affected by changes in the American business model, and the antimonopoly movement. For a while, the 1984 breakup of the telephone monopoly gave me new opportunities, but in the end the research model of Bell Labs and Bellcore could not be sustained. In 1992, I changed the direction of my work and I started to work on reliability of optical components. With the support of the International Telecommunication Union, I served as an expert adviser on optical devices in Brazil and India during my time at Bellcore.

In the course of my career, I contributed to the inventions in the field of ferrites and perovskite catalysts. Also, it was very exciting being part of the discoveries at the beginning of optical fiber development and ceramic superconductors. In 1996, I retired from Bellcore and started working as Distinguished Member of Technical Staff in Lucent Technologies—Bell Labs working on government projects. During this period (about 10 years in duration), I worked on projects that did not permit publishing. In 2006 I retired from Lucent, but I continue to perform as a reviewer for Journal of Non-Crystalline Solids, Journal of the American Ceramic Society, Physics and Chemistry of Glasses, and NRC and NSF proposals in the Division of Materials Research.

In 2008, I joined the New School's Institute for Retired Professionals located in New York City where I strive to bring the joy of science to nontechnical audience.

3 Most Cited Publications

Title: Oxygen and rare-earth doping of the 90-K superconducting perovskite $YBa_2Cu_3O_{7-x}$
Author(s): Tarascon, JM; McKinnon, WR; Greene, LH; Hull, GW; Vogel, EM
Source: Physical Review B; volume: 36; issue: 1; pages: 226–234; published: July 1, 1987
Times Cited: 560 (from Web of Science)

Title: Observation of spatial optical solitons in a nonlinear glass wave-guide
Author(s): Aitchison, JS; Weiner, AM; Silberberg, Y; Oliver, MK; Jackel, JL; Leaird, DE; Vogel, EM; Smith, PWE
Source: Optics Letters; volume: 15; issue: 9; pages: 471–473; published: May 1, 1990
Times Cited: 337 (from Web of Science)

Title: Nonlinear optical phenomena in glass
Author(s): Vogel, EM; Weber, MJ; Krol, DM
Source: Physics and Chemistry of Glasses; volume: 32; issue: 6; pages: 231–254; published: December 1991
Times Cited: 323 (from Web of Science)

Challenges

One of the challenges of my early career was the attendance at the large technical meetings. I did not have a network of former professors and friends from college or graduate school. The presence of a young woman in the social gatherings after the technical sessions raised uncomfortable questions in those days. It took a few years until there were more women attending the conferences before they included me in their social circle.

On being a woman in this field . . .

In 2014 there was still a debate going on in American press with the headlines "Why Women Still Can't Have It All." I believe nobody can have it all, but most of us can have a lot more than we did years ago. Forty years ago in corporate or academic America, we faced overt sexism of a kind I see only when watching *Mad Men*. Women were paid substantially less than their male colleagues. Only after data were provided to substantiate this inequality, Bell Labs management adjusted our salaries across the board. The women took an active and collective move to focus on this issue. Finally, as more women choose science, I hope the world will view us as scientists, not women scientists. Some of us succeed and some of us fail, but we have choices to be in science and that is wonderful.

Words of Wisdom

I had a pleasure to be a mentor and an adviser to a number of young women scientists. One question was always asked—when is the best time to start a family and my answer was whenever you are ready, there is never a best time. The biggest difference in your experience to manage work and family life is the partner that you choose to share your life with. To have a family on your own and a full career is really hard. To be excited about your work, you have to love science. You can enjoy your work even if you are not a superwoman. You have a choice to define what success is for you—teach a science class, synthesize a new compound, or be a CEO of a company. Do not have other people define your success.

(Betty) Noemi Elisabeth Walsöe de Reca

Director of CINSO
Centro de Investigaciones en Sólidos
UNIDEF (MINDEF-CONICET)
Juan B. de La Salle 4397 Villa Martelli
Pcia. Buenos Aires
Argentina

e-mails: walsoe@citedef.gob.ar,
            bettywalsoe@gmail.com
telephone: +54 11 4709 8158

N.E. Walsöe de Reca
(at 28 years old) receiving
doctoral degree (1966).

Birthplace
Ciudad Autónoma de Buenos Aires (CABA),
Argentina

Born
February 7, 1937

Publication/Invention Record
>225 publications: h-index 14 (Web of Science)
16 patents, 9 book chapters

Tags
❖ Academe
❖ Domicile: Argentina
❖ Nationality: Argentinian
❖ Caucasian
❖ Children: 3

Proudest Career Moments (to date)

I would like to mention several moments: (1) In 2007, I was honored as a corresponding member of the ANCEFyN (Academia Nacional de Ciencias Exactas, Físicas y Naturales-Argentina). (2) Since 2009, I have been the Vice President of the AQA (Asociación Química Argentina); in 2013, I was re-elected for a new period. (3) Since 2013, I have served as an advisor of the Organizing Commission of Doctoral Thesis in Chemical Engineering of the UTN (Universidad Tecnológica Nacional).

Academic Credentials

Equivalence of a Ph.D. (1966) Chemistry, University of Buenos Aires-UBA, Buenos Aires City, Argentina.

Successful Women Ceramic and Glass Scientists and Engineers: 100 Inspirational Profiles. Lynnette D. Madsen.
© 2016 The American Ceramic Society and John Wiley & Sons, Inc. Published 2016 by John Wiley & Sons, Inc.

N.E. Walsöe de Reca nominated as Vice President of AQA with
the Ex-President of AQA, Dr. Carlos Azize (2013).

Licenciate Degree (1960) Chemistry, University of Buenos Aires-UBA, Buenos Aires
City, Argentina.

Research Expertise: nanoscale ceramics (synthesis, characterization and applications
to gas sensors and solid oxide fuel cells, nanostructured metallic oxides for ultraviolet
detectors, narrow gap II–VI semiconductors (for infrared, UV, gamma, and X-ray
radiation detectors)

Other Interests: organizing of national and international schools and meetings,
organizing commission of doctoral thesis-UTN (National Technological University)

Key Accomplishments, Honors, Recognitions, and Awards

- Distinction: "Women Leaders of the Global Chemistry Enterprise" given by the
  American Chemical Society (ACS)" to Dr. N.E. Walsöe de Reca, during the 248th
  ACS Annual Meeting and Exposition, San Francisco, USA, 2014
- Award: "Technology Leader in Chemistry 2014" given by Foundation for the
  Interaction of Production, Education, Science and Technology Systems (FUNPRECIT)
  to Dr. N.E. Walsöe de Reca, for her trajectory as researcher and technologist, 2014
- Award: "Dr. Horacio Damianovich" in Inorganic Chemistry given by the Argentine
  Chemical Society (Asociación Química Argentina-AQA) to Dr. N.E. Walsöe de Reca,
  2014
- The Chemistry Engineering Ph.D. Thesis of Eng. Ana María Martínez, directed by
  Dr. N.E. Walsöe de Reca and Dr. Alicia Trigubo, was awarded the Prize on Physics-

Chemistry "Enrique Herrero Ducloux" given by the AQA (Argentine Association for Chemistry), 2013

- Drs. N.E. Walsöe de Reca, Aldo Craievich, Diego Lamas, and Marcia Fantini received UNESCO Integration Mercosur 2010 Prize for the project "Nanostructured Materials: Synthesis, Synchrotron Light Studies, Properties and Applications," 2010

- Physics Ph.D. Thesis by Lic. Eugenio OTAL, directed by Dr. Horacio Canepa and Dr. N.E. Walsöe de Reca, was awarded "Best Ph.D. Thesis" by the Instituto Jorge Sabato, 2010

- Ph.D. thesis by Ulises Gilabert, directed by Dr. Alicia Trigubo and Dr. N.E. Walsöe de Reca, won the Prize on Physical Chemistry "Enrique Herrero Ducloux" given by the AQA (Argentine Association for Chemistry), 2008

- Ph.D. thesis by Martín Bellino, directed by Dr. Diego Lamas and N.E. Walsöe de Reca, earned Special Mention Prize "J.J. Giambiaggi" given by the AFA (Argentine Physics Association), 2007

- Ph.D. Thesis of Valeria Messina, directed by Dr. N.E. Walsöe de Reca and Dr. Andrea Biolatto, won prize by AATA for "Best Ph.D. Thesis on Food," 2007

- Keynote Speaker at the XV International Materials Research Congress in Cancun, México, 2006

- First Mention of Repsol-YPF 2004 Award to the Best Innovative Project in Science and Technology given to a Research Project by Drs. Diego Lamas and N.E. Walsöe de Reca, 2005

- Award for Best Research Work of the International Conference "Monitoring Systems and Novel Technologies for Detection/Removal of Pollutants in/from Ecosystems" organized by the Surrey University (UK) and the European Community, held at Buenos Aires, 2004

- First Honorable Mention of the DuPont-CONICET Award, Assistance Program to the Scientific and Technologic Development, to the project: "Fuel Cells of the SOFC Type to be Operated at Medium Temperatures, Employing Biogas as Fuel," Buenos Aires, 2004

- Part of research team for "Integration"—Mercosur—UNESCO Award to the project: "Solid Oxide Fuel Cells to Generate Electrical Energy Operated at Intermediate Temperatures with Methane–Air Mixtures as Fuel," 2004

- "Bernardo Houssay 2003 Award in Scientific and Technological Research" by SECyT (National Secretary of Science and Technology), 2003

- Team member of "Repsol-YPF Award" for the Best Innovative Project in Science and Technology, 2003

- National Academy of Exact, Physical and Natural Sciences "Simón Delpech Award in Materials Science," 2001

- National Delegate to the High Temperatures Commission and Solid State Chemistry of International Union for Pure and Applied Chemistry (IUPAC)—England. Afterward Argentine delegate in Solid State Commission, 1993–2000

- Team member, Defense Ministry Award (CITEFA), A-Category for "The Best Applied Research Work," Document Number 07/98, signed 1998

- "International Course of Ophthalmological Centers and the Cleveland Clinic Foundation" award for research work, 1997
- Award "The Best Experimental Scientific Work of PPMSS'97" of the Ukrainian State Academy of Sciences, given during the Second International School Conference on Physical Problems in Materials Science of Semiconductors (PPMSS), Chernivtsii, Ukraine, for research work, 1997
- Member of the Scientific Advisory Committee Exterior Relations Ministry to study the Basic Scientific Agreements between Germany and Argentina for Scientific Research and Technological Development Cooperation. SECYT Document Number 103/96, signed 1996/1997
- Invited by the Brazilian and French governments to participate in ÍEcole Franco-latinoamericaine d'Hiver sur la Diffusion dans les Matériaux and to dictate lectures on "La diffusion dans les Semiconducteurs II–VI"—Ouro Preto M.G. Brasil–1994
- Alicia B. Trigubós Ph.D. Thesis directed by Dr. N.E. Walsöe de Reca received a special mention in Concourse "Jorge Kittl" for the Best Research Work, given by SAM (Argentinean Society of Materials), Bahía Blanca, 1994
- Japanese Government Invitation to participate as member of the International Board—DIMAT'92—Second Iketani Conference (International Conference on Diffusion in Solids) held at Kyoto, Japan, 1992
- SAM (Argentine Society of Materials) Award to "Young Researchers" received by the Lic. M.A. Prokopek for her Physics Thesis under the direction of Dr. N.E. Walsöe de Reca, 1989
- NATO (Advanced Studies Institute) Invitation to assist the Diffusion in Solids School to expose on "PRINSO-CONICET-CITEFA Research Activities on Diffusion in Ionic Solids and Semiconductors," Aussois, France, 1988
- Award, Condecoration des Palmes Academiques du Gouvernement Français dans la Ordre de Chevalier des Palmes Academiques given to Dr. N.E. Walsöe de Reca by the First Minister of the French Republic as proposed by the French Education Ministry, 1987
- Award "Silver Clover" (in Science) by International Rotary Club to Seven Women, detached in Sports, Education, Journalism, Science, Social Assistance, Art and Medicine, 1986
- Invitation and fellowship of the Pennsylvania University to assist the Gordon Conference on Solid Electrolytes and to present "Studies on Ionic-Conducting Compounds at the CINSO-CONICET-CITEFA," New Hampshire, USA, 1986
- NATO Advanced Study Institute Invitation and Fellowship and CONICET Grant to assist the NATO School on Solid State Batteries—Alcabideche (Portugal) and to visit the Instituto de Química-Física—"Rocasolano"—Madrid, Spain. In both cases, to attend the Materials Science courses and to present "CINSO-CONICET-CITEFA Research Activities in the Area of Solid State Batteries," 1985
- Invitation of Austrian Government to dictate conferences on "II–VI Semiconductors" and "Research Activities at CINSO-CITEFA-CONICET in the Semiconductors Area" at Linz, Graz, and Wien Universities, Österrreich, 1983

- Invitation of the French Government to participate in the 4th International Conference on Solid-State Ionics, Grenoble, France—two research works were presented—and to visit the University of Rennes I to integrate the Jury of a Doctoral Thesis and to visit the Solid State Laboratory at Meudon, CNRS—Bellevue (cooperation work), 1983
- NATO Invitation to be professor in the Ecole d' Eté on "Transport Atomique dans les Solides," held at Lannion, Francia, 1981
- Invitation Tohoku University to assist the 3rd International Meeting on Solid-State Ionics, Tokyo, Japan, 1980
- Drs. N.E. Walsöe de Reca and Juan Franco have obtained twice the Research Projects Concourse given by the "Alberto J. Roemmers" Foundation, 1979 and 1980
- Drs. N.E. Walsöe de Reca and Juan Franco have won the Scientific Publication Concourse for the book *Primary Batteries*, published by FECYC (National Fund for Education, Science and Culture), CONICET, 1979
- Invitation of the Materials Science Department, University of Tohoku (Sendai), Japan, to dictate several conferences on "Diffusion in Metals" and "Diffusion in III–V Semiconductors," 1978
- Invitation of the Argentine Cardiologic Society to join the panel on "New Batteries for Pacemakers" at the International Cardiology Congress, Buenos Aires, 1977
- O.E.A. invitation to participate in the Latin-American Physics School (ELAF), Caracas, Venezuela, 1976
- Invitation of the Brazilian National Council of Research to participate in the IV Congreso Iberoamericano de Cristalografía, Campinas, Brazil, 1975
- ILAFA (Latin-American Institute of Iron and Steel) Award: First Prize for Students' Research Direction: "Selective Separation of Associated Oxides in the Red Earth of Misiones" given to Dr. N.E. Walsöe de Reca and UTN Students, 1969
- Junior Chamber of Buenos Aires: One of "Ten Outstanding Young People of Argentina" Scientific Award, 1968

Biography

Early Life and Education

N.E. Walsöe de Reca's father was Norwegian and mother was Italian. She has one brother (who is an architect). She grew up in Argentina and received an early education (primary and secondary) in the Public School (Escuela Normal de Maestros, Number 9 "Sarmiento") at Buenos Aires, where she received the degree of teacher for primary school. She then undertook university studies at the Faculty of Exact and Natural Sciences (FCEN), Buenos Aires University (UBA), where she received a Chemistry (Licenciate) Degree (in 1960). After graduating in Chemistry (at the age of 21), she married Raúl Martín Reca, a medical doctor.

N.E. Walsöe de Reca obtained her chemistry doctoral degree (Ph.D.) from FCEN-UBA (in 1966) for her thesis on "Self-Diffusion in β-Phase Titanium and Hafnium." In these early years, she lived in Germany with her husband and her eldest son Gustavo. Her second son, Guillermo, was born a year later, about the same time that she received a fellowship of the International Atomic Energy Agency (IAEA), Wien, for post-degree studies in materials science, at Institut für Metallurgie und Metallkunde, Technische Hochschüle, München, and her husband, Raúl received a DAAD Fellowship to perform research in the Augenklinik der Universität. Afterward, they moved to France because they won fellowships from the French Government Technical Cooperation for post-degree studies. Betty worked in materials science at the Centre d' Etudes Nucleaires (Service de Recherche en Metallurgie Physique et Chimique), Saclay, and also on physics and materials science studies at the Faculté d' Orsay, while Raúl has performed research on glaucoma in the Cauchin Hospital in Paris.

Career History

Coming back from Europe, they continued their research careers in Argentina. N.E. Walsöe de Reca joined CITEFA (today CITEDEF) as a member of the RPIDFA (Researcher Career of the Armed Forces) and, at the same time, as a member of the Researcher Career—CONICET (National Council of Scientific and Technologic Research). From the scientific point of view, her career is divided into three distinct periods.

From 1960 to 1980, she worked in the area of metallurgy. During this period, she published several papers, filed patents, and directed others in their research work (fellowships holders, magister, and doctoral thesis).

The next period spanned from 1980 through 1992. During this time, she helped to create the CINSO-CONICET-CITEFA in 1980 with two research projects: "Functional Materials for Solid State Batteries" (covering studies on solid electrolytes and electrode materials, organic-iodide charge transfer complexes, and intercalation compounds) and "IV–VI Semiconductors for the Infrared Radiation Detection." All these studies enabled her to develop original solid-state cells and file patents such as *solid-state cells for cardiac pacemakers*, *thin-film cells* (for photocameras, calculators, and memories), *batteries for surveillance systems* that can be used at extreme temperatures ($-40°C$ through $+60°C$).

Research on *IV–VI semiconductors* was performed taking into account their synthesis, characterization, and study of structural, electrical, and optical properties, these materials are used for *photoconductive infrared (IR) detectors*, operated at room temperature. Surface activation of detectors was accomplished by an original oxidation method patented at CINSO. IR detectors were used in devices to measure cereals, textiles, or papers humidity; to detect electric short paths, for devices measuring the CO_2 (g) emission, and early fire detection in woods and silos. *Infrared photovoltaic detectors* have also been built with the *II–VI semiconductor* $Hg_{1-x}Cd_xTe$ (MCT), the compound exhibits a high strategic and commercial value useful for IR detection at 8 and 14 µm. For this work and due to a very small research budgets, it was necessary to build a considerable part of infrastructure (furnaces to grow single crystals, temperature regulators, high precision systems to move the specimens, etc.). MCT has been used in devices for surveillance, air and satellite surveillance of natural resources, and industrial and medical applications. In this second period, numerous research results have been published in refereed national and international scientific reviews or communicated in meetings; invention patents were also obtained and educational work has continued intensively.

The final period began in 1992 with her first doctoral thesis on ceramic nano-materials. This marked the beginning of an intensive period on the subject, which is still ongoing today. Ceramic nanosemiconductors were synthesized and characterized. Several of their surprising properties enabled *gas sensors* to be built for CO (g), VOCs, H_2, and NH_3. Sensors built with nanoceramics exhibit a higher sensitivity (30–37%) and lower operation temperature in comparison with those built with the same materials but microstructured (several patents have been obtained). In this case, research enabled to control food quality (olive oils, orange juices, garlic, effect of packing on food, etc.), study medicinal plants, monitor the environmental contamination, the residual effect of insecticides, or the evolution of biosolids and particularly to be applied to electronic noses. Nanostructured ceramics were successfully synthesized, character-ized, and used for IT-SOFC (solid oxide)-type fuel cells, operated at intermediate operation temperatures (550–650°C) in comparison with working temperature of usual SOFC fuel cells (800–1200°C) and using hydrocarbons, natural gas, or biogas (coming from garbage) or even hydrogen as fuels were built for electrical energy generation. Research on new nanoceramics for solid electrolytes and electrodes was performed (synthesis, characterization, and properties study) and new design of cells (two and single chambers) was proposed. This project is still being developed with important technologic applications in the energy area.

In 1972 and for nearly 20 years, she was a Materials Science Professor at the UTN (National Technology University) and the FI-UBA (Engineering Faculty, University of Buenos Aires). She also delivered many lectures in other universities in Buenos Aires, in different provinces of Argentina, and in foreign countries (e.g., Spain, Brazil, Venezuela, France, United Kingdom, and Japan).

At present, she is a senior member of CITEDEF and of CONICET Researcher Careers. She has participated in many international research collaborations and has received both foreign and local honors and awards. She has a long list of achievements, including the development of commercial products, several patents, many publications, participation in meetings, and direction of many doctoral students and fellows. She has also had many invitations to conferences in foreign countries. She is at present writing with some colleagues/friends a book *Ceramic Nanostructured Mate-rials: Synthesis, Characterization and Applications*.

N.E. Walsöe de Reca opens the First Meeting on Computation in Chemistry held at AQA (1987).

After 38 years of being happily married, her husband Raúl Martin passed away. N.E. Walsöe de Reca has three sons: Gustavo who is also a Medical Doctor, Guillermo who is an artist (painter and musician), and Federico who is a graphic design expert. She also has five grandchildren: Sebastian, Ana, Miranda, Martina, and Martin.

3 Most Cited Publications

Title: Enhanced ionic conductivity in nanostructured, heavily doped ceria ceramics
Author(s): Bellino, MG; Lamas, DG; de Reca, NEW
Source: Advanced Functional Materials; volume: 16; issue: 1, pages: 107–113; doi: 10.1002/adfm.200500186; published: January 5, 2006
Times Cited: 106 (from Web of Science)

Title: Synthesis of nanocrystalline zirconia powders for TZP ceramics by a nitrate–citrate combustion route
Author(s): Juarez, RE; Lamas, DG; Lascalea, GE; et al.
Source: Journal of the European Ceramic Society; volume: 20; issue: 2; pages: 133–138; doi: 10.1016/S0955-2219(99)00146-6; published: February 2000
Times Cited: 77 (from Web of Science)

Title: Comparison between two combustion routes for the synthesis of nanocrystalline SnO_2 powders
Author(s): Fraigi, LB; Lamas, DG; de Reca, NEW
Source: Materials Letters; volume: 47; issue: 4–5; pages: 262–266; doi: 10.1016/S0167-577X (00)00246-9; published: February 2001
Times Cited: 52 (from Web of Science)

Challenges

One important challenge is continuing the education efforts at UNIDEF-MINDEF-CONICET to train others to carry out these research and education traditions. Another challenge is to convey to both researchers and students the importance of encouraging scientific curiosity and innovation in research, increasing true scientific collaboration and friendship among the group members, and trying to ensure the best work and study environment.

On being a woman in this field . . .

Being a very young woman (working and studying hard) in the 1960s was easier at CNEA (today Institute Jorge Sabato) in Argentina than it was in Germany, indeed. There were few or no women in science in the 1960s in Metallkunde Lehrsthul, Technische Hochschule-München, and I was afraid of thinking that some colleagues regarded me "as a rare insect coming from the Pampas." It was sometimes a little depressing; my husband was always very supportive. Besides, time and goodwill were good medicines to surmount difficulties and the last time in Germany it was both productive and pleasant. Then, working in France was easier still. I returned to Argentina in 1969 and a new research group was created at CITEFA—the Argentine Armed Forces Research Center (today CITEDEF). It was necessary to contribute to the scientific research organization in a well-known institute devoted to service and technology. It was not easy indeed, particularly for a young woman in a military institute having, at that moment, no grants or infrastructure. It was necessary to build everything from the start. Five professional women constituted the group. With hard work and relying on the advice and teaching from their authorities or professors—Admiral Fernando Millia (CITEFA President), Dr. Sonia Nasif, Prof. Jorge Sabato, and Dr. Antonio Rodriguez (CONICET president), everything became easier. I was often the only woman in physics or chemistry meetings. Soon that changed and I was glad to see that the number of women studying in Argentina was increasing every year. Today in Argentina, in several careers, the number of women students and professors or those occupying faculty positions is higher than for men. Women fought to defend their salaries, so now it is only in a few places that women still receive lower salaries than men do. Unfortunately, misogynous groups remain not only in Argentina but also everywhere and their members usually bear resentments; education is a central part of the solution.

Words of Wisdom

I do not feel capable to advise anybody, but I have several words of wisdom for myself that I repeat like "mantras," which I am willing to share with you:

- Rarely do things turn out as we expect, be prepared to surmount the resulting difficulties.
- If you do, you will succeed.
- As my grandmother often said: "A woman must work outside her home (as well as in her home) and find equilibrium between salary and necessities, so yoúll be neither poor nor rich, but yoúll be happy."
- Life happiness consists in having always somebody to love and something to do, to learn, and to hope.
- Work must provide a simple and lasting happiness and fun, so that you would wait anxiously for each Monday.
- To feel a deep down peace is the most evident sign of having found the truth.
- What you are doing today for your family, your research work, your community, or your country is important since you are performing all this with a day of your life.
- The best judge of your behavior is your own conscience.
- The scientist has a double social responsibility with the community to give back to the society something of what s/he received from the community (grants, fellowships, studies, and positions). Then the acquired knowledge and "know-how" must be transferred to the other institutes or to the local industry.
- If knowledge is necessary for children and uneducated people, it must be communicated in a simple, pleasant, and useful way. Teaching, explaining for the different education levels, and training are important ways to give back the community what the scientist has received.
- Happiness consists in loving what you have (or do) and not in wishing what you have not.

Profile 96

Ellen D. Williams

Director of Advanced Research Projects
Agency—Energy (ARPA-E)
U.S. Department of Energy
1000 Independence Ave. SW
Washington, DC 20585
USA
and
Distinguished University Professor
Department of Physics and Institute for Physical
Science and Technology
University of Maryland
College Park, MD 20742-4111
USA

Williams in her research
laboratory at University of
Maryland around 1985.
Source: Family photos.

e-mails: ellen.williams@doe.gov; edw@umd.edu

Birthplace
Oshkosh, Wisconsin, USA
Born
December 5, 1953

Publication/Invention Record

>200 publications: h-index 47

Tags

❖ Academe
❖ Administration and
  Leadership
❖ Domicile: USA
❖ Nationality: American
❖ Caucasian
❖ Children: 2

Proudest Career Moment (to date)

It is an honor to be nominated by President Obama to be the Director of U.S. Department
of Energy's Advanced Research Projects Agency—Energy (ARPA-E). I have been
privileged with opportunities to serve my community and science in various capacities—
as a professor, at a major corporation, and at the ARPA-E. In all of these roles, one of my
greatest sources of pride has been working with and supporting dedicated scientists and
engineers who apply their talent and creativity to generate new technical value through
innovation.

Academic Credentials

Ph.D. (1981) Chemistry, California Institute of Technology, Pasadena, California, USA.
B.S. (1976) Chemistry, Michigan State University, East Lansing, Michigan, USA.

Successful Women Ceramic and Glass Scientists and Engineers: 100 Inspirational Profiles. Lynnette D. Madsen.
© 2016 The American Ceramic Society and John Wiley & Sons, Inc. Published 2016 by John Wiley & Sons, Inc.

Research Expertise: surfaces at the atomic scale, thin films, low-dimensional inter-faces and graphene, quantitative use of scanned probe microscopy in the statistical mechanics of surfaces and new materials

Other Interests: energy sustainability, enhancing science education, engaging women and minorities in physics and related fields, science policy development

Key Accomplishments, Honors, Recognitions, and Awards

- Cheetham Award Lecture, University of California, Santa Barbara, 2014
- Women in Science, Technology, Engineer-ing and Production Award, Manufacturing Institute, 2013
- Supported and supervised 23 doctoral disser-tation students through the award of their Ph.D., 1986–2012
- Chair, National Academy of Sciences Com-mittee for Review and Update of Technical Issues Related to the Comprehensive Nuclear Test Ban Treaty, 2009–2012
- Founding Director, NSF Materials Research Science and Engineering Center, University of Maryland, 1996–2010
- Director, UMD-NIST Cooperative Research Agreement: Nanoscience and Technology Research for Nano-Metrology in Materials, Electrical and Mechanical Engineering, and Physics, Chemistry, and NanoBiology and NanoMedicine, 2006–2010
- Member: Congressional Commission on the Strategic Posture of the United States, 2008–2009

Williams presenting a talk at the MIT Energy Initiative Symposium in 2012. Source: https://mitei.mit.edu/news/ bp-chief-scientist-ellen-williams-energy-sustainability-rests-innovation-and-collaboration. (See insert for color version of figure.)

- Member, National Security Panel of University of California President's Council, 2000–2009
- Elected to National Academy of Sciences, 2005
- Co-Organizer National Nanotechnology Initiative Grand Challenges Workshop and Report (DOE), 2004
- Materials Research Society—David Turnbull Award, 2003
- Elected to American Academy of Arts and Sciences, 2003
- American Physical Society—David Adler Lectureship Award, 2001
- Chairman of the Gordon Conference on Thin Films and Crystal Growth, 2001
- Meeting Chair, Materials Research Society, 1999

- American Physical Society Centennial Speaker, 1998–1999
- Principal Investigator: NSF Materials Research Group, 1991–1996
- E.W. Mueller Award Lecturer, University of Wisconsin, Milwaukee, 1996
- Japan Society for the Promotion of Science Fellow, 1996
- Fellow of the American Vacuum Society, 1993
- Fellow of the American Physical Society, 1993
- Co-Organizer, Workshop on New and Emerging Techniques for Imaging Surfaces, Washington, DC, 1993
- Co-Chairman, Local Arrangements Committee, Physical Electronics Conference, 1990
- American Physical Society Maria Goeppert Mayer Award, 1990

Biography[1]

Early Life and Education

Although Ellen Williams was born in Wisconsin, she moved during her childhood with her family to Livonia, Michigan, a sprawling suburb northwest of Detroit. Her father was one of many workers for the Ford Motor Company. Since he was an engineer, he discouraged his daughter from studying the subject. In the 1930s, engineers then were on the production floor dealing with tough guys, and so from his perspective, it was no life for a woman. Instead, he encouraged Ellen to study computers and computer programming, a prescient observation for the early 1970s.

While attending Franklin High School in Livonia, she found an after-school course to explore computer programming. With the help of outstanding high school teacher David Danes, she developed a strong interest in chemistry. "[He] helped me see a logical train of argument in the reactions." Upon graduation from high school, she traveled to East Lansing, home of Michigan State University, and majored in chemistry. Her advisor, Frederick Horne, taught her the fundamentals of the statistical analysis of spectroscopic signals. She was encouraged to take a graduate-level statistical mechanics class as an undergraduate. It was in that class where she explored the physical basis for the laws of thermodynamics, including the effects of entropy. She found the course compelling, and the experimental tenets of "stat mech," as the field is called by its practitioners, have formed the basis of her research.

She continued her education at the California Institute of Technology—only a few years after women were allowed admittance as undergraduates—to pursue a graduate degree in chemistry. She worked in the laboratory of her thesis advisor, surface chemist W. Henry Weinberg, studying the atomic-scale mechanisms important to catalysts. She developed calluses on her hands from cranking down the bolts that sealed shut the ultra-high-vacuum chamber used in her electron diffraction studies, designed to measure the atomic structure of well-ordered surfaces. Her first paper, published while she worked in

---

[1] Adapted from her profile in the Proceedings of the National Academy of Sciences: http://www.pnas.org/content/105/43/16415.full.

Weinberg's lab, combined her interest in then-emerging computational capabilities with the mechanics of order and disorder.

At Caltech, she also met her husband, astro-physicist Neil A. Gehrels. He now works at NASA's Goddard Space Flight Center (Green-belt, MD) and, like Williams, was elected to both the American Academy of Arts and Sciences and the National Academy of Science. She counts him as "one of the most influential people in [her] scientific career." Williams says that "it's phenomenal interacting with someone at home who is also actively pursuing fundamental sci-entific questions."

Williams and husband Neil Gehrels at the 2012 Olympics. Source: Family photos.

Williams with her children Thomas and Emily in 1990. Source: Family photos. (See insert for color version of figure.)

Career History

After completing her Ph.D. in 1981, Williams left California and drove across the country for a postdoctoral fellowship at the University of Maryland at College Park, working in the research group of Professor Robert Park, and then moving on to become a physics professor at the university's Physics Department and in the Institute for Physical Science and Technology. Early in her faculty career, her interest in how atoms move and organize to form interfaces intersected with a new scientific tool, the scanning tunneling microscope (STM). Williams recognized the transformational power of being able to directly observe how atoms move and organize into structures, so she developed an experimental program to pursue this vision. Her research group explored how new experiments could be designed to draw information out of the new types of data available from the STM, and uncovered a wealth of unexpected physical behaviors. Recognizing the importance of theory to understand these new observations, she pulled together a team of researchers in a National Science Foundation (NSF)-sponsored Materials Research Group. This was followed in 1996 by the formation of the University of Maryland College Park Materials Research Science and Engineering Center (UMD-MRSEC) in which she led until 2010.

In parallel with her research activities, Williams actively engaged in work with professional societies, with the extensive program of Educational Outreach to Pre-College students that was an important feature of the UMD-MRSEC, and with government advisory work. The latter led to engagement with questions of national security and Williams was called to serve on the Congressional Commission on the Strategic Posture of the United States in 2008, and to lead a National Academy study on technical issues relating to the Comprehensive Test Ban Treaty in 2009.

She took a leave of absence from the University of Maryland beginning in 2010 to serve as Chief Scientist at BP—the third largest oil company and the fifth largest corporation on the planet. In this position, she works on how developments in science and technology can contribute to sustainable, secure, and environmentally responsible energy.

In November 2013, Williams was nominated to be the Director of the Advanced Research Projects Agency—Energy (ARPA-E).

Williams and her BP colleagues Chief Bioscientist John Pierce and Chief Chemist Mike Desmond at University of Manchester in 2013. Source: International Centre for Advanced Materials.

3 Most Cited Publications

Title: Atomic structure of graphene on SiO_2
Author(s): Ishigami, M; Chen, JH; Cullen, WG; et al.
Source: Nano Letters; volume: 7; issue: 6; pages: 1643–1648; doi: 10.1021/nl070613a; published: June 2007
Times Cited: 706 (from Web of Science)

Title: Charged-impurity scattering in graphene
Author(s): Chen, J-H; Jang, C; Adam, S; et al.
Source: Nature Physics; volume: 4; issue: 5; pages: 377–381; published: May 2008
Times Cited: 615 (from Web of Science)

Title: Steps on surfaces: experiment and theory
Author(s): Jeong, HC; Williams, ED
Source: Surface Science Reports; volume: 34; issue: 6–8; pages: 171–294; doi: 10.1016/S0167-5729(98)00010-7; published: 1999
Times Cited: 462 (from Web of Science)

Challenges

Scientific research is challenging—in a productive and exciting way. However, I also accept new challenges by serving in leadership and policy positions; these roles are a tremendous responsibility—and each one helps me to grow in a different way and prepares me for the next challenge.

On being a woman in this field . . .

I began my graduate studies at Caltech in 1976, shortly after undergraduate women were allowed admittance. Although women had not been forbidden as graduate students in the past, it had been unusual, and Caltech was just starting to regularly accept women as graduate students. Somewhat surprisingly, in my entering class a large fraction of the graduate students in chemistry were women and we forged a bond as pioneers in Caltech's experiment in co-education. We felt a lot of pride in demonstrating that we were tough enough to compete on an equal footing.

Words of Wisdom

- My family, friends, and colleagues have provided guidance, love, and support and this serves as the foundation for all my efforts.
- Be open to new experiences; there are many ways to bring your lifetime of experience (e.g., in scientific research and in supporting the application of cutting-edge technology) to meet pressing social needs.

Wanda Wieslawa Wolny

Retired, Meggitt PLC/Ferroperm Piezoceramics A/S
Porthusvej 4
DK-3490 Kvistgaard
Denmark

e-mail: wanda.wolny@gmail.com

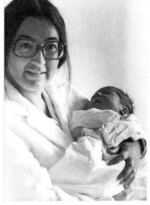

Wanda in 1977 with
one of her children.

Birthplace
Katowice, Poland
Born
December 7, 1945

Publication/Invention Record
>20 publications: h-index 6 (Web of Science)
2 patents, 1 book chapter

Tags
❖ Industry
❖ Domicile: Denmark
❖ Nationality: born Polish, Danish citizen
❖ Caucasian
❖ Children: 2

Proudest Career Moments (to date)

(1) Development of Ferroperm Piezoceramics from two to three employees' local business to an internationally acknowledged business leader with high-quality piezoceramics and a supplier to medical, aerospace, and defense system manufacturers. (2) Playing an active part in the development of non- (or minimally) invasive medical ultrasound, such as ultrasonic treatment of Parkinson disease.

Academic Credentials

M.Sc. (1969) Electronics and Solid State Physics, Warsaw University of Technology, Warsaw, Poland.

Successful Women Ceramic and Glass Scientists and Engineers: 100 Inspirational Profiles. Lynnette D. Madsen.
© 2016 The American Ceramic Society and John Wiley & Sons, Inc. Published 2016 by John Wiley & Sons, Inc.

Research Expertise: ferroelectrics, piezoceramics, piezoelectric technologies, including lead-free and high-temperature piezoelectric ceramics, piezoelectric thick films, piezoelectric transducers and sensors

Other Interests: assisting end users in product development process

Key Accomplishments, Honors, Recognitions, and Awards

- Member of Steering Committee—European Technology Platform for Advanced Engineering Materials and Technologies (EUMaT) and leader of Working Group (WG) "Materials for Information and Communication Technologies" (ICT), 2009–2014
- Member, Ultrasonic Industry Association (UIA) Board, 2010–2014
- President of The Piezoinstitute Association Internationale Sans But Lucratif (AISBL), 2008–2013
- Member of Strategic Advisory Board, European Union (EU) Commission, 2008–2012
- Member of NMT Eranet (New Generation of Titanium Biomaterials) Programming Board, 2010–2011
- Member of European Commission's Enterprise Policy Group, Business Group, 2001–2010
- Coordinator of Multifunctional & Integrated Piezoelectric Devices (MIND) project 2005–2009, resulting in establishment of The Piezoinstitute AISBL, 2008
- Coordinator of Polecer Network with 80 academic and industrial members (Polar Electroceramics), 2000–2005
- Agnes and Betsy Award (Danish Female Engineer of the Year), 2004
- Convenor of WF Standardisation Committee of CENELEC, resulting in 3 standards on piezoelectric materials

Wanda Wolny receiving the Female Engineer of the Year Award (with the Minister of Education), 2004.

Biography

Early Life and Education

From 1964 to 1969, Wanda Wolny studied at Warsaw University of Technology. In 1969, she was awarded a Master's degree in Electronics and Solid State Physics. Midway through this process (in February 1967), she married Ewald R. Wolny, a Polish gynecologist.

Career History

They came to Denmark, a month or two after her graduation in July 1969, as political refugees and she started to work at Ferroperm 3 months later. Early in her career, they had two sons, the first was born in 1976 and the second in 1982. She was promoted to R&D Manager of the Piezoceramics Department in 1986. A few years later, in 1998, she became the R&D Manager of Ferroperm Piezoceramics A/S. From 1999 to 2001, she served as the Ferroperm Piezoceramics A/S Director and she became the Managing Director in 2001. In 2005, she received another promotion to Managing Director R&D and then in 2008 to Executive Director R&D and at this time Ferroperm became Meggitt. From 2010 to 2013,

Wanda Wolny, 2014. Credit: Meggitt.

she served as the Director of Business Development for Materials Science. In 2014, she turned to being a Consultant of Business Development, and then at the start of 2015, she retired.

3 Most Cited Publications

Title: Liquid-phase sintering of PZT ceramics
Author(s): Corker, DL; Whatmore, RW; Ringgaard, E; et al.
Source: Journal of the European Ceramic Society; volume: 20; issue: 12; pages: 2039–2045; published: November 2000
Times Cited: 77 (from Web of Science)

Title: European approach to development of new environmentally sustainable electroceramics
Author(s): Wolny, WW
Source: Ceramics International; volume: 30; issue: 7; special issue: SI; pages: 1079–1083; published: 2004
Times Cited: 40 (from Web of Science)

Title: Extrinsic contribution to the non-linearity in a PZT disc
Author(s): Perez, R; Albareda, A; Garcia, JE; et al.
Source: Journal of Physics D: Applied Physics; volume: 37; issue: 19; pages: 2648–2654; article number: PII S0022-3727(04)81688-7; published: October 7, 2004
Times Cited: 16 (from Web of Science)

Challenges

- To learn Danish.
- To combine private life as a mother and wife and professional career. Not to give in to the pressure from well meaning "friends."
- To travel overseas on business or to conferences leaving small children at home.

On being a woman in this field . . .

There is no intellectual difference neither different professional abilities between women and men, no other limits than those we put on ourselves. Good friends will confirm that for you whenever you are in doubt. I was extremely lucky with having friends such as Maricka (Prof. Marija Kosec) and Nava (Prof. Nava Setter).

Wanda Wolny combining work and pleasure in Hawaii;
taking a few days of holidays after the ISAF2000
conference.

Words of Wisdom

My favorite quote is from musical and movie "Annie Get Your Gun": "Anything you can do, I can do better!"

Profile 98

Jackie Y. Ying

Executive Director
Institute of Bioengineering and Nanotechnology
31 Biopolis Way
The Nanos, 138669
Singapore

e-mail: jyying@ibn.a-star.edu.sg
telephone: +65-6824-7100

Prof. Jackie Ying, Executive Director
of the Institute of Bioengineering
and Nanotechnology. Credit:
Institute of Bioengineering and
Nanotechnology.

Birthplace
Taipei, Taiwan
Born
April 30, 1966

Publication/Invention Record

>325 publications: h-index 63 (Web of Science)
>150 patents/patent applications

Tags

❖ Administration and Leadership
❖ Domicile: Singapore
❖ Nationality: American
❖ Asian
❖ Children: 1

Proudest Career Moment

Returning to Singapore in March 2003 to set up the Institute of Bioengineering and Nanotechnology (IBN) was a tremendous career opportunity. At that time, Singapore was trying to establish biomedical sciences as a new pillar of its economy, and become a global R&D hub. It was challenging to create a research niche in bioengineering and nanotechnology, recruit talented people from all over the world, and build up the infrastructure from scratch.

Today, IBN is a multidisciplinary research institute at the interface of science, engineering, and medicine, with over 180 researchers conducting research in nano-medicine, cell and tissue engineering, biodevices and diagnostics, and green chemistry and energy. We have contributed significantly to the growth and visibility of Singapore's R&D sector with over 1000 publications in top journals and about

Successful Women Ceramic and Glass Scientists and Engineers: 100 Inspirational Profiles. Lynnette D. Madsen.

Prof. Ying with some nanomedicine group members who developed green tea-based nanocarriers that can kill cancer cells more effectively. Credit: Institute of Bioengineering and Nanotechnology. (See insert for color version of figure.)

340 active patents/patent applications. We are excited that our innovations have been commercialized through 84 patents that were licensed to multinational and start-up companies. We have also established nine start-up companies to commercialize our technologies. We hope that these new technologies would impact the society through better health care.

Academic Credentials

Ph.D. (1991) Chemical Engineering, Princeton University, Princeton, NJ, USA.
M.A. (1988) Chemical Engineering, Princeton University, Princeton, NJ, USA.
B.E. (1987) Chemical Engineering, The Cooper Union, New York, NY, USA.

Research Expertise: synthesis of advanced nanostructured materials for biomedical and catalytic applications

Other Interests: I am active in the local education community as I would like to encourage more young people to pursue careers in scientific research. One of the things we need to do as scientists is to convey to the general public what we are doing and the impact of our work, and I regularly go to the schools to give talks on my research. IBN started a Youth Research Program in October 2003 to provide hands-on research experience and mentorship. The program has reached out to over 80,400 students and teachers from 290 schools through various activities such as open houses, career talks, science camps, and workshops. More than 2200 students and teachers have also completed research internships at IBN.

Key Accomplishments, Honors, Recognitions, and Awards

- American Institute for Medical and Biological Engineering (AIMBE) Fellow, 2015
- The Brunei Crown Prince Creative, Innovative Product and Technological Advancement (CIPTA) Award—Crown Prince Grand Prize; First Prize in ASEAN New Invention and Innovation Category, 2015
- Singapore Ministry of Education Service to Education Awards, 2011 and 2015
- Royal Society of Chemistry Fellow, 2014
- The Singapore Women's Hall of Fame, 2014
- Materials Research Society Fellow, 2013
- Named as one of the "World's 500 Most Influential Muslims" by The Muslim 500: 2012, 2013, and 2014
- International Union of Biochemistry and Molecular Biology (IUBMB) Jubilee Medal and Lectureship, 2012
- Asian Innovation Silver Award, Wall Street Journal Asia, 2011
- Singapore National Institute of Chemistry-BASF Award in Materials Chemistry, 2010
- World Economic Forum Young Global Leader, 2004–2009
- Named as one of the "One Hundred Engineers of the Modern Era" by American Institute of Chemical Engineers (Centennial Celebration), 2008
- Great Woman of Our Time Award for Science and Technology, Singapore Women's Weekly, 2008
- Honorary Professorship, Sichuan University, China, 2006

Prof. Jackie Ying (second from right) with Sir John Major (far right) at the Asian Innovations Awards 2011 organized by The Wall Street Journal Asia. Credit: Institute of Bioengineering and Nanotechnology.

- Elected Member of the German National Academy of Sciences, Leopoldina, 2005
- Best of Small Tech Advocate Award Finalist, Small Times Magazine, 2005
- Peter V. Danckwerts Memorial Lectureship, Chemical Engineering Science and World Congress of Chemical Engineering, 2005
- Linsay Lectureship, Texas A&M University, 2004
- Gouq-Jen Su Distinguished Lectureship, University of Rochester, 2002
- Honorary Professorship, Jilin University, China, 2002
- American Institute of Chemical Engineers Allan P. Colburn Award, 2000
- Technology Review Inaugural TR100 Young Innovator Award, 1999
- Union Carbide Innovation Recognition Award, 1998 and 1999
- Ernest W. Thiele Lectureship, University of Notre Dame, 1997
- American Chemical Society Faculty Fellowship Award in Solid-State Chemistry, 1997
- Camille Dreyfus Teacher–Scholar Award, 1996
- David and Lucile Packard Fellowship for Science and Engineering, 1995
- Royal Academy of Engineering ICI Fellowship for Young Academic Chemical Engineers, 1995
- American Ceramic Society Ross C. Purdy Award for the Most Valuable Contribution to the Ceramic Technical Literature, 1995
- Office of Naval Research Young Investigator Award, 1995
- 3M Innovation Fund Award, 1993–1994
- National Science Foundation Young Investigator Award, 1992
- Alexander von Humboldt Research Fellowship, 1992
- National Science Foundation, NATO Postdoctoral Fellowship, 1991
- AT&T Bell Laboratories Ph.D. Scholarship, 1989–1991
- Hoechst Celanese Excellence in Engineering Award, 1990
- General Electric Graduate Fellowship, 1987–1988
- The Cooper Union H. M. Siegel Memorial Prize for Scholastic Excellence, 1987
- New York State Regents Scholarship, 1983–1987
- Tau Beta Pi National Honor Society of Engineering
- Iota Sigma Pi National Honor Society for Women in Chemistry

She serves on many editorial and advisory boards, and also board of directors (these are not listed here).

Biography

Early Life and Education

Jackie Y. Ying was born in Taipei, and raised in Singapore and New York. She graduated with B.E. *summa cum laude* in Chemical Engineering from The Cooper Union in 1987.

As an AT&T Bell Laboratories Ph.D. Scholar at Princeton University, she began research in materials chemistry, linking the importance of materials processing and microstructure with the tailoring of materials surface chemistry and energetics. She pursued research in nanocrystalline materials with Prof. Herbert Gleiter at the Institute for New Materials, Saarbrücken, Germany, as an NSF-NATO Post-Doctoral Fellow and Alexander von Humboldt Research Fellow.

Career History

Prof. Ying joined the Chemical Engineering faculty at Massachusetts Institute of Technology (MIT) in 1992, and was promoted to Associate Professor in 1996 and then to Professor in 2001. She is currently the Executive Director of the Institute of Bioengineering and Nanotechnology (IBN), Singapore. IBN is a multidisciplinary national research institute founded by Prof. Ying in March 2003 to advance the frontiers of engineering, science, and medicine.

Prof. Jackie Ying with her team members from the Biodevices and Diagnostics group who have recently invented a rapid diagnostics kit that can detect a key dengue antibody from saliva within 20 min. Credit: Institute of Bioengineering and Nanotechnology.

Prof. Ying's research is interdisciplinary in nature, with a theme in the synthesis of advanced nanostructured materials for catalytic and biomaterial applications. Her laboratory has been responsible for several novel wet-chemical and physical vapor synthesis approaches that create nanocomposites, nanoporous materials, and nanodevices with unique size-dependent characteristics. These new systems are designed for applications ranging from biosensors and diagnostics, targeted delivery of drugs and proteins, generation of biomimetic implants and tissue scaffolds, pharmaceuticals synthesis to green chemistry and energy. Prof. Ying has authored over 340 articles, and presented over 410 invited lectures on this subject at international conferences.

Prof. Ying is an Honorary Professor of Jilin University (China) and Sichuan University (China), and an Adjunct Professor of National University of Singapore, Nanyang Technological University (Singapore), and King Saud University (Saudi Arabia). She was a founding member of the Board of Directors of Alexander von Humboldt Association of America. Prof. Ying has been recognized with a significant number of research awards (including one from the American Ceramic Society), and she has been elected as a member of the German National Academy of Sciences, Leopoldina. She was appointed by the U.S. National Academy of Engineering in 2006 to serve on the blue-ribbon committee that identified the grand challenges and opportunities for engineering in the 21st century.

Prof. Ying is the Editor-in-Chief of *Nano Today*. Under her leadership, *Nano Today* underwent a successful transition from a magazine to a journal, witnessing major increases in the impact factor. In addition, Prof. Ying serves on the editorial board for dozens of other prestigious journals. She serves on the Scientific Advisory Boards of Molecular Frontiers (a global think tank that promotes molecular sciences), King Abdullah University of Science and Technology Catalysis Center, and National University of Ireland Galway Centre for Research in Medical Devices (CÚRAM). Prof. Ying has over 150 primary patents issued or pending, and has served on the Advisory Boards of six start-up companies and two venture capital funds. One of the start-up companies that she cofounded, SmartCells, Inc., has developed a technology platform that is capable of autoregulating the release of insulin therapeutic depending on the blood glucose levels. Merck acquired SmartCells, Inc. in 2010, with potential aggregate payments in excess of $500 million to further develop this technology for clinical trials.

3 Most Cited Publications

Title: Synthesis and applications of supramolecular-templated mesoporous materials
Author(s): Ying, JY; Mehnert, CP; Wong, MS
Source: Angewandte Chemie International Edition; volume: 38; issue: 1–2; pages: 56–77; published: 1999
Times Cited: 1673 (from Web of Science)

Title: Role of particle size in nanocrystalline TiO_2-based photocatalysts
Author(s): Zhang, ZB; Wang, CC; Zakaria, R; et al.
Source: Journal of Physical Chemistry B; volume: 102; issue: 52; pages: 10871–10878; published: December 24, 1998
Times Cited: 937 (from Web of Science)

Title: Synthesis of hexagonally packed mesoporous TiO_2 by a modified sol–gel method
Author(s): Antonelli, DM; Ying, JY
Source: Angewandte Chemie International Edition in English; volume: 34; issue: 18; pages: 2014–2017; published: October 2, 1995
Times Cited: 847 (from Web of Science)

ResearcherID (A-8402-2012)

Challenges

When I was starting out my career at MIT, I was a triple minority; I was young, female, and Asian. I had to work twice as hard as everyone else, but I enjoyed what I was doing and I set my own standards. I was very fortunate to be in a department that was very supportive.

The challenge I face now is to create and drive cutting-edge research projects that can make a significant impact on society. These projects involve addressing complex problems, and require a multidisciplinary approach. IBN has fostered strong collaborations and team work between scientists, engineers, and medical doctors to tackle such research problems.

On being a woman in this field . . .

As a woman in research, I am in a position to make a difference and become a positive role model, to pass on what I have received and learned to the future generations. Certain fields are male dominated, but it is a matter of having enough women going into an area to form a critical mass. When I was going through university and graduate school, I had only one female professor in chemical engineering. So I hope to be a role model to other young scientists.

Words of Wisdom

- It is important to find meaningful things that you want to focus on, and enjoy doing.
- Dare to work on challenging problems, and contribute in your own way to make an impact on society.
- It is not possible to please everyone, you just have to set your own goals, persevere, and try your best to make a difference.

Profile 99

Maria Magdalena Zaharescu

Senior Researcher and Head of Department
Ilie Murgulescu Institute of Physical Chemistry
Romanian Academy
Bucharest
Romania

e-mail: mzaharescu@icf.ro
telephone: +4021-3167912

Maria Zaharescu at about
30 years on the terrace of her
house, in the 1970s.

Birthplace

Cluj-Napoca, Romania

Born

January 20, 1938

Publication/Invention Record

>250 publications: h-index 22 (Web of Science)
3 patents, 8 book chapters

Proudest Career Moments (to date)

Tags

❖ Academe
❖ Domicile: Romania
❖ Nationality: Romanian
❖ Caucasian
❖ Children: 1

In 1964, I moved from Cluj-Napoca to Bucharest, Romania, to the Center of Inorganic Chemistry of the Romanian Academy, getting the position of assistant researcher after competing with two male candidates.

I was appointed Head of Laboratory when I was 33 years old, only three months after I defended my Ph.D. thesis.

I have received the "Gheorghe Spacu" award of the Romanian Academy for "Phase equilibria studies in oxide systems" based on papers I published by 1971. One of these papers was: S. Solacolu, R. Dinescu, M. Zaharescu, Die thermischen Phasengleichge-wichte des Systems $BaO–TiO_2–V_2O_5$, Rev. Roum. Chim., 15, 401–408 (1970). The results communicated in the paper mentioned above were also included in the "Phase

Successful Women Ceramic and Glass Scientists and Engineers: 100 Inspirational Profiles. Lynnette D. Madsen.
© 2016 The American Ceramic Society and John Wiley & Sons, Inc. Published 2016 by John Wiley & Sons, Inc.

Equilibria Diagrams," NIST Standards Reference Database 31, National Institute of Standards and Technology, USA (Fig. 06654 and EC-448). As of today, I have nine phase equilibrium diagrams included in the same database.

I was elected corresponding member of the Romanian Academy in 2001 and full member in 2015. The Romanian Academy, founded in 1866, is Romania's highest cultural forum. It covers the scientific, artistic, and literary domains. The Academy has 181 acting members who are elected for life. At the present moment, only four women are full members of the Romanian Academy.

Academic Credentials

Ph.D. (1971) Institute of Chemistry, Cluj-Napoca, Romania.
B.Sc. (1959) Chemistry (with Merit Award), Babes-Bolyai University, Cluj-Napoca, Romania.

Research Expertise: physical chemistry of oxide systems (reaction mechanisms, thermal phase equilibria, structure–properties correlations), vitreous oxide systems with special properties (thermally and chemically stable), oxide nanostructures obtained by chemical routes, especially by the sol-gel method

Other Interests: teaching physical chemistry of oxide systems, serving as a doctoral supervisor

Key Accomplishments, Honors, Recognitions, and Awards

- Member of the Editorial Board of 4 Romanian journals (Revue Roumaine de Chimie, Revista de Chimie, Optoelectronics and Advanced Materials—Rapid Communications, and Romanian Journal of Materials), 1995 to present
- Member of the Editorial Board of the International Journal of the Sol-Gel Science and Technology, 2005 to present
- Member of the Chemistry Commission of the National Council for the Recognition of the University Degrees, Diplomas, and Certificates, 2005 to present
- Full member of the Romanian Academy, 2015
- Vice President of the Romanian Ceramic Society, 1993–2014
- Member of National Counsel of Scientific Research, Materials Science Division, 2010–2012
- Coordinator of the Section of Chemical Science of the Romanian Academy, 2006–2009
- President of Balkan network COSENT (Cooperation of South Eastern European (SEE) Countries in the Field of Nanotechnology), 2005–2008
- Honor Award and Gheorghe Spacu Medal, Romanian Society of Chemistry, 2007
- Corresponding Member of the Romanian Academy, 2001
- Scientific Order Third Degree, Romania, 1983
- Scientific Merit Medal, Romania, 1982
- Gheorghe Spacu Award, Romanian Academy, 1971
- Third Prize for Research of the Ministry of Education, Romania, 1967

Biography

Early Life and Education

Maria Magdalena Zaharescu was born in Cluj-Napoca, Romania. She was a single child. She attended school in her native town and graduated from high school in 1954 with a Diploma of Merit. She then studied at the Faculty of Chemistry of the Babes-Bolyai University of Cluj-Napoca where she received her Bachelor of Science degree in the field of inorganic and analytical chemistry in 1959. Together with two male colleagues, she graduated with a Merit Award, the highest academic achievement at that time. (In a turn of events that was unprecedented, more than 30 years after, all three former colleagues became members of the Chemical Science Section of the Romanian Academy.)

After receiving her bachelor's degree, she worked for 5 years (1959–1964) in the research and development department of the "Carbochim" abrasives factory in Cluj-Napoca, where she first started to study oxide materials and where she also had the opportunity to take a study trip to the present Institute of Silicate Chemistry of the Russian Academy of Science in St. Petersburg. Based on her work at Carbochim, three papers were published in national journals.

In 1964, she competed for and won the position of assistant researcher at the Center of Inorganic Chemistry of the Romanian Academy, for which she moved from Cluj-Napoca to Bucharest, Romania. In 1969, the Center of Inorganic Chemistry merged with the Center of Physical Chemistry of the Romanian Academy, keeping the latter name. In Bucharest, besides her duties as a researcher, Maria also studied toward her doctorate. She received her Ph.D. doctorate in June 1971. The topic of her Ph.D. thesis was the study of the ternary SiO_2–TiO_2–NiO system.

Career History

The topic of her research in the first years of activity at the Center of Inorganic/Physical Chemistry was oriented mainly toward establishing phase equilibrium diagrams in binary and ternary oxide systems containing TiO_2, V_2O_5, and alkaline earth oxides.

In 1974, she obtained a year-long postdoctoral scholarship at the Technical University of Denmark in Lyngby where she contributed to the elaboration of a new method of thermal analysis and did high-temperature X-ray diffraction studies.

Inspired by an article she read by chance in a professional journal, Maria started the sol-gel studies in Romania in 1975. She had the idea to combine her university focus on inorganic and analytical chemistry with her oxide field research at the time, in order to obtain reactive oxides that could enhance the reactions in solid states. In 1979, the first paper on the preparation of the La silicate by the sol-gel method was published in a Romanian journal. Her first sol-gel paper in an international journal (Silikattechnik) was published in 1986, concerning TiO_2 thin film preparation by the sol-gel method.

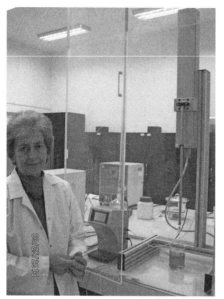

Maria Zaharescu in the laboratory of oxide compounds preparation by chemical methods, mainly sol-gel.

Maria Zaharescu at the Conference of Non-Crystalline Solids, Foz do Iguaçu, Brazil, 2009.

From 1971 to 1990, she worked on 61 applied national research contracts and was the coordinator for 50 of them. During this time, she participated in only one international collaboration, between the Romanian Academy and the Academy of the former Soviet Union, having as partner the present Institute of Silicate Chemistry of the Russian Academy of Science, St. Petersburg. She also attended some international conferences held in Hungary, Czech Republic, Poland, and the former Soviet Union.

After 1990, the possibilities of scientific collaboration were much expanded and she was involved in several types of international collaborations as director or co-director of the Romanian teams. These collaborations were funded by the U.S. National Science Foundation, European Union (FP5, FP6, COST Actions, M-ERA-NET), NATO, bilateral intergovernmental cooperation agreements with Greece, India, and Slovenia, or inter-academic cooperation projects with the Czech Republic, Bulgaria, Slovenia, France, and Italy. Maria visited and presented lectures at universities and prestigious research institutions such as University of California at Davis (USA), Rutgers University (USA), University of York (England), Université Claude Bernard, Lyon (France), Institute of Radio Engineering and Electronics of the Academy of Science of Czech Republic, Prague (Czech Republic), Indian Institute of Chemical Technology, Hyderabad (India), and Josef Stefan Institute, Ljubljana (Slovenia). She also had the opportunity to present invited lectures at numerous international conferences held in Bulgaria (1999, 2002), the United States (2003, 2007, 2010), Mexico (2004, 2005), Serbia (2005), Spain (2006), Poland (2007, 2009), France (2007, 2012), Germany (2007), South Korea (2007), Italy (2008), Malaysia (2008), Brazil (2009, 3 conferences), Chile (2009), Sweden (2011), South Africa (2011),

China (2012), New Zealand (2013), Morocco (2015), and Romania.

She served as a scientific reviewer for more than 20 prestigious peer-reviewed scientific journals (including Journal of the American Ceramic Society, Journal of the European Ceramic Society, Journal of Sol-Gel Science and Technology, Journal of Physical Chemistry, Journal of Non-Crystalline Solids, Journal of Alloys and Compounds, Journal of Materials Science and Engineering, Journal of Thermal Analysis and Calorimetry, and the Central European Journal of Chemistry) as well as all 4 Romanian journals.

Maria Zaharescu with Professor Alexandra Navrotsky at the MS&T Congress held in Columbus, Ohio, 2011 (with the permission of Prof. A. Navrotsky).

3 Most Cited Publications

Title: $TiO_2(Fe^{3+})$ nanostructured thin films with antibacterial properties
Author(s): Trapalis, CC; Keivanidis, P; Kordas, G; Zaharescu, M; Crisan, M; Szatvanyi, A; Gartner, M
Source: Thin Solid Films; volume: 433; issue: 1–2; pages: 186–190; doi: 10.1016/S0040-6090(03)00331-6; published: June 2, 2003
Times Cited: 134 (from Web of Science)

Title: Influence of the silica based matrix on the formation of iron oxide nanoparticles in the Fe_2O_3–SiO_2 system, obtained by sol-gel method
Authors: Jitianu, A; Crisan, M; Meghea, A; Rau, I; Zaharescu, M
Source: Journal of Materials Chemistry; volume: 12; issue: 5; pages: 1401–1407; doi: 10.1039/b110652j; published: 2002
Times Cited: 81 (from Web of Science)

Title: Comparative study of the sol-gel processes starting with different substituted Si-alkoxides
Authors: Jitianu, A; Britchi, A; Deleanu, C; Badescu, V; Zaharescu, M
Source: Journal of Non-Crystalline Solids; volume: 319; issue: 3; pages: 263–279; doi: 10.1016/S0022-3093(03)00007-3; published: May 15, 2003
Times Cited: 54 (from Web of Science)

ResearcherID (C-4185-2012)

Challenges

My most important challenge was to find the solution for an optimal balance between the scientific activity and my family. I hope I succeeded and I attribute this to my very understanding husband and a very intelligent daughter, both who accepted my work and frequent traveling abroad.

On being a woman in this field . . .

 . . . it depends on the period, place, and mentality of the people you meet. I did not feel any discrimination during my university studies, but I experienced it in Cluj-Napoca when I tried to change my job from applied work to academic research. That was the reason I moved from Cluj-Napoca to Bucharest, and it turned out to be a blessing in disguise. I was not discriminated against in the institutes in which I worked, belonging to the Romanian Academy. I was accepted and I was fairly promoted in competition with male candidates. As a matter of fact, due to the historical evolution of our country during for the last 40 years, today there are more women than men studying and working in the chemistry field in Romania.

Words of Wisdom

Research is not an ordinary job. In order to achieve really significant achievements in this field, you have to have a true passion for research.

When you are doing research, you have to be well informed and well connected with other research teams. Nothing can replace the personal contacts between scientists (not even the Internet, Facebook, Skype, etc.). Therefore, attending conferences and visiting other research institutes and/or universities is a duty for a scientist, not an option.

Academician Ilie Murgulescu, the founder of our institute, used to say: "if you are doing the best you possibly can in your work that means you are doing the right thing." I fully agree! He also used to say that a leader should be a generous person.

I consider that for a professor or a research group leader, the highest satisfaction is to have students or collaborators who have (or will have) achievements that surpass their own.

Profile 100

Jing Zhu

Professor
School of Materials Science & Engineering
Tsinghua University
Beijing 100084
P.R. China

e-mail: jzhu@mail.tsinghua.edu.cn
telephone: 86-10-62794026

Jing Zhu as a
student at Fudan
University when
she was about
20 years old.

Birthplace
Shanghai, China
Born
October 10, 1938

Publication/Invention Record
>300 publications: h-index 43 (Web of Science)
17 patents, editor of 2 books, 3 book chapters

Tags
❖ Academe
❖ Domicile: China
❖ Nationality: Chinese
❖ Asian
❖ Children: 3

Proudest Career Moment (to date)

In 1951, after completing primary school study I passed the Shanghai urban examination with high grades and subsequently was accepted by Shanghai Middle School, the most prestigious middle school in Shanghai. After passing the national university entrance exam in 1957, I was matriculated into Fudan University, one of the most famous; in that year the matriculation rate was less than 10% among middle school students in China. After the Cultural Revolution in 1977, I passed the national examination (math/physics/chemistry/English) for going abroad and then qualified as a Chinese Visiting Scholar.

In the early 1970s, I found reversible austenite in aging maraging steel by electron microscopy. I felt very excited by this finding. Through heat treatment control, further high strength and high toughness maraging steel was realized by aging strengthening of the reversible austenite and other compounds.

At the end of September 1980, about 40 days after I had arrived at Arizona State University, I was notified at Prof. J.M. Cowley's office that I may get the offer if I applied for the faculty research associate position announced in the University Job Opening. I was thrilled that my academic ability had been recognized. I felt gratitude to Prof. J.M.

Successful Women Ceramic and Glass Scientists and Engineers: 100 Inspirational Profiles. Lynnette D. Madsen.
© 2016 The American Ceramic Society and John Wiley & Sons, Inc. Published 2016 by John Wiley & Sons, Inc.

Cowley for giving me the opportunity and to all the people in China who nurtured and helped me to be competent for this position in the United States.

One day in 1981 I was doing an experiment in the HB-501 (a modified dedicated scanning transmission electron microscope) room of the Center for Solid State Science of Arizona State University. I observed superlattice diffraction spots splitting and the splitting pattern changing on the display screen attached to the HB-501 while the sample, ordered/disordered Cu_3Au thin film, was shifted. I believed this to be a new insight. After calculation and analysis, we concluded this to be the electron beam coherent wave effect of a defect with a single atomic plane, published in 1982 and co-authored with Prof. Cowley.

In 1995, I was voted in the first of 10 outstanding faculties by all people (more than 1500) voting in the Central Iron & Steel Research Institute of China. I was honored and proud to be recognized by my peers in a democratic election.

In 2009, I was invited to present a plenary talk in 12th International Meeting on Ferroelectricity and 18th IEEE International Symposium on Applications of Ferroelectrics. In this talk, it was emphasized how we use advanced electron microscopy to study the ferroelectrics. Three topics were presented: (1) Chemical ordered domain and polar nanoregion in $Pb(Mg_{1/3}Nb_{2/3})O_3$; (2) Hierarchical micro-/nanometer scale domain structure near morphotropic phase boundary in single crystal $Pb(Mg_{1/3}Nb_{2/3})O_3$–$PbTiO_3$; and (3) Aberration-corrected transmission electron microscopy (TEM) study of ferroelectric materials. This talk was received with warm applause.

Jing Zhu was one of the six plenary speakers in the 12th International Meeting on Ferroelectricity and 18th International IEEE International Symposium on the Applications of Ferroelectrics in 2009.

Academic Credentials

(1962) Physics, Fudan University, Shanghai, China.
During the period of 1949–1980, there was no academic degree system in China.

Research Expertise: electron microscopy of materials science, interfaces and surfaces, complex oxides, nanomaterials, ferroelectrics, multiferroics

Key Accomplishments, Honors, Recognitions, and Awards

- Plenary Lecture "Electron microscopy study of polar nanoregion in perovskite relaxor ferroelectrics" at the AMEC-8: The 8th Asia Meeting on Electroceramics, 2012
- Plenary Lecture "Analytical electron microscopy study of ferroelectrics with perovskite-type structure" at IMF-ISAF 2009 (12th International Meeting on Ferroelectricity and 18th IEEE International Symposium on Applications of Ferroelectrics), 2009
- Fellow, The Academy of Sciences for the Developing World (TWAS), 2007
- Award of Science & Technology Achievement, Ho Leung Ho Lee Foundation, 2004
- First author of the book "Nano-Materials and Devices," Tsinghua University Publishing House, in Chinese, 2003

- Author of "Chapter 11. Silica nanowires/nanotubes" in the book "Nanowires and Nanobelts—Materials, Properties and Devices, Volume 2: Nanowires and Nanobelts of Functional Materials," edited by Zhong Lin Wang, Kluwer Academic Publishers, Boston/Dordrecht/London, 2003
- Author of "Chapter D1. Study of integrated ferroelectric by electron microscopy" in the book "Progress in Transmission Electron Microscopy," edited by Hengqiang Ye and Yuanming Wang, Scientific Publishing House, in Chinese, 2003
- Invited Speaker, "Polar cluster in relaxor ferroelectric $Pb(Mg_{1/3}Nb_{2/3})O_3$" at the 15th International Conference of Electron Microscopy, 2002
- Author of "Possible center for polar cluster in lead magnesium niobate $Pb(Mg_{1/3}Nb_{2/3})O_3$" in the book "Grain Boundary Engineering in Ceramics" (Ceramic Transactions, Volume 118), pp. 191–199, The American Ceramic Society, 2000
- Invited Speaker, "Ordered domain and polar cluster in lead magnesium niobate $Pb(Mg_{1/3}Nb_{2/3})O$," JFCC International Workshop on Fine Ceramics 2000, "Grain Boundary Engineering in Ceramics—From Grain Boundary Phenomena to Grain Boundary Quantum Structure," 2000
- Invited Speaker, "The investigation of electronic structure by analytical electron microscopy," the Fifth IUMRS International Conference on Advanced Materials (IUMRS-ICAM'99), Beijing, China, 1999
- Author of "Chapter 3. Study of electronic structure at interface" in the book "Structure and Properties of Interface in Materials," edited by Hengqiang Ye, Scientific Publishing House, in Chinese, 1999
- International Advisory Board Member, the Third Pacific Rim International Conference on Advanced Materials and Processing (PRICM-3), 1998
- Plenary Lecture at the Second Pacific Rim International Conference on Advanced Materials and Processing (PRICM-2), 1995
- Member, Chinese Academy of Sciences, 1995
- The Science and Technology Progress 2nd Prize by National Education Ministry, 1995
- The Science and Technology Progress 2nd Prize by Metallurgical Industry Ministry, 1995
- The Science and Technology Progress 2nd Prize by Chinese Academy of Sciences, 1994

Jing Zhu presented a plenary lecture in the PRICM-2 held in Seoul in South Korea in 1995.

- The Science and Technology Progress 2nd Prize by Metallurgical Industry Ministry, 1993
- 1st Prize of National Award for Science and Technology Progress, 1992
- 1st Prize of the Science and Technology Progress Prize by Metallurgical Industry Ministry, 1989
- 4th Prize of the National Nature Science Prize (the coherent electron wave micro-diffraction effect of an individual planar fault was found and defined for the first time in the world in 1982 by her observation using the cold field emission gun scanning transmission electron microscope in Arizona State University (ASU) and theoretical calculation with diffraction theory), 1989
- First author of the book "High Spatial Resolution Analytical Electron Microscopy," Scientific Publishing House, in Chinese, 1987

Biography

Early Life and Education

Jing Zhu was raised in Shanghai by her adoptive parents. She has one biological elder sister and elder brother in her family as well as one elder sister in her adoptive family. Jing received her early education in the Shanghai public school system and also continued with her higher education in Shanghai. In 6 years of study at Shanghai Middle School, she received the best basic education in the fundamentals of mathematics, physics, chemistry, biology, and the knowledge of geography and history as well as language training (Chinese, Russian), music, and physical culture. In her 5 years of study at Fudan University, her focus included higher mathematics, methods of mathematical physics, general physics (mechanics, optics, electromagnetics, thermodynamics, atomic physics, and statistical physics), modern physics (quantum theory, electrodynamics, and principle of relativity), and specialized courses on electron physics and experimental physics. She spent the final term writing her dissertation.

It was after graduation from Fudan University that she began self-studying electron microscopy, the physics of metals, dislocations, metallography and heat treatment, material science, oxides, ferroelectrics, and multiferroics. Her background in physics provided the ability to self-study and acquire the knowledge to be successful in her research career.

Career History

After graduation, she was assigned to the Central Iron & Steel Research Institute in Beijing. At that time, the assignment system was performed by the government; there was no free job position application for common people in China. Her first project was to design an electron optic system of an electron microscope with 100 keV and making the 100 keV power supply with the 5×10^{-6} stability. Simultaneously, she modified and used a Russian ЭM-3 electron microscope to improve its stigmatism and brightness through a reconstructing of the condensed lens aperture.

Since 1964, she spent 13 years engaged in the installation and maintenance of electron microscopes and served as an electron microscopist for senior researchers.

From the end of 1960s to late 1980s, Jing and her colleagues investigated alloying principle and toughening pathway in high strength/high toughness steels. She proposed a rational heat treatment system and alloying principle that has been adopted in the alloy used as the material of the centrifugal machine for uranium in China.

After the Cultural Revolution (started in 1966), China opened the door to the world. In October 1977, Jing was qualified as a Chinese Visiting Scholar and was sent to study abroad through a national examination organized by the National Ministry of Education and National Ministry of Metallurgical Industry of China. She received 1 year of spoken English training and then proceeded with the very complicated procedures for going to the United States. On August 20, 1980, Jing arrived at the Phoenix airport to meet Prof. J.M. Cowley of ASU. Studying under his guidance was her first selection in her application form. Forty days later, Jing was employed as a faculty research associate in the Center for Solid State Science of ASU by Prof. Cowley's recommendation. There she started her research career in the United States and was eager to study and get back lost time due to the Culture Revolution in China. Her dream was to catch up with the world-class in the electron microscopy field. Following conferment (2 years to study abroad) with the Chinese government, Jing came back to China in September of 1982. She completed the calculation on microdiffraction from a single atomic plane defect immerged in a strain field of a solid in 1983. From 1984 to 1985, she accepted Prof. Cowley's invitation as a faculty visiting associate professor in the Department of Astronomy and Physics of ASU. The Center for Solid State Science and the Facility for High Resolution Electron Microscopy of ASU reached its zenith during the 1970s and 1980s. This position gave Jing a chance to work with and learn from many international senior and famous scientists there.

Jing went back to Arizona State University in 1993: (left) Jing Zhu, (middle) J.M. Cowley, and (right) Jianmin Zuo.

In 1996, Jing left the Central Iron & Steel Research Institute and accepted a position as professor in Tsinghua University. Since then she has focused her research on carbon materials, nanomaterials, ceramics, oxides, ferroelectrics, and multiferroics by advanced electron microscopy. From early in 1997 through to the beginning of 2007, she was Dean of the School of Materials and Engineering at Tsinghua University. She has been the

Director of the National Center for Electron Microscopy in Beijing since 2006. From July to September 2006, Jing also held a visiting professor position at Tohoku University in Japan.

3 Most Cited Publications

Title: Size dependence of Young's modulus in ZnO nanowires
Author(s): Chen, CQ; Shi, Y; Zhang, YS; Zhu, J; Yan, YJ
Source: Physical Review Letters; volume: 96; issue: 7; pages: 075505-1–075505-4; doi: 10.1103/PhysRevLett.96.075505; published: February 24, 2006
Times Cited: 464 (from Web of Science)

Title: Aligned single-crystalline Si nanowire arrays for photovoltaic applications
Author(s): Peng, KQ; Xu, Y; Wu, Y; Yan, YJ; Lee, ST; Zhu, J
Source: Small; volume: 1; issue: 11; pages: 1065–1067; doi: 10.1002/smll.200500137; published: November 2005
Times Cited: 443 (from Web of Science)

Title: Synthesis of large-area silicon nanowire arrays via self-assembling nanoelectrochemistry
Author(s): Peng, KQ; Yan, YJ; Gao, SP; Zhu, J
Source: Advanced Materials; volume: 14; issue: 16; pages: 1164–1167; doi: 10.1002/1521-095(20020816)14:16<1164:AID-ADMA1164>3.0.CO;2-E; published: August 16, 2002
Times Cited: 391 (from Web of Science)

Challenges

I was born during the second war. Before my birth, the Japanese army invaded my hometown. When I was only 1 month old, my parents left me to go to Chongqing city for the anti-Japanese war. When I look back, I was in a weak position. I had no bought toys or new clothing in my childhood. Since I was cognizant of my situation, I told myself that I should be strong and study hard to become a person who is useful for society and for constructing a rich and independent motherland. In my life journey, there have arisen many challenges that every time I bravely face. I am very lucky to have met so many good persons who through their instruction, teaching, and caring have given me the wisdom and ability to face challenges.

On being a woman in this field . . .

One thing I feel is not good is that there are less than 10% women among the academicians of Chinese Academy of Sciences.

Words of Wisdom

I would give the Tsinghua motto here as following:

- Self-Discipline and Social Commitment
- Actions Speak Louder than Words

AFTERWORD

When I was first asked to write an Afterword for this book on profiles of women in "ceramic and glass science and engineering," I was forced to go back to some very basic questions: what exactly are ceramics and glass; how does one become a scientist or engineer in this field; and how does this field connect to STEM in general, i.e., anything I might care about? But rather than looking for answers within the science and engineering, I decided primarily to explore these issues *and the science* through the eyes and the lives of an incredible group of women.

I was amazed at the terrain covered by their areas of study: from materials used in medicine and dentistry, to use in aerospace and telecommunications, to storage of nuclear wastes, to restoration of works of art. I just had never thought about this kind of range or the flexibility of these areas of study. Applying different lenses, they found opportunities in many places.

The careers of these women span over half a century: from the time of overt barriers, through the movement for women's rights, through to the present. There are more women in this field today than 30 years ago, but they still have the "opportunity" to be "the one" or one of a few in so many places where they go, such as in scientific conferences or programs. Some of these women have reached the top of their field; they are or have been chairs of departments or heads of centers or institutes, leaders within professional societies and companies, their own as well as major corporations, recipients of top prizes, and recognized as elected fellows and members of national and international academies.

Most have felt the difference that being *different* makes: many embraced the opportunities, even doing so while clearly seeing the barriers. As with women in so many fields of science and engineering where they are woefully underrepresented, they shared stories of being both very visible AND invisible. Some (but not all) remember being unheard, participating in groups where they shared thoughts and ideas that only got recognized and valued when later repeated by (and attributed to) a male colleague. Many recounted instances of being underestimated, where lower expectations existed for them, in the same space where the bar that got set for them might have been higher than for male colleagues. They also have felt the joy of success, for example, solving a previously intractable problem, when they felt the respect of their colleagues. Not every woman who shared her story reported being treated differently because she was a woman in a field that had few women. Some found those (still rather rare) places where they were welcome and their contributions were recognized and valued.

These scientists and engineers come from all over the world, nearly 30 countries. They shared their experiences in growing up and in getting their education. One woman recalled the story of not being able to attend the high school focused on science and mathematics in her city because, at that point in history, it was for males only. Some were educated and nurtured in women's colleges that have an outstanding track record of developing women for leadership in science and engineering. Some women attended the Seven Sister Colleges, while others were among the first to enter previously all-male technically focused institutions. Even as the doors to opportunity opened, some male faculty retained their pre-civil rights era attitudes: women not wanted here. One woman noted that attitudes are improving, particularly in the youngest generations.

Some of the women found their chosen areas of study quite early, while others came to them quite late; in at least one case, one woman "stuffed" all her science and mathematics in less than 2 years after having migrated over from a focus in the arts. Interestingly and importantly, these scientists and engineers have taken all these "detours" and meanderings into their lives and careers and have been enriched by them!

Rightfully, they advised putting excellent science first. In all too many cases where I have been in conversations with women in fields where they are a distinct minority, I have found that they chose not to get caught up in the workplace "drama," instead developing coping mechanisms and ignoring any slights and differential treatment.

As with many of the women in science and engineering with whom I interact, their journeys were very much shaped by the choices they have made relating to career–life integration, how to balance the career with having a family, how to support other life interests such as in music, art, or even windsurfing, or otherwise to have a multi-dimensional life (while doing twice as much on the science/engineering front). This is truly living a rich life!

For those who shared these aspects of their stories, we learn that success in this arena is often interlinked with choosing a life partner who is committed to **both** careers and who "walks the talk" in sharing the day-to-day responsibilities of childcare and family maintenance. Research tells us that having children, a very personal decision, has an impact on the arc of the career of a woman in science or engineering. While the data regarding the research output of women with children represent collective decision making by many women, the profiles here relate to individual decision making: a young single mother of two who goes to college to be able to support her children and winds up discovering her talent and creating a rewarding career; a scientist who became a mother in grad school urges women not to put off having children because of career concerns; a scientist with four children who wanted to make sure women got the message that children and career are not incompatible; a scientist who just couldn't see how she could make the juggling act work for her and decided accordingly. I wonder to what extent the male colleagues of these scientists went through the same kind of soul-searching about having (or trying to balance) a family and a career.

When beginning their careers, many of these women scientists and engineers were the first, and for some years, the only woman in their work group or area. They advanced through raw talent, hard work, determination, focus, and "overproducing," exceeding expectations even when the bar was higher. I saw within their stories willingness to take risks, to explore understudied areas (before they became too crowded), to pursue what they had been warned were "dead zones," and the value of broad interdisciplinary backgrounds as many of them "pieced together" the courses, experiences, opportunities, and work history that led them to their careers. Some of those profiled talk about being lucky to have been able to encounter the circumstances that led to their success; I see instead a playing out of the adage ascribed to Louis Pasteur, "Fortune favors the prepared mind."

I was also pleased to see the commitment to serve of those profiled. They expressed a sense of responsibility to their field, to the science enterprise, and to the larger global community—to give back in payment for all they received from others.

Far beyond traversing big chunks of America's civil rights history, the lives of women profiled here were also shaped by world history and social movements else-where—for example, the Holocaust, the Cultural Revolution in China, and the breakup of the Soviet Union. When a parent flees a homeland, for example, everyone in that family must adapt and make a new life, opening up new and different possibilities. This is a reminder that our many identities lead us to challenges and opportunities, shaping the people we become and the values we hold. I saw within many of these stories a deep commitment to women in science and engineering, to smoothing the educational paths for their students, and to other aspects of social justice.

I also want to point out what I did NOT see. Somehow the voices of American women of color are nearly absent here. Perhaps this is largely due to the relative rarity of African American, American Indian, and Hispanic American women within these fields. In addition, many women of color in the United States are younger or new entrants; this is also true for many women in Korea, Thailand, Malaysia, and even Gaza. I also did not see women from Africa. I do not believe that the absence of minority women is related to intentional exclusion of their narratives, but likely more a reflection of lack of development and priorities given to these areas of study. Perhaps the institutions that many attended early in their higher education (two-year colleges and/or minority serving institutions) do not have gateways into materials science and engineering. Likely this represents an opportunity for the fields, to reach out into and form partnerships with the institutions where these women of color are enrolled. It may also be that students from these groups aren't aware that such fields exist or that such lives and careers might be possible. Hopefully, this book will begin to fill in the knowledge gap and increase awareness of what is possible.

For women of color who read this book, I urge you to look beyond your absence and the small numbers that probably caused it; look instead at the interesting lives of purpose that the women profiled here have been able to build. Connect with organizations where you can find community within science and engineering. There are many of us who went first. As a woman of color who received her doctorate in ecology over 40 years ago, I have learned that being first just means making a path where one does not yet exist. While challenging, it is always interesting; it can also be rewarding and fun. The women in this volume give the reader the sense that they love what they do and are happy they have had the chance to explore the unknown and to advance their fields. Clearly, these are more than jobs to them; their careers are intimately integrated into their lives and are great sources of joy. Whether they choose to combine career and family (or not), these decisions felt right for them. What more can anyone ask for?

While I call out particular groups of women who should find this volume especially useful, I want to note that this book is valuable for **anyone** who wants to see the lives that these scientists and engineers have built around ceramics/glass. And because the profiles are those of women, the lives are seen in 3D: self; family/community; and career, fully integrated. I greatly appreciate the willingness of these women to share their personal narratives, not just their career progression. It is useful to see their pathways to the present. I want this to become a book that men in these fields will seek out to read. I hope this book will help them dispel any lingering notions they might have that a "woman's place" in these fields is lesser. And I also wish for them, in these respects, to be like these women—excellent in their science or engineering and multidimensional in their lives.

DR. SHIRLEY M. MALCOM
Head of Education and Human Resources Programs,
The American Association for the Advancement of Science

ACKNOWLEDGMENTS

I am thankful to my husband, Dr. Erik B. Svedberg, for his support in so many ways that encouraged me during the tough times, enabled me to write the book, and carried out lots of tasks (to help the progress of the writing, refinement of images, and also directly with the background research on the featured women).

I would like to thank my parents, Dawn and David Madsen, for their faith in me and their support of my many endeavors.

Anita Lekhwani at Wiley (in Hoboken, NJ) is acknowledged for her patience in dealing with a first-time book author and for enthusiastically supporting the writing, publishing, and marketing of this book. I appreciate the four initial reviewers of the book proposal. I would also like to thank Cecilia Tsai (Wiley, Hoboken, NJ) who dealt with my questions along the way, and Sue Joshua (at Wiley in Chichester) who provided legal guidance. Melissa Yanuzzi at Wiley assisted with the production, and Abhishek Sarkari at Thomson Digital worked on the copyediting.

I am grateful to Greg Geiger (and others) at The American Ceramic Society for support throughout the process. I am thankful to NSF for granting me moonlighting permission.

It is always a challenge to identify top people in a field, and in this task, I had much assistance for which I am grateful. Professor Gary Messing of Pennsylvania State University identified the women in the World Academy of Ceramics and provided me with details about each person. In addition, Gary reviewed my entire list of women and provided suggestions—I hasten to point out, if there is blame, it is all mine and not his. Prof. Yury Gogotsi helped in identifying women in Japan and Korea, Dr. Greg Exarhos helped to identify women at national labs, Dr. Olivia Graeve pointed out women in South America, Dr. Tania Paskova identified leading researchers in Germany, Dr. Guillermo Aguilar identified women in Mexico, Dr. Pearl Sullivan helped me navigate Malaysian universities, and Drs. Jane Chang of UCLA, C.T. Sun of Purdue University, and Masahiro Yoshimura of Japan all provided valuable assistance in identifying Taiwanese leaders in the field of ceramics. Finally, many of the women featured in this book provided suggestions—worst case, their suggestions were reassuring (since they were women already on my list); best case, they were helpful.

As with any book featuring accomplished, busy people, assistants of various types were patient and generous with their time in helping me assemble interesting and accurate profiles. Special thanks are extended to Read Schusky (Mildred Dresselhaus' assistant) who provided timely and thoughtful comments, Antonio Salinas (who assisted with María Vallet-Regí's profile), Stefan Wagner (who assisted with Ellen Ivers-Tiffee's profile), Stephanie Diaz (Maxine Savitz' assistant), Rebecca L. McDuffee (Laura Greene's assistant), and Profs. Jongheun Lee and Han-Ill Yoo (who offered to review Prof. Park's profile while she was traveling). Many thanks to everyone who helped me to make connections, including Prof. Alice M. Agogino, Ms. Mary Lee Berger-Hughes, Dr. Eric Wuchina, and Dr. Jonathan Mallett.

I am grateful to Ms. Xiaoyin Ren in China for providing information about a Chinese politician (Zhili Chen who was educated in ceramics and carried out research in the field before moving fully to politics). Additional people assisted with her profile: Profs. Susan Trolier-McKinstry (images), Long-Qing Chen (contact information), and Haiyan Wang (communication).

I am grateful to everyone who contributed to the profile of Prof. Marija Kosec who passed away in December 2012.

I would also like to thank Dr. Steve Freiman for doing background research on several of the women. I appreciate the careful editing and suggestions from Dr. Lisa Frehill and Mr. C. Bruce Wells. Dr. David Nelson, a former NSF colleague, periodically provided advice about the book. I am grateful to my NSF colleague, Dr. Sean L. Jones, for his enthusiasm and suggestions and for loaning me his personal collection of books on underrepresented groups in science.

I am grateful to Dean Cristina Amon for providing the Foreword and Dr. Shirley Malcom for writing the Afterword. I also appreciate the remarks from Drs. Rita Colwell, Yury Gogotsi, Olivia Graeve, and Rainer Waser.

Finally, I give my thanks to the women who helped complete their profiles and for their encouraging comments. Their input was essential to the quality, depth, and beauty of the book. I am grateful to Dr. Julia Phillips who holds the record time for a profile—from my first point of contact to a completed profile (with images and a signed release); it was less than 12 hours—if only all the profiles were so fast and easy! A special thank you goes to Prof. Alex Navrotsky who supplied the very first complete profile within one week. Within hours of receiving her input, Prof. Nava Setter sent me the second profile (nearly complete with one image missing). I am particularly grateful to them because they showed faith in me and provided the first full glimpse of the richness of the book. I really appreciate Prof. Jennifer Lewis for her enthusiasm—she wrote to me early in 2014 with these comments: "Great!! This is really going to be fun!"—remarks like these really helped to keep my spirits high.

CONCLUSION

At a Rosalind Franklin Society Board Meeting in 2012, Millie Dresselhaus explained she considered a career as a primary school grade teacher. Well, this is not too surprising since she was born in the 1930s and came of age at a time when women who had entered science and engineering fields during WWII were pushed out of the post-war workforce in the 1950s. It is disheartening, though, to hear from others, e.g., Laura Greene, who was born more than 20 years later, that she was also only encouraged to pursue a teaching certificate in the 1970s, a time when the U.S. Women's Liberation Movement was ramping up. This book demonstrates that women's career choices have broadened in the past half-century, and that science has benefited from women's increased presence.